AF869287

Date Palm Genetic Resources and Utilization

Jameel M. Al-Khayri • Shri Mohan Jain
Dennis V. Johnson
Editors

Date Palm Genetic Resources and Utilization

Volume 2: Asia and Europe

Springer

Editors
Jameel M. Al-Khayri
Department of Agricultural Biotechnology
King Faisal University
Al-Hassa
Saudi Arabia

Dennis V. Johnson
Cincinnati
Ohio
USA

Shri Mohan Jain
Department of Agricultural Sciences
University of Helsinki
Helsinki
Finland

ISBN 978-94-017-9706-1 ISBN 978-94-017-9707-8 (eBook)
DOI 10.1007/978-94-017-9707-8

Library of Congress Control Number: 2015934366

Springer Dordrecht Heidelberg New York London
© Springer Science+Business Media Dordrecht 2015
This work is subject to copyright. All rights are reserved by the Publisher, whether the whole or part of the material is concerned, specifically the rights of translation, reprinting, reuse of illustrations, recitation, broadcasting, reproduction on microfilms or in any other physical way, and transmission or information storage and retrieval, electronic adaptation, computer software, or by similar or dissimilar methodology now known or hereafter developed. Exempted from this legal reservation are brief excerpts in connection with reviews or scholarly analysis or material supplied specifically for the purpose of being entered and executed on a computer system, for exclusive use by the purchaser of the work. Duplication of this publication or parts thereof is permitted only under the provisions of the Copyright Law of the Publisher's location, in its current version, and permission for use must always be obtained from Springer. Permissions for use may be obtained through RightsLink at the Copyright Clearance Center. Violations are liable to prosecution under the respective Copyright Law.
The use of general descriptive names, registered names, trademarks, service marks, etc. in this publication does not imply, even in the absence of a specific statement, that such names are exempt from the relevant protective laws and regulations and therefore free for general use.
While the advice and information in this book are believed to be true and accurate at the date of publication, neither the authors nor the editors nor the publisher can accept any legal responsibility for any errors or omissions that may be made. The publisher makes no warranty, express or implied, with respect to the material contained herein.

Printed on acid-free paper

Springer is part of Springer Science+Business Media (www.springer.com)

Preface

Date palm (*Phoenix dactylifera* L.) trees bear edible fruits under adverse conditions of high temperature, limited water resources and high soil and water salinity. Dates have contributed to the emergence and survival of ancient societies in desert regions worldwide. Now they are being rediscovered as a nutritional health food due to richness in fructose, a low glycemic index sugar, vitamins and dietary fibers. Dates have diverse medicinal properties due to high content of phytochemicals including carotenoids, polyphenols (e.g. isoflavons, lignans and flavonoids), tannins and sterols. These compounds are known for antioxidant activity, cholesterol-lowering properties and chemoprevention of cancer, diabetes and cardiovascular diseases.

Date palm plantations worldwide are estimated to have over 150 million trees, distributed between 10° and 39°N Lat. but thrive mostly between 24° and 34°N Lat. in the Middle East and North Africa. Date production has steadily increased from (in thousands) 2,535 mt in 1980 to 7,549 mt in 2012 (FAOSTAT database apps.fao.org). Although modern agricultural practices have contributed to increase in production, expansion in agricultural land and utilization of natural resources has played an important role. Sustainable date palm cultivation in the twenty-first century will require implementation of a set of principles drawn from existing research results and farm-level experiences. Despite the existing body of research, realization of the full potential of the date palm is being hindered by the lack of comprehensive studies of key areas: (a) Comprehensive assessment of date palm genetic resources and their conservation, including the role of seedling palms; (b) Broad evaluation and identification of cultivars and the role of DNA fingerprinting; (c) Development of a set of best-cultivation practices to enhance yield and reduce post-harvest loss for both monoculture and mixed oasis farming systems; (d) Conservation and utilization of germplasm resources and agroforestry in light of climate change; (e) Appraisal of the role of tissue culture according to the scale of production; and (f) Evaluation of date palm products and marketing. This book fills the gaps in the preceding key areas for major date-producing countries worldwide. It provides an assessment of how to achieve these key principles across the date-producing countries with reference to the local environmental conditions and socioeconomic practices.

The book is published in two separate volumes; each consists of 15 chapters. Volume 1 addresses date-producing countries within Africa, North America and South America, whereas, Volume 2 pertains to Asia and Europe. Volume 2 begins with an introductory chapter highlighting modern scientific discoveries and the main book features. Chapters 2, 3, 4, 5, 6, 7, 8, 9, 10, 11, 12, 13, and 14 provide an assessment specific to an individual date-producing Asian countries including Iran, Saudi Arabia, Iraq, Pakistan, Oman, Yemen, Israel, Kuwait, Qatar, Bahrain, Syria, Palestine and India. Two other date-producing Asian countries not included in this volume, due to unavailability of authors, are the United Arab Emirates, a major producer, and Jordan, a minor producer. Europe, represented by Spain, is addressed in Chap. 15. The volume ends with appendixes summarizing available information on each country. The appendixes provide surveys of characteristics and distribution of major date palm cultivars, commercial sources of dates, offshoots and in vitro plants; and research institutes and scientific societies concerned with date palm research and production.

In this book volume, each chapter describes the current status and prospects on: cultivation practices and agroforestry; genetic resources focusing on threats and research progress in genetics, breeding, conservation, and germplasm banks; role and importance of cultivar classification and identification based on morphological and molecular techniques; importance of micropropagation technology and research progress toward scale-up production; and advances in dates processing, novel products and innovative utilizations and marketing. In addition, each chapter describes the main characteristics of important date palm cultivars including growth requirements, cultivar distribution shown in a country map, production statistics and economics, nutritional aspects as well as morphological descriptors supported with color photos. The chapter concludes with imperative recommendations for further development of date palm sector in the respective country.

This volume presents, for the first time, a comprehensive assessment of date palm genetic resources and utilizations in date palm–growing countries of Africa and the Americas. The book is a valuable resource for students, researchers, scientists, commercial producers, consultants, and policymakers interested in agriculture, particularly in the date palm industry. Individuals and companies throughout the date industry would benefit greatly from the adoption of the policies and practices discussed in this book to enhance production and expand industrialization and marketing of traditional and potentially new date products. The chapters were graciously contributed by prominent scientists of each country. All manuscripts were critically reviewed and revised to ensure accuracy and quality. We greatly appreciate all contributory authors for their contribution towards the success and quality of this book. We are grateful to Springer for giving us an opportunity to compile this book.

Al-Hassa, Saudi Arabia — Jameel M. Al-Khayri
Helsinki, Finland — Shri Mohan Jain
Cincinnati, OH, USA — Dennis V. Johnson

Contents

Contributors

Adel A. Abul-Soad Botany Department, Horticulture Research Institute, Agricultural Research Center, Orman, Giza, Egypt

Hassan Abu-Qaoud Department of Plant production and Protection, Faculty of Agriculture, An-Najah National University, Nablus, Palestine

Osman E. Ahmed Department of Agricultural Affairs, Ministry of Environment, Doha, Qatar

Talaat A. Ahmed Department of Biological and Environmental Sciences, Qatar University, Doha, Qatar

Abdulaziz M. Al-Bahrany Department of Agricultural Biotechnology, College of Agriculture and Food Sciences, King Faisal University, Al-Hassa, Saudi Arabia

Francisco Alcaraz Department of Plant Biology, Faculty of Biology, University of Murcia, Murcia, Spain

Salah M. Aleid Date Palm Research Center of Excellence, King Faisal University, Al-Hassa, Saudi Arabia

Department of Food and Nutrition Sciences, King Faisal University, Al-Hassa, Saudi Arabia

Ibrahim J. Al-Jboory Plant Protection Department, College of Agriculture, University of Baghdad, Baghdad, Iraq

Mohammed A. Al-Khalifa Department of Agriculture Affairs, Ministry of Municipalities Affairs and Urban Planning, Budaiya, Kingdom of Bahrain

Jameel M. Al-Khayri Department of Agricultural Biotechnology, College of Agriculture and Food Sciences, King Faisal University, Al-Hassa, Saudi Arabia

Thuraya A. Almansoori Department of Biology, College of Science, University of Bahrain, Sakhir, Kingdom of Bahrain

Lateefa Al-Sabah Biotechnology Program, Environment and Life Sciences Research Center Kuwait Institute for Scientific Research, Safat, Kuwait

Mohammed S. Al-Yafai Department of Agricultural Affairs, Ministry of Environment, Doha, Qatar

Rashid Al-Yahyai Department of Crop Sciences, College of Agricultural and Marine Sciences, Sultan Qaboos University, Al-Khod, Oman

Asunción Amorós Department of Applied Biology, Escuela Politécnica Superior de Orihuela, Miguel Hernández University of Elche, Orihuela, Spain

Jawad Ashkanani Biotechnology Program, Environment and Life Sciences Research Center Kuwait Institute for Scientific Research, Safat, Kuwait

Saeed Ba-Angood Department of Plant Protection, University of Aden, Khormaksar, Aden, Yemen

Saleh M. Bader West Asia Research Program, The International Center for Agricultural Research in the Dry Areas (ICARDA), Baghdad, Iraq

Encarna Carreño Department of Plant Biology, Faculty of Biology, University of Murcia, Murcia, Spain

Yuval Cohen Department of Fruit Tree Sciences, Institute of Plant Sciences, Volcani Research Center, Bet Dagan, Israel

Gisela Díaz Department of Applied Biology, Torreblanca, Miguel Hernández University of Elche, Elche, Spain

Baruch (Buki) Glasner Agricultural Manager, Department of Agriculture, HADIKLAIM, Tel Aviv, Israel

Nadia Haider Department of Molecular Biology and Biotechnology, AECS, Damascus, Syria

Saeed Hajian Department of Plant Breeding, Date Palm and Tropical Fruits Research Institute of Iran, Ahvaz, Iran

Zohreh Hamidi-Esfahani Department of Food Science and Technology, Faculty of Agriculture, Tarbiat Modares University, Tehran, Iran

Kadhim M. Ibrahim Department of Biotechnology, College of Science, Al-Nahrain University, Baghdad, Iraq

Shri Mohan Jain Department of Agricultural Sciences, University of Helsinki, Helsinki, Finland

Dennis V. Johnson Consultant, Cincinnati, OH, USA

M. Mumtaz Khan Department of Crop Sciences, College of Agricultural and Marine Sciences, Sultan Qaboos University, Al-Khod, Oman

Hussam S.M. Khierallah Date Palm Research Unit, College of Agriculture, University of Baghdad, Dawoodi, Baghdad, Iraq

Emilio Laguna Centro para la Investigación y Experimentación Forestal – Servicio de Vida Silvestre, Generalitat Valenciana, Valencia, Spain

Shaimaa M. Mahdi Tropical Fruit Department, Central Agricultural Pesticides Laboratory (CAPL), Agricultural Research Center, Al Giza, Egypt

Ghulam S. Markhand Bioassay Department, Date Palm Research Institute (DPRI), Shah Abdul Latif University, Khairpur, Sindh, Pakistan

Abdulaziz M.A. Mohamed Department of Life Sciences, Arabian Gulf University, Manama, Kingdom of Bahrain

Asunción Morte Department of Plant Biology, Faculty of Biology, University of Murcia, Murcia, Spain

Nazir H. Muhammed Department of Agricultural Affairs, Ministry of Environment, Doha, Qatar

Concepción Obón Department of Applied Biology, Escuela Politécnica Superior de Orihuela, Miguel Hernández University of Elche, Orihuela, Spain

Sunil Pareek Department of Horticulture, Rajasthan College of Agriculture, Maharana Pratap University of Agriculture and Technology, Udaipur, Rajasthan, India

Diego Rivera Department of Plant Biology, Faculty of Biology, University of Murcia, Murcia, Spain

Chellan Sudhersan Biotechnology Program, Environment and Life Sciences Research Center Kuwait Institute for Scientific Research, Safat, Kuwait

Jibi Sudhersan Biotechnology Program, Environment and Life Sciences Research Center Kuwait Institute for Scientific Research, Safat, Kuwait

Chapter 1
Introduction: Date Production Status and Prospects in Asia and Europe

Dennis V. Johnson, Jameel M. Al-Khayri, and Shri Mohan Jain

Abstract Successful future development of date palm depends largely on evaluating, utilizing, and conserving genetic resources; assessing the value of present and potential cultivars; promoting the best cultivation, processing, and marketing practices; and broadening the number of fruit and tree products. The date palm was domesticated at the dawn of agriculture, new technologies were adopted for its cultivation, the palm was dispersed widely by humans and it has become a major world tree crop with the introduction of modern plantations. France, the USA, and international organizations have contributed to date palm development and more recently domestic programs in the larger producing countries. Creating a new international date palm organization to bring together research and development efforts and to serve as a resource center would be beneficial. Date cultivation originated in Iraq and that country has always been a major producer; Iran, Saudi Arabia, Pakistan, Oman, and the United Arab Emirates also are all primary world producers. These countries present a mixed pattern of old and new date cultivation which receive federal government support for research and development. In recent decades, cultivation in the Arabian Peninsula has expanded significantly with new plantations using tissue-cultured plants. Date production in Pakistan and India comes mostly from seedling date palms under traditional cultivation. Modern date cultivation is being established to meet the large domestic demand. Israel's date production is unique coming entirely from modern plantations. Spain's very small date production from seedling dates is a novelty and likely to remain so. Because they are not included in country chapters, summary accounts of date cultivation in Jordan and the United Arab Emirates are included.

D.V. Johnson (✉)
Consultant, Cincinnati, OH, USA
e-mail: djohn37@aol.com

J.M. Al-Khayri
Department of Agricultural Biotechnology,
College of Agriculture and Food Sciences,
King Faisal University, Al-Hassa, Saudi Arabia
e-mail: jkhayri@kfu.edu.sa

S.M. Jain
Department of Agricultural Sciences, University of Helsinki,
PL-27, Helsinki 00014, Finland
e-mail: mohan.jain@helsinki.fi

© Springer Science+Business Media Dordrecht 2015

J.M. Al-Khayri et al. (eds.), *Date Palm Genetic Resources and Utilization: Volume 2: Asia and Europe*, DOI 10.1007/978-94-017-9707-8_1

Keywords Dispersal and distribution • Domestication • Food value • Modern plantations • Production statistics • Research and development

1.1 Introduction

The date palm (*Phoenix dactylifera* L.) is one of the most valuable domesticated fruit trees because of ritual significance in human societies, health benefits, productive capacity in harsh semiarid and arid environments, and the range of subsistence products from its fruits and other parts of the large palm. In planning this book, the editors strived for complementarity with their earlier book, *Date Palm Biotechnology* (Jain et al. 2011), as well as two books by other authors, *Dates*: *Production, Processing, Food and Medicinal Values* (Manickavasagan et al. 2012) and *Dates*: *Postharvest Science, Processing Technology and Health Benefits* (Siddiq et al. 2014). Each of these three books is composed of individual chapters, on a broad range of topics. None of them deals with the date palm status of genetic resources and production in each of the date palm-growing countries.

Additionally in this book, four major themes were identified as exemplified in the book's title. Genetic resources of the date palm are of great importance in the future of the crop, for two reasons. First, the diversity present in the entire *Phoenix dactylifera* gene pool represents an invaluable resource for conventional and molecular breeding of future date palms of high fruit quality, high fruit yield, tolerance to abiotic stresses, and resistance to major pests and diseases. Genetic erosion as well as genetic sedimentation imperil the date palm. Second, the strong and increasing trend toward cultivation of known elite cultivars, so evident in nearly every new date palm plantation, is being carried on at the expense of a narrowing of the overall cultivar diversity as traditional cultivars are less likely to be propagated and unevaluated seedling date palms (Johnson et al. 2013) are eradicated as if they were weeds. Cultivar assessment, for all but a few well-known elite types, is at a very rudimentary stage of knowledge. Because the unassessed cultivars are often native to a single country, studies of promising cultivars must be organized at the national level, with active collaboration among neighboring countries which may harbor close relatives; in certain instances the same cultivar may be present in adjoining areas within a country or in adjacent countries and bear different vernacular names.

Cultivation practices exhibit considerable variability across the date-growing world. Certain accepted practices have been adapted to local environmental conditions, such as early picking and postharvest ripening because of climatic limitations preventing full maturity of the fruits on the trees. Best management practices have yet to be perfected for date palm growing under a basic set of environmental factors. The country-level practices described in this book provide a point of initiation to begin developing guidelines for growers to assure them with maximum return for their investment and labor. The novel products part of the book title was included to prompt individual countries to examine the products derived from date fruits not marketed as fresh or dried products. Descriptions of derived date fruit products in the various countries demonstrate the cottage-level and small- and large-scale industrialization in existence. These descriptions make possible a cross-fertilization of ideas from country to country to valorize date fruit production to the highest level possible, which represents a potential enhancement of income to producers and manufacturers. Although the novel products described in the book are almost exclusively focused on the fruits, other parts of the date palm tree have a range of other subsistence and industrial products such as those derived from the leaves, petioles, edible heart, and stem wood. These nonfruit date products merit a fresh in-depth assessment to update the earlier publication *Date Palm Products* (Barreveld 1993).

1.2 Agrobotany, Domestication, and Dispersal of Date Palm

1.2.1 Agrobotanical Description

A mature date palm can reach 30 m in height and is the tallest of the *Phoenix* species. The stem has a diameter of 40–50 cm, bearing suckering offshoots at its base during the early years of growth. The leaves are pinnate and erect with numerous stiff leaflets; the rachis is armed with broad sharp spines on each side. An adult palm has a moderately dense crown of 100 or more leaves 3–6 m in length; each leaf has a life span of 3–7 years; about 12 new leaves are produced each year. The date palm is dioecious, bearing male and female flowers on separate trees. Dioecy in date palm is an impediment to seed propagation and traditional crossbreeding experimentation as equal numbers of male and female plants are produced and their gender not apparent in the field until flowering begins. Propagated by seed, it reaches sexual maturity at about 5–7 years of age, significantly sooner if by offshoots. Flowering is normally annual. Fruit development from pollination to maturity is reached in about 200 days and is variable in shape, color, and size, 4–7 × 2–3 cm, larger than in any other *Phoenix* species. Yields are highly variable, 20–100 kg per adult tree, and depend upon cultivar, environmental conditions, and cultivation practices. The Medjool cv. produces the largest commercial fruits. Fruits contain a single elongate seed, variable in shape and size, 20–30 × 5–8 mm, containing a useful oil. On average, a plantation has an economic life of about 50 years, although the trees will continue to produce fruit beyond that age, albeit at lower yields. When trees become very tall, they are more expensive to care for and harvest, which negatively affects the production economics.

The true date palm, *Phoenix dactylifera* L., is one of 14 recognized species of the genus (Barrow 1998). It is most closely related to the Cretan date palm (*P. theophrasti* Greu.) of the Eastern Mediterranean and to the Canary Islands date palm (*P. canariensis* Chab.), native to that island group, and the sugar date palm (*P. sylvestris* (L.) Roxb.) which is native to South Asia. Where *Phoenix* species distributions overlap in the wild, natural hybrids occur; likewise *Phoenix* palms grown in botanical gardens readily hybridize (Gros-Balthazard 2013). This facility makes artificial crossbreeding of species easily achieved. The date palm has 36 chromosomes ($n = 18$; $2n = 36$).

1.2.2 Domestication

The date palm is one of the world's very early domesticates and a traditional Old World fruit. Date palm, olive tree (*Olea europaea* L.), common fig tree (*Ficus carica* L.), and wine grape (*Vitis vinifera* L.) together are a seminal quartet of ancient fruit crops closely associated with the earliest stages of agriculture. The first domestication of the date palm is believed to have taken place at least 6,000 years ago in Mesopotamia, the lands between the Tigris and Euphrates rivers, in what is today Iraq. The area is often referred to as the *cradle of agriculture*. The early history of date palm as a cultivated palm is known only from archaeological records and ancient cuneiform clay tablets. It is presumed that wild date palms were selected for propagation on the basis of their possessing desirable fruit characteristics.

Over an indeterminate period of time following the first planting of date palms, several simple but important innovations took place in Mesopotamia which fostered the development of date palm cultivation. In roughly presumed chronological order, the following took place:

(a) Female and male palms were distinguished.
(b) Organized plantings made with fixed spacing.
(c) Palms under cultivation segregated by gender.
(d) Irrigation and water management.
(e) Artificial pollination.
(f) Separation of offshoots for propagation.
(g) Stages of fruit development recognized and named.

From the latter are derived the five sequential stages of fruit growth which were named hababauk, kimri, khalal, rutab, and tamar. Somewhere along the way, the first cultivar was given a name, which may have been Zahidi in the Basrah area of present-day Iraq, a cultivar prominent in that area to the present. The foregoing is summarized from Dowson (1921, 1923, 1982), Popenoe (1973), and Pruessner (1920).

Since the date palm was initially domesticated, new cultivars have repeatedly been created as a result of seedling propagation and the selection of those of superior fruit quality subsequently propagated by offshoots and assigned individual names. Cultivar names typically derive from either the particular fruit characteristics, the locality where the palm originated, or the name of the farmer who selected it.

1.2.3 Dispersal and Geographic Distribution

From its presumed homeland in Mesopotamia, over succeeding millennia, date palms were dispersed by humans to the east, west, and south and took hold wherever climatic conditions and water sources were favorable for fruit production. Initial dispersal was clearly by means of seed. Even when abandoned, dates persisted and became naturalized where conditions allowed. Munier (1973), in a general study of date growing, described dispersal of date palm in two major historic routes. One route is from Mesopotamia south into the Arabian Peninsula and eastward to undivided India. A second route begins in Egypt (itself the location of an independent domestication of date palm) across North Africa west to Morocco. The Caravan trade and the Nile River may have been the means of spreading date palms southward into the Sahelian countries and Sudan in the early days. From North Africa, dates were carried to Spain by the Moorish invasion in the eighth century, or possibly earlier. Seedling dates were grown successfully for fruit in southwestern Spain, although the climate is marginal for carrying fruits to maturity on the tree. In the sixteenth century, the Spaniards successfully introduced date seeds to the Americas, initially to Peru and later to Mexico, and achieved modest fruit production. In the mid- to late 1800s, there are reports of seedling dates being grown in Southern Africa, Australia, and New Caledonia (Johnson 2010). In the past 100 or so years, as offshoots have successfully been transported over long distances, modern date palm plantations employing elite cultivars have emerged or are emerging in countries such as the UAE, Oman, Kuwait, USA, Mexico, India, Israel, Jordan, Namibia, and Australia, as well as in several Sahelian countries in Africa.

1.3 The Rise of Modern Date Palm Plantations

Scientific literature on date palm published before the twentieth century consists mostly of brief accounts of date growing in a few countries or entries in reference volumes; an exception is the book *The Date Palm in India* (Bonavia 1885), which brings together information about date growing in the Persian Gulf and British India.

The US Department of Agriculture created a program in the 1890s to promote an irrigated commercial date industry in the American Southwest and in 1904 established a date experiment station in Mecca, California, later moved to Indio. Offshoots of recognized cultivars were imported from North Africa and the Middle East and provided to pioneer farmers. Commercial date production began in 1912. Needing a source of technical information, in 1924 the date growers created the Date Growers' Institute, which organized technical meetings each year and published an annual review (DGIAR 1924–1979). The USDA published a number of technical studies on date growing in the USA (e.g., Nixon and Carpenter 1978; Swingle 1904) as well as bulletins on date cultivation in other countries such as Egypt (e.g., Mason 1923). Three general books have helped to popularize dates in the USA (Dunham

1948; Paulsen 2005; Simon 1978). US scientific publications on date palm have made a major contribution to modern date growing. However, scientific research on date palm in the USA was greatly reduced after cessation of the date growers' annual meetings in 1979 and closure of the date experiment station a year later.

The French also made important contributions in the twentieth century to modern date cultivation in their North African colonies by organizing the first two international date palm conferences. In 1931 a weeklong meeting was held in Biskra, Algeria, with presentations on broad climatic, agronomic, and economic aspects of date production. One of the recommendations was to promote the Deglet Noor cultivar (Sem Dattier 1931). It was not until 1950 that a second such conference took place in Tunis, Tunisia. Among the recommendations made were to conduct scientific research on date palm agronomy and adopt uniform fruit export standards (Cong Int Datte 1951). French scientists produced numerous articles and technical reports on dates in Africa, broadening the knowledge base. Pereau-Leroy (1958) authored the first national study of dates in Morocco and Munier (1973) the first general scientific study of date cultivation.

International backing for data palm development was provided by FAO in the second half of the twentieth century. This was initiated by sponsorship of three technical conferences on date palm: one in Tripoli, Libya, in 1959 and two in Baghdad, Iraq, in 1965 and 1975. No proceedings were published, but the presentations by leading date specialists from around the world generated background materials for the issuance by FAO of a series of important books on date palm (Dowson and Aten 1962; Dowson 1982; Barreveld 1993; Zaid 1999, 2002). Another contribution made to date palm science was *The Date Palm Journal* (FAO 1981–1988).

A recent contributor to date palm development is the International Center for Agricultural Research in the Dry Areas (ICARDA), recently relocated from Aleppo, Syria, to Beirut, Lebanon. One of its projects focuses on increasing date production among smallholder farmers in Iraq (Mazid et al. 2013), ravaged by war; another involves enhancing date palm production systems in the GCC (Gulf Cooperation Council) countries (ICARDA 2011). This project is being led by an ICARDA date palm specialist based in Oman. A study of date harvest and postharvest handling was a part of the project (Kader and Hussein 2009).

With the major contributions outlined above from the USA, France, FAO, ICARDA, and numerous lesser but important efforts, a firm scientific foundation has been established for the cultivation and processing of dates, upon which the future of the crop is being built.

1.4 Research and Development

Apart from the international development efforts described above which have fostered modernization of date palm production to varying degrees in individual countries, national programs also have played an important role. Major common problems shared by all date-producing countries are associated with cultivation practices, pests and diseases, soil and water conditions, harvest and postharvest practices, processing facilities, technical support and extension, the agricultural labor force, as well as domestic and international marketing. The best example of differences is pests and diseases. Remote areas of date growing in southern Africa and the Americas have few such problems to contend with, whereas Morocco and Algeria are experiencing significant crop and tree losses given the occurrence of bayoud disease and the red palm weevil. Climatic factors present problems only in certain countries such as Pakistan and India where annual monsoon rain occurs before date fruits are mature; but in Saudi Arabia, for instance, climate presents no hindrance to date growing. Each date-growing country has a unique set of impediments to improving production, and the severity of each should set the agenda for domestic R & D.

This book provides information on the research centers carrying out research on date palm on various aspects, supported by mainly the federal government in academic institutions and private investment. A close link exists between the magnitude of date production and the funding to support date palm R & D. Egypt, Iran, Iraq, Saudi Arabia, and Pakistan, each ranking among the top six producers worldwide, have established research institutes devoted to date palm, which are funded by the federal governments. In other date palm-producing countries, including Algeria and Morocco, date palm research is carried out by the national agricultural research centers. In Syria and Palestine date palm currently is considered as a minor crop, with annual production of 3,986 and 3,600 mt, respectively. The paucity of federal government funding becomes a major obstacle in developmental efforts. Therefore, these countries must fall back on larger date-producing countries for importing improved date palm cultivars, technologies to control pests and diseases, and agro-food-based products and other by-products. Any important factor in domestic support for date palm research depends on the priority of the national governmental support in creating date palm-based agro-food industry and local and international marketing boosting local economy.

Throughout the date palm world in general, insufficient attention has been given to programs of genetic improvement employing emerging biotechnology procedures. This neglect may be attributed to the limitation of research funds, inadequate advance research infrastructure, and scarcity of experts in many of the date palm-producing nations. Interrelated issues include the need for precise knowledge of the genetic resources present in a country, including that contained in seedling date populations, and the effective means to conserve genetic resources through germplasm gardens and cryopreservation.

Given the problems shared by date-growing countries, it would be of great benefit to create an international date palm center. The advantages of such a center include the following:

(a) Avoid expensive duplication of effort in addressing major pest and disease problems.
(b) Enhance opportunities for collaborative research both at the bilateral and international levels.
(c) Develop and disseminate information on best practices in date cultivation, harvesting, postharvest handling, and marketing by developing an interactive

website. An excellent model for date palm exists in the Coconut Timeline (http://cocos.arecaceae.com/).

(d) Develop international industry descriptors and standards for fruit quality and packaging and marketing.
(e) Maintain a database of world date palm cultivars and their conservation status.
(f) Maintain information on date palm genetic diversity, genetic erosion, conservation, and utilization of germplasm.
(g) Develop programs on health benefits of dates and various commercial food products.
(h) Develop date palm functional genomics for studying useful genes leading to genetic improvement of date palm growing under climate change, enhanced fruit quality, and industrial products.

1.5 Date Fruit Production Statistics

During the planning of the present book, available data on date production were critically evaluated in order to determine the relative importance of the crop within a country based on national production in relation to global production quantities. The Food and Agriculture Organization of the United Nations (FAO) collects data from member nations and publishes annual agricultural production data on major crops. These data are the most comprehensive available and of primary concern in this book. In addition, FAO publishes data on the value of crops, as well as import and export quantities, but those data are of lesser importance given the focus of the book on production.

Domestic production statistics in most countries are incomplete or difficult to access. The exceptions are Saudi Arabia and Morocco which have published detailed accounts of their date palm cultivation which include some data on the numbers of palms being grown and area under cultivation, along with cultivar descriptions (INRA 2011; Min Agr Saudi Arabia 2006).

Aggregate world date fruit production for 2012 amounted to 7,548,918 mt, according to FAOSTAT. Table 1.1 lists reported producing countries in descending order of magnitude. The production figures given in Table 1.1 reflect the extent of date growing around the world. These data are a mixture of production figures provided by the various countries and FAO estimates. Burkina Faso, Ethiopia, Mali, and Senegal, discussed in this book, are not included in FAO data, apparently because commercial date production hardly exists yet.

There are some serious problems with the FAO data that have a bearing on this book and need to be discussed; date production reported for five countries is incorrect. The production reported for China of 150,000 mt represents the production of the red date fruit, rather than the true date palm. Red date or jujube is scientifically *Zizyphus jujuba* (L.) H. Karst., a tree or shrub of the Rhamnaceae or Buckthorn family. China has no production of true dates. A similar situation is found with respect to the reporting of 31,675 mt of date production in Turkey. In this instance, the figures represent production of the common fig. This misunderstanding arises from the same Turkish word being used to refer to dates and figs. Albania's reported

Table 1.1 Date fruit production by country, 2012, in mt

Ranking	Country	Production (mt)
1	Egypt	1,470,000
2	Iran	1,066,000
3	Saudi Arabia	1,050,000
4	Algeria	789,357
5	Iraq	650,000
6	Pakistan	600,000
7	Sudan (former)	433,500
8	Oman	270,000
9	United Arab Emirates	250,000
10	Tunisia	190,000
11	Libya	170,000
12	China*	150,000
13	Morocco	113,397
14	Yemen	55,181
15	Israel	42,866
16	Kuwait	34,600
17	Turkey*	31,765
18	USA	28,213
19	Mauritania	22,000
20	Qatar	21,843
21	Chad	20,000
22	Niger	17,000
23	Bahrain	15,000
24	Somalia	13,000
25	Albania*	12,935
26	Jordan	10,417
27	Mexico	6,012
28	Spain	4,000
29	Syria	3,986
30	Palestine	3,600
31	Benin	1,300
32	Kenya	1,100
33	Cameroon	600
34	Namibia	400
35	Peru	400
36	Swaziland*	330
37	Djibouti	86
38	Colombia*	30

Source: FAOSTAT (2012)
Countries which erroneously are reported to have date production are indicated with an asterisk

production of 12,935 mt of dates is a reflection of date imports from other countries being considered as national production; the country's climate is unsuited to commercial date production. Swaziland's reported production of 330 mt is incorrect, because there is no commercial date cultivation within the country

(McCubbin M, personal communication, 2014). The neighboring Republic of South Africa has some date growing focused on Medjool cv. in the Limpopo Region (McCubbin 2007), which does not appear in FAO statistics for the RSA. There are no FAO trade data on date imports and exports between the two countries, so it appears unlikely there is any connection. FAO has reported date production for Swaziland each year since 1999, all based on either an estimate or imputation methodology. The most plausible explanation is that the production reported as dates is for some other fruit. Finally, the statistic that Colombia produced 30 mt of dates is not from domestic production but rather from imports and exports. Climatic conditions in the country also are unsuited for commercial date growing. Inclusion of these five countries in world production totals resulted in an overstatement of 195,060 mt in 2012. India should be but is not included in Table 1.1 for unknown reasons; data are either not collected in the country or, if they are, not submitted to FAO. Whatever the case, according to data from a newspaper estimate cited in Chap. 14, India's date production was estimated in 2010/2011 to be at least 120,000 mt, which would place it among the top twelve world producers.

The quality and accuracy of national date production data vary, irrespective of the significance of the crop to the individual country's agricultural sector. Among the factors at play are the following:

(a) The mix of modern formal plantation cultivation and traditional date growing where densities vary and other crops are grown in association such as is typical in oases. Statistical data on the former are much easier to gather and are more accurate.
(b) Planting densities in some countries are highly variable; hence it is difficult to take data on area under cultivation and derive the actual number of trees being grown.
(c) Data from farmers may be understated in countries where taxes are assessed on the basis of total production and likewise the quantities of dates consumed by growers and their families or used as livestock feed are poorly known.
(d) The level of proficiency of the national agency collecting agricultural production data is again highly variable.
(e) Data from one producing country may be double counted. For example, Mexico's small date production takes place just south across the border from the Yuma, Arizona, and Bard, California area, one of the two most important production regions of the USA. An indeterminate quantity of Mexican dates is known to be shipped to Yuma, Arizona, and sold through the large cooperative processing facilities in that city. Other such examples likely exist in some African countries.

Regarding production data by date cultivar at the national level, data are nonexistent. National date production statistics in even the most advanced agricultural countries such as the USA are aggregated and not collected by cultivar. As indicated in the country chapter in these volumes, the most prominent cultivars are identified, but their relative ranking as regards total national production is unknown. The most reliable production data by date cultivar is derived from farm or research station records of per tree productivity.

The lesson to be drawn from this discussion is that date palm production statistics for each country must be appraised for their accuracy and reliability within the context of the conditions which are known to exist in that country.

1.6 Food Value of Dates

In the human diet, date fruits are rich sources of sugars, protein, fiber, minerals, and certain vitamins. A quick energy source because of the high sugar content, dates have been called *nature's candy*; they are an excellent snack food and currently being recommended as such. When the fruits reach maturity (tamar), sucrose inverts into glucose and fructose. Dates contain reasonable amounts of vitamin A, thiamin, riboflavin, and niacin and are a good source of minerals such as potassium, calcium, and iron, as well as of dietary fiber. The burgeoning natural food sector is promoting date consumption as a functional food and as possessing antioxidant qualities. Dates are included in the group of recommended healthy food choices for individuals suffering from type 2 diabetes. Each date cultivar has an individual profile of nutritional values. Table 1.2

Table 1.2 Nutritional value of Deglet Noor and Medjool cv. dates (nutrient values and weights are for edible portion)

		Value per 100 g	
Nutrient	Unit	Deglet Noor cv.	Medjool cv.
Proximates			
Water	g	20.53	21.32
Energy	kcal	282	277
Protein	g	2.45	1.81
Total lipid (fat)	g	0.39	0.15
Carbohydrate	g	75.03	74.97
Fiber, total dietary	g	8.0	6.7
Sugars, total	g	63.35	66.47
Minerals			
Calcium, Ca	mg	39	64
Iron, Fe	mg	1.02	0.90
Magnesium, Mg	mg	43	54
Phosphorus, P	mg	62	62
Potassium, K	mg	656	696
Sodium, Na	mg	2	1
Zinc, Zn	mg	0.29	0.44
Vitamins			
Vitamin. C, ascorbic acid	mg	0.4	0.0
Thiamin	mg	0.052	0.050
Riboflavin	mg	0.066	0.060
Niacin	mg	1.274	1.610
Vitamin B_6	mg	0.165	0.249
Folate, DFE	µg	19	15
Vitamin A, RAE	µg	0	.7
Vitamin A, IU	IU	10	149
Vitamin E	mg	0.05	–
Vitamin K	µg	2.7	2.7

Source: US Department of Agriculture National Nutrient Database for Standard Reference, Basic Reports 9087 and 9421. Accessed 18 Sept 2014 gives the nutritional values for two common commercial cultivars: Deglet Noor and Medjool. Ahmed et al. (2014) provide a good current summation of research of the composition and nutritional value of date fruits, and Vayalit (2014) of the bioactive compounds and functional properties of the fruit.

1.7 Date Production in Asia and Europe

Together the countries included in this volume account for just over one-half of the world's date production. Iran, Saudi Arabia, and Iraq consistently rank among the top five world producers.

An insect pest problem common to the region is the red palm weevil (*Rhynchophorus ferrugineus* Oliv.), of tropical South Asian origin, which has disseminated westward and spread throughout the Middle East and Mediterranean Basin. This destructive pest of date palm attacks the stem and if untreated can kill the tree. No highly effective control measures have yet been found.

The countries in this region are characterized by large populations of seedling dates, often grown under haphazard conditions, which have low fruit productivity. Compared to elite cultivars, seedling dates often exhibit greater tolerance to biotic and abiotic stresses; therefore they may harbor genetic resources of value to plant breeders and for that reason merit scientific assessment (Johnson et al. 2013).

The South Asian date-growing countries (Pakistan and India), Spain, and the Mediterranean parts of the North African countries share a problem of marginal climatic conditions for date growing either because of rainfall during the period of fruit maturation or an insufficient number of days with suitably high temperatures to fully mature fruit on the tree. Under such conditions growers elect to plant early-bearing cultivars which ripen before the annual monsoon rains or cooler weather, and/or the fruits are harvested at khalal or rutab stages and artificially ripened; the latter involving extra labor costs.

Two Asian countries, United Arab Emirates and Jordan, are not covered in this book, despite their significance and the concerted efforts by the editors to involve chapter authors from the two countries. Brief accounts of the date production status in the two countries, derived from available source materials, are given below.

1.7.1 United Arab Emirates

The UAE is the ninth largest date producer in the world, having ascended rapidly to that rank. From 1980 to 2012, the country experienced a five-fold increase in date production (51,157–250,000 mt), according to FAO data. Currently, it is estimated to have some 44 million date palms. In addition to its cultivation for fruit, the date palm is by far the most common ornamental plant in the UAE.

A considerable number of introduced and indigenous date cultivars are grown in the UAE. Ba-Angood and Ahmed (1984) studied the chemical composition of 64 major date cultivars grown in the UAE, introduced and native, but did not give their individual origins. A new study of genetic diversity of date palm in the UAE, also introduced and native, enumerated 72 different cultivars, with their countries of origin. In the study, 22 native cultivars were included: Bahlani, Khoor, Menaa, Ayn Baqar, Anwan, Jask Falk, Jash Jafar, Khad Alfaras, Shahil, Um AlQuareen, Um AlSalaa, Abu AlAuzook, Abu Kebal, Azmi, Dabbassi, Jash AlAlwan, Jash Samra, Khad Alasal, Khattery, Nawander, Salahee, and Um AlDihen (Chaluvadi et al. 2014). A comparison of the cultivar names in the two studies is challenging because of different spellings, but it appears about 14 are common to both. These figures suggest that overall about 120 date cultivars are grown in the UAE. Zaid (2003) also estimated the gene pool at approximately 120 cultivars. He also enumerated 14 new cultivar introductions: Khlas, AbouMaan, Hallawi, Khissab, Khnaize, Nabut Saif, Jabri, Hillali, Lulu, Chichi, Khadraoui, Sakii, Sultana, and Barhi.

Other date research specific to the UAE is a study by Dakheel (2003) on salinity and drought stress and another by Al Shakir (2003) on date postharvest technology.

Traditional date palm cultivation has existed in what is now the UAE for millennia. Salem (1998) described date production according to regions of the Emirates. The Abu Dhabi region is the largest, accounting for about 45 % of production, followed by the Middle Region (25 %) and the North and East regions each with 15 %. The cultural practices described are irrigation, fertilizers (manure), hand pollination, fruit thinning by bunch or strand removal, and pruning of leaves and spines. Major constraints included lack of irrigation water and salinity of water and soils, as well as marketing difficulties. A problem of older date farms is poorly performing cultivars and high planting densities which reduce yields while consuming large quantities of irrigation water. Otier (1997) also provides some information about small-farmer practices. Among the constraints to expansion of existing date growing cited by Zaid (2003) are water resources for irrigation, a shortage of farm workers, pests and diseases, and weak farm management practices. Among the pests, red palm weevil, which reached the UAE in 1985, is the most serious (Zaid 2003).

An earlier problem of a scarcity of planting material has been dealt with. The UAE now has ample sources of tissue-cultured date palms of leading cultivars, for both domestic and foreign date growers, produced by the Date Palm Tissue Culture Laboratory, UAE University, Al Ain; Green Coast Nursery, Fujairah; and Al Wathba Marionnet, Abu Dhabi.

The UAE has made significant contributions to date palm science through organizing and hosting five international date palm conferences, with published proceedings (First Int Conf 1998; Second Int Conf 2001; Zaid et al. 2007; Zaid and Alhadrami 2010; Fifth Int Conf 2014). Since 2000, the United Nations Office for Project Services (UNOPS) has supported a date palm research and development program at the UAE University in Al Ain. This program has provided technical support to the UAE, the Gulf Region, and other date-producing countries. The annual Khalifa International Date Palm Awards, created in 2007, with a secretariat in Al

Ain Abu Dhabi, is providing recognition to the contributions of both individuals and institutions in the advancement of date palm cultivation and production.

1.7.2 Jordan

Date production is highly concentrated in the Jordan Rift Valley, which extends along the entire western boundary of the country. Elevations below sea level in most of the valley provide a frost-free environment of higher than average temperatures for the latitude, conditions ideal for date growing. Reportedly, there are 335 date farms in the country containing about 250,000 trees. Date production has experienced a significant recent increase, climbing from 220 ha in 1995 to 2,497 ha in 2011. Production in 2012 amounted to 10,417 mt. According to a 2005 survey, in order of importance, the major cultivars grown are Medjool, Barhi, Khadrawy, Khalas, Deglet Noor, Zaghloul, Hayany, Zahidi, Mektour, and Ahmar Talal. The Jordanian Date Producers and Marketing Association was created in 2005 to market the expanding fruit production. The major constraint to increased date growing is a reliable supply of irrigation water which is not too salty. Red palm weevil is the leading insect pest. The foregoing information is drawn from (Rumman and Al-Zubi 2014).

1.8 Conclusions and Prospects

Chapters covering 13 Asian date-producing countries are included in this Volume 2. They range in importance from Iran and Saudi Arabia, each producing more than one million mt annually, to Syria and Palestine with production below 4,000 mt per year. The inclusion of Spain provides an exceptional instance of growing date palms in a climatic area that is marginal for quality fruit production, but where date growing is motivated, to a large degree, by a profound historic appreciation of their distinctive ornamental value in the southwestern region of the country.

Common to all of the Middle Eastern countries is the antiquity of date cultivation, including the homeland of domestication in the Tigris-Euphrates Valley. A history of cultivation reaching back some 6,000 years has produced a multitude of cultivars, many of which are of commercial importance today, such as Barhi, Dayri, Halawy, Khadrawy, and Zahidi, all historic Iraqi cvs. Seedling date populations exist throughout these countries, as well as in South Asia where the palm was introduced about 4,000 years ago (Pareek 2015). The Asian countries represent a rich storehouse of genetic variability of the date palm. Another shared characteristic of Asian date-growing countries is the number of prepared products from date fruit, such as syrup, sugar, vinegar, and alcohol, along with a host of artisanal objects made from the leaves, midribs, leaf bases, and stem wood.

Prospects for increased date fruit production in Asia are encouraging. Attempts are underway to rehabilitate the Iraqi date industry so ravaged by war. As discussed above, new plantations established in the UAE have advanced the country to be among the top ten world producers. Saudi Arabia has maintained its prominence as a world producer over recent decades. India has a high domestic demand for dates currently being met chiefly from imports, giving the country a pronounced incentive to increase production; private and public programs to that end are being pursued. Spain, the sole European country included in this volume, has some potential to improve local fruit production, but the dates grown are likely to remain an exotic gourmet food item.

References

Ahmed J, Al-Jasass FM, Siddiq M (2014) Date fruit composition and nutrition. In: Siddiq M et al (eds) Dates: postharvest science, processing technology and health benefits. Wiley Blackwell, Chichester, pp 261–283

Al Shakir S (2003) Date post-harvest technology in the United Arab Emirates. In: The date palm from traditional resource to green wealth. The Emirates Center for Strategic Studies and Research, Abu Dhabi, pp 171–198

Ba-Angood SA, Ahmed MS (1984) Chemical composition of major date cultivars grown in the United Arab Emirates. Date Palm J 3(2):381–394

Barreveld WH (1993) Date palm products. Agricultural services bulletin 101. FAO, Rome

Barrow S (1998) A revision of *Phoenix*. Kew Bull 53(3):513–575

Bonavia E (1885) The date palm in India. Thacker, Spink & Co., Calcutta

Chaluvadi SR, Khanam S, Aly MAM, Bennetzen JL (2014) Genetic diversity and population structure of native and introduced date palm (Phoenix dactylifera) germplasm in the United Arab Emirates. Trop Plant Biol 7:30–41

Cong Int Datte (1951) Congrès International de la Datte. Bull Infor 38, Office Tunisien de Standardisation, Tunis

Dakheel A (2003) Date palm and biosaline agriculture in the United Arab Emirates. In: The date palm from traditional resource to green wealth. The Emirates Center for Strategic Studies and Research, Abu Dhabi, pp 199–211

DGIAR (1924–1979) Date growers' institute annual report, vol 1–54. Coachella Valley, California

Dowson VHW (1921, 1923) Dates and date cultivation in the 'Iraq. Parts I, II, III. Heffer & Sons, Cambridge

Dowson VHW (1982) Date production and protection. Plant production and protection paper 35. FAO, Rome

Dowson VHW, Aten A (1962) Dates – handling, processing and packing. Agricultural development paper 72. FAO, Rome

Dunham WA (1948) It's a date. Publication Press, Pasadena

FAO (1981–1988) The date palm journal. Regional Project for Palm & Dates Research Centre in the Near East & North Africa, Baghdad

FAOSTAT (2012) Food and Agricultural Organization Corporate Statistical Database. FAO, Rome

Fifth Int Conf (2014) Fifth international date conference, Abu Dhabi. Proceedings forthcoming, 16–18 Mar 2014. Acta Hort forthcoming

First Int Conf (1998) Proceedings: the first international conference on date palms, UAE University, Al-Ain, 8–10 Mar 1998

Gros-Balthazard M (2013) Hybridization in the genus *Phoenix*: a review. Emir J Food Agric 25(11):831–842

ICARDA (2011) Development of sustainable date palm production systems in the Gulf Cooperation Council countries. Project report 1, International Center for Agricultural Research in the Dry Areas, Aleppo

INRA (2011) Atlas du palmier dattier au Maroc. Institut National de la Recherche Agronomique (In French and Arabic)

Jain SM, Al- Khayri JM, Johnson DV (eds) (2011) Date palm biotechnology. Springer, Dordrecht
Johnson DV (2010) Worldwide dispersal of the date palm from its homeland. Acta Hort 882:369–375
Johnson DV, Al-Khayri JM, Jain SM (2013) Seedling date palms (*Phoenix dactylifera* L.) as genetic resources. Emir J Food Agric 25(11):809–830
Kader AA, Hussein AM (2009) Harvesting and postharvest handling of dates. ICARDA, Aleppo
Manickavasagan A, Essa MM, Sukumar E (eds) (2012) Dates: production, processing, food, and medicinal values. CRC Press, Boca Raton
Mason SC (1923) The Saidy date of Egypt: a variety of the first rank adapted to commercial culture in the United States. USDA Bull 1125, Washington DC
Mazid A, Al-Hashimy MJ, Zwain A, Haddad N, Hadwan H (2013) Improved livelihoods of small-holder farmers in Iraq through integrated pest management and use of organic fertilizer. Socioeconomic and Agricultural Policy Report, Working Paper 11, ICARDA, Beirut
McCubbin MJ (2007) The South African date palm industry – strengths and weaknesses. Acta Hort 736:53–57
Min Agr Saudi Arabia (2006) The famous date varieties in the Kingdom of Saudi Arabia. Ministry of Agriculture, Riyadh (In Arabic and English)
Munier P (1973) Le palmier-dattier. Maisonneuve & Larose, Paris
Nixon RW, Carpenter JB (1978) Growing dates in the United States. USDA Agr Infor Bull 207, Washington DC
Otier M (1997) Phoenix dactylifera in the United Arab Emirates. Principes 41(1):29–35
Pareek S (2015) Date palm status and perspective in India. In: Al-Khayri JM et al. (eds) Date palm genetic resources, cultivar assessment, cultivation practices and novel products, vol 2. Springer, Dordrecht, p 441–485
Paulsen ME (2005) The amazing story of the fabulous Medjool date. Mark Paulsen Press, Tualatin
Pereau-Leroy P (1958) Le palmier dattier au Maroc. Ministère de l'Agriculture, Rabat
Popenoe PC (1973) The date palm. Field Research Projects, Coconut Grove, Miami
Pruessner AH (1920) Date culture in ancient Babylonia. Am J Sem Lang Lit 36:213–232
Rumman GA, Al-Zubi J (2014) Date palm in Jordan. Unpublished manuscript, Amman, Jordan
Salem J (1998) Date production and marketing in the United Arab Emirates. In: Proceedings: the first international conference on date palms, UAE University, Al-Ain, 8–10 Mar 1998, pp 531–543
Sec Int Conf (2001) Proceedings: the second international conference on date palms, UAE University, Al-Ain, 25–27 Mar 2001
Sem Dattier (1931) Semaine du dattier – compte rendu general. Imp Imbert, Algiers
Siddiq M, Aleid SM, Kader AA (eds) (2014) Dates: postharvest science, processing technology and health benefits. Wiley Blackwell, Chichester
Simon H (1978) The date palm: bread of the desert. Dodd Mead & Co., New York
Swingle WT (1904) The date palm: its utilization in the southwestern states. Bur Plant Ind Bull 53, USDA, Washington DC
Vayalit PK (2014) Bioactive compounds, nutritional and functional properties of date fruit. In: Siddiq M et al (eds) Dates: postharvest science, processing technology and health benefits. Wiley Blackwell, Chichester, pp 285–303
Zaid A (ed) (1999) Date palm cultivation. Plant production and protection paper 156, FAO, Rome
Zaid A (ed) (2002) Date palm cultivation, Rev. ed. Plant production and protection paper 156, FAO, Rome
Zaid A (2003) Date palm culture in the UAE: present situation and future potential. In: The date palm from traditional resource to green wealth. The Emirates Center for Strategic Studies and Research, Abu Dhabi, pp 151–169
Zaid A, Alhadrami G (eds) (2010) Proceedings of the fourth international date palm conference, March 15–17, 2010. Acta Hort 882:1–715
Zaid, A, Hegarty V, Al Kaabi HHS (eds) (2007) Proceedings of the third international date palm conference, February 19–21, 2006. Acta Hort 736:1–586

Part I
Asia

Chapter 2
Date Palm Status and Perspective in Iran

Saeed Hajian and Zohreh Hamidi-Esfahani

Abstract Date palm is the second most important horticultural crop of Iran after pistachio. It is mostly grown in a southern belt of the country. More than 400 cultivars, among 3,000 named cultivars worldwide, are found in Iran, thus providing a major global date palm gene pool. Recently, a comprehensive National Date Palm Program was established to provide support for all key aspects of the date production industry in Iran. Although most of the plantations rely on traditional practices, the growers have started to improve their old orchards and establish new commercial plantations. In the chain of cultivation practices, pollination is the most important. Accordingly, an electric pollinator has been recently developed to efficiently pollinate date palms. Bunch fading disorder has been the most important obstacle in recent years. Identification studies on domestic genetic resources of date palm began in the 1960s in Iran based on morphological characteristics leading to distinctiveness, uniformity, and stability (DUS) national indices for date palm cultivars. Moreover, scientists have recently begun to develop and employ molecular markers to identify and characterize some elite cultivars. The top ten cultivars are Piarom, Barhi, Zahidi, Sayer, Kabkab, Shahani, Dayri, Halawy, Haji Mohammadi, and Dehdar Moradi. Although the fruit is still harvested according to tradition, there have been rapid developments in postharvest operations. The fungus *Beauveria bassiana* as well as gamma rays were found as efficient alternatives to methyl bromide as a fumigant. Date products in Iran are categorized as semifinished, ready-to-use, date-derived, fermented products and by-products.

Keywords National Date Palm Program • Electric pollinator • Bunch fading disorder • Genetic biodiversity • Beauveria • Irradiation • Date products

---

S. Hajian (✉)
Department of Plant Breeding, Date Palm and Tropical Fruits Research Institute of Iran, 61355/16, Ahvaz, Iran
e-mail: saeed1551@yahoo.com.au

Z. Hamidi-Esfahani
Department of Food Science and Technology, Faculty of Agriculture, Tarbiat Modares University, Tehran, Iran
e-mail: hamidy_z@modares.ac.ir

© Springer Science+Business Media Dordrecht 2015

J.M. Al-Khayri et al. (eds.), *Date Palm Genetic Resources and Utilization: Volume 2: Asia and Europe*, DOI 10.1007/978-94-017-9707-8_2

2.1 Introduction

Iran is located in southwest Asia and borders the Gulf of Oman, Persian Gulf, and Caspian Sea. Its mountains have shaped both the political and the economic history of the country for centuries. Iran shares its northern borders with three post-Soviet states: Armenia, Azerbaijan, and Turkmenistan. Iran's western borders are with Turkey in the north and Iraq in the south, terminating at the Shatt al-Arab, which Iranians call the Arvand River. The Persian Gulf and Gulf of Oman littorals form the southern border. To the east lies Afghanistan on the north and Pakistan on the south. With a land area of 1,648,000 sq km (636,000 sq mi), Iran ranks eighteenth in size among the countries of the world (FAOSTAT 2014a).

Iran gains a wide range of climatic conditions. Winters are cold with heavy snowfall and below zero temperatures in the northwest during December and January. In these regions, spring and fall are relatively moderate, while summers are dry and hot. On the other hand, winters are temperate and the summers are very hot, having average daily temperatures in July exceeding 38 °C in the south. On the Khuzestan Plain (south west of Iran), summer heat goes along by high humidity. Iran commonly has a semiarid to arid climate in which most of the annual precipitation falls from October through April. Annual precipitation averages 250 mm or less in the most regions of the country. However, rainfall exceeds 1,000 mm annually and is relatively distributed evenly within the year in the western part of the Caspian Sea. In contrast, some basins of the Central Plateau usually receive 100 mm or less of precipitation annually. A wide range in climatic condition makes it possible to cultivate a diverse variety of crops such as cereals (wheat, barley, rice, and maize (corn)); fruits (dates, nuts such as pistachios, olives, figs, pomegranates, melons, and grapes); cotton; sugar beets and sugarcane; spices, e.g., saffron; raisins; tea; tobacco; berberis (barberry); and medicinal herbs. Nearly, 2,000 plant species are domestically grown in Iran; of these, date palm could be considered as one the most historic one (Flora of Iran 2011).

Date palm is the second most important horticultural crop of Iran after pistachio. It is well adapted to poor soil and water conditions as well as the hot and dry climatic conditions in the southern belt of Iran; hence dates play a prominent role in development and environmental sustainability of oasis landscapes and can contribute to desert rehabilitation.

The exact origin of the date palm has been lost in history; however it is clear that it was cultivated in old Iran as early as 4,000 B.C. in the area which today straddles the boundary between Iran and Iraq. Date palm is appreciated and cultivated because of its high nutritional value, and it serves as an important survival food resource, when the country experiences natural calamities and political strife.

Based on the latest reports of the Ministry of Agriculture (Jihad-e-Keshavarzi), nearly 98 % of date palm plantations are located in the southern belt of Iran. Date palm is grown in 13 of the 31 provinces, among them, Khuzestan, Hormozgan, Bushehr, Fars, Kerman, and finally Sistan and Baluchestan, with more than 95 % of total annual production, are the main producing provinces (Fig. 2.1) (Hajian 2005b).

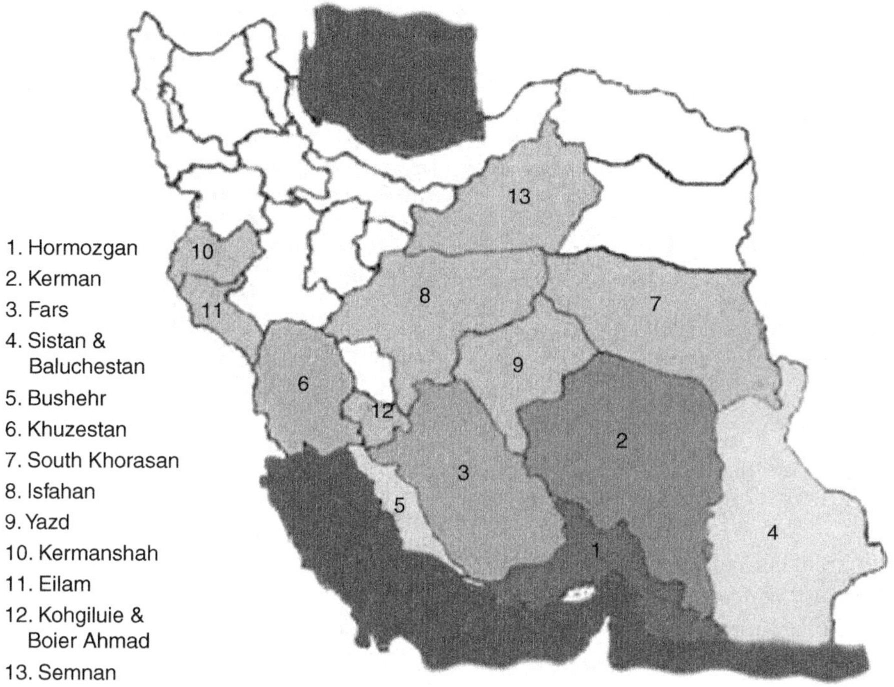

Fig. 2.1 Major (1–6) and minor (7–13) cultivated date palm provinces in Iran (*Source*: Hajian (2005b))

Table 2.1 Ranking of Iranian date industry in the world in 2011

Index	Quantity/value	World rank
Cultivated area	256,000 ha	1st
Harvested area	154,274 ha	3rd
Annual production	1,016,608 mt	3rd
Yield	6.6 t/ha	2nd
Export (quantity)	112, 030 mt	3rd
Export (value)	160.3 million USD	2nd
Germplasm diversity	>400 cultivars	1st

Source: FAOSTAT (2014b)

More than 400 date palm cultivars, among some 3,000 named worldwide, are found in Iran; therefore it occupies the first position in the global date palm gene pool (Zaid and de Wet 2002). In 2011, Iran produced 1,016,608 mt of fresh dates from 154,274 ha harvested; it was the third largest world date-producing country after Egypt and Saudi Arabia (FAOSTAT 2014b). Exports amounted to 112,030 mt of fruit generating a cash return of USD 160.3 million (FAOSTAT 2014b). Table 2.1 reflects Iran's standing as a world date producer. In spite of large-scale cultivation, high production, and a wide diversity of cultivars, the quantity of date fruit exports

Fig. 2.2 Date Palm and Tropical Fruits Research Institute of Iran, Ahvaz City, Khuzestan Province (Photo by Saeed Hajian, Jun 2012)

has been unsatisfactory in recent years. Traditional cultivation and harvesting practices, use of low-yielding cultivars, expensive production costs, insufficient cold storage facilities, traditional packaging systems, high marketing costs, scarce technical knowledge of producers and exporters, insufficient investment by the private sector, competitive global markets, and finally ineffective advertising methods have been key limiting factors in Iran's date industry in the last decades (Hajian 2011).

Research programs on dates commenced 50 years ago within Iran in different disciplines including breeding, protection, irrigation, nutrition, plant cultural improvement, and postharvest technology. The Date Palm and Tropical Fruits Research Institute of Iran was established in 1993 (Fig. 2.2) to coordinate and support all national research programs, with headquarters in Ahvaz City, the capital of Khuzestan Province in the southwest. The institute supports seven agricultural research centers in Hormozgan, Kerman, Bushehr, Fars, Iranshahr, Zabol, and Jiroft, as well as ten field stations at Omm-e-Tomeir, Bahookalat, Jahrom, Hajiabad, Zahak, Shabankareh, Shahdad, Azizabad, Aliabad, and Minab. The headquarters has six research departments: Plant Breeding, Agro-technical Improvement, Plant Nutrition and Irrigation, Plant Protection, Processing and By-Products, and Research and Technical Services.

One of the recent achievements of the institute has been the formulation of a comprehensive National Date Palm Program. The program strongly supports key aspects of the date production industry in Iran. To do this, more than 90 private and governmental organizations are collaborating as major stakeholders, among them the following:

- Agricultural Production Deputy, Ministry of Agriculture
- Agricultural Extension Deputy, Ministry of Agriculture
- Plant Protection Organization
- National Center of Agricultural Mechanization Improvement

- National Agricultural Products Insurance Organization
- Customs Organization
- Shahid Chamran University
- Shiraz University
- Nut Exporters Union of Khuzestan Province
- Commercial Date Producers and Exporters

Formulation of the program began in July 2006 by organizing seven technical groups affiliated with the head office. The program was formulated by consultation and presented to all affiliated parties by the Minister of Agriculture in August 2009. The program is subdivided in the following eight categories: Plant Breeding, Plant Agro-technical Practices, Plant Protection, Irrigation, Nutrition, Mechanization, Processing and Packaging Systems, and Policy Management (Hajian 2011).

In conjunction with the comprehensive plan, new date plantations were encouraged to cover key components including feasibility studies, suitability of the site, selection of cultivars to be planted, site preparation, irrigation systems, and technical practices, while financial establishment will highlight the establishment and operational costs and the cash flow statement.

2.2 Cultivation Practices

Most Iranian date palm plantations have been cultivated following traditional practices and with a high density of trees per hectare. Also they are cultivated along with other fruit crops, mostly citrus species. However, in recent years most of the growers have begun to improve their old orchards and establish new commercial plantations based on modern scientific principles (Pezhman 2002). Growers usually perform cultivation practices of pollination, thinning, bunch management, bunch bagging, and pruning.

Pollination is the most important of the agro-technical practices and has been improved in recent years. Pollination development has focused on five key factors: pollinator cultivar, pollination time, minimum quantity of pollen required, pollen density, and pollination method. According to research results, the best male date palm cultivars recommended as pollen sources for some of the commercial cultivars are listed in Table 2.2.

Growers are able to pollinate the flowers 1 week before natural splitting of the spathes until 10 days after it. The effective pollination period (EPP), for instance, varies from 7 days before natural spathe splitting to 10 days after it in Hayani and Medjool cvs., respectively. However studies revealed that the best time for pollination of most commercial cultivars is 1–2 days before natural spathe splitting up to 3 days after it in Iran (Table 2.3).

A few date palm plantations, especially in the south and southeast, are still pollinated by wind, bees, and insects, so-called *natural pollination*. All these regions are characterized by their 100 % seedling date palm populations with about 50 %

Table 2.2 Recommended male date palm pollen sources for pollination of some domestic cultivars

Female cultivars	Recommended male cultivars	Region (province)	Reference
Sayer	Khekri, Ghanami, Verdi, Samesmavi	Ahvaz (Khuzestan)	Eata (1991)
Zahidi	Towarz, Zahidi	Jahrom (Fars)	Zargari (2000b)
Shahani	Shahani	Jahrom (Fars)	Zargari (2000b)
Kabkab	Zahidi	Jahrom (Fars)	Zargari (2000b)
Mozafati	Pollen of Jiroft male stocks	Aziz Abad (Bam)	Ehsani (1986)
Halili	Pollen of stock no. 1002	Minab (Hormozgan)	Samavi (1999)
Haliaei	Khekri, Samesmavi, Sabzparak, Sorkhparak, Zard Parak	Jiroft (Kerman)	Abazarpoor (1998b)

Table 2.3 The optimal time for pollination of some of the commercial cultivars

Female cultivar	Effective pollination period	Region (province)	Reference
Sayer	From spathe splitting to next 4 days	Ahvaz (Khuzestan)	Sayahpoor (1998)
Mozafati	1–3 days after spathe splitting	Jiroft (Kerman)	Abazarpour (1998c)
Shahani	2–4 days after spathe splitting	Jahrom (Fars)	Zargari (2000c)
Kabkab	2 days after spathe splitting	Bushehr (Bushehr)	Farashbandi (2000)

males that lead to low efficiency of fruit production per unit area. Commercial date production, however, needs artificial pollination to ensure desired fertilization and overcome disadvantages of dichogamy. Artificial pollination practices are performed by traditional hand and mechanical methods in Iran. Most growers are accustomed to pollinate their date palms by the following methods:

2.2.1 Placement of Male Strands in Female Inflorescence

The male strands are cut from a freshly cut male spathes and 4–5 strands placed in a lengthwise inverted position, between the strands of the female inflorescence (Hajian 2005a). It is recommended some pollen be shaken over the female inflorescence after placement of the strands. It is also recommended to tie the pollinated female clusters about 10 cm from the outer end in order to keep the male strands in place and avoid entanglement with the female cluster strands during their rapid growth (Fig. 2.3a). This method could be done by a long wood stick for tall trees (Fig. 2.3b). Both the above cases are very common in Khuzestan, Kerman, and Hormozgan provinces.

Fig. 2.3 Pollination by putting few male strands between the strands of female inflorescence in young (**a**) and tall (**b**) trees

2.2.2 *Shaking Dried Pollen Grains over Female Inflorescence Using a Fine Cloth Bag*

Pure dried pollen grains are shaken over the female clusters using a fine cloth bag (like a stocking) 48–72 h after the male inflorescence splits. The bag is filled and tied to a long wooden pole, then shaken over the inflorescence of tall trees. Depending upon the height of the tree, it is recommended that shaking be repeated 2–3 times to obtain the highest fruit set (Hajian 2005a).

2.2.3 *Pollen Bearing Cotton Pieces Between Strands of Female Inflorescence*

This is not very common in Iran but is used by few growers. In this method, dried pollen is dusted on a spherical piece of cotton about the size of a golf ball and 1–3 pieces placed between the strands of female inflorescences (Hajian 2005a).

All the above traditional methods of artificial pollination are based on the climbing of the taller palms which is slow and costly operation and needs considerable labor and time to complete. Accordingly, mechanical pollination has become more popular especially in Kerman Province (Jiroft, Bam, and Kahnuj regions) in recent years (Pezhman 2001). The mechanical device is similar to a manual back sprayer (Fig. 2.4).

By repeatedly pressing the air pump, the air is compressed in the air tank. About half of pollen hopper is filled by dry pure or mixed pollen grains, and when the handle is pressed, the compressed air is released and moves through connecting

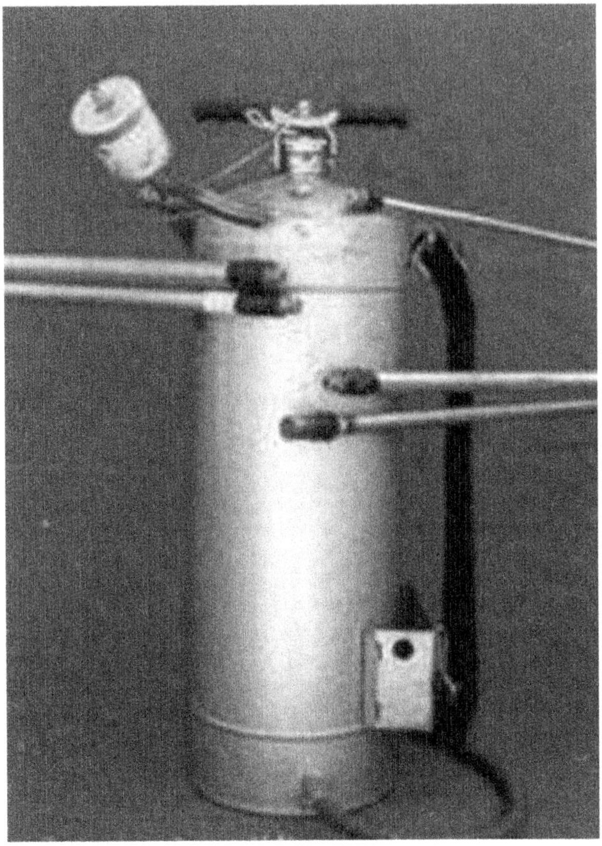

Fig. 2.4 Mechanical pollinator for date palm

hose. The grains move from the pollen hopper toward the nozzle through connected pollinator pipes and are expelled over female inflorescence. It is recommended that mechanical pollination be repeated twice with a 2–3 day interval. Mechanical pollinators eliminate the need to climb the palms and save labor and decrease accidents. Also, the growers are able to better manage their time for pollination practices; however, they suffer from inconvenience due to the design. It is necessary that two workers perform the operation.

Recently, an electric pollinator has been designed and developed in Iran to facilitate pollination practices in date palms. The dispersing system is completely different from the previous designs. It can be easily operated by a remote control (Fig. 2.5a). Thus the pollination feasibility and controllability have been enhanced for the operation. The size and weight of the tool as well as operation cost and time reduction with the electric pollinator is superior in comparison with former mechanical pollinators (Fig. 2.5b). Preliminary evaluation of the tool performance on Barhi cv. in the Ahvaz region showed no significant difference between the fruit set of trees pollinated by either traditional or mechanical methods. Mean fruit set attained by the developed tool, mechanical pollinator, and traditional method was 68.12,

Fig. 2.5 Electric pollinator with small remote-controlled dispersing system (**a**) facilitates pollination practices in date palm plantations (**b**)

Table 2.4 Recommended pollen density for pollination some of the commercial cultivars

Female cultivars	Recommended density	Region (province)	Reference
Sayer	20 % pollen + 80 % pollard	Ahvaz (Khuzestan)	Eata (1988)
Kabkab	20 % pollen + 80 % pollard	Kazerun (Fars)	Zargari (2000a)
Shahani	20 % pollen + 80 % pollard	Jahrom (Fars)	Zargari (2000a)
Mozafati	10 % pollen + 90 % pollard	Jiroft (Kerman)	Abazarpoor (1998a)

62.04, and 64.94 %, respectively (Mostaan et al. 2010). The electric pollinator can hold 200 cm^3 of the pollen mixture sufficient to pollinate about 120 palms (about 1 ha) and can be done by a single worker.

Although the growers consider 15–20 male trees/ha in traditional date plantations of Iran to provide naturally enough pollen grains, research over the past 20 years has showed that it could be decreased to five male trees/ha if collected pure dried pollens are mixed with an inert filler substance (Table 2.4).

The particle size of inert filler materials (such as talcum powder, wheat pollard, and ground date bunch remains) must be similar to the pollen grains with no harmful effect on pollen viability or their germination on female stigma.

2.2.4 Pests, Diseases, Disorders, and Weeds

Bunch fading disorder has been the most important problems of Iran date palm plantations in recent years. This disorder was first reported from Kahnodge City in Kerman Province on Mozafti cv. in 1988 and gradually spread to other provinces such as Hormozgan, Khuzestan, Bushehr, and Sistan and Baluchistan on Mordaseng, Khasoui, Kabkab, and Mozafti cvs., respectively.

Fig. 2.6 Symptoms of date palm fading disorder on bunches of cvs. (**a**) Khasoui and (**b**) Mozafti

Fig. 2.7 Covering of bunches with aluminum foil (**a**) and wicker baskets (**b**)

This disorder has been observed mostly on soft fruit cultivars during the mid-ripening stage. Sudden wilting of the fruits that finally leads to fruit and bunch drying which are the most important symptoms of this disorder (Fig. 2.6). The disorder takes place so fast (1–2 weeks) usually at khalal stage (Pezhman 2002). Due to economic impact of this disorder, many projects have been implemented in different fields in order to identify its causes as well as control methods.

Based on preliminary studies, several fungi, especially the asexual forms of *Ceratocystis* spp., have been isolated from infected fruit bunches in different regions; however, their pathogenic effects have not been proven (Karampour 2001a). It was observed that the environmental factors have more influence on the incidence and spread of this disorder than the other factors. It seems that a sudden drop in relative humidity (<20 %) and high temperature (>43 °C) along with hot and dry winds during fruit growth development (from khalal to rutab) have significant effects on higher incidence and development of this disorder (Karampour 2001b). Also, poor attention to appropriate agro-technical practices in date palm plantations plays a key role in exacerbating the level of damage (Mirzaee and Saei 2001).

According to the latest findings, covering of bunches with aluminum foil or a wicker basket reduces the damage (Fig. 2.7) (Pezhman 2002). In addition, cultivation of sorghum or alfalfa in infected plantations significantly improved damaged fruits (Darini and Ezadi 2001; Davoodian et al. 2001).

Table 2.5 The most important pests, diseases, disorders, and weeds of date palm in Iran

Pests	Diseases	Disorders	Weeds
Oligonychus afrasiaticus	*Mauginella scattae*	Bunch fading	*Imperata cylindrica*
Ommatissus binotatus	*Graphiola phoenicis*	Black nose	*Cynodon dactylon*
Batrachedra amydraula		Bastard offshoot	*Alhagi camelorum*
Arenipses sabella		Top bending	*Prosopis* spp.
Oryctes elegans			*Cyperus rotundus*
Parlatoria blanchardi			*Convolvulus arvensis*
Pseudophilus testaceus			*Glycyrrhiza glabra*
Microcerotermes diversus			*Sorghum halepense*
Rhynchophorus ferrugineus			*Amaranthus retroflexus*
Vespa orientalis			*Chenopodium album*
Polistes hebraeus			*Sonchus oleraceus*
Rattus rattus			*Xanthium strumarium*
Nesokia indica			*Malva sylvestris*
Mus musculus			*Solanum nigrum*

Source: Pezhman (2002)

The results of implemented research on this disorder revealed that the level of damage could be decreased significantly from 45 down to 5 % if growers perform integrated agro-technical practices including sufficient nutrition (particularly K_2O spraying with 0.5 % solution at 2, 4, 10, and 15 weeks after pollination), regular irrigation (4–7 day intervals from fruit set to the middle of rutab stage), and bunch pruning (cutting one third of bunch tip) (Pezhman et al. 2005). Other important Iranian pests, diseases, disorders, and weeds in date palm plantations are listed in Table 2.5.

2.3 Genetic Resources and Conservation

Plant genetic resources, water, and suitable land (soil) could be considered as fundamental to support sustainable development agricultural programs in Iran. Research activities have focused largely on prominent genetic resources due to limitations in suitable water and soil conditions in the country.

Investigations of domestic genetic resources of date palm began in the 1960s in Iran. Date cultivars have been developed over thousands of years by seedling selection and propagation of those palms possessing desirable fruit characteristics. Date palm cultivars number around 3,000 worldwide; meanwhile there are more than 400 different cultivars in Iran, placing the country first in the world in that regard.

Accurate identification and conservation of new date palm cultivars has always been of research interest to assess their value as genetic resources. The first Iranian field collection of date palms was established in Khuzestan Province, southwestern

Table 2.6 New identified and collected female cultivars of date palm in Iran

No. of identified female cultivars	No. of collected female cultivars	Province
114	25	Hormozgan
67	130 (Bam) and 150 (Jiroft)	Kerman
85	40	Fars
115	30	Bushehr
33	14	Sistan and Baluchestan
65	105 (Ahvaz)	Khuzestan

Source: Hajian et al. (2011)

Table 2.7 Major date palm cultivars in Iran

Province	Cultivars
Hormozgan	Piarom, Mordaseng, Khasi, Almehtari
Kerman	Mozafti, Kalute, Mordaseng
Fars	Shahani, Kabkab, Khasi, Zahidi, Gantar
Bushehr	Kabkab, Shahabi, Zahidi
Sistan and Baluchestan	Mozafti, Rabbi
Khuzestan	Sayer, Barhi, Zahidi, Dayri, Halawy, Khazrawi, Braim, Khasi, Kabkab, Gantar, Haj Ghanbari, Shekar

Source: Hajian et al. (2011)

Iran, in the early 1960s; three additional collections were created in the early 1980s within the Ahvaz, Bam, and Jiroft agricultural research stations (Hajian and Pezhman 2004). The number of newly identified and collected female cultivars of date palm is shown in Table 2.6. In addition, the major domestic cultivars by province are listed in Table 2.7 (Hajian et al. 2011).

The first foreign cultivars, including Deglet Noor, Medjool, and Thoory, were imported in 1986. These cultivars have been monitored for 15 years for the uniformity of plantlets (propagated by tissue culture), and their quantitative and qualitative fruit characteristics. Results revealed that Thoory cv. is tolerant of high relative humidity and could be recommended for the southern belt of the country (Hormozgan Province), which experiences high relative humidity throughout most of the fruiting season.

Medjool has exhibited desirable fruit quality and quantity; however, Deglet Noor demonstrated low-quality fruit production, probably due to special pollen requirements (Hajian and Pezhman 2004). Other imported cultivars are listed in Table 2.8.

Jarvis no. 1 and Fard no. 4 male cultivars as pollinators were imported at the same time. They are being studied at the Ahvaz (Khuzestan), Minab (Hormozgan), and Zahak (Sistan and Baluchestan) agricultural research stations.

Although the selection and conservation of elite cultivars by creating traditional genetic resources collections, to support rehabilitative programs of old date palm plantations as well as establishment of new ones, these collections do not provide sufficient quantities of propagation materials. Traditional genetic conservation of

Table 2.8 Distribution of imported commercial cultivars in agricultural research stations of Iran

Cultivars	Agricultural research stations (province)
Ashrasi	Khuzestan, Hormozgan, Sistan and Baluchestan
Koosh Zabad	Kerman, Hormozgan, Sistan and Baluchestan, Fars, Bushehr
Abu Narenja	Hormozgan, Sistan and Baluchestan, Fars, Bushehr
Deglet Noor	Khuzestan, Hormozgan, Hormozgan, Kerman
Helali	Kerman, Hormozgan, Fars, Bushehr, Sistan, Baluchestan
Abu Moaan	Hormozgan, Sistan and Baluchestan, Fars, Hormozgan, Bushehr
Hayani	Khuzestan, Bushehr
Medjool	Khuzestan, Hormozgan, Hormozgan, Kerman
Fard	Kerman, Hormozgan, Sistan and Baluchestan, Fars, Hormozgan
Nabat Seif	Hormozgan, Sistan and Baluchestan, Fars, Bushehr
Raana Tala	Kerman, Hormozgan, Sistan and Baluchestan, Fars, Bushehr
Shi Shi	Khuzestan, Kerman, Hormozgan, Sistan and Baluchestan, Fars, Bushehr
Thoory	Khuzestan, Hormozgan, Sistan and Baluchestan, Fars, Bushehr

Source: Hajian and Pezhman (2004)

date palms is threatened by such factors as nonagricultural land development in centers of origins, desertification, climate change, and salinization.

The high maintenance cost of traditional field collections represents a continuous challenge to conserving genetic diversity of date palm; an alternative is to establish in vitro collections to facilitate access and utilization of date palm germplasm. Preliminary conservation efforts have been initiated in recent years in Iran (Hajian 2005c); however, advanced studies are required to identify, characterize, and evaluate the date palm germplasm. This is vital to achieving a rational and successful utilization of date palm genetic resources. Meanwhile, date growers and research centers should still be encouraged to continue conserving elite cultivars traditionally in the oasis cultures.

2.4 Plant Tissue Culture

Mass propagation of date palms from mature specimens is impossible due to the limited number of offshoots produced. Indeed, offshoot production is limited to the early period in a date palm's life span. Moreover, seed and offshoot propagation is impractical. Since the early 1980s, many Iranian governmental initiatives have been put forward to restock the lost groves by the private and public sectors. Scientists and governmental policymakers believe that tissue culture should be considered as the best method for mass propagation of date palm.

The first successes with tissue culture were achieved at the Seed and Plant Improvement Research Institute, affiliated with the Ministry of Agriculture, in 1988–1990. Implemented researches mostly focused on Sayer and Kabkab cvs. by embryo culture (Majidi et al. 1991; Nazeri et al. 1993), meristematic tissues (shoot tips and buds), and highly differentiated somatic tissues (leaf, stem, inflorescence, and root sections). Somatic embryogenesis, based on callus production and multiplication, followed by growth and elongation of somatic embryos, has been mostly applied (Davoodi et al. 2002; Eshragi et al. 2005). Up to now, this technique had shown to be genotype independent with a high rate of multiplication along with a high survival rate upon transfer to soil. Nevertheless, there is always a question among date growers and scientists about the true-to-typeness of plants produced in vitro. At the present time, three groups of tissue culture-derived plants can be found in Iran. On average, their bearing age is 15, 9, and 6 years. Those in the first group were micropropagated and acclimatized abroad and brought into the country at 1985. The second group of plantlets was micropropagated abroad and acclimatized in Iran; those in the third group were micropropagated and acclimatized within the country.

Subsequently, a private joint venture agreement was signed by the government with Date Palm Developments Limited, a UK company, to facilitate the transfer of date palm tissue culture technology to Iran over a period of 3 years. The technology transfer was successfully achieved by 1996 and certified by the parent company. The private sector subsequently acquired whole ownership of the technology and has been producing large numbers of tissue-cultured date palms for over 10 years. It is worth mentioning that tissue culture-derived plants of a few cultivars are subject to somaclonal variation in particular and to genetic variations in general. These are unlike the epigenetic variations observed and reported in Iran, which are at the physiological level. Iranian date growers have experienced various reproductive and vegetative abnormalities on young palms since 2007. Results of a national comprehensive research project revealed that fruit set failure was the leading abnormality throughout the country.

Barhi was found to be the most sensitive cultivar to propagate by tissue culture. Also, younger palms generally showed more reproductive and vegetative abnormalities. According to recent findings, the technique used for propagation (protocol), nature of mother plant (chimera), type of growth regulators used, type of explant used (ploidy gradients: apex to root), age of subcultures (>1 year), medium composition, and incubation conditions are key factors causing the variations (Hajian 2009). In addition, studies by means of enzymatic systems revealed that there were some differences in the isozyme banding pattern of offshoot and micropropagated plants (Zivdar et al. 2008). The polymorphism of peroxidase (PRX) and shikimate dehydrogenase (SHD) enzymatic systems confirmed that the differences distinguished had an epigenetic basis.

Most commercial laboratories in the world are doing their best to ensure the true-to-typeness of the produced date plant material. Various techniques such as histocytology, isoenzyme, restriction fragment length polymorphism (RFLP), and random amplified polymorphic DNA (RAPD) are used to produce and certify the conformity of the plants. In most cases, fingerprinting is the technique actually used, but it is believed that field response is the only reliable test to confirm that palms derived from tissue culture are true to type to the mother plant (Zaid and de Wet 2002). However, the private sector in Iran which is using tissue culture for mass propagation of date palm has not revealed how they certify the true-to-typeness of their produced plants.

Furthermore, genetic transformation in date palm by means of particle bombardment has been studied since the late 2000s in Iran. Studies have involved plant-related parameters including osmotic conditioning of explants, before and after bombardment, type of explants (embryogenic callus and somatic embryo) as well as bombardment parameters like acceleration pressure, bombardment distance and gold particle size (Habashi et al. 2008; Mousavi et al. 2009, 2014). Further efforts to optimize transformation conditions and the integration of useful genes for the improvement of date palm should be considered as prime targets in Iran.

2.5 Cultivars Identification

The biodiversity of date palms has always been one of the most important aspects of the date palm industry in Iran. The large number of cultivars or landraces among date plantations has made it difficult to identify and characterize all cultivars; however, initial studies commenced in early 1960s on the basis of morphological characteristics. Iranian scientists have attempted to identify and characterize the required plant cultivars for cultivation and breeding programs as well as for cultivar-right-protection purposes. All attempts have been implemented on the basis of phenotypic variation to the early 2000s.

Dawson and Kashani, in cooperation with other Iranian experts, first identified 400 different cultivars throughout the country in the 1950s (Kashani 1979). They established the first field collection which included 175 elite cultivars, in Ahvaz. This involved observations of the morphological and physiological characteristics of an unknown cultivar to provide an official description by comparison with appropriate reference cultivars, which can be used to establish its distinctness, uniformity, and stability (DUS). Subsequent studies mostly focused on fruit characteristics such as color, shape, weight, ripening stage, TSS (total suspended solids), pH, etc. Morphological findings based on the International Union for the Protection of New Varieties of Plants (UPOV), system of intellectual property protection, assisted scientists to determine national DUS indices for date palm in 2007 (Sadeghian 2008). Accordingly, about 20–25 of the over 400 cultivars, as the commercial cultivars, were recommended for cultivation in different regions of Iran. Some of them are briefly described in Table 2.9 (Pezhman 2002).

In spite of that, there are increasing technical problems in achieving distinctness from a large number of cultivars in Iran. Phenotypic plasticity is still an obstacle in identification of date palm biodiversity within the country. Hence, the potential of molecular systems for producing reliable cultivar descriptions, which are largely unaffected by the environment, has become increasingly attractive since the mid-2010s. The potential of molecular systems for application in cultivar registration is therefore under careful and active consideration within UPOV. However, it is clear that the power of molecular techniques could potentially allow discrimination between cultivars down to the very small differences of a few base pairs within the genome. Therefore, before their possible introduction, it has been necessary to

Table 2.9 Fruit characteristics and growth locations of important commercial Iranian date palm cultivars

Cultivar	Color (tamar)	Harvest time	Fruit type	Consumption stages	Total sugar (%)	Flesh to seed ratio	Growth province
Piarom	Brown	Oct–Nov	Semidry	Khalal-rutab-tamar	66.5	7.7	Hormozgan
Zahidi	Yellow to light brown	Oct	Semidry	Tamar	63	8.4	Khuzestan
Dayri	Brown	Oct	Dry	Rutab-tamar	58.5	6.6	Khuzestan
Sayer	Light brown	Aug–Sep	Semidry	Rutab-tamar	63	12.6	Khuzestan
Mordaseng	Light brown	Aug	Soft	Rutab-tamar	47.5	8.8	Hormozgan and Kerman
Halawy	Yellow to light brown	Aug	Semidry	Rutab-tamar	58	5.6	Khuzestan
Shahani	Light brown	Sep–Nov	Soft	Khalal-rutab-tamar	50.5	12.5	Fars
Kabkab	Light brown	Sep	Soft	Rutab-tamar	55.5	11.5	Bushehr
Mozafati	Brown	Aug–Sep	Soft	Rutab-tamar	50	13.0	Kerman
Almehtari	Brownish to yellow	Jun	Soft	Rutab-tamar	54	5.4	Hormozgan
Khasoui	Light brown	Oct	Soft	Khalal-rutab-tamar	55	10.4	Fars
Barhi	Light brown	Sep	Soft	Khalal-rutab-tamar	60	10.5	Khuzestan

consider how to ensure that sufficient genetic distance between varieties is maintained to be able to give certainty of continuing protection (Camlin 2003). To mitigate this, Iranian scientists at the Agricultural Biotechnology Research Institute of Iran (ABRII) have recently begun to develop and employ molecular markers, statistical tests, and software to identify and characterize ten date palm cultivars including Sayer, Piarom, Shahani, Kabkab, Mazafati, Barhi, Gantar, Dayri, Zahidi, and Rabbi. They identified specific molecular keys using seven SSR markers. The results also proved that there was different genetic background within Dayri, Rabbi, and Piarom mother trees (Mardi et al. 2010).

2.6 Cultivars Description

Over 400 various cultivars are known in Iran. The wide variation of date palm biodiversity makes it difficult to describe all of them. However, fruit and tree morphology has been used to distinguish the cultivars since the 1960s. Morphological characteristics, especially the color variation during fruit development, supplemented with leaf morphology have shown promising results to individualize the cultivars. From fruit set to full ripened stage, the fruits pass through distinct changes of development beginning with hababouk and ending at tamar. Color variation during these stages is the best diagnostic marker for Iranian scientists to distinguish the cultivars. In addition, shape and size of fruits are good indicators for cultivar identification, although the size is mostly influenced by agro-technical practices. Some Iranian botanists consider the position of the fruit cap (perianth) to characterize a cultivar. The fruit cap entirely covers the fruit base in some cultivars, whereas it covers only the central portion in others. Moreover, seed characters such as shape, size, margin of ventral furrow, and pulp-to-seed weight ratio are used to identify the cultivars. Date palm botanists also consider biochemical composition of fruits such as TSS and pH to determine cultivar description (Hajian 2007). Accordingly, the top ten superior date palm cultivars of Iran are illustrated in Fig. 2.8. In addition, their descriptors are shown in Table 2.10.

2.7 Dates Production and Marketing

Over one million mt of fresh dates currently are produced each year in Iran. Statistics show that 55–60 % of annual production is consumed domestically; 12–16 % is exported, leaving 24–33 % considered as surplus or wasted production. Most date fruits are consumed directly with little or no processing, although the quantity of processed date products is growing rapidly.

According to export statistics of FAO, Iran has typically been the second largest date producer and exporter in the world over the decade 2002–2011. Although export quantities have fluctuated during this period (Fig. 2.9a), the unit value has risen significantly (Fig. 2.9b).

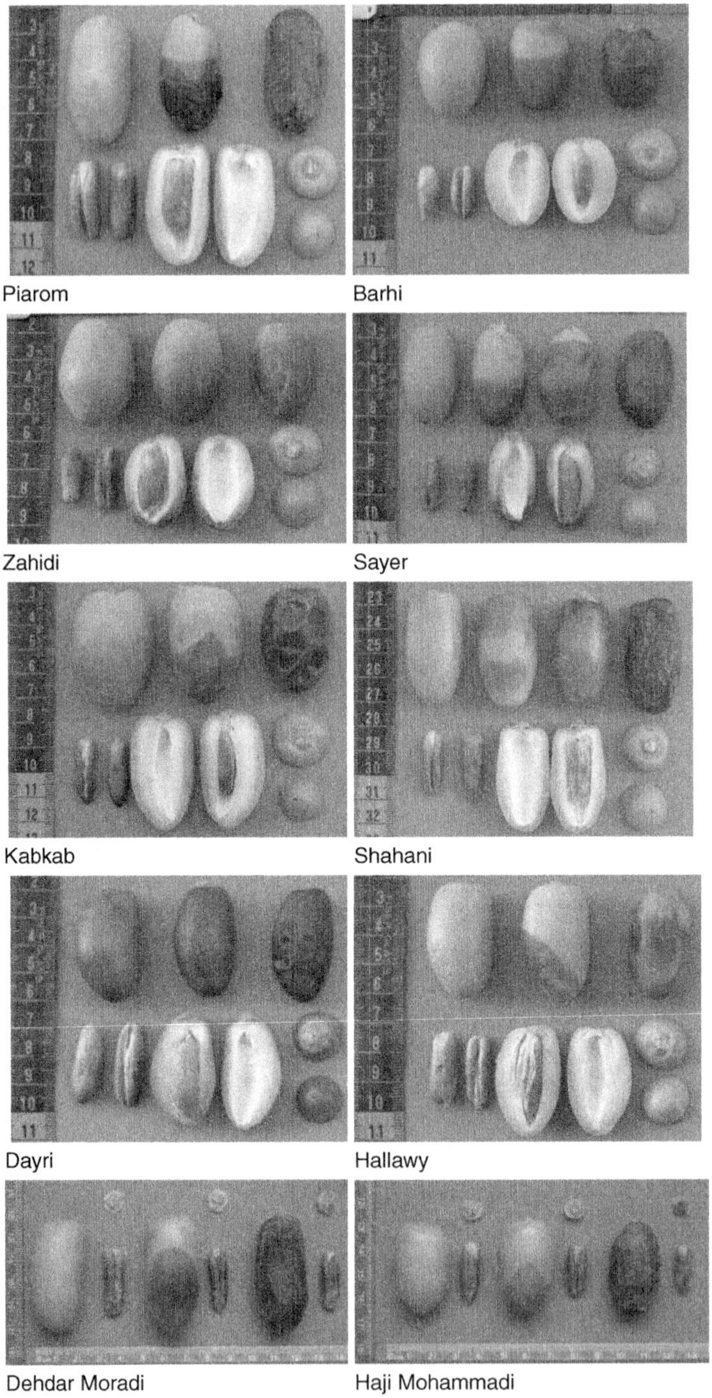

Fig. 2.8 Popular date palm cultivars grown in Iran

Table 2.10 Descriptors of elite date palm cultivars grown in Iran

Cultivar	Average yield (kg/tree)	Max recorded yield (kg/tree)	Tree height[a]	Fruit length (mm)	Fruit diameter (mm)	Fruit weight (g)	Fruit type	Sweetness[b]	Ripening time
Barhi	145 (khalal)	350 (khalal)	Dwarf	35	25	9.8	Soft	Moderate (khalal)	June–July (khalal), Sept (tamar)
Dehdar Moradi	88	110	Tall	51	22	9.7	Semidry	Moderate	July–Aug
Dayri	45	110	Dwarf	35	16	6.5	Dry	Very high	Sept–Oct
Haji Mohammadi	60	85	Tall	43	22	8.8	Soft	Very high	June–July
Halawy	65	135	Semi-tall	41	22	6.9	Semidry	High	June–July
Kabkab	65	95	Semi-tall	33	22	14.8	Soft	Very high	Aug–Sep
Piarom	50	95	Tall	41.5	17	8.7	Semidry	Moderate	Oct–Nov
Sayer	47	100	Semi-tall	40	20	8.0	Semidry	Very high	Aug–Sep
Shahani	91	175	Tall	41	15	11.5	Soft	Moderate	Sept–Nov
Zahidi	75	130	Dwarf	35	20	7.7	Semidry to dry	Very high	Aug–Oct

[a]Height of tree (m): dwarf 2.45–3.64, semi-tall 3.65–4.84, tall 4.85–9.64, very tall >9.65

[b]Sweetness of fruit (total sugar %): low < 67.4, moderate 67.5–74.4, high 74.5–77.9, very high 78.0–88.4

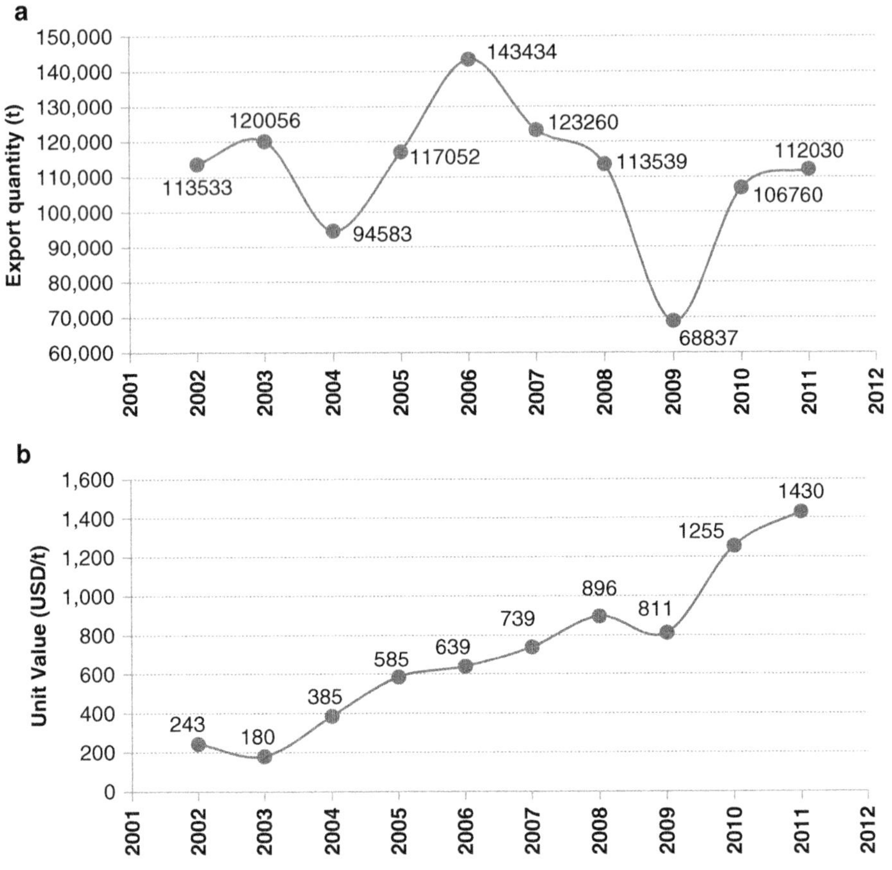

Fig. 2.9 Quantity (**a**) and unit value (**b**) dates export from Iran during the decade 2001–2011 (*Source*: FAOSTAT (2014b))

The knowledge of Iranian date growers about production as well as preservation and storage of dates has improved in recent decades. Since the early twenty-first century, reports indicate that the potential and viability for a modern date industry exists in Iran. Most of these reports focus on economics and marketing of the crop at the national or regional level. Although the fruits are still harvested by traditional methods, there is rapid development in postharvest operations.

Postharvest processing of dates includes sorting, pitting, disinfestation of insects, washing, grading, heat treatment, adjusting moisture content, coating, pasteurization, and packaging. Although most date producers follow the processing steps, the first fully mechanized sorting, disinfesting, and packaging pilot plant was established in Ahvaz in 2010. It was supplemented by global standard cold storage rooms to enhance technological knowledge of both producers and exporters. The pilot plant includes different peripheral devices such as a washing and dryer machine (Fig. 2.10a) and date paste maker (Fig. 2.10b).

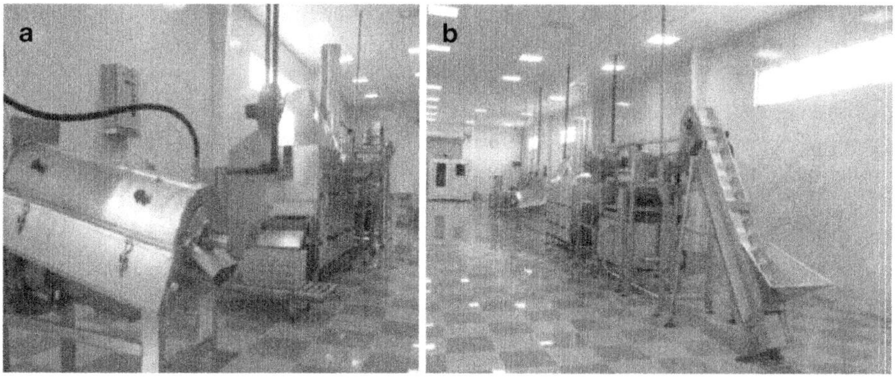

Fig 2.10 Date paste maker (**a**) and automatic washing and drying machine (**b**) in the first modern pilot date processing in Ahvaz, Iran

New alternatives for fumigation of fruits by methyl bromide must be considered as one of the recent achievements in postharvest technology in the Iranian date industry. Novel research findings revealed that application of the fungus *Beauveria* (*Beauveria bassiana*) is able to effectively control adult *Oryctes elegans* (Coleoptera: Scarabaeidae) in storage rooms (Latifian and Rad 2012). Moreover, radiation of Sayer date fruits by gamma ray (0.25 KGY) prolonged shelf life of the fruits in storage rooms for up to 9 months. Findings proved that *Plodia interpunctella* and *Oryzaephilus surinamensis* were efficiently controlled by this method of irradiation (Garshasebi 2010).

Nevertheless, growing mixed cultivars, traditional harvesting, a high level of wastes in the plantations, expensive production, low standards of packaging and storage systems, high marketing cost, and ineffective advertising methods should be considered as the key obstacles of postharvest technology of Iran's date industry (Hajian 2011).

2.8 Processing and Novel Products

Industrial producers often sort date fruits in the field just after harvesting. Sorting is usually done manually by workers who cull damaged fruits as well as remove foreign materials. Date processing mostly involves removing the cap and in some cases seeds of the dates for better marketing and reducing transfer costs (Barreveld 1993). Removing date seeds may be done by crushing and sieving the fruits or, more sophisticatedly, by pushing the seed out of the fruit. The next stage involves fumigation and sterilizing the fruit to prevent pest damage when they are stored. In Iran, the major techniques to prevent insect infestation are fumigation, heat treatment, cold storage, and irradiation, of which fumigation is the most common. Heat treatment and cold storage are rather beneficial when applied to dates for other reasons, and irradiation is an effective but not yet common. Other fumigants are also used such as carbon disulfide, hydrocyanic acid, Phostoxin, and ethylene oxide, but methyl bromide is the most prevalent and effective technology. Nevertheless, it is worth mentioning that the current technique of disinfestation using methyl bromide has been prohibited by the year 2015 (UNEP 1998). Washing is usually done by automatic machines to remove dust or other foreign materials using water sprays. After washing, dates are subjected to flowing warm air to remove the moisture from the surface of fruits (Ashraf-Jahani 2002).

At the next stage, dates are graded according to size, color, and moisture content, after which heat treatment of 60–65 °C as a partial pasteurization is applied to limit the activity of microorganisms, enzymes, and insects. Date drying is carried out by using drying tunnels or solar drying. Dates are subjected to deterioration because of fermentation, mold formation, darkening, and aroma and flavor loss. All forms of deterioration increase with higher water content. Air drying should result in a moisture content of 20 % or less to prevent the activity of molds and yeasts. If dates do not have the recommended minimum water content, artificial hydration may be used. Hydration is carried out for dates that have been dried by remaining on the tree for a long time or in long-term storage. The most common method of hydration is using low-pressure saturated steam. To remove the stickiness of the date surface, starch solution or methyl cellulose is used. Edible oil, water, BHA (butylated hydroxyanisole), and BHT (butylated hydroxytoluene) are used to eliminate the roughness of date surface and make it shiny. Then, pasteurization at 66 °C for 30 min is conducted to complete the process of microbial reduction. Finally, dates are packaged in suitable containers, labeled, and controlled for weight and toxic metals. Subsequently, dates are transferred to cold storage chambers to retard chemical and biological processes (Ashraf and Hamidi-Esfahani 2011).

Many products, such as date syrup, alcohol, animal feed, date powder, different types of bread, marmalade, candy, chocolate, and date paste, can be obtained from dates. Other parts of the tree are also important in the rural economy. For example, the stem is used for making wooden boats and rafts as well as roof coverings for rural houses in addition to serving as raw materials for fiber, paper and wood industries. The foliage is used for making handicrafts such as fans and straw hats. It is estimated that more than 70 various products and by-products can be produced from different parts of date palm.

Date products in Iran are categorized as semifinished date products, ready-to-use date products, derived products from date fruits, fermented products, and by-products from processing (Ashraf and Hamidi-Esfahani 2011).

2.8.1 Semifinished Date Products

Date Paste The production of date paste is one of the most interesting products in the date food industry. It leads to reduced transportation and storage costs, since the seeds (10–20 % of the total fruit weight) are removed during processing. Moreover the availability of date paste for the food industry is ensured throughout the year. Date paste makes it possible to convert dates of even inferior quality into an intermediate value-added product for the date processing industry. Date paste is usually made from culled fruits unsuitable for marketing. For the preparation of date paste, pitted dates (tamar stage) are either soaked in hot water (95 °C) for 5–15 s or steamed for 3 min at 69 kPa. Dates are turned into date paste by grinding. Citric or ascorbic acid (0.2 %) may be added to lower the pH of date paste for improved shelf life and to maintain a desirable color. Date paste can also be used as a partial substitute for flour in bakery and confectionary products. The use of 4–8 % date paste in bread formulation results in significant improvements in the dough's rheological properties, delays gelatinization, improves gas production and retention, extends the shelf life, retards staling, and improves the crumb and crust characteristics. Using date paste in cookies results in a higher spread ratio increasing with amounts added up to 20 %. It prevents the crystallization of sucrose in cookies during the cooling off period immediately after baking.

Extruded Date Seeded crushed dates are extruded through 0.6 cm holes and shaped into rolls, cut into 1.3–2.5 cm lengths, and coated with dextrose or barley flour to prevent them from sticking together. To improve marketing, it is dried and solidified by exposure to air.

Diced Date This product is made from cutting seeded dates into pieces with a dicer. The diced date pieces are coated with dextrose or oat flour to prevent the pieces from sticking together. This product is usually consumed with cereals, cooked products like cake, different kinds of bread, and sweetmeats. Adding date pieces (10 %) to ice cream slightly reduces the overrun.

Date Powder After dilution of date paste with water, it is spread on metal trays and dried using a tunnel or cabinet dryer until moisture is less than 5 %. The dried material is milled and sold based on different granular sizes. This product is used in confectionary and baby foods as a sweetening agent.

2.8.2 Ready-to-Use Date Products

Some preserved products such as pickles, chutney, jam, jelly, dates in syrup, date butter, candy, and confectionary products are prepared from date. Dates at the kimri and khalal stages of maturity are most suitable for preparing pickles in oil and for chutney. Pickles in oil are prepared using pitted, sliced kimri fruit with various spices, condiments, and mustard oil. Chutney is a generic term for a condiment composed of fruit sugar, citric acid or vinegar, vegetables, and hot spices. Derived from an Indian word, chutney is probably best known when mango is used as the basic fruit, but other fruits such as dates can be utilized. Brine and salt-stock pickles are other popular products that could be prepared from kimri dates.

Ripe dates with a high sugar content are suitable for jam making. A sugar-date pulp ratio of 55:45 is used for jam making with 65 % sugar content, 1 % pectin, and pH of 3.0–3.2. For jelly making, date juice-sugar at a ratio of 1:1 is used, and the finished product has total soluble solids content of 73° Brix and pH of 3.57.

Also, dates in tamar stage containing high sugar content are appropriate for date butter, which is similar to peanut butter in usage. It is similar to jam making, except the pH is adjusted to 4.7, the total soluble solids content of 75° Brix, and a sugar-date pulp ratio of 40:60.

Dates in syrup is prepared from peeled, seeded whole dates at the khalal stage. In this product, the syrup is concentrated to about 75–80° Brix with the pH adjusted to 2.8–3.0 with citric acid.

Date candy is prepared using date paste, roasted groundnuts, and coconut. The use of date paste and nuts in a 60:40 ratio coated with chocolate gives a good sensory quality for this type of candy. Plain date bars prepared from date fruit, almonds, coconut, groundnuts, and pistachios can be coated with chocolate and fortified with sesame, skim milk powder, and oat flakes.

Different desserts prepared from date fruits are ice cream, pudding, date sherbet, and fruit yogurt. Other products include macerated date and fiber-filled dates, with the former having medical uses. Other ready-to-use date products are different sauces such as steak sauce. Steak sauce is a traditional product, used as a flavor with meat dishes; it contains 10 % ground dates.

2.8.3 Derived Products from Date Fruit and Stem

In this type of production, second grade or culled fruits from large-scale date packing operations are typically used. The product list includes date spread, sherry, medicinal alcohol, arrack (spirits), vinegar, date wine, date syrup, soft drinks, liquid date sugar, and high-fructose liquid date sugar. Dates are mixed with water in all these products to facilitate the separation of undesirable materials. In some products, added water remains in the final product, as in beverages. However, in other products, water is added to create a suitable medium for the next process such as fermentation.

Other products from date syrup are yeasts (rich in protein), vitamins, and also baker's yeast. Fat can be produced from date juice and syrup by means of some microorganisms like *Penicillium lilacinum*, *Pen soppi zaluski*, and *Aspergillus nidulans*. Date pulp and/or date syrup can be utilized as a substrate for oxytetracycline formation by some suitable mutants of *Streptomyces rimosus*. Date seed lipids and hydrolysate can serve as carbon and nitrogen sources, respectively, in the fermentation medium for the formation of oxytetracycline by *Streptomyces rimosus*. Higher amounts of antibiotics were found with date substrates than with glucose and urea.

Date honey is concentrated date juice with colloidal compounds from which the major part of its pigments has been removed. It is utilized in beverages, chocolate, ice cream, marmalade, and confectionary. A kind of syrup can be obtained from the stem tissue sap of some date palm cultivars from which sugar is extracted. Caramel coloring made from date syrup can be used in a variety of foods, such as beverages, bakery, confectionary, and meat products. Another product is Tarooneh arrack (Tarooneh distillate), which is made from the chopped male or female spathe; it smells pleasant and has medicinal benefits.

Sago (white in color) and *lagmi* (date palm sap) are other products obtained from the date palm stem. Date palm sap (tapped from the stem) has a high sugar content (mainly sucrose), protein, ash, and phenolics. Its surface activity and foaming power are due to the presence of proteins. Because of its antioxidant activity and nutritional value, this natural juice could be used as a functional food. Sap extracted to produce sweet *lagmi* must be done away from sunlight to prevent fermentation.

2.8.4 By-Products from Date Processing

Three major by-products result from date fruit processing facilities: low-grade rejected date fruits, date seeds, and date press cake. Date fruits, unfit for fresh consumption or in derived products, are commonly utilized as animal feed or as a substrate for fermentation products because of their nutritive components such as sugars, proteins, and minerals. Date seeds possess good nutritional value based on their dietary fiber content, making them suitable for the preparation of fiber-based foods and dietary supplements (Habibi-Najafi 2011). Date seeds are also referred to as pits, pips, stones, and kernels. Date seeds represent on average 10 % of fruit weight. Because they are rich in oil, proteins, minerals, and fiber, they can serve as valuable raw materials in animal feed.

Press cake is the by-product of date juice extraction. Depending on the method used for juice extraction, date flesh with remaining sugar, with or without seeds, remains. Press cake is wet (about 70 % moisture) and bulky (it forms 30 % of date weight) and deteriorates quickly; it can create a disposal problem if unutilized. Investigations have revealed that the majority of protein exists in the press cake itself and the majority of fat in the date seed. Press cake which includes seeds is a rich source of dietary fiber and high levels of phenolics and antioxidants. Press cake is utilized for animal feed as well as microbial conversions.

2.9 Conclusions and Recommendations

Iran can be considered as one of the oldest date-producing countries in the world. Historic indigenous knowledge of date growing and large-scale cultivation, high annual production, high yields, and a wide diversity of cultivars have provided inimitable conditions to further develop the date industry. Over one million mt of fresh dates are presently produced annually in Iran. Nearly 55–60 % of annual production is consumed in domestic markets, around 12–16 % is exported, and the remaining 24–33 % can be considered as surplus and wasted production. The same pattern could be found in global domestic consumption, export, and surplus and wasted

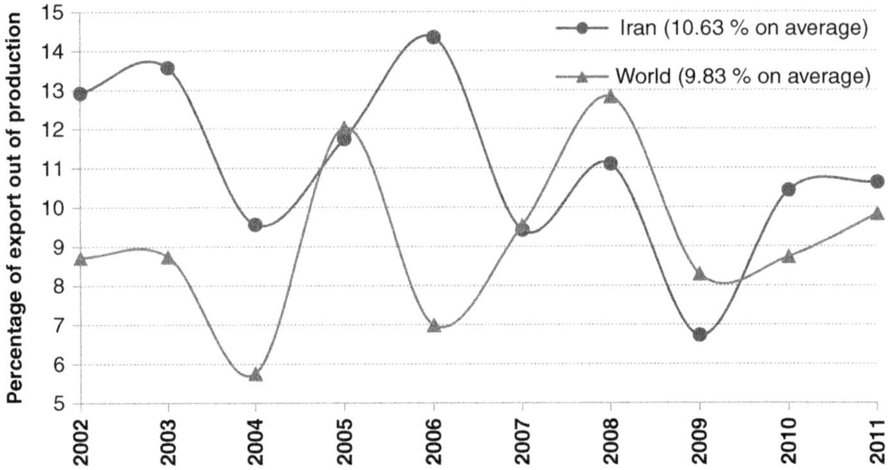

Fig. 2.11 Percentage of date fruits export out of production in Iran and the world

production, according to recent FAO statistics (FAOSTAT 2014b). Fig 2.11 illustrates that 10 % of annual production has been on average exported from Iran and worldwide during decade 2002–2011. Similarly, around 90 % of annual production has been globally consumed (including surplus and wasted production) in domestic markets. Therefore, global market demands are consuming no more than 10 % of annual production.

Efficient and innovative marketing methods must be employed to attract new date consumers. Indeed, the potential value of eating dates is not clearly defined in the daily human diet; most consumers do not consider dates as a fruit in the way they do apples, oranges, or bananas. Dates are also not accepted as a nut snack such as pistachios or peanuts. Moreover, the poorer segments of societies do not consider dates as a desirable food. Per capita consumption in Iran, as well as in the world, confirms the low status of dates in daily human diet. The latest statistics illustrate that average annual fresh date fruit consumption amounted to 11.6 and 0.9 kg per capita in Iran and in the world, respectively, in 2009 (FAOSTAT 2014b). The statistics also revealed that Iran occupies fifth place in per capita consumption of dates after Saudi Arabia, Algeria, Libya, and Egypt with 36.6, 16.3, 16.3, and 14.2 kg per capita per year, respectively. Providing new incentives to consumers by innovation in primary products and by-products can create new markets.

The broad patterns of date cultivation and harvesting, use of higher-yielding cultivars, efficient agro-technical practices, adequate cold storage facilities, modern packaging and transport systems, and inducement of the private sector to invest in the date industry have to be considered as contributory approaches to develop the industry within Iran. On the other hand, programs should be focused on gradual replacement of cultivars with higher-yielding palms, increasing of product prices to financially support the growers and exporters, effective use of mechanization to decrease production expenses, broadening the array of by-product to decrease waste, and finally creating vigorous private (or governmental) organizations to define high-level policies for marketing targets. Date palm research must focus on quality improvement in production by adopting new and efficient agro-technical practices as well as utilization of biotechnology to achieve objectives in the areas of gene banks and transgenic plants.

References

Abazarpoor M (1998a) Investigation and determination of the most suitable pollen density for mechanical pollination of the Mozafti cultivar (Final report of research project – No. 77/131). Date Palm and Tropical Fruit Research Institute Press, Ahvaz (in Farsi)

Abazarpoor M (1998b) Investigation on the effects of pollen of different cultivars on quality and quantity of the Halili cultivar fruit (Final report of research project – No. 77/139). Date Palm and Tropical Fruit Research Institute Press, Ahvaz (in Farsi)

Abazarpour M (1998c) Investigation and determination of the best time of pollination in Mozafti cultivar (Final report of research project – No. 77/144). Date Palm and Tropical Fruit Research Institute Press, Ahvaz (in Farsi)

Ashraf Z, Hamidi-Esfahani Z (2011) Date and date processing: a review. Food Rev Int 27: 101–133

Ashraf-Jahani A (2002) Date, a life fruit. Agricultural Science Publications, Tehran

Barreveld W (1993) Date palm products. FAO agricultural services bulletin 101, Rome

Camlin M (2003) Plant cultivar identification and registration – the role for molecular techniques. Acta Hort 625:37–47

Darini A, Ezadi M (2001) Effects of inter-planting of sorghum and alfalfa on date palm fading disorder (Final report of research project – No. 80/226). Date Palm and Tropical Fruit Research Institute Press, Ahvaz (in Farsi)

Davoodi D, Majidi E, Khoshkam S (2002) Some morphological and anatomical aspects of date palm (*Phoenix dactylifera* L.) somatic embryogenesis in tissue culture. J Agric Sci Tech 4:63–71

Davoodian A, Darini A, Ezadi M (2001) Effects of different bunch covers on date palm fading disorder (Final report of research project – No. 77/362). Date Palm and Tropical Fruit Research Institute Press, Ahvaz (in Farsi)

Eata M (1988) Determination of the best pollen density for pollination of the Sayer cultivar (Final report of research project – No. 67/014). Seed and Plant Improvement Institute Press, Karaj (in Farsi)

Eata M (1991) Determination of the best pollen for pollination Sayer cultivar (Final report of research project – No. 70/093). Seed and Plant Improvement Institute Press, Karaj (in Farsi)

Ehsani A (1986) Investigation on the effects of different pollens on quality and quantity properties (Final report of research project – No. 65/414). Date Palm and Tropical Fruit Research Institute Press, Ahvaz (in Farsi)

Eshragi P, Zarghami R, Mirabdulbaghi M (2005) Somatic embryogenesis in two Iranian date palm cultivars. Afr J Biotech 4(11):1309–1312

FAOSTAT (2014a) Food and Agriculture Organization; Resources. Available via faostat.fao.org. Accessed 18 Apr 2014

FAOSTAT (2014b) Food and Agriculture Organization; Crop Production. Available via faostat.fao.org. Accessed 28 Nov 2013

Farashbandi H (2000) Determination of the best time of pollination for Kabkab cultivar (Final report of research project – No. 79/106). Date Palm and Tropical Fruit Research Institute Press, Ahvaz (in Farsi)

Flora of Iran (2011) Geo-botany of Iran. Available via http://www.flora-iran.com. Accessed 18 Apr 2014

Garshasebi M (2010) Effects of irradiation date fruits (Sayer cv.) on storage pest control and fruit shelf life (Final report of research project – No. 89/789). Date Palm and Tropical Fruit Research Institute Press, Ahvaz

Habashi AA, Kaviani M, Mousavi A, Khoshkam S (2008) Transient expression of β-glucuronidase reporter gene in date palm (*Phoenix dactylifera* L.) embryogenic calli and somatic embryos via microprojectile bombardment. J Food Agr Env 6(2):160–163

Habibi-Najafi MB (2011) Date seeds: a novel and inexpensive source of dietary fiber. International conference on food engineering and biotechnology, Singapore, 28–30 Sept 2011

Hajian S (2005a) Fundamentals of pollination in date palm plantations in Iran. Paper presented at the first international conference on mango & date palm, University of Agriculture, Faisalabad, 20–23 June 2005

Hajian S (2005b) Introducing of international date cultivars situation in Iran. Paper presented at the first international conference on mango & date palm, University of Agriculture, Faisalabad, 20–23 June 2005

Hajian S (2005c) Identification of date palm germplasm in Iran (Final report of research project – No. 89/936). Date Palm and Tropical Fruit Research Institute Press, Ahvaz (in Farsi)

Hajian S (2007) Mapping of date palm biodiversity in Iran (Final report of research project – No. 89/937). Date Palm and Tropical Fruit Research Institute Press, Ahvaz (in Farsi)

Hajian S (2009) Evaluation of date palm abnormalities propagated by tissue culture technique in Iran. Paper presented at the 6th biotechnology conference. Iranian Biotechnology Society. Tehran, 13–15 Aug 2009

Hajian S (2011) Abstract of the national date palm program of Iran. Kerdegar Publication, Ahvaz

Hajian S, Marashi S, Torahi A et al (2011) Plant breeding sub program. In: Hajian S (ed) National date palm program of Iran. Kerdegar Publication, Ahvaz (in Farsi)

Hajian S, Pezhman H (2004) Situation of date palm and tropical fruits researches in Iran (approaches, obstacles and abilities). National Anniversary Research Week, Shahid Chamran University, Ahvaz, 21–27 Apr 2004 (in Farsi)

Karampour F (2001a) Study on causes of date palm fading disorder (Final report of research project – No. 80/612). Date Palm and Tropical Fruit Research Institute Press, Ahvaz, (in Farsi)

Karampour F (2001b) Study on fungal diseases on date palm fading disorder (Final report of research project – No. 80/619). Date Palm and Tropical Fruit Research Institute Press, Ahvaz (in Farsi)

Kashani M (1979) Identification of date palms in Iran (technical bulletin). Seed & Plant Improvement Institute Press, Karaj (in Farsi)

Latifian M, Rad B (2012) Pathogenicity of the entomopathogenic fungi *Beauveria bassiana* (Balsamo) Vuillmin, *Beauveria brongniartii* Saccardo and *Metarhizium anisopliae* Metsch to adult *Oryctes elegans* Prell and effects on feeding and fecundity. Int J Agric Crop Sci 4(12):811–817

Majidi E, Shakib A, Modiri M et al (1991) Study of callus induction from in vitro culture of different parts of date palm. Seed Plant J 7(1/2):9–13

Mardi M, Torahi A, Kavand A (2010) Application of microsatellite markers for identification and registration of date palm cultivars (Final report of research project). Agricultural Biotechnology Research Institute Press, Karaj (in Farsi)

Mirzaee M, Saei M (2001) Effects of date palm orchards management on date bunch fading disorder (Final report of research project – No. 80/305). Date Palm and Tropical Fruit Research Institute Press, Ahvaz (in Farsi)

Mostaan A, Marashi S, Ahmadizadeh S (2010) Development of a new date palm pollinator. Abstracts of the 4th international date palm conference, Abu Dhabi, 15–17 Mar 2010

Mousavi M, Mousavi A, Habashi AA, Arzani K (2009) Optimization of physical and biological parameters for transient expression of uidA gene in embryogenic callus of date palm (*Phoenix dactylifera* L.) via particle bombardment. Afr J Biotech 8(16):3721–3730

Mousavi M, Mousavi A, Habashi AA et al (2014) Transient transformation of date palm via Agrobacterium-mediated and particle bombardment. Emir J Food Agric 26(6):528–538

Nazeri S, Khoshkam S, Afshari M, Shakib AM, Majidi E (1993) Somatic embryogenesis in date palm varieties Estamaran and Kabkab. Seed Plant 8(3/4):16–20
Pezhman H (2001) Hand book of date (cultivation, practices and harvesting). Agricultural Education Press, Karaj (in Farsi)
Pezhman H (2002) A view on date palm situation and its research programs in Iran. Establishment meeting (conference) of date palm global network, Al Ain, 7–9 Apr 2002
Pezhman H, Roshan V, Rahkhodaei E (2005) Advanced study on the causes of date palm bunch fading and wilting disorder (Final report of research project – No. 84/912). Date Palm and Tropical Fruit Research Institute Press, Ahvaz (in Farsi)
Sadeghian Y (2008) National standards for distinctness, uniformity and stability of date palm. Seed and Plant Certification and Registration Institute Press, Karaj (in Farsi)
Samavi H (1999) Study on the effects of different pollen of male date palm trees on quantity and quality of Halili cultivar fruits (Final report of research project – No. 78/067). Date Palm and Tropical Fruit Research Institute Press, Ahvaz (in Farsi)
Sayahpoor H (1998) Determination of the best time of pollination of the Sayer cultivar (Final report of research project – No. 77/051). Date Palm and Tropical Fruit Research Institute Press, Ahvaz, (in Farsi)
UNEP (1998) Assessment of alternatives to methyl bromide. In Montreal protocol on substances that deplete the ozone layer, United Nations Environment Program (UNEP), Nairobi
Zaid A, de Wet P (2002) Date palm propagation. In: Zaid A (ed) Date palm cultivation. Food and Agriculture Organization, Rome
Zargari H (2000a) Determination of the best pollen density for pollination of Kabkab and Shahani cultivars (Final report of research project – No. 80/636). Date Palm and Tropical Fruit Research Institute Press, Ahvaz (in Farsi)
Zargari H (2000b) The best pollen of male date palm trees for pollination of Zahidi, Shahani and Kabkab cultivars (Final report of research project – No. 80/654). Date Palm and Tropical Fruit Research Institute Press, Ahvaz (in Farsi)
Zargari H (2000c) Determination of the best time for pollination of Shahani cultivar (Final report of research project – No. 80/695). Date Palm and Tropical Fruit Research Institute Press, Ahvaz (in Farsi)
Zivdar S, Mousawi M, Alemzadeh-Ansari N (2008) Genetic stability in date palm micropropagation. Asian J Plant Sci 7(8):775–778

Chapter 3
Date Palm Status and Perspective in Saudi Arabia

Salah M. Aleid, Jameel M. Al-Khayri, and Abdulaziz M. Al-Bahrany

Abstract Date palm is a major crop significantly contributing to the agroecosystems in Saudi Arabia. There has been vast support and development in the agricultural sector in general and the date palm sector in particular during the last decades in Saudi Arabia. However, there are several challenges facing date cultivation, processing, and marketing. Such difficulties relate to water scarcity and salinity, soil erosion and desertification, insect pest infestations and disease, insufficient processing and packaging facilities and technologies, as well as a decline in date fruit demand. There is an enhancement in investment in world-class facilities and operations by leading date producers in Saudi Arabia. Large farms are prevalent in Saudi Arabia, and they are making substantial investments in new or replacement trees not in production signifying a further increase in production capacity; however, cultivar selection is generally not based on technical analyses such as yield, disease resistance, or water and fertilizers requirements. The aim of this chapter is to provide an overview of major recent aspects of the date palm in Saudi Arabia, a hot and arid region. Subjects discussed include fruit production, propagation, molecular and genetics descriptions, field management practices, entomology and pathology, post-harvest handling, value-added products, commercial value, and nutrition and health benefits. Also presented are ways of minimizing the usage of nonrenewable natural resources as well as recent findings from research and development activities.

Keywords Cultivation • Dates • Germplasm • Marketing • Molecular identification • Processing • Propagation • Research • Tissue culture

---

S.M. Aleid (✉)
Date Palm Research Center of Excellence, King Faisal University,
420, Al-Hassa 31982, Saudi Arabia

Department of Food and Nutrition Sciences, King Faisal University,
420, Al-Hassa 31982, Saudi Arabia
e-mail: seid@kfu.edu.sa

J.M. Al-Khayri • A.M. Al-Bahrany
Department of Agricultural Biotechnology, College of Agriculture and Food Sciences,
King Faisal University, Al-Hassa, Saudi Arabia
e-mail: jkhayri@kfu.edu.sa; ambahrany@kfu.edu.sa

© Springer Science+Business Media Dordrecht 2015
J.M. Al-Khayri et al. (eds.), *Date Palm Genetic Resources and Utilization: Volume 2: Asia and Europe*, DOI 10.1007/978-94-017-9707-8_3

3.1 Introduction

3.1.1 Importance to Saudi Arabia Agriculture

Date palm (*Phoenix dactylifera* L.) is a unisexual fruit tree native to the hot arid regions of the world, mainly grown in the Middle East and North Africa. Since ancient time, this majestic plant has been recognized as the *tree of life* because of its integration into human settlement, well-being, and food security in hot regions of the world, where only a few plant species can flourish (Al-Khayri 2007). Saudi Arabia has a land area of 2,250,000 km^2, of which less than 0.2 % is presently under cultivation. The country is characterized by an arid and semiarid climate. The cultivated land is estimated at 525,000 ha. However, three to four times as much land is arable but not yet put to use. The significant feature of Saudi agriculture lies in the scarcity of water, with rainfall averaging about 100 mm per year in most regions except the southwest, which averages 250 mm or more (Abo-Hassan 1981). Temperatures are generally high, reaching 50 °C at times in some places in summer, when the relative humidity is also high in the eastern Saudi Arabia. Soils are fragile and subject to erosion by wind and water, as well as degradation through salinization. Over 95 % of the total land area in the Arabian Peninsula suffers from some form of desertification, of which 44 % is severe to very severe; wind and water erosion account for over 60 % of the desertification (Erskine et al. 2004).

The number of named date palm cultivars distributed all over the world was reported to be approximately 5,000 (Bashah 1996). Different date cultivars taste different from place to place, and even within one date grove, there is a range of qualities in the same cultivar.

Date palm is the most widely grown fruit tree in areas with long dry summers and mild winters, such as the predominating climate of Saudi Arabia. It thrives in desert and oasis where temperatures are high but with groundwater table that is close to the soil surface. Date fruit production is dependent on the availability of certain heat requirements, according to cultivar. Most dry fruit cultivars are found in the dry areas, whereas soft and semidry cultivars are confined to the humid and semidry areas. Date palms in Saudi Arabia can grow in different types of soil, but the best productions are recorded in light deep soils. It can tolerate high levels of salinity, i.e., some cultivars can survive salinity levels up to 22,000 ppm, but their growth and fruit productivity would be affected (Erskine et al. 2004). Salinity stress exerts negative influence on date palm growth and chlorophyll content (Al-Abdoulhadi et al. 2012).

Saudi Arabia occupies the first rank in the world in terms of average per capita consumption of dates, reaching 34.8 kg/year in 2003 (Al-Shreed et al. 2012).

3.1.2 Production Statistics and Economics

World date palm cultivation has expanded rapidly in recent years. The global production of date was estimated at 7.75 million mt, worth USD 3.82 billion in 2010. Saudi Arabia is ranked as the second largest date producer in the world after Egypt

Table 3.1 Estimated area and production of dates by region in Saudi Arabia in 2011

Region	Production (mt)	Area (ha)
Riyadh	248,327	42,208
Qassim	187,561	39,301
Eastern	147,305	13,625
Madinah	139,924	18,502
Hail	109,229	16,187
Jouf	46,898	5,471
Makkah	44,882	8,068
Tabuk	20,679	2,966
Asir	20,119	4,297
Najran	19,526	3,070
Baha	6,618	1,095
Jazan	326	288
Northern	152	40
Total	991,546	155,118

Source: Ministry of Agriculture (2011)

(FAOSTAT 2012). In 2011, date production in Saudi Arabia reached 991,546 mt, from 3.7 million trees which covered an area of 155,118 ha (Ministry of Agriculture 2011). Table 3.1 shows estimated area and production of dates by region in Saudi Arabia. The date palm orchards are distributed in various regions. Riyadh in the Central Region is considered the highest producer of dates (25.04 %) followed by Qassim region and Eastern region (Al-Hassa and Qatif) which represent 18.91 % and 14.85 %, respectively. While there are about 400 date cultivars grown in four main regions in the Kingdom, only about 50–60 cvs. are commercial (Mikki 1998).

Because of surplus date production in Saudi Arabia, domestic marketing of the product needs to be enhanced; moreover, exports have not reached the expected level. Also, the per capita consumption of dates in Saudi Arabia has significantly declined over the years, due to competition from other fresh fruits and sweets, in addition to changes in family size and eating habits, while the production has increased. This has impacted date production negatively, in spite of support from the government. The Kingdom currently has 64 date processing factories (Al-Shuaiby and Ismael 2007; Anonymous 2006, 2009).

3.1.3 *Current Agricultural Problems*

Presently, date palm development in Saudi Arabia is of great governmental concern, while project implementation concerns the private sector. After the ambitious expansion of developing programs for new date palm plantation, the overall picture is now a series of continuous new endeavors aimed at improving the date growing areas through the introduction of labor-saving methods in cultivation and modern irrigation systems, improved packaging, industrialization of fruit processing, and product diversification, for example, a better utilization of lignocellulosic residues of the date palm tree (Mikki 1998). However, there are constraints facing the development of date palm in Saudi Arabia. Date palm cultivation in the country has a long history, yet the efforts exerted on research and development, although significant, are still insufficient and fall below expectations. In general, product quality is still low, field and postharvest losses are high, and date product and by-product utilization needs improvement. To address these constraints, the Saudi Arabian government ranked date palm as having the highest research priority in setting priorities for agricultural research (Erskine et al. 2004).

3.2 Cultivation Practices

3.2.1 Current Cultivation Practices

Date palm cultivation practices in Saudi Arabia can be summarized in the following steps: fertilization, irrigation, pollination, fruit thinning, bending and bagging of bunches, pesticide control, and harvesting. Most date farmers practice manual pollination techniques and are generally reluctant to adopt innovative pollination techniques. Proper training programs for the use of mechanical fertilization methods need to be implemented. The three highest expenses of date farmers in Saudi Arabia are labor, chemicals, and energy to pump water. Brown (1983) showed that among cultural operations, harvesting, pollination, and pruning are the most labor-intensive work, accounting for more than 80 % of the total production costs. Most of palm trees in the old historic date palm orchards in Saudi Arabia such as Al-Hassa, Kharj, and Madinah have been planted densely with only 2–5 m between trees, which represents an impediment to mechanization.

In Saudi Arabia, flood irrigation is primarily practiced in Madinah and Al-Hassa. However, most date palm farms have adapted modern irrigation techniques such as drip and sprinkler irrigations. However, some old farms still practice flood irrigation. Farmers do not keep records of water consumption; date palms are typically overwatered, resulting in a waste of critical water resources. Al-Amoud et al. (2000) studied the effect of different irrigation methods on date palm tree yields in Saudi Arabia. The results demonstrated that the maximum average yield was obtained for trees irrigated with trickle methods, followed by basin-irrigated trees. An annual water volume of 108 m^3 per tree (1.08 m/ha) would be sufficient to produce the highest water use efficiency. The trickle irrigation method has been shown to be the best for water use efficiency, followed by the basin and then the bubbler methods (Al-Amoud et al. 2000).

Timing between applications of fertilizers varies considerably among growers in Saudi Arabia, ranging from every year to 3 years. Fertilizers are often applied without soil sampling, with limited use of chemical fertilizers in some regions such as Al-Hassa and Madinah. In Saudi Arabia, responses to fertilization are inconsistent and probably depend upon cultivar, soil type, and other factors. Manure has traditionally been used in date production, but in many instances inorganic fertilizers are used. Many producers, even in industrialized production, consider manure to be superior to inorganic fertilizer. Cover crops are also often grown in date groves (Abdul-Baki et al. 2002). In a study by (Almadini and Al-Gosaibi 2007), evaluating the effect of the use of organic fertilizer for palm trees on soil properties and fertility, the results indicate that applying organic amendments led to improved physical and chemical characteristics of the soil and greater soil fertility, indicating the importance of adding organic fertilizers to enhance soil quality and sustain the productive capacity of fruit quantity and quality. Al-Kahtani and Soliman (2012) reported that agricultural waste plus 40 % sheep manure gave the highest initial fruit set and retained fruit, bunch weight, yield fruit weight, flesh weight, flesh thickness, fruit volume, fruit dimensions, total soluble solids, nonreducing sugars, and total sugars, as compared with the other treatments over two seasons.

Soil testing can be a useful tool to assess the fertilizer needs of date palms. Of the 17 essential elements, N, P, and K are commonly deficient in soils. Therefore, field fertilizer trials predominantly involve these three elements. There is a direct relationship between soil test values and date palm yield, justifying the need to improve fertilizer. Elprince and Alsaeedi (2007) estimated fertilizer requirements of date palm from site variables including soil testing, water, and leaf analysis. The results warrant the establishment of an extension service for fertilizer recommendations in the Al-Hassa region. Presently, there is a tremendous opportunity to increase tree yield and fruit quality simply through the adoption of improved cultivation practices.

3.2.2 *Pollination and Fruit Quality*

3.2.2.1 Pollination

In most date palm growing countries, including Saudi Arabia, seedling males are used for pollination. These seedling males are highly variable and differ greatly in their growth, vigor, spathe characteristics, and pollen quality (Al-Baker 1972). In some date cultivars better fruit set results from pollen of some males than others, apparently due to compatibility of male and female cultivars (Nixon 1969). The principal characteristics desired in a male palm are fertility, flowering early in season, and a large quantity of flowers. At room temperature, pollen will retain its viability for 2–3 months. Soliman and Al-Obaid (2013) reported significant differences in pollen grain morphology among 11 Saudi date palm males.

Al-Ghamdi et al. (1988) evaluated eight date palm males and their effects on fruit character of three female cultivars grown in Al-Hassa. They observed that males varied considerably with respect to vegetative and reproductive characters. Fruit characteristics such as total weight, pulp weight, moisture content, length, and width were significantly affected to a degree, influenced by the genotype of both male and female trees. In general, the male designated M1 appeared to be superior to the other males tested based on the fruit characteristics evaluated.

3.2.2.2 Fruit Quality

Quality profile of dates involves an evaluation of four aspects: (a) color, shape, size, taste, texture, pit/flesh ratio, and uniformity in color and size of the fruit; (b) moisture, sugar, and fiber content; (c) defects of the fruits, which may include discoloration, broken skin, sunburn, blemishes, and shrivel deformity; and (d) presence of insect infestation, foreign matter, pesticide residues, mold, and decay (Saleem 2005). Horticultural practices, in particular, have direct effect on yield, on fruit quality, and ultimately on the sustainable utilization of all resources within the oasis agroecosystem, including date palm cultivars (Siebert et al. 2007). Some traditional horticultural practices may adversely impact date yield and quality. These include narrow spacing between trees, minimal fertilizer inputs, and inadequate irrigation due to limited resources, in addition to inadequate pollination, fruit thinning, and protection from biotic and abiotic stress as well as inefficient pest and disease control (Jaradat 2013). Hand pollination and fruit thinning guarantee adequate fruit set and optimal fruit quality (Mukhtar et al. 2011).

3.2.3 Pest and Disease Control

The most important pests and diseases affecting date palm in Saudi Arabia are the red palm weevil (RPW), *Rhynchophorus ferrugineus* (Olivier), and Al-Wijam disease. A phytoplasma pathogen is suspected to cause Al-Wijam; its main symptoms are leaf stunting, yellow streaking, and a marked reduction in fruit and stalk size, which leads to failure in fruit production at harvest (Alhudaib et al. 2007). The RPW has been identified by FAO as a *category-1* insect pest of date palm in the Arabian Gulf region and is a key pest in Saudi Arabia. Close spacing of palms and open flood irrigation favor increased attack by RPW, probably due to higher relative humidity in the plantations resulting from these practices (Sallam et al. 2012). In Saudi Arabia, RPW was reported in Qatif in the Eastern Province then spread throughout the country, mainly through transport of infested planting material (Al-Abdulmohsin 1987). In Saudi Arabia, the annual loss due to eradication of severely infested palms is put at 1–5 %, with estimated value to range from USD 1.74–8.69 million, respectively, at a fixed eradication level of 20 % of infested palms (El-Sabea et al. 2009).

Early detection of infestation is the key to the success of RPW management in the field. Infested palms, if not detected early and treated, usually die after harboring several overlapping generations. However, palms in an early stage of attack respond to insecticide treatment by stem injection (Faleiro 2006). Recently El-Shafie et al. (2011) assessed the efficacy of trap-free *attract and kill* pheromone technology for managing RPW in date plantations, while Faleiro et al. (2010) developed models to validate area-wide RPW-IPM programs in the date plantations of Saudi Arabia.

Stored dates are attacked by insects, birds, rodents, mites, and microorganisms including fungi. Fumigation is the most commonly important alternative to pest control of stored dates (see Sect. 3.7.4.1 for more details).

Treatment of dates on the bunch during April to June will effectively control the lesser date moth, date stone beetle, and greater date moth. This will affect positively the control of other ripe fruit moths and beetles, particularly the sap beetle during August to late October. Management of date pests during this time of the cropping season will guarantee clean undamaged and uninfected dates in the storehouse.

Precautions have to be taken to prevent reinfestation after storage and immigration of insects from outside the storage facilities. International CODEX food safety standards for dates list the absence of live insects and insect eggs and mites as an important index for quality. Therefore, to reduce the quantitative and qualitative losses due to insect pests, and to meet the international marketing standards, an integrated stored product management program (ISPM) for dates should be established and strictly applied (Abo-El-Saad and Elshafie 2013).

3.2.4 Cultivation Challenges and Limitations

The date palm is a drought-resistant and salt-tolerant plant. The annual water requirement of a mature date palm ranges between 115 and 306 cubic meters (1.15–3.06 mt/ha) (Al-Baker 1972). Scarcity of water in Saudi Arabia contributes significantly to reducing date palm productivity and even to the death of the trees (El-Juhany 2010). Date palm cultivation requires plenty of water, and if the current trend of palm expansion in Saudi Arabia continues, a considerable amount of additional irrigation water will be required. However, due to the limited availability of water resources, the increase in using water conservation measures, such as modern irrigation systems (trickle and bubbler), is necessary (Al-Amoud et al. 2000). Moreover, there are several constraints facing the development of date palm in Saudi Arabia, such as: (a) the presence of low-quality cultivars; (b) poor farm management; (c) pests and diseases, and inadequate IPM control; (d) deficiencies in harvesting, processing, and marketing practices; (5) shortage of qualified and national trained staff and laborers; and (e) insufficient research and development activities (Erskine et al. 2004). Although the cultivation of date palms has developed considerably, and great attention has been given by the Government of Saudi Arabia; nonetheless, the level of date productivity is low compared with other date-producing countries (Al-Obaid 1996). The main causes of low date productivity are the increase in the large number of the older trees, the existence of many low-quality and undesirable cultivars, lack of sufficient offshoots to establish new orchards or even renew the old ones, and the increase in cost of offshoots of good-quality cultivars (Al-Sakran and Muneer 2006).

3.3 Genetic Resources and Conservation

Date palm genetic diversity is the foundation of an effective crop improvement program to develop new cultivars resistance to current abiotic and biotic limitations. This highlights the importance to preserve date palm germplasm by utilizing all available means. In Saudi Arabia, limited research was encountered related to date palm in vitro cryopreservation. General conservation efforts are evident from the establishment of two field germplasm banks to conserve national and international cultivars, discussed below. Moreover, to protect germplasm against degradation threats due to the spread of pests, the authorities have implemented quarantine regulations on date palm international exchange and transport within the country.

3.3.1 Threats and Degradation

Date palm germplasm biodiversity and productivity in Saudi Arabia is threatened by a number of factors common to most date palm growing countries. In addition to urbanization, desertification, salinization, and genetic erosion, date palm is threatened by numerous diseases and insects as well as environmental stress factors including drought, salinity, and temperature extremes that may be exacerbated as a result by global climate change.

In Saudi Arabia, red palm weevil (RPW) (El-Sabea et al. 2009) and Al-Wijam disease (Alhudaib et al. 2007) are the most serious pests for date palm cultivation. Lack of efficient integrated pest management (IPM) has resulted in the spread of pests and diseases in different regions of the country (El-Juhany 2010). It is reported to attack mostly young date palm less than 20 years old. Reports indicated that 99 % of the infested date palms are less than 20 years old with 75 % being in the age group of 6–15 years which is the most vulnerable age of infestation (Anonymous 1998).

There are several factors that contribute to genetic erosion that also threaten date palm diversity and productivity. Krueger (2011) stated that modernization of traditional oasis culture can contribute to genetic erosion of date palms due to the importation of more profitable elite cultivars to replace old or dead date palms with elite foreign cultivars. Lately, it is noticeable that some Saudi framers are interested in the cultivation of certain foreign elite cultivars such as Medjool. Also, certain elite local cultivars comprise most of the cultivation areas in each growing regions, marking the decline of genetic diversity.

Water scarcity and salinity of water and soil, aggravated by typical lengthy rainless periods and drying of many water wells, restrict expanding date palm cultivation in most parts of the Arabian Peninsula, including Saudi Arabia (Jaradat and Zaid 2004).

Another important threat is the shortage of skilled labor to maintain date palm cultivation, which requires attention. Enhanced income through the opportunities presented by modern technologies has led the majority of the small farmers to abandon their date palm groves (El-Juhany 2010). Neglecting these aspects can cause the degradation of date palm germplasm and limit cultivation. It is important to strengthen research and conservation efforts to keep these factors in check in order to conserve date palm germplasm resources and sustain productivity.

In an effort to manipulate growth of date palm and examine the potential effects on various physiological parameters, several studies were conducted involving magnetic fields (Dhawi and Al-Khayri 2008a, b; 2009a, b, c; Dhawi et al. 2009) as well as x-ray irradiation (Al-Enezi and Al-Khayri 2012a, b; Al-Enezi et al. 2012). These studies involved exposing seeds collected from cv. Khalas in Al-Hassa, germinated and grown under greenhouse conditions, and chemical analyses based on seedling tissue. Changes were observed in germination and seedling growth as well as physiological responses including proline accumulation, ions concentration, DNA content, and photosynthetic pigments distribution. These studies provide guidelines for future studies aimed at inducing growth manipulation and possibly mutations that may contribute to enhancing the available genetic resources and expanding the genome of the date palm (Dhawi and Al-Khayri 2011).

3.3.2 *Cryopreservation*

Cryopreservation is currently considered an effective method for long-term conservation of date palm germplasm. Several studies to examine these factors include the effect of different components of cryoprotectant solution. In Saudi Arabia, the only relevant study encountered was reported by Al-Bahrany and Al-Khayri (2012a), which showed that the revived cell samples treated with 10 % DMSO in combination with 0.75 M sucrose showed the most colony formation, greatest callus growth, and highest embryo numbers. Accumulating fundamental knowledge of cryopreservation requirements may be utilized to develop a germplasm bank for the long-term conservation of date palm germplasm. This is of tremendous importance, especially under current predictions of global climate change and the potential abiotic and biotic threats that may challenge date palm cultivation in different regions of the world.

3.3.3 *Germplasm Banks*

3.3.3.1 Date Palm Germplasm Bank at King Faisal University

A date palm germplasm bank was established in 1982 at King Faisal University, Date Palm Research Center of Excellence, located at the Agricultural and Veterinary Experimental Station in southeastern Al-Hassa (25°15′36″ N lat.; 49°43′30″ E long., elev. 145 m). This collection contains national and international date palm cultivars as well as a section for trees produced by tissue culture. The national

Fig. 3.1 Views of the date palm germplasm bank at the Ministry of Agriculture, Date Palm Research Center, Al-Hassa, Saudi Arabia (Photo by: Kabeel M. Ghawas)

collection comprises 510 palms representing 31 Saudi cultivars collected from 7 major growing regions (Al-Hassa, Qatif, Madinah, Besha, Najran, Jouf, and Qassim). The international collection comprises 615 date palm trees representing 26 cultivars imported as offshoots from California, USA, and Iraq. In addition, 282 tissue culture-derived date palm trees representing 15 cultivars are obtained from the USA, England, France, and Saudi Arabia (Al-Ghamdi 2001).

This germplasm collection encountered adverse conditions which led to the decline of numerous trees. Degradation was mainly due to the decline of groundwater quality resulting in increased soil salinity, low soil fertility, and shallow soil depth, as well as the location of the germplasm field over an underground salt belt within the experimental station, high water table, and increased exchangeable sodium percentage. Consequently, a decision was made in 2010 to transfer some of the surviving trees to the Ministry of Agriculture Date Palm Germplasm Bank. Currently the relocation of the date palm germplasm bank to a more suitable site where soil conditions are more favorable is being considered.

3.3.3.2 Date Palm Germplasm Bank at the Ministry of Agriculture

The Ministry of Agriculture, Date Palm Research Center, located in northern Al-Hassa (25°27′ N lat.; 49° 33′ E long.), established a date palm germplasm bank in 2000 (Fig. 3.1). This was implemented through the technical cooperation between the Ministry of Agriculture and FAO to conserve elite cultivars. Subsequently, the collection was augmented with seedling date palms through collaboration with the Arab Center for Studies of Arid Zones and Dry Lands (ACSAD). The collection consists of 89 cultivars collected from major growing regions of Saudi Arabia and 5 cultivars from international sources. In addition, there are14 date palm seedling cultivars from the Zulfi region of central Saudi Arabia (26°18′ N lat.; 44°48′ E long.) and 2 seedling cultivars from the Al-Hassa region. Moreover, the collection contains elite male date palms collected from various regions of Saudi Arabia.

The main objectives of establishing this date palm germplasm bank were: (a) conserve genetic diversity, (b) study the morphological and physical characteristics of cultivars, (c) develop a germplasm repository for plant breeding and molecular work, and (d) study the reaction of the cultivars to major abiotic and biotic stresses.

3.3.4 Quarantine Regulations

Due to the threats posed by biotic factors on agriculture, numerous countries have imposed certain actions to minimize the phytosanitary risk associated with the inadvertent introduction and spread of pests and diseases. Saudi Arabia has approved implementation of Quarantine Law of the Gulf Cooperation Council of the Arab Gulf States through Royal Decree No. M/9 dated 8 March 2005 (Experts 2005). The aim of this law is to prevent the entry and spread of agricultural pests, protection of environmental and plant resources, and the facilitation of trade. Implementation of the Regulations of the Quarantine Law, Article 6 of Chapter II, prohibits import of consignments of plants or plant products or any other material of any palm species and their derivatives. Fruits of date palm and fiber-stripped coconuts are exempted (Ministry of Agriculture 2005).

Measures were taken in the early 1990s by the Ministry of Agriculture to restrict the spread of pests, especially red palm weevil (Committee 1993). These acts included (a) prohibiting transportation of date palm offshoots within Saudi Arabia among date palm growing regions, (b) obtaining official permits to transport any date palm offshoots within the Kingdom, and (c) strengthening extension services to increase farmers' awareness of the danger and prevention methods, as well as providing on-field technical assistance to deal with infested trees.

The transport of offshoots is restricted within the country mainly because of the lack of a protocol ensuring the phytosanitary condition of the plants. Recently, Al-Shawaf et al. (2013) developed an insecticide-based protocol that kills larval stages of RPW feeding inside date palm offshoots. This would allow transport of treated offshoots to a different growth region without the danger of spreading this devastating insect. The protocol involves dipping offshoots in 0.004 % fipronil 3.5 % (ThripGuard 35 SCTM) for 30 min.

3.4 Plant Tissue Culture

3.4.1 Importance and History

Propagation by offshoots, axillary buds growing from the base of the tree, is the preferred method as compared to seed propagation which normally produces off-type progeny. Although seeds are a rich resource of germplasm useful for genetic improvement (Johnson et al. 2013), seed propagation is not suitable for the multiplication of known date palm cultivars because genetic traits including fruit traits are not maintained. Offshoots are produced only in the early life of the date palm, the first 10–15 years, in very limited number, although this number varies among cultivars.

Offshoots must remain attached to the parent tree for 2–3 years to develop an adequate growth and their separation from the parent tree is often difficult and expensive, and a large number of offshoots do not survive, making this method inefficient particularly at commercial scale. Alternatively, micropropagation is gaining increased interest as it provides a relatively rapid means for mass clonal propagation (Al-Khayri 2005, 2007).

In the early 1980s, research on in vitro plant regeneration of date palm began in Saudi Arabia, mostly at the Date Palm Research Center, King Faisal University, (Abo El-Nil 1989a, b; Abo El-Nil and Al-Ghamdi 1989; Khalil et al. 1983; Khan et al. 1983) and continued to build in the 1990s (Al-Khayri et al. 1996; Al-Maarri and Al-Ghamdi 1995, 1997). This early work demonstrated successful micropropagation of a number of important Saudi date palm cultivars. Subsequently, research was focused optimizing in vitro factors (Al-Khayri 2013) and the utilization of tissue culture for physiological and genetic improvement studies including tolerance to drought (Al-Khayri 2002) and salinity (Al-Bahrany and Al-Khayri 2012b; Al-Khayri and Al-Bahrany 2004a), the most important stress factors limiting cultivation in this region.

3.4.2 Research and Development

Various research institutions which are actively engaged in date tissue culture include the Ministry of Agriculture, King Abdulaziz City for Science and Technology, King Saud University, and King Faisal University. Relevant research by Saudi scientists is chronologically presented in Table 3.2. This table reflects the studied factors to optimize somatic embryogenesis and organogenesis in various cultivars.

3.4.3 Scale-up Commercial Micropropagation

Two commercial tissue culture labs currently are involved in date palm micropropagation: SAPAD Tissue Culture Date Palm Co. (http://sapad.com.sa/en) in Dammam, Eastern Province, and Al-Rajhi Tissue Culture Laboratory (http://www.clonebiotech.com/) located in Riyadh, Central Province. They provide venues for local purchase as well as international export. Different sizes of in vitro plants are shipped depending on customers' specifications.

3.4.4 Micropropagation Protocol

A micropropagation protocol effectively used for date palm regeneration via somatic embryogenesis is fully described by Al-Khayri (2005, 2007). Briefly it involves culturing explants (sectioned lateral and apical buds of offshoots) on MS medium

Table 3.2 Major findings of research on date palm tissue culture in Saudi Arabia

Cultivar	Results	Reference
Helali	Callus and roots formed from various seed zones but no plant redifferentiation was observed	Khalil et al. (1983)
Not specified	NAA-containing medium resulted in highest callus formation in all explants. Plantlets were obtained by reculturing explants on a medium containing 1 mg L^{-1} NAA with 0.1 mg L^{-1} kinetin	Khan et al. (1983)
Not specified	SH and B5 media induced best callus induction, while MS medium supported best regeneration by organogenesis. Embryogenesis was best on ER and SH media. Best level of AC was 1.5–2 g L^{-1}	Abo El-Nil (1989a)
Khalas	Callus growth was enhanced by the inclusion of 0.5 mg L^{-1} thiamine and 2 mg L^{-1} biotin. This treatment also gave the highest number of embryos. Embryo elongation was greatest at 0.5 or 2 mg L^{-1} thiamine combined with 1 mg L^{-1} biotin	Al-Khayri (2001)
Barhi	The number of embryos increased in response to increasing $AgNO_3$ concentration in the absence of 2iP, but in the presence of 2iP, the opposite trend was observed. The number of resultant embryos was the highest on 25 μM $AgNO_3$ in the presence of 0.5 μM 2iP	Al-Khayri and Al-Bahrany (2001)
Khalas	The highest percentage of somatic embryos formed complete plantlets (86 %) occurred on half-strength MS containing 0.2–0.4 mg L^{-1} IBA	Al-Khayri (2003)
Barhi, Naboot Saif, Ruzeiz, Hilali, Khesab	Optimum callus growth occurred at 50 μM $AgNO_3$ in Barhi, Naboot Saif, and Ruzeiz, whereas Hilali grew best at 62.5 μM. Khesab maximum growth occurred at 62.5 μM. Best $AgNO_3$ concentrations for somatic embryogenesis were 62.5, 50, and 37.5 μM for Barhi, Hilali, and Ruzeiz, respectively; whereas 12.5 μM was best for Naboot Saif and Khesab	Al-Khayri and Al-Bahrany (2004b)
Sukury	The highest numbers of buds, number shoots, and shoot length were obtained on MS medium containing (in mg L^{-1}) 0.05 kinetin + 0.25 2ip + 0.025 BAP + 0.025 IAA + 0.025 NOA + 0.025 NAA	Al-Khateeb (2006a)
Sukkari	The results indicated that sucrose at a high level of 60 g improved production of embryos. The production of embryos was greatly enhanced by the addition of 5 % PEG combined with 60 g L^{-1} sucrose	Al-Khateeb (2006b)
Sukkari	Addition of 60 g L^{-1} sucrose produced the highest number of somatic embryos and longest shoot equal to that produced by 6 % date syrup. At 10 %, the syrup caused severe reduction in somatic embryogenesis	Al-Khateeb (2008a)
Khenaizy	Using 5 % of date syrup successfully induced buds and shoots formation; however, higher concentrations (10–20 %) resulted in browning and dryness of plant materials	Al-Khateeb (2008b)
Khenaizy	At 30 and 60 g L^{-1} of sugar, optimal shoot growth was obtained but abnormal growth was observed at 90 and 120 g L^{-1}. Fructose gave the highest dry weight. Rooting was enhanced by 60 g L^{-1} and higher sugar	Al-Khateeb (2008c)

(continued)

Table 3.2 (continued)

Cultivar	Results	Reference
Khesab, Naboot Saif	Callus weight and numbers of somatic embryos were directly proportional to CW concentration reaching optimum at 10 and 15 % CW for Khesab and Naboot Saif, respectively	Al-Khayri (2010)
Khesab, Barni, Barhi	The best callus growth was achieved in Khesab using W and WPM media, Barni using SH and NN medium, and Barhi using SH, W, and MS media. The highest regeneration percentage in Barni occurred on WPM, Khesab on W, and Barhi on W and WPM media	Al-Khayri (2011a)
Naboot Saif	Callus growth was best with 1 g L^{-1} of YE or CH. The best somatic embryo formation was seen on 1 g L^{-1} YE or 0.5 g L^{-1} CH	Al-Khayri (2011b)
Barhi	The growth pattern characterizing the exponential phase of cell suspensions commenced 4 weeks after initiation. The highest plating efficiency, 14.6 %, was obtained at cell density of 10,000 cells ml^{-1}	Al-Khayri (2012)
Naboot Saif	Increasing PEG concentration increased total somatic embryo numbers reaching a maximum at 10 % PEG. The highest percentage of medium-size embryos, 52 %, was obtained at 10 μM ABA, whereas the highest percentage of small embryos was obtained at 50–100 μM ABA	Al-Khayri and Al-Bahrany (2012)

Abbreviations: *2,4-D* 2,4-dichlorophenoxyacetic acid, *2iP* 2-isopentenyladenine, *ABA* abscisic acid, *AC* activated charcoal, *AgNO3* silver nitrate, *B5* Gamborg medium, *BAP* 6-benzylaminopurine, *CH* casein hydrolysate, *CW* coconut water, *ER* Eriksson medium, *HE* Heller medium (Heller Basal Salt Mixture), *IAA* indole-3-acetic acid, *IBA* indole-3-butyric acid, *kinetin* furfurylamino purine, *MS* Murashige and Skoog medium, *NAA* 1-naphthylacetic acid, *NOA* 2-naphthyloxyacetic acid, *PEG* polyethylene glycol, *SH* Schenk and Hildebrandt medium, *WH* white medium (White's basal nutrient medium), *YE* yeast extract

Table 3.3 Culture stages requirements for date palm micropropagation

	Hormones (mg L^{-1})				
Culture stage	2,4-D	2iP	NAA	Light*	Stage duration (weeks)
Initiation	100	3	–	Dark	9
Callus	–	30	10	Dark	3
Embryogenic callus	–	6	10	Light*	9
Callus multiplication	–	1.5	10	Dark	12
Embryogenesis	–	–	–	Light*	12
Rooting	–	–	0.2	Light*	6

*Light is provided at 16-h photoperiods of 50 $\mu mol.m^{-2}.s^{-1}$

(Murashige and Skoog 1962) supplemented with (in mg L^{-1}) $NaH_2PO_4 \cdot 2H_2O$ (170) glutamine (200), thiamine (5), nicotinic acid (1), and pyridoxine (1) in addition to 3 % sucrose and 0.7 % agar. Activated charcoal (1.5 g L^{-1}) is added to the medium during the first three culture stages. Hormonal and light requirements as well as the duration of culture stages are shown in Table 3.3 and some of the process is shown in Fig. 3.2.

Fig. 3.2 Date palm in vitro regeneration. (**a**) Callus induction from shoot tip explants, (**b**) multiplication of embryogenic callus, (**c**) development of somatic embryos, (**d**) in vitro plantlets

3.5 Cultivars Identification

3.5.1 Role and Importance

A reliable methodology for cultivar identification is important to assure the genetic identity and conservation of date palm germplasm. Local cultivar names may not necessarily be accurate because name duplication exists among date palm cultivars grown in various regions. Morphological characteristics of date palm are traditionally utilized to identify cultivars mainly based on fruit traits and vegetative parts; however, these traits can be influenced by environmental factors. Identification and evaluation of genetic diversity based on morphological markers of reproductive parts are difficult and restricted to after the onset of fruiting (El Hadrami and Al-Khayri 2012; El Hadrami et al. 2011).

The tools of morphological identification of date palm cultivars are being consistently refined. For example, Hamza et al. (2009) developed a new approach for cultivar identification based on six stable vegetative characters unaffected by cultivation conditions. They are the percentage of spined midrib part, apical divergence angle, maximal pinnae width of the top leaf, percentage of solitary spines, spine length at the middle, and maximal spine angle.

Alternatively, biochemical markers based on isozymes analysis have been adapted for cultivar identification, but they provide limited information for detecting genomic variations. Currently, molecular markers based on DNA analysis are the most reliable and accurate method for genomic studies including cultivar identification, evaluation of phylogenetic relationship, and analysis of genetic control of agronomic traits.

This section highlights the results of studies conducted in Saudi Arabia in relation to the utilization of these identification approaches in date palm, including morphological descriptors as well as molecular markers based on isozyme and nucleic acid analyses. It also provides an update of date palm genomic studies.

3.5.2 Morphological Descriptors

3.5.2.1 Cultivar Identification

Asif et al. (1989) studied popular date palm cultivars grown in Al-Hassa, which were classified into groups based on morphological characteristics of vegetative parts (leaf length, leaflet length and width) as well as fruit traits (color, weight, length and diameter of fruit, pulp weight, and seed weight) and season of production. Their classification consisted of three categories to which cultivars were assigned.

Al-Doss et al. (2001) documented variation among Saudi date palm cultivars based upon 21 vegetative traits, 13 flowering and yield characters, and 11 fruit properties at different stages of fruit development. They concluded that the principal component analysis can be effectively used for identification and description of date palm cultivars based upon morphological characteristics of leaves and fruits.

Using morphological characters of dates, Al-Khalifah et al. (2012) characterized 14 elite cultivars of date palm popularly cultivated in Saudi Arabia. They found that fruit morphological data demonstrated a high level of diversity in shape of the fruit, color variation during ripening, length-width ratio, percentage of area covered by the fruit cap, and fruit-base shape (Table 3.4). Fruit cap size and percentage of the fruit base covered by the cap are important morphological markers to distinguish cultivars where variation ranged at 25–90 % among cultivars.

Table 3.4 Morphological characteristics of fruit and ripening season of some Saudi date palm cultivars

Cultivar	Shape	Color variation during ripening			Length/width ratio	Fruit cap (%)	Base	Ripening
		Beser (late khalal)	Rutab	Tamar				
Barhi	Ovoid	Yellow	Apricot yellow	Light brown	1.21	40	Truncate	Mid-season
Deglet Noor	Ovoid elongated	Light red	Light brown	Light brown	2.62	90	Truncate	Late-season
Hilali	Ovoid	Yellow	Yellowish brown	Brown	1.1	90	Truncate	Late-season
Hulwa	Ovoid elongated	Red	Dark red	Dark	1.5	50	Shallowly cordate	Mid-season
Khalas	Ovoid elongated	Apricot yellow	Yellowish brown	Brown	1.46	30	Oblique	Mid-season
Maktoumi	Cylindrical	Yellow	Apricot yellow	Brown	1.46	60	Shallowly cordate	Mid-season
Moneifi	Ovoid elongated	Yellow	Apricot brown	Light brown	1.51	50	Truncate	Late-season
Nabtat Ali	Ovoid elongated	Yellow	Maroon	Maroon	1.44	25	Shallowly cordate	Mid-season
Um Khashab	Ovoid elongated	Light red	Maroon	Maroon	1.84	60	Truncate	Mid-season
Ruthana	Ovoid	Yellow	Yellowish brown	Yellowish brown	1.4	30	Deeply cordate	Early season
Sabaka	Ovoid elongated	Yellow	Golden brown	Light brown	1.5	50	Shallowly cordate	Mid-season
Shaqra	Ovoid elongated	Red	Dark red	Dark brown	1.32	33	Cordate	Mid-season
Sukkari	Ovoid elongated	Yellow	Yellowish brown	Brown	1.43	60	Cordate	Mid-season
Wannana	Ovoid elongated	Reddish yellow	Brown	Dark brown	1.44	30	Oblique	Mid-season

Source: Al-Fuhaid et al. (2011) and Al-Khalifah et al. (2012)

Two book resources providing fruit descriptors of Saudi date palm cultivars illustrated with high-quality photos were published by Al-Fuhaid et al. (2011) and Al-Khalifah et al. (2013). The latter reference also provides molecular identifiers.

3.5.2.2 Male Identification

In an effort to improve date palm production and fruits quality, research to identify superior pollinizers is essential. Most of date palm males available for pollinating different female cultivars are from seed propagation, resulting in many different local males that represent a source of genetic diversity (Soliman et al. 2013). Studying pollen grain characteristics is important for understanding distinguishing features, comparative morphology, ontogeny, as well as aspects of breeding systems and hybridization (Soliman and Al-Obeed 2013). Identifying national male date palms exhibiting superior quality including pollen viability, fertility, and compatibility is considered the first step towards establishing germplasm banks for conservation superior pollen compatible with local cultivars (Al-Ghamdi et al. 1988).

In order to identify highly potent male palms to propagate standard male cultivars, several studies were conducted. Nasr et al. (1989) evaluated 600 males grown in the Central Region of Saudi Arabia based on morphological characteristics relevant to time of flowering, spathe, strands, and weight of pollen grain per spathe. The researchers found significant differences in these parameters among various males and were successful in identifying superior pollinizers based on these characteristics.

Al-Ghamdi et al. (1988) studied date palm males cultivated on Al-Hassa farms and classified them on the basis of morphological characteristics of vegetative growth including number of leaves, leaf length, number of leaflets per leaf, length of leaflet, width of leaflet, length of spines, width of spines, as well as flowering characteristics relevant to spathe number, length, width, and fresh weight as well as pollen viability.

Similarly, Soliman et al. (2013) evaluated 10-year-old seedling date palm males which originated from various popular cultivars at the Research and Agriculture Experiment Station, Dirab, King Saud University, Riyadh. The evaluation was based upon leaf morphological characters (palm crown, pinnae color, pinnae distribution, pinnae end, pinnae bent, and leaf basal color), spathe properties (length, diameter, strand length and weight, strands number/spathe, flowers number/strand and total number flowers) and pollen traits (pollen grain weight and germinate percentage). Their results indicated that as compared with other males, Sukkari cv. exhibited the greatest leaf length, pinnae number, pinnae length, spine numbers, spathe length, strand length, and pollen grains weight, whereas Khalas and Sukkari male cvs. recorded the highest total number of flowers. In contrast, Sefri male cv. followed by Sukkari, Meneifi, and Sari male cvs. germinated at the highest rate.

To investigate pollen morphological traits as a tool to facilitate the identification of date palm males, Soliman and Al-Obeed (2013) examined morphological difference among males based on scanning electron microscopy (SEM) images reflecting surface properties of pollen. Significant variations were detected in terms of shape, pore frequency, and exine patterns of pollen grains proved.

3.5.3 Biochemical Descriptors Based on Isozyme Analysis

In the late 1980s, a study involving cultivar identification was conducted by Al-Helal (1988) using sodium dodecyl sulfate-polyacrylamide gel electrophoresis (SDS-PAGE), which separated amylase isoenzymes for fruit extracts of 13 date palm cultivars. He found two bands representing α-amylase and β-amylase activities and concluded that the amylase protein patterns appeared to be cultivar specific. Subsequently, Al-Helal (1992) demonstrated that date palm pollen isoenzyme patterns may be useful to investigate genetic variation among date palm males. He based his observation on electrophoretic patterns of soluble protein, nonspecific esterase (EST), glutamate-oxaloacetate transaminase (GOT), and glutamate dehydrogenase (GDH) of extracts of pollen grains from different date palm males. Qualitative variations were observed based on EST and GOT isoenzymes among males; whereas, three isoenzymes of GDH were present in the pollen extracts but showed no electrophoretic variation.

Early sex determination of seedlings is important for expediting breeding processes and conserve resources required to maintain trees until fruiting stage at which sex can be determined based on flowering morphology. Al-Fredan (2013) investigated the potential use of isozymes to differentiate males and female date palm trees. The genotypes showed specific reliable mobility values for three tested peroxidase isozymes. Results revealed different leaf isozyme patterns for the two female date palm cvs. Ghur and Khenaizy. Moreover, the presence or absence of peroxidase (PRO I) was identified as a reliable marker which distinguished between date palm genders. The most anodal loci PRO I was heterozygous present only in males and PRO II only in females whereas PRO III was found in both genders.

To determine variations within a cultivar, Al-Issa (2013) assessed different trees of cv. Khalas, grown at ten locations within the Al-Hassa oasis, using SDS-PAGE to analyze leaf extracts. Dendrogram analysis revealed three clusters comprising different locations: Cluster 1 occupies a distinct place in the dendrogram composed of Oyuni, Omran, Taraf, and Shoa'bah location samples reflecting monomorphism; cluster 2 included Aqair and Ain Merjan location samples that distinguished each by a unique band; and cluster 3 included Battalia, Gurain, Ulaijalah, and Mutairfi. Consequently, Al-Issa (2014) reported polymorphisms in isoenzyme patterns among cv. Khalas trees growing in the preceding locations with two to seven bands of nonspecific esterase (EST) and two to four bands of glutamate-oxaloacetate transaminase (GOT).

3.5.4 Molecular Descriptors Based on DNA Analysis

Molecular work involving date palm in Saudi Arabia began in the early 2000s (Al-Mssallem et al. 2004). Currently various research institutions are involved in molecular identification and phylogenetic and genomic studies on date palm.

3.5.4.1 Cultivar Identification and Phylogeny

Several studies have focused on molecular analysis of the genetic diversity among Saudi date cultivars. Most of these studies used random amplified polymorphic DNA (RAPD), but other techniques were also used like inter-simple sequence repeat linked simple single repeat (ISSR) for cultivar identification and unweighted paired group method of arithmetic mean (UPGMA) for cluster analysis of various cultivars (Table 3.5). Recently, Al-Khalifah et al. (2013) published a book presenting RAPD markers for the identification of 100 cultivars of date palm grown in Saudi Arabia.

3.5.4.2 Genome Mapping

King Abdulaziz City for Science and Technology (KACST) and Beijing Institute of Genomics, Chinese Academy of Science (BIG/CAS), cooperatively initiated a Date Palm Genome Project (DPGP) in 2008, with research conducted at KACST, Riyadh. The DPGP project is a comprehensive genome research project aimed at sequencing the date palm genome including organelles (chloroplast and mitochondria) as well as the nuclear genome (Zhang et al. 2011).

Table 3.5 Molecular markers research for the analysis of genetic diversity among Saudi date cultivars

Method	Results	Reference
RAPD	Found two clusters: (A) cvs. Shahal, Um Kbar, Ajwa, Um Hamam, and Bareem with 0.59–0.89 Nei and Li's similarity coefficient and (B) cvs. Rabeeha, Shaishi, Nabtat Saif, Segae, Sukkari Asfar, Sukkari Hamra, and Nabtat Sultan with a 0.66–0.85 coefficient	Al-Khalifah and Askari (2003)
RAPD	Found two clusters: (A) cvs. Barhi and Ajwa and (B) cvs. Nabtat Ali, Ruthana, and Sukkari	Al-Moshileh et al. (2004)
RAPD	Found two clusters: (A) cvs. Mouakil and Khalas with 0.73 Nei and Li's similarity coefficient and (B) cvs. Maktoumi and Nabtat Ali with 0.70 Nei and Li's similarity coefficient (Fig. 3.3)	Al-Khalifah (2006)
RAPD	Found high level of genetic similarity (96.3 %) between Sefri and Makfazi, followed by (92.3 %) between Khalas and Maktoumi. The similarity between Sefri, Makfazi, and Khalas and Maktoumi was 90 %, while Sukkari similarity to the other cvs. was the lowest (85 %)	El-Tarras et al. (2007)
RAPD	Found two clusters: (A) red Sukkari and yellow Sukkari with 0.81–0.87 similarity coefficient and (B) naptet Sukkari with 1.0 coefficient	El-Rayes (2009)
RAPD, ISSR	Found two clusters: (A) Sukkari with Nei and Li's coefficient equal to 0.55 and (B) cvs. Medjool, Segae, and Khalas with 0.66–0.85 similarity	Abdulla and Gamal (2010)
RAPD	Analysis of the morphological fruits and RAPD markers of 14 cvs. revealed similarity correlation. In addition, it was shown that fruit shape is most influenced by genetic variation (Fig. 3.4)	Al-Khalifah et al. (2012)

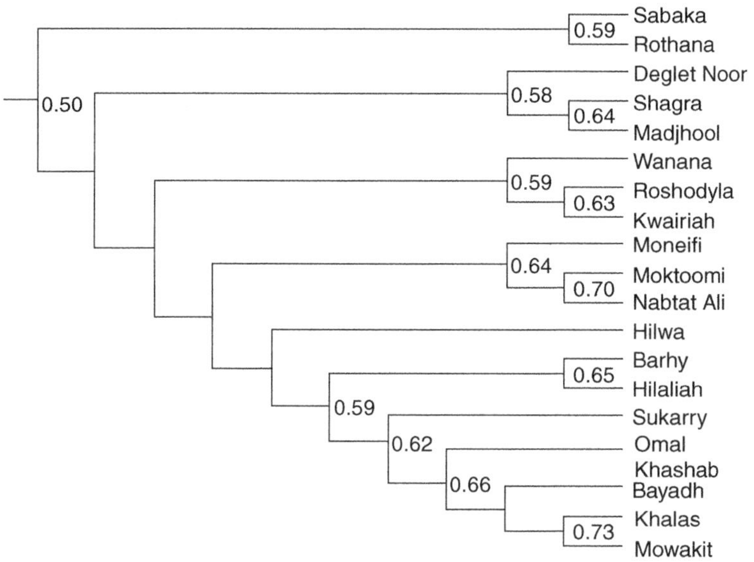

Fig. 3.3 Dendrogram of phylogenetic relationship among 19 date palm cultivars based on RAPD analysis (*Source*: Al-Khalifah et al. (2012))

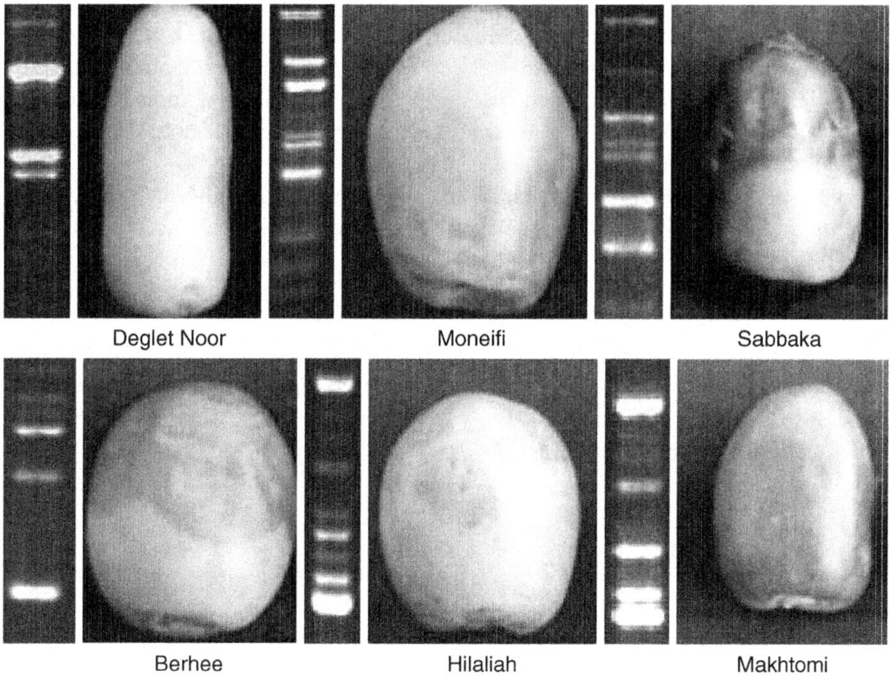

Fig. 3.4 Fruit morphology and DNA profiles of some date palm cultivars produced by A-06 primer (OPERON Tech) (*Source*: Al-Khalifah et al. (2012))

Sequencing and assembling of the date palm chloroplast genome using the pyrosequencing technology was recently achieved by Yang et al. (2010). The chloroplast genome was found to consist of a 158,462 bp double-stranded circular DNA molecule, with a typical quadripartite structure of the large (LSC, 86,198 bp) and small single-copy (SSC, 17,712 bp) regions separated by a pair of inverted repeats (IRs, 27,276 bp). Fang et al. (2012) assembled the mitochondrial genome of date palm, cv. Khalas, by combining data from two next-generation sequence platforms, pyrosequencing (Roche GS FLX) and ligation-based sequencing (Life Technologies SOLiD). They found that the mitochondrial genome of date palm consists of a circular molecule of DNA consisting of 715,001 bp of which 6.5 % (46,770 bp) encodes for 38 proteins, 30 tRNAs, and 3 ribosomal RNAs and the remaining 93.5 % comprises non-coding sequences. Based om the transcriptomic data of the mitochondrial genes expression levels in male flower, root, bud, and female flower, as compared to tissues of seed, fruit, and leaf. In addition, they identified 120 of single nucleotide polymorphisms (SNPs) among cvs. Khalas, Fahal, and Sukkari in which 7 SNPs were found in the coding sequences.

To further understand gene expression during various developmental states of date palm fruit, Yin et al. (2012) conducted a study, using Roche/454 GS FLX instrument, to determine sequencing-based gene profiling of eight stages according to the number of days after pollination. Stages studied included early and late stages of kimri, khalal, and beser (late khalal) as well as rutab and tamar stages. This study identified 7.6 million sequence tags from various fruit developmental stages, excluding tamar stage. Based on gene ontology categorization and pathway analysis, 10 core cell division genes, 18 ripening-related genes, and 7 starch metabolic enzymes were identified. It was concluded that most genes are highly expressed in early stages of development, whereas late stages are critical for expression of fruit ripening and metabolism-associated genes.

The most significant finding related to date palm genomic studies in Saudi Arabia was reported by Al-Mssallem et al. (2013) who successfully sequenced the nuclear genome (cv. Khalas) using pyrosequencing reads. Their results revealed that the gnomic size is 605.4 Mb which covers 90 % of the genome (approx. 671 Mb) and 96 % of its genes (approx. 41,660 genes).

Recent work by Sabir et al. (2014) resulted in sequencing of the mitochondrial and chloroplast genome, using Illumina HiSeq 2000 platform, of nine Saudi cultivars. Based on SNP analysis, this study revealed heteroplasmy and close phylogenetic relationships among cultivars.

3.6 Cultivars Description

3.6.1 Growth Requirements

While different cultivars of dates vary somewhat in their growing seasons, the primary date season in Saudi Arabia runs from March, when the green buds first appear, to the middle of October, when the mature fruits are ready for harvest. Saudi

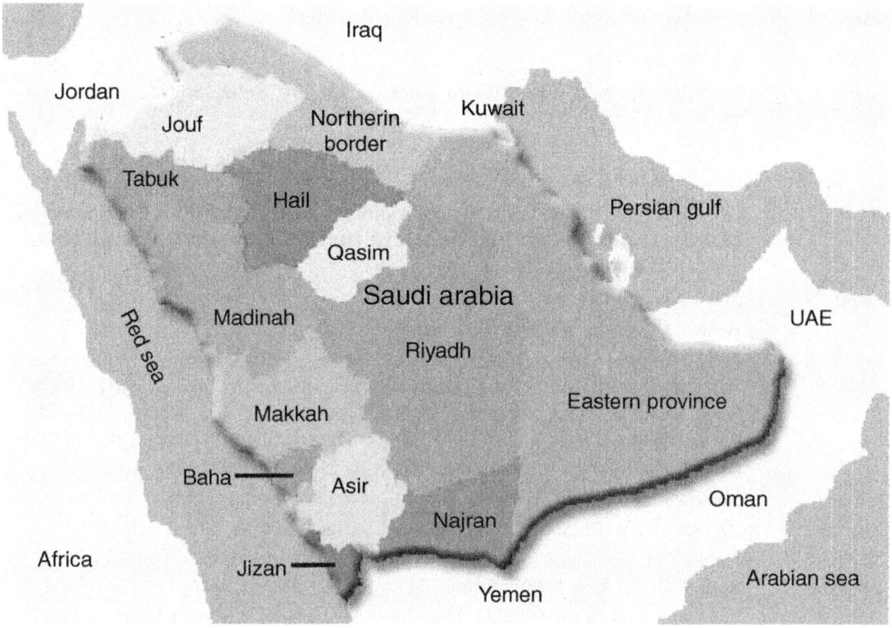

Fig. 3.5 Saudi Arabia map illustrating 13 geographical regions growing different cultivars of date palm (see Table 3.6) (*Source*: www.google.com)

Arabia is located in southwest Asia between 16 and 32 N lat. and 35–65 E long. Most of the regions are within the dry tropical zone. The climate is characterized by hot, dry, long summers where temperatures in certain areas can reach 50 °C during the months of June to August. Generally the average temperature during summer is around 35 °C; but in the winter season falls during the months of December, January, and February in certain areas to below zero (Ministry of Agriculture 2011). This vast range in temperature between winter and summer affords an optimum climate for date palm cultivars in Saudi Arabia. Each region has its own land race cultivars adapted to its agroecosystem.

3.6.2 *Cultivar Distribution and Production*

Saudi Arabia has an estimated 24 million date palms spread over the 13 different provinces of the Kingdom (Fig. 3.5). Each province is characterized by certain date palm cultivars. Table 3.6 shows the most famous date cultivars grown in each region of Saudi Arabia (Ministry of Agriculture 2006).

The total area under date palm cultivation reached 155,118 ha in the 2011 growing season. There was a marginal increase in the area under date cultivation attributed to incentives provided by the government to small farmers in terms of subsidies such as fertilizers, short-term loans, pesticides, fixed prices, and market facilities.

Table 3.6 Famous data palm cultivars by region in Saudi Arabia

Region	Famous cultivars
Asir	Sefri, Barni, Shakl, Sari, Bidaira, Ruthana, Khodry, Sullaj, Sagae
Baha	Sefri
Eastern Region	Khalas, Ruzeiz, Shaishi, Shebedi, Khenaizy, Wesaili, Ghur, Hatmi, Tayyar, Hilali, Shahal, Um Raheem, Khesab, Barhi, Bakkeria, Maktoumi
Hail	Hulwa, Kasba, Rikhairni, Majhoola, Hamra, Deglet Hamood, Deglet Shewayish, Miskani, Deglet Moftah, Fankha, Swiria, Hijria, Kdhiria, Dubia, Safran, Balga, Saddrah, Qurain, Khodhreya
Jazan	Sefri, Khodry
Jouf	Hulwa, Bewayda, Kathmaa, Hessaineya
Madinah	Ajwa, Anbara, Safawi, Ruthana, Rabeaa, Shabli, Hulwa, Barni Al Madina, Barni Al Ulaa, Beid, Barni Al Ais, Ruthana Alsharg, Sukkarat Alsharg, Hulya, Barhi, Suwaida Al Mashouk, Mutaban, Wannana, Sabah, Sukkarat Albeid, Al Arous, Looat Musaed, Gawj, Khoshaimi, Jaffari, Khodry, Shaqri, Qatara, Um Khashab, Lubana, Jebaili, Huiwat Beida, Sukkarat Al Madina, Maktoumi
Makkah	Moshwaq, Mutalbia, Alluban, Alhamri, Rabeaa, Khodry, Zafran
Najran	Bayad, Sieqat, Rutab, Barni, Mouakeel, Safraa, Hamraa, Ulooq, Khodair
Northern Boarders	Hulwa, Sukkari, Fankha, Safra, Khalas, Segae, Deglet, Kasba, Um Khashab, Sullaj, Nabtat Saif, Nebtat Sultan, Maktomi
Qassim	Sukkari, Asfar, Barhi, Shaqra, Um Hamam, Sukkaria Ahmar, Khalas, Nabtat Ali, Ruthana, Hulwa, Um Khashab, Nabtat Rasheed, Wannana, Rushodia, Maktoumi, Aseela, Lahrnia, Segae, Nabtat Saif, Um Kabar, Berairni, Hushara, Muneifi Ahmer, Muneifi Asfar, Hulwa Wasit, Khodry, Fankba, Motwah, Quatarah, Salmia
Riyadh	Sullaj, Meneifi, Nabtat Saif, Nabtat Sultan, Khodry, Makfazi, Miskani, Dekhaini, Sari, Sefri, Khalas, Barhi, Segae, Sukkari,Hilali, Rezeiz, Shaishee, Maktomi, Aseela, Ghur, Nabtat Ali, Ruthana, Um Hamam, Haqaqi, Um Raheem, Halawa, Shabal, Sabaka, Rushodia, Um Kbar, Barni, Khesab, Sukkaria Hamra, Khashram, Um Kouz, Degl, Hhawar, Um Aldbur
Tabuk	Hulwa, Barni

Source: Al-Fuhaid et al. (2011)

Total date production reached 991,546 mt in the 2011 growing season. The modest increase in total date fruit production could be due to the slight increase in total area brought under date palm cultivation (Alshuaibi 2011).

Growing citrus with dates, quantity sold, and the price received were significant variables in the econometric analyses of date production cost. Date production and its prices are related to cultivar, cost of production, and intercropping in the orchards, as well as optimal scale of operations, all of which influence the economic efficiency (Alseleem 1998). Osman and Al-Besher (1989) reported that labor was the most important input in date production. Many of the seasonal workers are hired and may not be under the farmer's control. Part-time recruitment necessities throughout the harvest season represent a key challenge for date farmers. The Ministry of Agriculture advises the labor office to issue foreign work visas at a ratio of one laborer per 300 palms.

3.6.3 Description of Date Cultivars

Genotype identification of date palm is commonly based on morphological characters. Female cultivars are recognized by their fruit traits like size, shape, color, and taste, along with vegetative morphology of the tree. Date fruits pass through four distinct stages of maturity, i.e., kimri, beser (also referred to as khalal at early stage), rutab, and tamar (Al-Ghamdi 1996). Young green fruits are termed kimri. The beginning of ripening marks the beser stage, the half-ripened stage is called rutab, and the fully ripened, soft-textured stage is called tamar. Descriptions of Saudi date cultivars are shown in Table 3.4. There is a wide variation among date palm cultivars in terms of fruits maturation in early, mid, or late season. Some cultivars are consumed only in beser or in tamar stage, whereas others are can be consumed in both beser and tamar stages. Fruit appearance for some Saudi cultivars is shown at rutab stage (Fig. 3.6) and at tamar stage (Fig. 3.7).

3.6.4 Nutritional Aspects and Health Benefits of Dates

The contribution of dates to the daily intake of energy and other nutrients in Saudi Arabia is declining. This might be due to the decrease in per capita consumption because of the competition with chocolate and candy. There should be more attention called to the presence of value-added products from dates so that consumers can eat dates in an indirect manner. Applied research culminating in feasibility studies is needed to promote new products fortified with dates, such as extruded baby foods, breakfast cereals, and snacks for youth, to market with the help of food industry in Saudi Arabia. Advertising programs endorsing dates as a healthy nutritional food need to be launched by the government. Certain studies show that the Kingdom occupies the first rank in the world in terms of average per capita consumption of dates per year, which reached 34.8 kg/year in 2003.

Dates are composed of water (10–22 %), invert sugar (62–75 %), protein (2.2–2.7 %), fat (0.4–0.7 %), ash (3.5–4.2 %), total acidity (0.06–0.2 %), and ascorbic acid (30–50 mg/100 g) on dry weight basis (FAO 1992). Total dietary fiber content of dates fluctuates from 6.26 to 8.44 g/100 g, of which 84–94 % is insoluble fiber and 278–301 kcal/100 g as energy expressed on a fresh weight basis, due to the high carbohydrate content (Al-Farsi et al. 2007). Dates are rich in potassium and low in less favorable components like fats, cholesterol, and sodium.

Dates are a good source of antioxidants. Studies on antioxidant activity and phenolic content of various date fruits demonstrated a linear relationship between antioxidant activity and the total phenolic content of date fruit extract (Alliath and Abdalla 2005). Fruits of different cultivars have different total phenolics content and antioxidant activity (Al-Farsi et al. 2007; Al-Turki et al. 2010). The antioxidant properties of date fruits vary depending on their phenolic components and vitamins C and E, carotenoids, and flavonoids (Al-Farsi et al. 2007; Mansouri

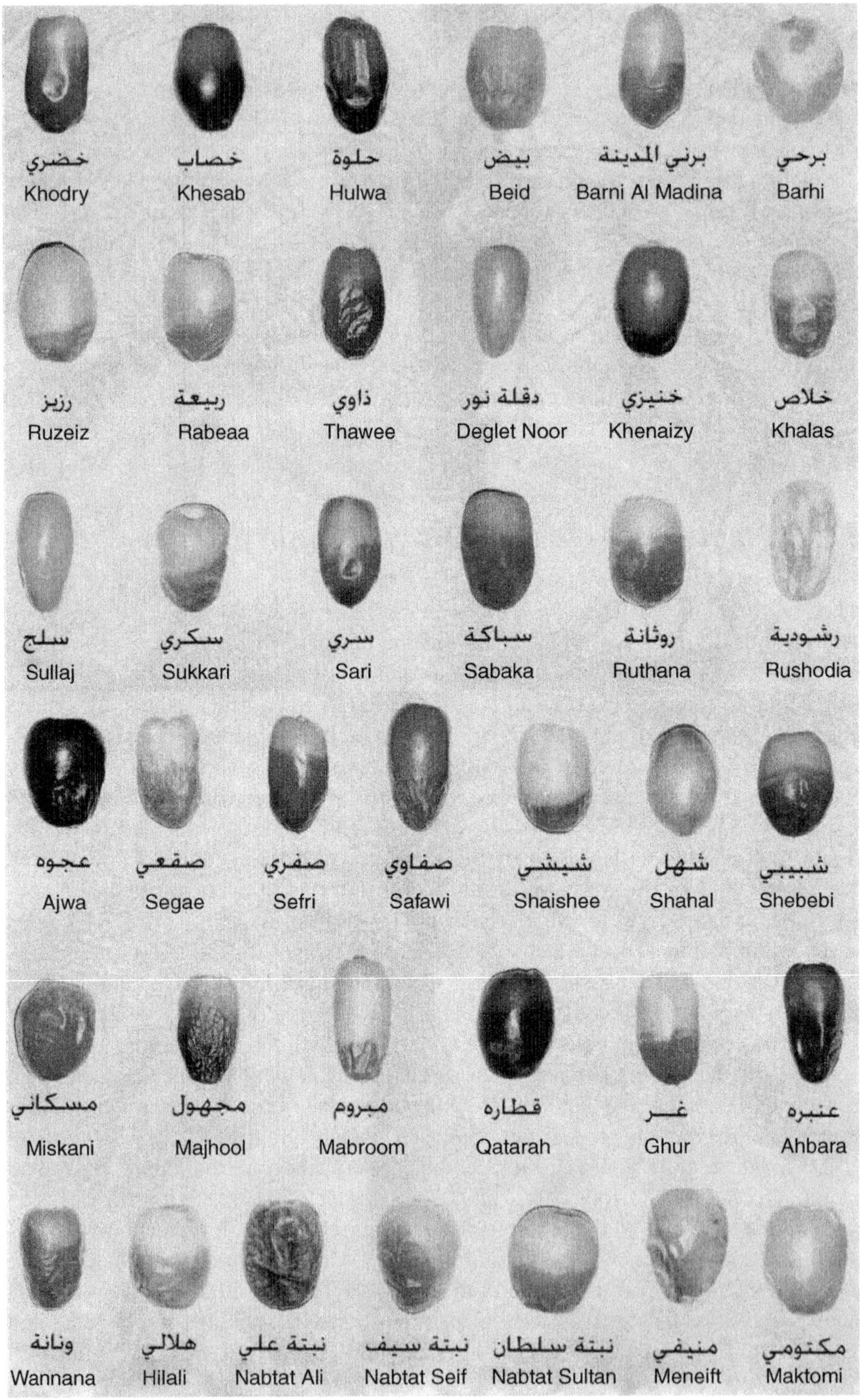

Fig. 3.6 Morphology of some Saudi date cultivars at rutab stage (*Source*: Al-Fuhaid et al. (2011))

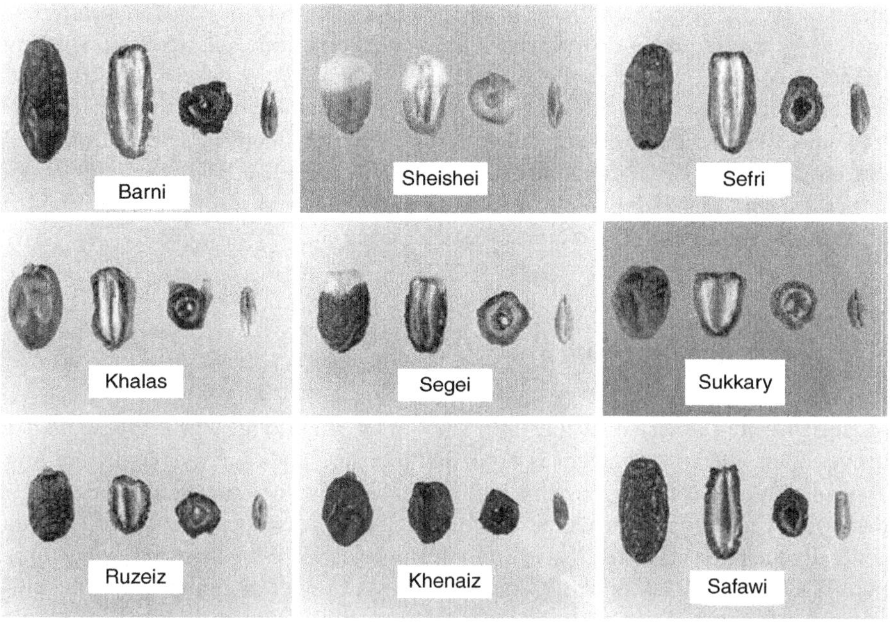

Fig. 3.7 Morphology of some Saudi date cultivars at tamar stage (*Source*: Al-Khateeb and Dinar (2002))

et al. 2005). All date cultivars are a good source of natural antioxidants and could potentially be considered as a functional food or functional food ingredient (Al-Farsi et al. 2007).

3.7 Dates Production and Marketing

There is a vast diversity in date palm cultivars in Saudi Arabia; the most notable are Khalas, Sukkari, Ajwa, Anbara, Ruthana, Segae, Barhi, and Ruzeiz. Although the cultivation of date palms has developed considerably and great attention has been given to date production in Saudi Arabia, nonetheless, the level of productivity is low compared with other producing countries (Al-Obaid 1996). The main causes for this decrease in date productivity are the increase in the number of the overage trees, the existence of many low-quality and undesirable cultivars, the lack of sufficient offshoots to establish new orchards or renew old ones, and the increase in cost of offshoots of good-quality cultivars (Al-Sakran and Muneer 2006; Bashah 1999). In date production and marketing, primary consideration should be given to quality. Consumer perception of date quality is mainly based on value attributes, which include date cultivar, taste, texture, size, color, crust cohesion, freshness, no skin fracture, and free of insects. Recently, there are increasing consumer concerns about date quality in the Saudi market. Accordingly, product differentiation strategies are increasingly used in date markets to attract the interest of consumers for various date characteristics. Differentiation is achieved through their distinct attributes and communicated through labeling (Aleid et al. 2014; Al-Kahtani et al. 2011).

Competitiveness of Saudi dates requires an evaluation of consumers' desired attributes. The recognized value may originate from product characteristics as well as from conditions of production. These values can be measured by examining consumer behavior in existing markets or by consumer interviews to assess perceptions of price and product quality (Carlsson et al. 2005).

3.7.1 Practical Approaches

Most dates are harvested at the fully ripe tamar stage when the fruits exhibit a deeper color, the sugar content is high, and moisture and tannin contents are low. Typical production in the primary cultivation regions does not exceed 20–40 kg/palm per year, although production inputs (i.e., fertilizers, pesticides) are low and generally the palms are too closely spaced. However, in well-organized date plantations, yields may reach over 100 kg/palm per year under favorable growth conditions. At 30 years of age, date palms reach their high-yield period. In general, an average well-managed palm can produce about 60–70 kg of fresh dates per year (Organic 2002). The decline in date yields in Saudi Arabia is attributed to inadequate cultural practices coupled with outbreaks of pests and diseases. Pollination and fruit thinning are critical processes in date palm production. Pollen origin in dates affects fruit quality, yield, and annual productivity. Different pollen sources affect fruit size, flesh and seed development, and time to fruit maturation (Al-Ghamdi et al. 1988).

3.7.2 Optimization of Yield

Date palms require regular fertilization for high stable yield (Al-Rawi 1998). El-Shurafa (1984) reported that date palm soils lose considerable amounts of macro- and micronutrient elements through leaf pruning and fruit harvest; therefore, fertilization is necessary. Al-Dekaili and Al-Dejaili (1989) affirmed the need for fertilizers, especially nitrogen to obtain high numbers of leaves with long green pinnae and high yield. Moreover, appropriate fruit thinning gives the remaining fruits a better chance to develop to a larger size and a better quality by reducing compaction among fruits on the bunch. It also helps to promote good flowering in the following year (Marzouk et al. 2007). Such results could be attained either by reducing the number or fruits per bunch or by reducing the number of bunches per palm. Al-Obeed et al. (2005) found that the 15 % shortening of strands at pollination time led to a reasonable yield with fruit quality. Soliman and Harhash (2012) reported that Sukkari cv. was thinned by removing 15 and 30 % of the total number of strands from the center of each bunch, while the control strands were used as reference. Although thinning caused significant decrease in yield (based on both bunch and palm weight), significant improvement in fruit quality was observed. Thinning 30 % at 4 weeks after pollination will lead to optimum fruit quality.

3.7.3 Harvest Mechanization

Harvesting is labor intensive because dates are handpicked. On large date plantations, platforms may be used to raise the workers to harvest fruit. But in most cases, workers must climb the palms to reach and pick the fruit (Organic 2002). There are two methods of date palm cultural operations, traditional and mechanical; the latter is still not a common practice in Saudi Arabia. The most difficult part of the traditional date palm cultural operations involves workers ascending to the crown of the tree; if the trees are very tall, the worker risks falling and injury or even death. Using the tree trunk leaf bases to climb the tree in Saudi Arabia is typical of cultural operations (Nixon 1969). Al-Kiady (2000) reported that skilled specialized laborers were becoming rare and more expensive causing a serious problem in the date production industry.

Mechanized cultivation methods can improve date quality and is essential to optimize this industry (Albozahar 2003). Al-Suhaibani et al. (1988) made a date service machine in Saudi Arabia which was designed at Silsoe College, UK. They carried out a survey on 19 orchards in Saudi Arabia and measured certain physical properties such as tree spacing, height, trunk circumference, bunch spacing, and ground profile to consider in their design. Shamsi (1990) designed a sprocket-type climbing machine to harvest dates, and he later designed and developed a tree-climbing date harvesting test rig at the same college (Shamsi 1998). This machine climbs up the tree trunk with a worker to reach the fruit bunches. Nicklin (1993) designed a tree-climbing rig to lift a man up into the palm.

Fadel (2005) developed a tractor-mounted machine for date palm service based on physical properties. The machine had the capability to lift a worker and required tools to the crown zone as high as 4.5 m. Mazloumzadeh and Shamsi (2007) designed and developed a light tractor-mounted date palm service machine that could reach to a maximum height of 10 m with payload up to 130 kg.

Attempts to mechanically harvest date fruits have not been successful due to inadequate efficiency and the availability of a low cost labor in Saudi Arabia. Date fruits have to be firm enough and resistant to mechanical damage. The developed mechanical harvesting system needs to compensate for the differences in soil compactness taking into account the characteristics of the date palm tree. Moreover, the machine must be manufactured at a commercial scale and made available to farmers at a reasonable cost. Similarly, mechanization of fruit sorting and sizing systems also requires further development.

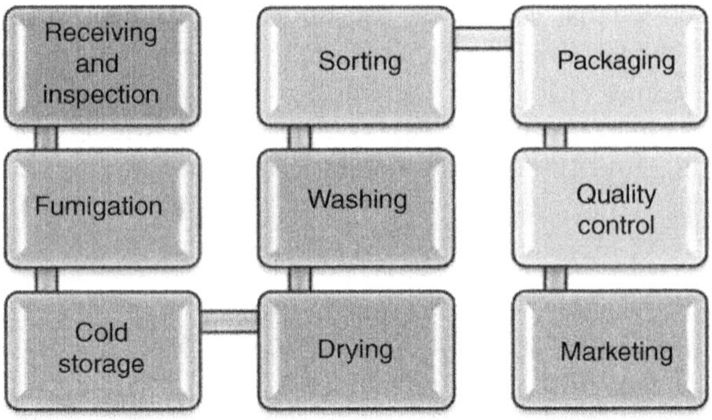

Fig. 3.8 A flowchart of common processes for dates processing

3.7.4 Postharvest Operations

Hygienically processed and properly packed dates harvested in Saudi Arabia have substantial potential in both domestic and international markets. Processing involves fumigating, washing, sorting, grading, glazing, weighing, and packaging (Sindh 2010). A flowchart of date processing steps in Saudi factories is shown in Fig. 3.8.

3.7.4.1 Fumigation

Fumigation is the first step after harvesting to protect fruits from infestation. The date industry suffers substantial losses from the ravages of insects. Infestations of dates with moths (almond moths, meal moths), fig moth *Ephestia cautella*, Indian meal moth *Plodia interpunctella*, beetles (sap beetles, saw-toothed grain beetles, flour beetles), rats, mice, and ants result in contamination and loss of volume (Glasner et al. 2002). Methyl bromide is highly effective for controlling insects in stored products. However, methyl bromide emissions have been found to have deleterious effects on the atmosphere and present a hazard to human health. Therefore, in accordance with the Montreal Protocol, its production and use will be eliminated by the end of 2015 worldwide (Council 1985). Alternatives to methyl bromide include the following: (a) phosphine, the principal alternative; (b) a controlled atmosphere high in carbon dioxide; and (c) physical control methods such as heating or cooling regimes, active oxygen (ozone or hydrogen peroxide), and irradiation. However, some of these methods are very costly. In organic date palm production, carbon dioxide is used instead of methyl bromide (Glasner et al. 2002). Heat treatment is a common alternative for date disinfestation replacing fumigation (Belarbi et al. 2001).

3.7.4.2 Cold Storage

Dates in Saudi Arabia may need to be stored for long periods, as much as 10–12 months. Temperature is the most single important environmental factor affecting shelf life and quality of fresh dates. FAO has developed several approaches to prolonging date storage using refrigeration or freezing. The process is mainly based on slowing down fruit maturation. Most dates are harvested at full maturity (tamar stage) when the color and sugar content characteristic of the cultivar has fully developed. Also, tamar stage fruit has the longest potential storage life (many months) compared to rutab or khalal (several days to weeks maximum). Fresh dates harvested at tamar stage and stored at −18 °C provide for maximum economical extension of shelf life and preserve edible product quality (FAO 2008).

3.7.4.3 Sorting

Sorting of dates is done manually. A chain conveyor for sorting dates is usually used. During this step, workers sort and remove dates with any indication of infestation as well as other particles and damaged dates (Organic 2002).

3.7.4.4 Washing

Date processors generally rely on wash-water sanitizers to reduce microbial counts to maintain quality and extend shelf life (Gil et al. 2009). Dates are typically washed in a circular washer with sprinklers and dried using a hot air blower system in a hygienic environment (Sindh 2010). Washing with sanitizers is important to remove dirt and debris and for water disinfection to avoid cross-contamination between clean and contaminated products. Most sanitizing solutions achieve higher microbial reductions immediately after washing compared to water washing; however, after storage, epiphytic microorganisms grow rapidly, reaching similar levels. Chlorine-based sanitizers are among the most effective and efficient sanitizers when appropriate doses are used (Gil et al. 2009).

3.7.4.5 Drying

Air-drying is designed to result in fruit moisture content of 20 % or below to prevent growth of molds and yeasts (Organic 2002). Temperatures of 55–65 °C for drying of soft dates are generally used (Barreveld 1993). The hot air blast removes excess water and the dates are then loose- or press-packed, sealed, wrapped, and placed in cold storage (Mikki et al. 1989).

3.7.4.6 Packaging

Fancy dates are usually packaged at the processing plants in Saudi Arabia. This involves vacuum packaging of compressed or non-compressed whole dates in flexible sealed plastic coverings (Alhamdan and Hassan 1999) or polyethylene-polyamide (PE-PA) bags. Vacuum packaging is a useful technique for reducing darkening of dates during lengthy storage Mohsen et al. (2003). Mutlak and Mann (1984) reported that both enzymatic and nonenzymatic browning occurred in dates, increasing with higher moisture content and higher temperatures, and inhibited at low oxygen potentials. However, for packing loose dates, cleaned and graded dates are weighed and packaged in cardboard boxes. Generally, the weight of these packages may be 1–20 kg, depending on domestic or international customer requirements.

3.7.5 Survey of Commercial Producers and Major Farms

Date palms in Saudi Arabia are grown primarily in groves, such as found in Riyadh, Qatif, Qassim, Madinah, Bishah, and Al-Hassa, the latter the world's largest oasis, in the Eastern Province. Al-Hassa's famous springs and the extensive irrigation network they feed make it an ideal area for date growing. As far back as 4000 B.C., there is evidence of date cultivation in what is now the Eastern Province of Saudi Arabia (http://www.saudiembassy.net). The government supports establishing modern and well-managed date farms, enhancing date production efficiency, and optimizing use of natural resources such as land and water. Table 3.7 highlights business domains and marketing practices for some active modern date palm farms in Saudi Arabia.

3.7.6 Current Export and Import

In view of the increasing cultivation and surplus date production in the Kingdom, there is an emphasis on exports (Al-Abbad et al. 2011). Saudi Arabia exports dates to France, Germany, and India; Tunisia and Algeria are the potential competitors (Al-Shreed et al. 2012). Saudi Arabia exports only 48.8 mt of dates annually valued at a unit price of USD 0.83 per kg. In general, the potential strengths of exporting Saudi dates into international market can be summarized in the following: (a) existence of modern date palm plantations that can control the quality of dates at the upstream and traceability, (b) existence of cultivars with low demand in the Saudi market that meet international market requirements, (c) possibility to upgrade the exporting packing house to be more competitive with customer needs, (d) capacity of companies to launch new products (value addition) on the international market (e.g., frozen dates and pastries from dates), and (e) spiritual benefits of Saudi dates within the Muslim populations worldwide (Al-Shreed et al. 2012).

There are several technical aspects to be considered for exported dates. For example, for exporting packed Saudi dates to the USA, exporters should consider the following: (a) types of dates to be exported, (b) specification (sugar and moisture content) as well as physical properties, (c) date packaging (size and weight of the carton), (d) number of cartons and their weight to fill a 6 m long container of approximately 8.6 m^3, (e) requirements and specifications for exporting dates from Saudi Arabia to USA, and (f) export mechanism. Some date cultivars have export potential, due to their physical and chemical properties, as well as their abundant availability, including cvs. Sukkari, Khodry, Sefri, Ruzeiz, and Khalas.

Table 3.7 Some major date palm farms in Saudi Arabia

Farm	Business identity	Marketing practices
Al-Aseel Dates	250,000 palms with annual production of 3,000 mt. Has the largest research and development center in private sector to promote dates framing according to international standards	Retail marketing and distribution outlets. Existence in international markets
Al-Rajhi Endowments (Awqaf)	Al-Baten (200,000 palms) and the Darmaa (50,000 palms). Guinness record as largest palm plantation worldwide (2005). ECCOCERT certificated organic (2007). Farm returns are donated to charities	Part of production sold while fruits on the heads of palms. Another part is marketed in sorting and packaging stores. The rest is marketed through direct sales in local date markets. Fancy organic dates are marketed in international markets
Kingdom Dates	Multifarms in various places in Saudi Arabia cultivating more than 300 varieties. Supplying market needs through two integrated and equipped factories, with washing, processing, and packing devises for dates in Qassim	Kingdom Dates is a wholesaler, retailer, and exporter maintaining national and international standards having more than 100 branches
Al-Butain Agricultural Cooperative Association	Nonprofit organization in Qassim with a cold storage complex, financed by Saudi Agricultural Development Fund, with a capacity of 10,000 mt. Equipped with lines to sort, refine, fumigate (ultralow oxygen system), and pack	International marketing. Care of producer and manufacturer of dates in Qassim region
Dohayan Dates	Farms are based in Qassim. Provides a large selection of dates from all regions of Saudi Arabia	Supplies market's needs and delivers a variety of sweets that are derived from dates as raw material
Bin Zaid Dates	Farms and processing plant is located in Al-Hassa Oasis, an area classified as the most productive of high-quality dates in the northeast of Saudi Arabia	Supplying local and international markets with fresh fancy dates, dates stuffed with almonds and cashew nuts, as well as date paste and date syrup
Zadna Trading Company	Providing in the vicinity of consumers a good-quality of dates around the year with reasonable packed quantity	Supplying local and international markets with fresh dates, paste, and syrup. Ma'moul (shortbread pastries filled with date paste covered with chocolate). Dates stuffed with nut (almonds, cashew nuts, and walnuts)
Al-Mohamadia Dates	Farm and dates processing plant located in Kharj. The farm with an area of 18 km^2. HACCP certified 2004	Supplying local and international markets with a variety of date products. Considered one of the major producers of date paste
Yogi and Yousef Dates	Dates produced and packaged	Distribute dates to end retail, such as luxury food stores, organic shops, and boutique hotels in European and American markets

Source: Saudi Dates (2013)

3.8 Processing and Novel Products

3.8.1 Industrial Processing Activities

The date industry in Saudi Arabia is highly supported by government industrial programs. The support for well-established date processing facilities is the initiative of the Agricultural Development Fund through interest-free long-term loans which have advanced the spread of date processing facilities. The most important cultivars used for manufacturing processing in Saudi Arabia are Khalas (16 %), Ruzeiz (21 %), Khodry (15 %), Sukkari (10 %), and others (38 %). About 90 % of manufacturing activities in Saudi Arabia focus on the packaging process, instead of producing value-added manufactured date products. About 56 % of date processing factories produce value-added date products such as paste, jam, syrup, and vinegar. There are many sources of dates as raw material in Saudi Arabia. The best source for quality dates is at wholesale directly from farmers or through intermediate traders. Average and low-quality dates usually go to wholesale markets. About 70 % of dates delivered to processing factories comes from farmers, 25 % from traders, and 5 % from wholesale markets (Aleid 2013a, b; Elsabea and Aleid 2012). Two important products, date paste and date syrup, currently represent the date industry in Saudi Arabia.

3.8.1.1 Date Paste

Date paste typically is used as a filling in pastries and biscuits in the baking industry in Saudi Arabia, as well as a sugar substitute in cereals, puddings, breads, cakes, cookies, ice cream, and confectionaries. Modern industrial processing lines were installed within some large date processing facilities in Saudi Arabia. In date paste production, dates are steamed, seeded, macerated, and converted to a semisolid paste (Alhamdan and Hassan 1999). The extruded date paste is usually packed in high- or low-density polyethylene or polypropylene packaging. Yousif et al. (1991a, b) reported that date paste storage involves many challenges, including hardening, microbial spoilage, and darkening in color. Most date sugars are invert sugars, which increase the softness of bread and cookies (Mikki et al. 1983). Aleid (2009) found that the incorporation of date paste (Khalas cv.) into an Arabic bread as a source of 20 % of the sugar in the recipe was recommended.

3.8.1.2 Ma'moul

Ma'moul is a Middle Eastern pastry usually filled with dates and walnuts. Recipes for ma'moul are geographical and cultural dependent (Fig. 3.9). The two major components of ma'moul are dough and date filling. Major ingredients of the dough are semolina, flour, butter, sugar, yeast, and water; often, rosewater is substituted for

Fig. 3.9 Ma'moul filled with date paste (Prepared by Maryam M. Aleid, photographed by Nasreen N. Alsarra)

part of the water. The filling usually is made of date paste, fine pieces of walnuts, and butter. A specific amount of the filling usually is placed inside a piece of dough and then shaped using a specific type of wooden molding tool. The dough is then baked and cooled and may be dusted with fine sugar.

3.8.1.3 Date Syrup

Consumption of date syrup is very popular in Saudi Arabia, as a sweetener for a number of traditional food dishes. Low-quality dates can be used as raw material for producing date syrup, which is a good source of glucose and fructose (Aleid 2012). Date syrup is rich in sodium, potassium, calcium, magnesium, and iron and has a low sodium-to-potassium ratio (1:8.3). This may be of dietetic importance, particularly for those with restricted sodium intake (Aleid et al. 1999). Moisture content strongly impacts storage stability of syrups, since it directly affects the likelihood of undesirable fermentation (Aleid et al. 2007). Date syrup also has antioxidant activity related to total phenolic content.

Several date processing premises in Saudi Arabia have an automated date syrup production line on a commercial scale. In date syrup extraction, the dates are placed in a vessel and the required amount of water is added. The temperature of the mixture is adjusted and it is mixed for a given time period (El-Shaarawy et al. 1989). The mixture of the sugar solution and date fibers is filtered to separate the solid materials and obtain a clear solution. The dilute solution obtained has a Brix of 12–15. This solution can be heated under vacuum and concentrated to a Brix of 70, which is appropriate for long shelf life. Evaporation is normally carried out at 55–60 °C to prevent burning of the sugars. From 1 kg of dates, approximately 600 g of date syrup with a Brix of 70 can be obtained. Date syrup can also be produced as an incidental by-product when bagged humid dates are heaped for several months and syrup oozes out by the force of their own weight (FAO 1996).

Aleid et al. (2007) utilized second-grade dates of the Ruzeiz cv. to extract syrup through high hydraulic pressure at room temperature. The sugar content of the syrup obtained was 70–85 % dry weight; most of the sugars were reduced sugars. This process gave a yield of 40–50 %. The color of the syrup was light brown compared with syrup from a heat extraction experiment (85–95 °C) which had a dark color and a caramelized flavor due to a high extraction temperature. However, the extraction time for the heat extraction process was relatively short, with a higher yield than the cold extraction process (Aleid et al. 2007). Efficient clarification of date syrup has been achieved using an ultrafiltration membrane with a molecular weight of 20,000 to obtain clear syrup that can be used as a flavoring agent for processed foods. Aleid (1998) reported on the incorporation of date syrup into bread dough recipes as a sugar substitute. Breads with 6 and 9 % date syrup, at 85 and 95 % extraction, exhibited a high total bread score compared with no sugar or 4 % sucrose treatments. Replacement of sugar with date syrup up to 6 % was strongly recommended.

3.8.2 Commercial Dates Processers and Packaging Plants

Date processing factories are licensed in six areas of Saudi Arabia, namely, Riyadh, Madinah, Qassim, Eastern region, Mecca, and Asir. This industry is characterized by geographical concentration. Labor and invested capitals are the most important factors determining the economic feasibility of date factories (Al-Tamimi 2005). In general, date processing factories depend mainly on packing activity and packing for others (farmers and private citizens) for a charge of about USD 0.5 per kg. Table 3.8 represents some major date packaging factories and their activities. One of the most important dates processing factories is the government factory in Al-Hassa. The government established this factory with annual capacity of 25,000 mt to support farmers and to ensure a minimum limit for date prices. It packs dates and donates them to international food programs (Elsabea and Aleid 2012; Mikki 1998).

3.8.3 Baker's Yeast from Dates

Baker's yeast could be produced from dates in Saudi Arabia and such an undertaking could be highly feasible for several reasons. There is a daily need for baker's yeast but no commercial domestic production as yet. Such a value-added product will benefit the date sector in general and thus reduce price volatility. Baker's yeast production is a clean process not harmful to the environment. All fresh yeast consumed in Saudi Arabia is imported. Local dates of low quality available at low

Table 3.8 Some major date processing plants in Saudi Arabia

Region	Processing plant name	Processing activity
Al-Hassa	Al-Hassa Company for Food Industry Date Processing Plant	Packing, paste, syrup, sweets
	Al-Jawharia Dates Factory	Packing, paste, syrup
	Al-Jazirah Dates and Food Factory	Packing
Riyadh	Amal Al-Khair Dates Packing and Packaging Factory	Packing, syrup
	Bateel Sweets and Chocolates Factory	Packing, syrup
	Nakhail Al-Watan Dates Factory	Packing
Kharj	Al-Babtain Factory for Dates Filling and Packing	Packing
	Al-Faisaliah Dates Factory	Packing
	Al-Mohamadia Dates Company	Packing, paste, syrup
	Al-Rajhia Dates Factory	Packing
	Al-Yamamah Dates Packing Factory	Packing
	Kharj Dates Factory	Packing
	Madaen Star Dates Factory	Packing
Qassim	Al-Salehaiah Dates Factory	Packing, syrup, sweets
	Kingdom Dates Factory	Packing, syrup, sweets
	Nadheed Dates Factory	Packing, paste, syrup
	Qassim Agricultural Company Dates Packaging Factory	Packing, syrup
Madinah	Al-Ahli Ideal Dates Packing Factory	Packing
	Al-Ansar Factory for Dates Packing and Wrapping	Packing
	Barakat Al-Madinah Dates and Sweets Factory	Packing
	Madinah Al-Munawarah Dates Company	Packing, stuffed, chocolate coat
	Quba Dates Factory	Packing
	Taiba Dates Packing Factory	Packing

Source: Saudi Industrial Development Fund (2010)

prices are a potential raw material for baker's yeast production. Research projects for the production of baker's yeast are important despite the high cost of the equipment required. Most countries in the world which produce bread yeast use molasses as a raw material.

The use of dates has several advantages over molasses, the traditional raw material. These advantages include saving the cost of some minerals and vitamins not existing in molasses but are naturally present in dates and giving the dates a comparative advantage for the production of baker's yeast. The search for yeasts prefix of high quality is the decisive element in the production process. The production of baker's yeast will increase the economic returns from dates and absorb part of the surplus dates which helps to maintain the balance between production and consumption, which in turn helps to stabilize prices for both producers and consumers (Aleid 2011).

Extensive research on baker's yeast production from dates has been carried out at King Faisal University in Saudi Arabia, leading to awarding of a patent on the process registered at the US Patent Office (US Patent No. 8,323,717B2 dated Dec 4, 2012). This patent has been one of the few pioneering works focusing on development of a fermentation process suitable for the production of baker's yeast from date substrates. The know-how obtained from this study could be applied to commercial production. All calculated financial and economic criteria indicate the feasibility of production of dry yeast, providing that this could be a productive activity as an extension line to an existing date processing factory and not as a stand-alone project. Almost all of the baker's yeast is produced from molasses using *Saccharomyces cerevisiae* (Barnett et al. 2000). The sugars in dates are mainly glucose and fructose, which are easily assimilated by most microorganisms (Sawaya 1986). Substrates from pure date syrup and pure molasses for the propagation of the baker's yeast strain *S. cerevisiae* were examined (Aleid 2009). They found that the overall biomass yield from pure date syrup substrate was significantly lower than that from pure molasses. Despite the fact that lower biomass yield using date syrup was obtained compared to molasses as a substrate, the use of dates as a substrate for baker's yeast production in Saudi Arabia as an alternative to the traditional substrate of molasses imported from external markets will help in finding new profitable ways for the utilization of this important, locally produced food crop and to reduce dependency on imported foods.

3.8.4 Bioenergy

The Saudi Arabian Basic Industries Corporation (SABIC) exclusively exports synthetic ethanol produced from petrochemical feedstocks. The product is exported in crude form, mostly to the USA and South Korea. In Saudi Arabia, ethanol is mainly used for the production of perfumes, cosmetics, medicines, cleaners, paints, and vinegar (Capital 2004). Ethanol derived from biomass has the potential to be a substitute of fossil fuel which is renewable, nontoxic, biodegradable, and more eco-friendly. The three major classes of feedstocks for ethanol production are sugars, starches, and lignocelluloses (Gupta and Kushwaha 2011). Raw biological materials that can be used as feedstock for ethanol production are numerous. The Date Palm Research Center of Excellence, King Faisal University, is currently studying date fruits, by-products, and tree wastes as raw materials for the production of ethanol.

3.9 Conclusions and Recommendations

Agricultural production in Saudi Arabia depends mainly on irrigation from groundwater and, to a lesser extent, on rainfall during the winter season (November to February). Date palm plantations are constrained by severe biotic and abiotic stresses including heat and salinity. Date crops in the Saudi Arabia suffer from many difficulties including high postharvest losses due to fermentation, insect infestation, birds, and mechanical damage. The dates produced should comply with international regulations and standards necessary for export (Kader and Hussein 2009). Several problems and constraints might affect the future of date palm industry in Saudi Arabia, including (a) low-quality cultivars; (b) poor farm management; (c) losses from pests and diseases and inadequate IPM control; (d) deficiencies in harvesting, processing, and marketing; (e) shortage of qualified national trained staff and laborers; (f) insufficient research and development activities.

High marketing margins and low farmer share indicate local marketing inefficiency for dates in Saudi Arabia. Moreover, a great fluctuation in prices is considered the most serious local marketing problem facing dates farmers in Saudi Arabia. Minimizing the usage of nonrenewable underground water through greater efficiencies, as well as restricting the production of low-quality date cultivars, is highly recommended. Also of importance is the conservation of genetic diversity. As a leading date-producing country with plentiful resources, Saudi Arabian scientists and concerned officials ought to consider developing a global strategy to conserve date palm germplasm utilizing all conservation technologies including in vitro cryopreservation approaches.

There are numerous research studies in Saudi Arabia aimed at finding new ways for economic utilization of high-quality dates and date surpluses, by-products, and wastes. Liquid sugar, date syrup, carbonated and non-carbonated drinks based on ultra-filtered date syrup, single-cell protein, baker's yeast, industrial alcohol, date paste, and date powder are some of the potential value-added products. Successful industrial producer and processor optimization and standardization for such value-added products will contribute to expanding the date industry. Conducting pre-feasible studies which include market structure and financial evaluation of industrial-scale production for such products from dates is highly recommended.

Governmental policy should focus on programs that limit future expansion of date palm plantations and economize on the use of nonrenewable groundwater, as well as limit the production of high-quality commercially feasible date palm cultivars. Improvement in tree yield and production quality as well as alignment of production capacity with demand is highly recommended. Launching a program to stimulate national and international demands for Saudi dates is highly recommended.

Moreover, a long-term comprehensive plan for applied research and development activities needs to be established. The main rationale for such plan is to accommodate the date palm sector's research, training, and outreach needs according to stakeholders' priorities and to provide measures for monitoring progress, evaluating outcomes, and assessing impacts. Furthermore, the operational activities for the plan should focus on the following: (a) attracting the most qualified researchers and staff; (b) generating funds; (c) improving operations and utilization of resources; (d) creating effective outreach programs to the date farming community, including the date palm industry and the society at large; (e) developing integrated research programs of high impact nationally and internationally; (f) enhancing multidisciplinary date palm research; and (g) establishing collaboration among leading research centers nationally and internationally.

References

Abdul-Baki A, Aslan S, Linderman R et al (2002) Soil, water, and nutritional management of date orchards in the Coachella Valley and Bard, 2nd edn. California Date Commission, Indio

Abdulla M, Gamal O (2010) Investigation on molecular phylogeny of some date palm (*Phoenix dactylifera* L.) cultivars by protein, RAPD and ISSR markers in Saudi Arabia. Aust J Crop Sci 4(1):23–28

Abo El-Nil M (1989a) Refining methods for date palm micropropagation. In: Proceedings of the 2nd symposium on the date palm, vol 1. King Faisal Univ, Al-Hassa, 3–6 Mar 1986, pp 29–41

Abo El-Nil M (1989b) The effects of amino acid nitrogen on growth of date palm callus. In: Proceeding of 2nd symposium on the date palm, vol 1. King Faisal Univ, Al-Hassa, 3–6 Mar 1986, p 59–65

Abo El-Nil M, Al-Ghamdi AS (1989) Stimulation of growth and tissue culture of date palm axillary buds by injection of offshoot with a cytokinin. In: Proceeding of the 2nd symposium on the date palm, vol 1. King Faisal Univ, Al-Hassa, 3–6 Mar 1986, pp 43–49

Abo-El-Saad M, Elshafie H (2013) Insect pests for stored dates and their management. In: Siddiq M, Aleid SM, Kader AA (eds) Dates: postharvest science, processing technology and health benefits. Wiley, Ames, pp 81–104

Abo-Hassan AA (1981) Rangeland management in Saudi Arabia. Rangelands 3(2):52–53

Al-Abbad A, Al-Jamal M, Al-Elaiw Z et al (2011) A study on the economic feasibility of date palm cultivation in the Al-Hassa Oasis of Saudi Arabia. J Dev Agric Econ 3(9):463–468

Al-Abdoulhadi IA, Dinar HA, Ebert G, Büttner C (2012) Influence of salinity stress on photosynthesis and chlorophyll content in date palm (*Phoenix dactylifera* L.) cultivars. Afr J Agric Res 7(22):3314–3319

Al-Abdulmohsin AM (1987) First record of red date palm weevil in Saudi Arabia. Arab World Agr 3:15–16

Al-Amoud AI, Bacha MA, Al-Darby AM (2000) Seasonal water use of date palms in the central region of Saudi Arabia. Agric Engin J 9(2):51–62

Al-Bahrany AM, Al-Khayri JM (2012a) Optimizing in vitro cryopreservation of date palm (*Phoenix dactylifera* L.). Biotechnology 11(2):59–66

Al-Bahrany AM, Al-Khayri JM (2012b) In vitro responses of date palm cell suspensions under osmotic stress induced by sodium, potassium and calcium salts at different exposure durations. Am J Plant Physiol 7(3):120–134

Al-Baker A (1972) The date palm. Ministry of Higher Education, Baghdad

Albozahar A (2003) Design, development and evaluation of a hand-held date palm dehorning device. Annual report No: 82/Nakhl/119, Ahwaz, Date Palm and Tropical Fruits Research Institute of Iran, Ahvaz, Iran

Al-Dekaili AA, Al-Dejaili JA (1989) Fruits production. Ministry of Higher Education Press, Mousil

Al-Doss AA, Aly MA, Bacha MA (2001) Morphological and agronomical variations among some date palm cultivars grown in Saudi Arabia using principal component and cluster analysis. King Saud Univ Agr J 13(1):3–18 (In Arabic)

Aleid SM (1998) Substitution of sugar with date syrup in bread making. Egypt J Appl Sci 13(11):271–285

Aleid SM, Aljaser MS (2009) Effect of Date Paste on Arabic Bread Quality. Journal of Saudi Society for Food and Nutrition. King Saud University 4(1):1–13

Aleid SM (2011) Industrial biotechnology: date palm fruit applications. In: Jain SM, Al-Khayri J, Johnson DV (eds) Date palm biotechnology. Springer, Dordrecht, pp 675–709

Aleid SM (2012) Dates production, storage and processing. In: Siddiq M (ed) Tropical and subtropical fruit processing and packaging. Wiley, Ames, pp 171–202

Aleid SM (2013a) Date fruit processing and processed products. In: Siddiq M, Aleid SM, Kader AA (eds) Dates: postharvest science, processing technology and health benefits. Wiley, Ames, pp 171–202

Aleid SM (2013b) Innovative processing technologies for processing dates. In: Siddiq M, Aleid SM, Kader AA (eds) Dates: postharvest science, processing technology and health benefits. Wiley, Ames, pp 203–231

Aleid SM, El-Shaarawy ML, Mesallam AS, Al-Jendan SI (1999) Chemical composition and nutritional value of some sugar and date syrups. Minufiya J Agric Res 24(2):577–587
Aleid SM, Aljaser MS, El Neshwey AA et al (2007) Utilization of high hydraulic pressure in extraction of date syrup. In: Proceeding of the 4th symposium on the date palm in Saudi Arabia, King Faisal Univ, Al-Hassa, 5–8 May 2007, pp 2151–2160
Aleid SM, Hassan BH, Almaiman SA et al (2014) Microbial loads and physicochemical characteristics of fruits from four Saudi date palm tree cultivars: conformity with applicable date standards. Food Nutr Sci 5:316–327
Al-Enezi NA, Al-Khayri JM (2012a) Alterations of DNA, ions and photosynthetic pigments content in date palm seedlings induced by X-irradiation. Int J Agric Biol 14:329–336
Al-Enezi NA, Al-Khayri JM (2012b) Effect of X-irradiation on proline accumulation, growth, and water content of date palm (*Phoenix dactylifera* L.) seedlings. J Biol Sci 12(3):146–153
Al-Enezi NA, Al-Bahrany AM, Al-Khayri JM (2012) Effect of X-irradiation on date palm seed germination and seedling growth. Emir J Food Agric 24(5):415–424
Al-Farsi M, Morris A, Baron M (2007) Functional properties of Omani dates (*Phoenix dactylifera* L.). Acta Hort 736:479–487
Al-Fredan MA (2013) Peroxidase activity in male and female plants of date palm (*Phoenix dactylifera* L.) growing in Al-Hassa, Saudi Arabia in vitro. El-Minia Sci Bull 24(1):37–55
Al-Fuhaid KM, Al-Afalique A, Oihabi A et al (2011) The famous dates varieties in the Kingdom of Saudi Arabia. Ministry of Agriculture, Alahsa, Saudi Arabia
Al-Ghamdi AS (1996) True-to-type date palm (*Phoenix dactylifera* L.) produced through tissue culture techniques: inflorescence and pollen grain evaluation. In: Proceedings of the 3rd symposium on the date palm, vol 1. King Faisal Univ, Al-Hassa, 17–20 Jan 1993, pp 93–103
Al-Ghamdi AS (2001) Date palm (*Phoenix dactylifera* L.) germplasm bank in King Faisal University, Saudi Arabia. Survival and adaptability of tissue cultured plantlets. Acta Hort 450:241–244
Al-Ghamdi AS, Al-Hassan GM, Jahjah M (1988) Evaluation of eight seedling date palm (*Phoenix dactylifera* L.) males and their effects on fruit character of three female cultivars. Arab Gulf J Sci Res 6(2):175–187
Alhamdan AM, Hassan BA (1999) Water sorption isotherms of date pastes as influenced by date cultivar and storage temperature. J Food Eng 39:301–306
Al-Helal AA (1988) Amylase isoenzymes and protein of date palm (*Phoenix dactylifera* L.) fruit. Bot Bull Acad Sin 29:239–244
Al-Helal AA (1992) Electrophoretic analysis of three selected isoenzymes of date palm pollen grains. Bot Bull Acad Sin 33:241–246
Alhudaib K, Arocha Y, Wilson M, Jones P (2007) Al-Wijam: a new phytoplasma disease of date palm in Saudi Arabia. Bull Insectol 60(2):285–286
Al-Issa AM (2013) Electrophoretic analysis of protein patterns in date palm "Khalas" cultivar leaflets among different locations of Al-Ahsa. Am J Agr Biol Sci 8(4):343–349
Al-Issa AM (2014) Identification of original "Khalas" cultivar date palm by using of electrophoretic analysis of isoenzymes. Int J Agric Sci Res 4(2):79–88
Al-Kahtani SH, Soliman SS (2012) Effects of organic manures on yield, fruit quality, nutrients and heavy metals content of Barhy date palm cultivar. Afr J Biotech 11(65):12818–12824
Al-Kahtani SH, Ismaiel SM, Aleid SM, Hassan BH (2011) The prospects and possible electronic trade of Saudi dates. Final report. Date Palm Research Center of Excellence. King Faisal University, Alahsa, Saudi Arabia, pp 1–191
Al-Khalifah NS (2006) Micropropagation and DNA fingerprinting of date palm trees of Saudi Arabia. Association of Agricultural Research Institutions in the Near East and North Africa, ICARDA West Asia Regional Program, Amman
Al-Khalifah NS, Askari E (2003) Molecular phylogeny of date palm (*Phoenix dactylifera* L.) cultivars from Saudi Arabia by DNA fingerprinting. Theor Appl Genet 107:1266–1270
Al-Khalifah NS, Askari E, Khan AES (2012) Molecular and morphological identification of some elite varieties of date palms grown in Saudi Arabia. Emir J Food Agric 24(5):456–461
Al-Khalifah NS, Askari E, Shanavaskhan AE (2013) Date palm tissue culture and genetical identification of cultivars grown in Saudi Arabia. National Center for Agriculture Technologies, King Abdulaziz City for Science and Technology, Riyadh. 207 p. Available online at: http://www.kacst.edu.sa/en/about/publications/

Al-Khateeb AA (2006a) Role of cytokinin and auxin on the multiplication stage of date palm (*Phoenix dactylifera* L.) cv. Sukkari. Biotechnology 5(3):349–352
Al-Khateeb AA (2006b) Somatic embryogenesis in date palm (*Phoenix dactylifera* L.) cv. Sukkari in response to sucrose and polyethylene glycol. Biotechnology 5(4):466–470
Al-Khateeb AA (2008a) Comparison effects of sucrose and date palm syrup on somatic embryogenesis of date palm (*Phoenix dactylifera* L.). Am J Biotech Biochem 4(1):9–23
Al-Khateeb AA (2008b) Enhancing the growth of date palm (*Phoenix dactylifera*) in vitro tissue by adding date syrup to the culture medium. Sci J King Faisal Univ Basic Appl Sci 9(1):71–85
Al-Khateeb AA (2008c) Regulation of in vitro bud formation of date palm (*Phoenix dactylifera* L.) cv. Khenaizy by different carbon sources. Biores Tech 99:6550–6555
Al-Khateeb AA, Dinar HM (2002) Date palm in Saudi Arabia: cultivation, production and processing. King Faisal University Press, Al-Hassa (in Arabic)
Al-Khayri JM (2001) Optimization of biotin and thiamine requirements for Somatic embryogenesis of date palm (*Phoenix dactylifera* L.). In Vitro Cell Dev Biol Plant 37:453–456
Al-Khayri JM (2002) Growth, proline accumulation, and ion content in NaCl-stressed callus cultures of date palm (*Phoenix dactylifera* L.). In Vitro Cell Dev Biol Plant 38:79–82
Al-Khayri JM (2003) In vitro germination of somatic embryos in date palm: effect of auxin concentration and strength of MS salts. Curr Sci 84:101–104
Al-Khayri JM (2005) Date palm *Phoenix dactylifera* L. In: Jain SM, Gupta PK (eds) Protocols of somatic embryogenesis in woody plants. Springer, Berlin, pp 309–318
Al-Khayri JM (2007) Micropropagation of date palm *Phoenix dactylifera* L. In: Jain SM, Haggman H (eds) Protocols for micropropagation of woody trees and fruits. Springer, Berlin, pp 509–526
Al-Khayri JM (2010) Somatic embryogenesis date palm (*Phoenix dactylifera* L.) improved by coconut water. Biotechnology 9:477–484
Al-Khayri JM (2011a) Basal salt requirements differ according to culture stage and cultivar in date palm somatic embryogenesis. Am J Biochem Biotech 7:32–42
Al-Khayri JM (2011b) Influence of yeast extract and casein hydrolysate on callus multiplication and somatic embryogenesis of date palm (*Phoenix dactylifera* L.). Sci Hort 130:531–535
Al-Khayri JM (2012) Determination of the date palm cell suspension growth curve, optimum plating efficiency, and influence of liquid medium on somatic embryogenesis. Emir J Food Agric 24(5):444–455
Al-Khayri JM (2013) Factors affecting somatic embryogenesis in date palm (*Phoenix dactylifera* L.). In: Aslam J, Srivastava PS, Sharma MP (eds) Somatic embryogenesis and genetic transformation in plants. Narosa Publishing House, New Delhi, pp 15–38
Al-Khayri JM, Al-Bahrany AM (2001) Silver nitrate and 2-isopentyladenine promote somatic embryogenesis in date palm (*Phoenix dactylifera* L.). Sci Hort 89:291–298
Al-Khayri JM, Al-Bahrany AM (2004a) Growth, water content, and proline accumulation in drought-stressed callus of date palm. Biol Plant 48:105–108
Al-Khayri JM, Al-Bahrany AM (2004b) Genotype-dependent in vitro response of date palm (*Phoenix dactylifera* L.) cultivars to silver nitrate. Sci Hort 99:153–162
Al-Khayri JM, Al-Bahrany AM (2012) Effect of abscisic acid and polyethylene glycol on the synchronization of somatic embryo development in date palm (*Phoenix dactylifera* L.). Biotechnology 11(6):318–325
Al-Khayri JM, Al-Ghamdi AS, Bautista CM (1996) Potential of in vitro leaf explants for date palm micropropagation. Dar Zahran for Publishing & Distribution, Amman, Jordan, In Vitro 32:106A
Al-Kiady HK (2000) Date palm tree: processing and agricultural technology-mechanical and harvesting of dates. Dar Zahran for Publishing & Distribution, Amman, Jordan
Alliath A, Abdalla AA (2005) In vitro evaluation of antioxidant activity of different extracts of *Phoenix dactylifera* fruits as functional foods. Deutsche Lebensmittel Rundschan 101:305–308
Al-Maarri K, Al-Ghamdi AS (1995) Effect of culturing date on in vitro micropropagation of date palm (*Phoenix dactylifera* L.) cv. Hillaly. Arab Univ J Agric Sci 3:151–167
Al-Maarri K, Al-Ghamdi AS (1997) Micropropagation of five date palm cultivars through in vitro axillary buds proliferation. Damascus Univ J Agric Sci 13:49–66

Almadini AM, Al-Gosaibi AM (2007) Impacts organic fertilizers for date palm trees on soil properties in Al-Hassa Oasis, Saudi Arabia. In: Proceedings of the international conference on date palm, King Saud Univ, Qassim Branch, Qassim, Saudi Arabia, p 69

Al-Moshileh AM, Motawei MI, Al-Wasel A et al (2004) Identification of some date palm (*Phoenix dactylifera* L.) cultivars in Saudi Arabia using RAPD fingerprints. Agril Marine Sci 9(1):1–3

Al-Mssallem IS, Al-Bahrany AM, Al-Khayri JM (2004) Genetic engineering techniques as a tool to identify some economical date palm (*Phoenix dactylifera* L.) cultivars in Saudi Arabia. Project AT-18–72 final report. King Abdualziz City for Science and Technology, Riyadh, Saudi Arabia

Al-Mssallem IS, Hu S, Zhang X et al (2013) Genome sequence of the date palm *Phoenix dactylifera* L. Nat Commun 4:2274. doi: 10.1038/ncomms3274. p 1–9

Al-Obaid AA (1996) Economies of dates in Saudi Arabia. In: The Extension Bulletin for Date Palm Trees. Extension Centre, Faculty of Agriculture, King Saud University, Riyadh, Saudi Arabia

Al-Obeed RS, Harhash MA, Fayez NS (2005) Effect of bunch thinning on yield and fruit quality of Sukkari date palm cultivar grown in the Riyadh region. J King Saud Univ Agric Sci 2:235–249

Al-Rawi AAH (1998) Fertilization of date palm tree (Phoenix dactylifera) in Iraq. In: Proceedings of the 1st international conference on date palm, 8–9 Mar 1998, Al-Ain Hilton Hotel, Al-Ain, pp 320–327

Al-Sakran MS, Muneer SE (2006) Adoption of date palm tissue culture technology among date palm producers in the central region of Saudi Arabia, Res Bull No. 145, Agricultural Research Center, Faculty of Food Sciences and Agriculture, King Saud University, Riyadh, Saudi Arabia

Alseleem YA (1998) Econometric analysis of dates cost functions in the Kingdom of Saudi Arabia. J King Saud Univ Agric Sci 10:34–42

Al-Shawaf A, Al-Shagag A, Al-Bagshi M et al (2013) A quarantine protocol against red palm weevil *Rhynchophorus ferrugineus* (Olivier) (Coleoptera: Curculionidae) in date palm. J Plant Prot Res 53(4):409–415

Al-Shreed F, Al-Jamal M, Al-Abbad A et al (2012) A study on the export of Saudi Arabian dates in the global markets. J Dev Agric Econ 4(9):268–274

Alshuaibi AM (2011) The Econometrics of investment in date production in Saudi Arabia. Int J Appl Econ Fin 5:177–184

Al-Shuaiby AM, Ismael M (2007) Economic alternatives of dates as animal feed in Al- Hassa and Al- Qatif. In: Abstract of the 4th symposium of date palm, King Faisal Univ, Al-Hassa, 5–8 May 2007, p 197

Al-Suhaibani SA, Babier AS, Kilgour J, Flynn JC (1988) The design of a date palm service machine. J Agric Eng Res 40(2):143–157

Al-Tamimi AA (2005) Evaluation of economic performance for date factories in Saudi Arabia. Master Thesis. Department of Agricultural Economics. King Saud Univ, Riyadh

Al-Turki S, Shahba MA, Stushnoff C (2010) Diversity of antioxidant properties and phenolic content of date palm (*Phoenix dactylifera* L.) fruits as affected by cultivar and location. J Food Agric Environ 8(1):253–260

Anonymous (1998) Final report (part A) – Red palm weevil control project, submitted by the Indian Technical Team to the Ministry of Agriculture and Water

Anonymous (2006) A review of the processing of dates in the Kingdom of Saudi Arabia: 1997 to 2005. Ministry of Agriculture, Riyadh, Saudi Arabia

Anonymous (2009) Statistics on processing of dates in the Kingdom of Saudi Arabia. Department of Studies, Planning and Statistics, Ministry of Agriculture, Riyadh, Saudi Arabia

Asif MI, Al-Ghamdi AS, Al-Tahir OA, Latif RAA (1989) Studies on the date palm cultivars of Al-Hassa oasis. In: Proceedings of the 2nd symposium on the date palm, vol 1. King Faisal Univ, Al-Hassa, 3–6 Mar 1986, pp 405–413

Barnett JA, Payne RW, Yarrow D (2000) Yeasts: characteristics and identification. Mycopathologia 149:159–160

Barreveld WH (1993) Date palm products. FAO Agricultural Service Bulletin No. 101, Food and Agricultural Organization of the United Nations, Rome

Bashah MA (1996) Date variety in the Kingdom of Saudi Arabia. Guide booklet for palms and dates. King Abdulaziz Univ Press, Jeddah, Saudi Arabia, pp 1225–1319

Bashah MA (1999) Propagation of date palm trees in Saudi Arabia. Ministry of Agriculture and Water, Administration of Extension and Agricultural Services, Riyadh. Saudi Arabia Agr Mag 30(1):24–40

Belarbi A, Aymard CH, Hebert JP (2001) Evolution of Deglet Noor date quality on it heat treatments (color and texture). In: Proceedings of the 2nd international conference on the date palms, Al-Ain, 25–26 Mar 2001, p 86

Brown GK (1983) Date production and mechanization in USA. In: Proceedings of the 1st symposium on the date palm, King Faisal Univ, Al-Hassa, 23–25 Mar 1982, pp 2–13

Capital Advisory Group (2004) Pre-feasibility study: medical alcohol and vinegar from dates. Riyadh, Saudi Arabia

Carlsson F, Frykblom P, Lagerkvist CG (2005) Consumer preferences for food product quality attributes from Swedish agriculture. Roy Swed Acad Sci, Ambio: J Hum Env 34(4):366–370

Committee for Red Palm Weevil Control in Qatif (1993) Extension leaflet on control of red palm weevil. Department of Plant Protection, Agricultural Extension Services Unit, Ministry of Agriculture, Maramar Press, Riyadh (in Arabic)

Council on Radiation Application (1985) Atomic Industrial Forum, National Council on Radiation Protection and Measurements. Bethesda, MD, USA

Dhawi F, Al-Khayri JM (2008a) Proline accumulation in response to magnetic fields in date palm (*Phoenix dactylifera* L.). Open Agric J 2:80–83

Dhawi F, Al-Khayri JM (2008b) Magnetic fields induce changes in photosynthetic pigments content in date palm (*Phoenix dactylifera* L.) seedlings. Open Agric J 2:121–125

Dhawi F, Al-Khayri JM (2009a) Magnetic fields-induced modification of DNA content in date palm (*Phoenix dactylifera* L.). J Agric Sci Tech 2:6–9

Dhawi F, Al-Khayri JM (2009b) The effect of magnetic resonance imaging on date palm (*Phoenix dactylifera* L.) elemental composition. Comm Biomet Crop Sci 4:11–18

Dhawi F, Al-Khayri JM (2009c) Magnetic field increase weight and water content in date palm (*Phoenix dactylifera* L.). J Agric Sci Tech 2:23–29

Dhawi F, Al-Khayri JM (2011) Magnetic field induced biochemical and growth changes in date palm seedlings. In: Jain SM, Al-Khayri JM, Johnson DV (eds) Date palm biotechnology. Springer, Dordrecht, pp 287–309

Dhawi F, Al-Khayri JM, Essam H (2009) Static magnetic field influence on elements composition in date palm (*Phoenix dactylifera* L.). Res J Agric Biol Sci 5:161–166

El Hadrami A, Al-Khayri JM (2012) Socioeconomic and traditional importance of date palm. Emir J Food Agric 24(5):371–385

El Hadrami A, Daayf F, El Hadrami I (2011) Date palm genetics and breeding. In: Jain SM, Al-Khayri JM, Johnson DV (eds) Date palm biotechnology. Springer, Dordrecht, pp 479–502

El-Juhany LI (2010) Degradation of date palm trees and date production in Arab countries: causes and potential rehabilitation. Aust J Basic Appl Sci 4(8):3998–4010

Elprince AM, Alsaeedi AH (2007) National fertilizer program for date palm. In: Proceeding of the 4th symposium on the date palm in Saudi Arabia, King Faisal Univ, Al-Hassa, 5–8 May, pp 80–104

El-Rayes DA (2009) Characterization of three date palm cultivars based on RAPD fingerprints and fruit chemical composition. Env Arid Land Agric Sci 20(2):3–20

Elsabea AMR, Aleid SM (2012) Marketing and industrial problems for the most important kinds of dates in KSA. Project no.10139 final report. Deanship of Scientific Research. King Faisal Univ, Alahsa, Saudi Arabia

El-Sabea AMR, Faleiro JR, Abo El Saad MM (2009) The threat of red palm weevil *Rhynchophorus ferrugineus* to date plantations of the Gulf region of the Middle East: an economic perspective. Outlook Pest Manag 20:131–134

El-Shaarawy MI, Mesallam AS, El-Nakhal HM, Wahdan AN (1989) Studies on extraction of dates. In: Proceedings of the 2nd symposium on the date palm, vol 2. King Faisal Univ, Al-Hassa, 3–6 Mar 1986, pp 259–271

El-Shafie HAF, Faleiro JR, Al-Abbad AH et al (2011) Bait-free attract and kill technology (Hook TM RPW) to suppress red palm weevil *Rhynchophorus ferrugineus* (Coleoptera: Curculionidae) in date palm. Fla Ent 9(4):774–778

El-Shurafa MY (1984) Studies on the amount of minerals annually lost by way of fruit harvest and leaf prunings of date palm tree. Date Palm J 3(1):277–286

El-Tarras A, Al-Tawatti N, Al-Malki F (2007) Genetic fingerprint of some KSA date palm cultivars using modern biotechnological techniques. Biotechnology 6(2):263–267

Erskine W, Moustafa AT, Osman AE et al (2004) Date palm in the GCC countries of the Arabian Peninsula. In: Proceedings of the regional workshop on the date palm development in the Arabian Peninsula, Abu Dhabi, 29–31 May 2004. (http://www.icarda.org/aprp/datepalm/introduction/introduction.htm)

Experts Bureau Saudi Arabia (2005) Quarantine law of the Gulf Cooperation Council in Saudi Arabia. Downloaded 3 Apr 2014. Available online at: http://www.boe.gov.sa/ViewSystemDetails.aspx?lang=arandSystemID=144andVersionID=147

Fadel MA (2005) Development of a tractor-mounted date palm tree service. Emir J Agric Sci 17(2):30–40

Faleiro JR (2006) A review of the issues and management of red palm weevil *Rhynchophorus ferrugineus* (Coleoptera: Rhynchophoridae) in coconut and date palm during the last one hundred years. Int J Trop Insect Sci 26:135–154

Faleiro JR, Ben Abdallah A, Kumar JA et al (2010) Sequential sampling plan for area-wide management of *Rhynchophorus ferrugineus* (Olivier) in date palm plantations of Saudi Arabia. Int J Trop Insect Sci 30(3):145–153

Fang Y, Wu H, Zhang T et al (2012) A complete sequence and transcriptomic analyses of date palm (*Phoenix dactylifera* L.) mitochondrial genome. PLoS One 7:e37164

FAO (1992) Dates. Food and agriculture organization of the United Nation, Rome

FAO (1996) Production yearbook 50. Food and Agriculture Organization of the United Nations, Rome

FAO (2008) Technical assistance for the cold storage needs for palm dates projects in Morocco. IOS Partners, Inc. 2008, USAID. Ministry of Agriculture, Rabat, Morocco

FAOSTAT (2012) Crop production 2010. Food and Agriculture Organization of the United Nations, Rome

Gil MI, Selma MV, López-Gálvez F, Allende A (2009) Fresh-cut product sanitation and wash water disinfection: problems and solutions. Int J Food Microbiol 134(1–2):37–45

Glasner B, Botes A, Zaid A, Emmens J (2002) Date harvesting, packinghouse management and marketing aspects. In: Zaid A (ed) Date palm cultivation, FAO plant production and protection paper no.156. Food and Agriculture Organization of the United Nations, Rome, pp 237–267

Gupta N, Kushwaha H (2011) Date palm as a source of bioethanol producing microorganisms. In: Jain SM, Al-Khayri JM, Johnson DV (eds) Date palm biotechnology. Springer, Dordrecht, pp 711–727

Hamza H, Rejili M, Elbekkay M, Fercichi A (2009) New approach for the morphological identification of date palm (*Phoenix dactylifera* L.) cultivars from Tunisia. Pak J Bot 41:2671–2681

Jaradat AA (2013) Date palm production. In: Siddiq M, Aleid SM, Kader AA (eds) Dates: postharvest science, processing technology and health benefits. Wiley, Ames, pp 29–55

Jaradat AA, Zaid A (2004) Quality traits of date palm fruits in a center of origin and center of diversity. Food Agr Env 2(1):208–217

Johnson DV, Al-Khayri JM, Jain SM (2013) Seedling date palms (*Phoenix dactylifera* L.) as genetic resources. Emir J Food Agric 25(11):809–830

Kader AA, Hussein A (2009) Harvesting and postharvest handling of dates. ICARDA, Aleppo

Khalil MS, Khan MA, Al-Kahtani MS (1983) In vitro embryo culture of date palm (*Phoenix dactylifera* L.). In: Proceedings of the 1st symposium on the date Palm, King Faisal Univ, Al-Hassa, 23–25 Mar 1982, pp 142–150

Khan MA, Khalil MS, Al-Kahtani MS (1983) In vitro culture of different tissues of date palm (*Phoenix dactylifera* L.) offshoot. In: Proceedings of the 1st symposium on the date palm, King Faisal Univ, Al-Hassa, 23–25 Mar 1982, pp 152–157
Krueger RR (2011) Date palm germplasm. In: Jain SM, Al-Khayri JM, Johnson DV (eds) Date palm biotechnology. Springer, Dordrecht, pp 313–336
Mansouri A, Embarek G, Kokkalou E, Kefalas P (2005) Phenolic profile and antioxidant activity of the Algerian ripe date palm fruit (*Phoenix dactylifera*). Food Chem 89:411–420
Marzouk HM, El-Salhy AM, Abdel-Galil HA, Mahmoud AE (2007) Yield and fruit quality of some date palm cultivars in response to some flower thinning rates. In: Proceedings of the 4th symposium on the date palm, King Faisal Univ, Al-Hassa, 5–8 May 2007, pp 312–321
Mazloumzadeh M, Shamsi M (2007) Evaluation of alternative date harvesting methods in Iran. Acta Hort 736:463–469
Mikki MS (1998) Present status and future prospects of dates and dates palm industries in Saudi Arabia. In: Proceedings of the 1st international conference on the date palm, Al-Ain, 8–11 Mar, Riyadh, Saudi Arabia, pp 469–513
Mikki MS, Al-Tai WF, Hamodi ZS (1983) Industrialization of dates and development of new products. 1- canning of date pulp and khalal dates. Proceeding of the 1st symposium on the date palm, King Faisal Univ, Al-Hassa, 23–25 Mar 1982, pp 520–533
Mikki MS, Hegazi AH, Abdul Aziz AA, Al-Taisan SM (1989) Stability of major Saudi date cultivars for commercial handling and packaging. In: Proceedings of 2nd symposium on the date palm, vol 2. King Faisal Univ, Al-Hassa, 3–6 Mar 1986, pp 9–25
Ministry of Agriculture (2005) Implementation procedures of the quarantine law for the Gulf Cooperation Council of the Arab Gulf States. Available at: http://ecm.moa.gov.sa/cs/allContent/el_ministry/animal_qur/qur.htm. Accessed 3 Apr 2014
Ministry of Agriculture (2006) Agricultural statistical year book, vols 17 and 19. Department of Economic Studies and Statistic, Riyadh
Ministry of Agriculture (2011) Twenty fourth statistical year book. Total estimated of agricultural crops area and production in Saudi Arabia: estimated area and production of dates crop by region in Saudi Arabia. Department of Economic Studies and Statistics, Ministry of Agriculture, Riyadh, Saudi Arabia, pp 68–70
Mohsen A, Amara SB, Salem NB et al (2003) Effect of vacuum and modified atmosphere packaging on Deglet Nour date storage in Tunisia. Fruits 85:205–212
Mukhtar M, Rasool KG, Parrella MP et al (2011) New initiative for management of red palm weevil threat to historical Arabian date palms. Fla Entomol 94:733–736
Murashige T, Skoog F (1962) A revised medium for rapid growth and bioassays with tobacco tissue cultures. Physiol Plant 15:473–497
Mutlak HH, Mann J (1984) Darkening of dates, control by microwave heating. Date Palm J 3(1): 303–316
Nasr TA, Shaheen MA, Basha MA (1989) Evaluation of date palm males used in pollination in the central region, Saudi Arabia. In: Proceedings of the 2nd symposium on the date palm, vol 1. King Faisal Univ, Al-Hassa, 3–6 Mar 1986, pp 337–346
Nicklin CD (1993) A review of past and current date harvesting mechanization to investigate the concept of designing a palm tree climber. B.Sc. Thesis. Harper Adams Agricultural College, UK
Nixon RW (1969) Growing dates in the United States. Agricultural research service, USDA, Washington
Organic Farming in the Tropics and Subtropics (2002) Exemplary description of 20 crops. Date palm, 1st edn. Naturland e.V., Germany, 19 p. Available online at: http://www.naturland.de/fileadmin/MDB/documents/Publication/English/date_palms.pdf
Osman GES, Al-Beshr AA (1989) Econometric analyses of date palms production cost in Al-Hassa. In: Proceedings of the 2nd symposium on the date palm, vol 1. King Faisal Univ, Al-Hassa, 3–6 Mar 1986, pp 523–526
Sabir JSM, Arasappan D, Bahieldin A et al (2014) Whole mitochondrial and plastid genome SNP analysis of nine date palm cultivars reveals plastid heteroplasmy and close phylogenetic relationships among cultivars. PLoS One 9(4):e94158. doi:10.1371/journal.pone.0094158

Saleem SA (2005) Aspects of ripening of Dhakki dates (*Phoenix dactylifera* L.) and post harvest stability employing hurdle technology. PhD Dissertation, Department of Food Science and Technology, Gomal Univ

Sallam AA, El-Shafie HAF, Al-Abdan S (2012) Influence of farming practices on infestation by red palm weevil *Rhynchophorus ferrugineus* (Olivier) in date palm: a case study. Int Res J Agric Sci Soil Sci 2(8):370–376

Saudi Dates (2013) Brochure, Saudi Arabian Pavilion, For FRUIT LOGISTICA 2013. Berlin, pp 1–18

Saudi Industrial Development Fund (2010) National factories directory, 12th edn. Ministry of Finance, Riyadh, Saudi Arabia, pp 4–6

Sawaya WN (1986) Dates of Saudi Arabia. Safir Press, Riyadh

Shamsi M (1990) Design and theoretical calculation of a tree climbing machine. Scientific and technologic research organization, Unpublished project report. Kerman

Shamsi M (1998) Design and development of a date harvesting machine. Dept Agric Bio-Syst Eng, School Agric Food Env, Cranfield Univ, p 180. https://dspace.lib.cranfield.ac.uk/.../1/Mohsen_Shamsi_Thesis_1998.pdf

Siebert S, Nagieb M, Buerkert A (2007) Climate and irrigation water use of a mountain oasis in northern Oman. Agric Water Manage 89:1–14

Sindh Board of Investment (2010) Dates processing plant. Sindh Board of Investment, Government of Sindh, Karachi, Pakistan

Soliman SS, Al-Obeed RS (2013) Investigations on the pollen morphology of some date palm males (*Phoenix dactylifera* L.) under Saudi Arabia conditions. Aust J Crop Sci 7(9):1355–1360

Soliman SS, Harhash MM (2012) Effects of strands thinning on yield and fruit quality of Sukkari date palm. Afr J Biotech 11(11):2672–2676

Soliman SS, Al-Obeed RS, Omar AA, Ahmed MA (2013) A Comparative study of the morphological characteristics of some seedling date palm males. J Appl Sci Res 9(7):4463–4468

Yang M, Zhang X, Liu G et al (2010) The complete chloroplast genome sequence of date palm (*Phoenix dactylifera* L.). PLoS One 5:99–106

Yin YX, Zhang X, Fang Y et al (2012) High-throughput sequencing-based gene profiling on multistage fruit development of date palm (*Phoenix dactylifera* L.). Plant Mol Biol 78:617–626

Yousif AK, Morton ID, Mustafa AI (1991a) Processing, evaluation and water relation of date paste. Trop Sci 31:147–158

Yousif AK, Morton ID, Mustafa AI (1991b) Effect of storage and packaging on the chemical and physical properties of date paste. Trop Sci 31:159–169

Zhang X, Tan J, Yang M et al (2011) Date palm genome project at the Kingdom of Saudi Arabia. In: Jain SM, Al-Khayri JM, Johnson DV (eds) Date palm biotechnology. Springer, Dordrecht, pp 427–448

Chapter 4
Date Palm Status and Perspective in Iraq

Hussam S.M. Khierallah, Saleh M. Bader, Kadhim M. Ibrahim, and Ibrahim J. Al-Jboory

Abstract Iraq is the birthplace of the date palm, and historically it was the domestication center of this crop. Moreover, for some years, Iraq was the largest producer of dates in the world. Many factors negatively have affected both the production and natural genetic diversity of the crop. However, efforts are being made by the Iraqi authorities and researchers alike to compensate for the serious damage the date palm sector has experienced over the past 30 years. New approaches have been introduced including biotechnology, grove management, pest control, and industrial practices. Production limitations have been diagnosed and constraints are on their way to be resolved. Date palm plantations are under stress from many biotic and abiotic factors including key insect pests like dubas bug, lesser date moth, trunk and stalk borers, and Old World date mite. Date palm diseases cause serious damages to date palm trees especially where stress factors are present such as palm weakness, soil salinity, high water table, borers, and tree aging. The use of plant tissue culture to support propagation by offshoots is necessary and started in the early 1980s. Both direct organogenesis and callus induction with subsequent asexual embryo formation protocols were achieved. Approximately, 600 date palm cultivars were grown in Iraq before 1980; however, currently their number is reduced to 500. Morphological

H.S.M. Khierallah (✉)
Date Palm Research Unit, College of Agriculture, University of Baghdad, 28339, Dawoodi, Baghdad, Iraq
e-mail: khierallah70@yahoo.com

S.M. Bader
West Asia Research Program, The International Center for Agricultural Research in the Dry Areas (ICARDA), Baghdad, Iraq
e-mail: knak100@yahoo.com

K.M. Ibrahim
Department of Biotechnology, College of Science, Al-Nahrain University, Baghdad, Iraq
e-mail: kadhimm2003@yahoo.co.uk

I.J. Al-Jboory
Plant Protection Department, College of Agriculture, University of Baghdad, Baghdad, Iraq
e-mail: ijboory@yahoo.com

© Springer Science+Business Media Dordrecht 2015
J.M. Al-Khayri et al. (eds.), *Date Palm Genetic Resources and Utilization: Volume 2: Asia and Europe*, DOI 10.1007/978-94-017-9707-8_4

traits have been used to describe genetic variation in these cultivars which are mainly related to the fruit, leaf, trunk, and other parts of the tree. DNA marker analysis in Iraqi date palm is at the developmental stage and began in 2000. Tree management and fruit handling improvements are required urgently for better quality production. Date trading has to be reassessed in Iraq in order to overcome the outmoded market chains.

Keywords Cultivar characterization • Diseases and pests • Genetic diversity • History • Micropropagation • Molecular markers • *Phoenix dactylifera*

4.1 Introduction

4.1.1 Historical and Current Agricultural Aspects

The date palm (*Phoenix dactylifera* L.) is one of the oldest cultivated fruit trees and was well known in Babylon, Iraq, 4000 B.C. The palms were celebrated during that period for their strength and majesty. People were heavily dependent on this tree to provide food and wood for making tools, furniture, and baskets. Figure 4.1 shows the blessed date palm during the Sumerian era.

While date palm trees numbered 32 million in the mid-twentieth century, this figure dropped dramatically to approximately 12 million by 2000. The decline was a result of numerous military conflicts in the date-producing regions since 1980. Iraq used to be the main exporter for dates, and dates represented the second largest export revenue after petroleum. The outskirts of Basrah, the country's main commercial hub in the south and the most fertile land for date palm growing, are now littered with trunks, dubbed a *palm tree desert*, which is a legacy of the 1980–1988 war with Iran. Just fewer than two million trees have survived. The embargo that followed in 1990 deprived farmers of modern agricultural equipment leading to poor management, while water became increasingly scarce as a result of a dam-building in neighboring Turkey, and Iran, which reduced the flow of the Tigris and Euphrates rivers on the Iraqi side. During the war of 2003, the date palm industry was substantially damaged. Now, Iraqi officials are promoting replanting programs to rehabilitate the country's date palms. The government is supporting a USD 150 million investment to triple the number of date palm trees by 2021. The program, which initially started in 2005, involved planting around 30 date palm farms and the Ministry of Agriculture established a new plant tissue culture laboratory for date palm micropropagation. Also, the government is encouraging the private sector to rehabilitate old date palm plantations, and private investors have been encouraged to cultivate additional sites in the Iraqi deserts. Overall, the Iraqi government program aims to increase the number of trees to 40 million in the next 10 years.

Fig. 4.1 The blessed date palm in the Sumerian era

Table 4.1 Date production and numbers of date palm trees grown in Iraqi governorates in 2012

Governorate	Production (mt)	Average yield per tree (kg)		Total number of date palms
		In production stage	Productive	
Kirkuk	115	65.6	65.6	3,904
Diyala	83,792	63.6	63.6	2,498,128
Al-Anbar	43,196	82.1	84.2	618,943
Baghdad	93,782	67.5	67.5	2,555,207
Babil	100,343	78.6	78.6	1,753,364
Karbala	72,217	58.5	70.7	1,496,632
Wasit	43,649	87.4	87.8	739,496
Salah- Al-Deen	19,734	90.6	90.6	443,620
Al-Najaf	32,304	68.0	71.0	560,717
Al-Qadisiyah	31,436	75.1	75.1	710,536
Al-Muthanna	25,275	57.9	61.1	822,052
Thi Qar	40,086	60.8	64.3	1,026,675
Maysan	8,967	63.9	65.2	181,529
Basrah	60,554	60.4	64.9	1,353,793
All females	655,450	68.3	71.1	14,764,596
Males	–	–	–	523,814
Total	–	–	–	15,288,410

Source: Central Statistical Organization (CSO) (2012), Iraq

4.1.2 Importance to Iraqi Agriculture

The date fruit itself is a high energy food or feed item for people and livestock. Since ancient times, the date palm has been an important food source for Iraqis. Dates have proved to be the best resource to insure food security during food shortages and crises. The most commonly used parts of the date palm are its fruits, trunks, and leaves, for commercial and, in the case of fruits, for medicinal applications. Date fruits contain more than 70 % sugar at maturity. Dates are also processed into paste and date palm syrup (*dibis*) which are used in certain prepared food recipes. The tree is a spectacular palm for landscaping large areas. It provides shade, prevents soil degradation and desertification, and thus protects the environment.

Date palm groves are found in 14 of the 18 Iraqi governorates; tree numbers in governorates and date yields are given in Table 4.1.

4.1.3 Current Agricultural Problems

Iraq is considered the date palm's birthplace and the tree is symbolic of Iraq. More than 30 million trees existed in the early 1960s. As a result of wars, the number of trees declined by more than one-half. Additionally, more than 600 cultivars were known to exist; however, their numbers were reduced about 500. The decline in date palm plantations was caused by: (a) drought and water scarcity; (b) soil and water salinity, and loss of fertility, especially in the southern region of Iraq where groves are widespread; (c) insect pests and the shortage of spraying equipment; (d) poor management; (e) reduction in numbers of skilled laborers; (f) high input cost vs low output value; (g) urbanization; (h) low investment in the date palm sector; (i) no effective rehabilitation for overaged orchards; (j) problems associated with date processing and marketing; and (k) absence of an effective agricultural extension service. Those impediments have caused a significant decline in date palm production and industry.

4.2 Cultivation Practices

Iraqi farmers have long mastered the cultivation of date palm trees. They are skilled professional laborers in this field. New technologies have been introduced for tree management including pollination machinery although still at a limited scale. Bunch bending, fruit thinning, and harvesting practices are being conducted manually. Protection and control of pests are carried out as routine work using appropriate pesticides. Date processing and packaging is at an infancy stage; modern techniques are limited to a few investors. Iraq is ambitious to launch a policy for competitive production and export of high-quality dates.

4.2.1 Research and Development

Research work on date palm started early in Iraq and elite cultivars were identified and dispersed to other regions of the world. Al-Baker (1972) stated that the numbers of date palm trees in Iraq were the largest in the world at the mid of 1970s. Despite the decline in tree numbers which has occurred for many reasons, but mainly because of the continuous wars that destroyed palm groves, efforts are going on to restore plantations. Development of date palm tree management has attracted many Iraqi scientists to carry on research work in this vital subject (Hussain 2002; Ibrahim 1979, 1995, 1998, 2008, 2010; Khalaf 2002; Sahi 1986; Shalash and Hamood 1989).

Research has concentrated on date palm tree management, micropropagation, and molecular markers, in addition to industrial applications. A plan by the Ministry of Agriculture to conserve cultivars is already established in Iraq to propagate them in vitro. Experiment stations concerned with date palm production have been established in all date-producing governorates. A Date Palm Research Unit was recently established in the College of Agriculture, Baghdad University, for conservation and micropropagation of economically important cultivars.

4.2.2 Pollination

One of the important date palm characteristics is the metaxenia phenomenon that affects fruit morphology and quality. It has been observed that Iraqi date palm male cultivars differ in their pollen grain size ranging from large to small as follows: Khakri Kratley, Khakri Adi, Khakri Smasmi, Ghanami Ahmer, Khakri Wardi, and Ghanami Ahmer. It is worth mentioning that Ghanami Ahmer is superior in bunches and quality of pollen grains per cluster. Differences in fruit flesh and seed weight were recorded in Khadrawy cv. whereas no influence appeared on Maktoom cv. fruits. Pollen grains taken from Resasi males led to enhanced maturity in Khadrawy females; however, no effect was noticed in Maktoom fruits. Percentages of maturity in Halawy cv. were raised when trees were pollinated with Ghanami Ahmer, but this was not the case in Sukar cultivar (Ibrahim 2008).

4.2.3 Pest and Disease Control

The gradual decline in the number of date palm trees and low productivity in Iraq is due to a combination of factors, including the biotic factors represented by pests and diseases and other abiotic factors such as increased salinity, drought and climate changes, as well as the geopolitical events which have occurred in the last three decades. Among the most important pests directly responsible for the damage are dubas bugs, Old World date mite, lesser date moth, and borers, along with rot diseases. The production of dates has been reduced by more than 50 % due to the abovementioned factors (Al-Jboory 1999, 2001).

The increase in date palm services costs and the low prices in the Iraqi market, as well as the lack of institutions to handle date exports, have led to the farmers neglecting their orchards and the gradual deterioration of tree health due to pests and physiological factors. This case was clearly demonstrated in Basrah governorate where the date palm numbers plunged from ten producing palms to three million weak and low production trees. The Ministry of Agriculture took note when the drastic decrease in date palm numbers dropped to 50 % of the previously recorded number of 32 million trees. It was obligatory for the relevant authorities to implement a national program to increase, improve, and protect the present date palm orchards in Iraq.

In September 2000, the National Program for the Propagation and Improvement of Date Palm was approved, covering five research lines. Date palm protection was among the research topics adopted in this program which includes determining an effective solution to pest problems, replacing pesticides or reduction of their use to the minimum, and adopting biological, physical, and agricultural control methods. A team of experts targeted Integrated Pest Management (IPM) components as follows:

(a) Identify date palm pests and determine their economic importance.
(b) Survey and identify biological agents (parasites, predators, entomopathogenic agents).

(c) Monitor the crop and forecast the pest appearance by using pheromones, food, and light traps.
(d) Determine the economic threshold for the key pests.
(e) Use of appropriate pesticides and their formulations.
(f) Follow pesticides resistance management program.
(g) Utilize correct regulations with exact use of pesticide dosages and direct-spray techniques.

Based on the abovementioned points, the program has achieved significant results; however, this success unfortunately has not been continued due to the collapse in the political and administration structure since 2003. As a consequence, progressive deterioration led to a decrease in the productivity and numbers of date palms. In order to avoid this deterioration, the Iraqi government should invest maximum energies in a joint effort between the private sector and the owners of date palm plantations and also introduce regional or international expertise to restore the value of Iraqi date palms.

Studies of date palm pests began in 1912 when so-called aphids producing heavy honeydew were observed on date palms in the Baghdad area. This insect was identified later as the dubas bug (Anonymous 1912). Buxton (1920), Dutt (1922), Dowson (1936), Memarian (1947), Rao (1921), and Rao and Dutt (1922) all studied date palm pests in Iraq, focusing on dubas bug, *Ommatissus binotatus* Fieb., which appeared first in Basrah and later in all date-growing areas of the country.

El-Haidari and Al-Hafidh (1986) and Hussein (1974) categorized the insect and noninsect pests according to their taxonomic orders: Homoptera (dubas bug, mealy bug, scale insects), Lepidoptera (lesser date moth, greater date moth), Coleoptera (stem borers, fruit stalk borers, frond borers, and others), Hymenoptera (yellow and oriental red wasp), Orthoptera (locusts), Isoptera (termite), Thysanoptera (thrips), Acarina (Old World date mite, flat mite, eriophyid mite), and stored date pests.

A field survey of date palm pests in Iraq determined the most important pests and their severity. Table 4.2 includes the key pests attacking date palm in Iraq with reference to the degree of severity. In terms of the importance of borers and their great damage on date palm in the middle and southern provinces of Iraq, the infection percentage in Babel, Kerbela, and Basrah reached 90–100 %. Extensive field and laboratory work has been undertaken in Babil governorate, Mahaweel region (Al-Jboory and Saleh 2001b).

Dubas bug was investigated by Hasoon (1988) and Hasan et al. (2003). The biological and ecological parameters of Old World date mite (Ghobar mite) were studied by Al-Sewidi (2003) and Al-Sewidi and Al-Jboory (2006). Al-Bahili (2004) studied the biological and chemical control of longhorn stem borer in Basrah, while Al-Rubaee (2003) developed a local hydraulic apparatus to inject chemicals inside the trunk to control dubas bug and borers. Ghali (2001) studied the impact of *Chalara paradoxa* fungus on date palm productivity and various factors influencing the infection severity.

Various research groups have provided very important facts regarding the population dynamics of bores, in a survey of living creatures inside and surrounding the date

Table 4.2 Common insect pests infecting date palm trees in Iraq

Name	Scientific name	Order and family	Degree of important
Longhorn date palm stem borer	*Jebusaea hammerschmidti* Reich	Coleoptera, Cerambycidae	+++
Fruit stalk borer	*Oryctes elegans* Prell	Coleoptera, Scarabaeidae	++
Frond borer	*Phonapate frontalis* Fahraeus	Coleoptera, Bostrichidae	+
Dubas bug (Old World date bug)	*Ommatissus lybicus* de Berg.	Homoptera, Tropiduchidae	+++
Lesser date moth	*Batrachedra amydraula* Meyrick	Lepidoptera, Momphidae	++
Greater date moth	*Arenipses sabella* Hampson	Lepidoptera, Pyralidae	+
Termite	*Microcerotermes diversus* Silvestri	Isoptera, Termitidae	+ (+)[a]
Parlatoria date scale	*Parlatoria blanchardii* Targioni	Homoptera, Diaspidae	+
Ghobar mite (Old World date mite)	*Oligonychus afrasiaticus* (McGregor)	Acari, Tetranychidae	++ (+)[a]

[a]The importance of termite infection depends on the health condition of the date palm, while Ghobar mite infestation is higher during the dust storms which blow during summer associated with drought and low rainfall

palms. The borer population density during the period from 4/10/2000 to 10/10/2001 indicated that the longhorn stem borer is dominant in date palm trees, causing a significant damage compared to the fruit stalk borer. Borers carry pathogens either externally or in their excrement, entering into the palm trunk through the borer holes (Abbas and Muhee 1991). The longhorn date palm borer is mostly found at the growth point, the base of young leaves, and fruit bunch bases, while the stalk borers are present either at low density attacking the bunch stalk or inside the trunk. Khalaf and Al-Taweel (2014) mentioned that six different borer species attack date palm in Iraq causing economic damage; among them four belongs to *Oryctes* species.

In order to achieve the second point of the IPM component, the following bioagents have been collected and identified, and some tested either in the field or in the laboratory:

(a) Entomopathogenic nematodes on longhorn date palm trunk borer and date palm stalk borer in Iraq (Al-Jboory 2001)
(b) Survey and classification of mite species found in/on date palm (Al-Jboory and Saleh 2001b)
(c) New record of *Megaselia* sp., a parasite on the female of longhorn stem borer (Al-Jboory and Saleh 2002b)
(d) Isolation and identification of date palm borer pathogens, *Beauveria bassiana* (Al-Jboory et al. 2002, 2006), Poxviridae virus from longhorn stem borer (Al-Jboory and Saleh 2002a), and *Oryctes*-like virus from date palm stalk borer (Mohamed and Al-Jboory 2001)

Table 4.3 Pathogens infected date palm orchards in Iraq

Pathogen	Common name	Infected parts	Economic importance
Mauginiella scaettae	Inflorescence rot	Inflorescence, spathe	+++
Thielaviopsis paradoxa *Chalaropsis radiciola*	Black scorch (fool's disease)	Rachis, leaflets, growth point, trunk, inflorescence	+++
Diplodia phoenicum	Diplodia disease	Leaves, growth tip	++
Graphiola phoenicis	Graphiola leaf spot	Rachis, leaflets	++
Phytophthora sp.	Belaat disease	Frond, growth point rot	++
Alternaria spp.	Brown leaf spot	Leaflets, rachis, spines	++

New species of a dubas bug eggs parasitoid, *Pseudoligosita babylonica*, has been identified (Hasan et al. 2003) and many other predators on dubas bug and Old World date mite. This research activity unfortunately ceased after 2003, and many of the unique bioagents collected such as the two viruses and the nematodes were lost due to the electricity network collapse. Some trials are now proceeding on date palm borers, directed by the Ministry of Science and Technology and also at the Organic Agriculture Center, Ministry of Agriculture, testing *Bacillus thuringiensis* and in addition mass rearing *Trichogramma* parasitoid against the lesser date moth. Light traps and neem oil have been used recently as an alternative to conventional insecticides against borers and dubas bug (Hama et al. 2010).

Iraqi universities play a crucial role in conducting trials on date palm pests including dubas bug (Al-Abbassi 1987, 1988; Al-Dhamin 2002, 2008; Al-Rubeai et al. 2010; Hamad 2005; Hasoon 1988; Jassim 2007), on lesser date moth (Al-Delamy 2004; Al-Jorany and Al-Delamy 2010b, 2013; Al-Fahdawi 1988; Aziz 1990, 2005), on borers (Al-Ali and Ismail 1987), and on date palm diseases (Al-Asady 2003; Juber and Al-Mohamadawi 2010). Unfortunately there is no investment for such valuable applied results in the practical implementation.

Extensive investigations were conducted during 1992 when date palm disorders were observed in Baghdad and other date palm-growing areas in Iraq. The Ministry of Agriculture appointed a committee and together with FAO experts (Saaidi 1992) described four important types of diseases of date palm in Iraq: (1) neck bending disease and date palm destruction caused by four fungi, *Thielaviopsis paradoxa* (Al-Hassan and Abbas 1987), *Chalaropsis radicicola* (Abbas and Abood 1996; MOA 1992), *Chalara paradoxa* (Ghali 2001), and *Chalara radicicola* (Al-Hamdany et al. 2011); (2) dry rot and growth stunting caused by the same fungi as in the first; (3) frond malformation and dwarfing caused by *Alternaria* sp., *Fusarium* sp., and *Thielaviopsis* sp.; and (4) heart wet rot associated with many fungi, e.g., *Thielaviopsis* sp., *Fusarium* sp., *Helicocephalus* sp., *Sphaeropsidales* sp., *Thaminidium* sp., and *Helminthosporium* sp. The trunk and stalk borers were among the most important disposition factor for pathogen entrance (Abbas and Muhee 1991).

Common diseases infecting date palm trees in Iraq are listed in Table 4.3 with their economic importance (Abdullah et al. 2010; Al-Ani et al. 1971; Al-Badran 2008; Al-Beldawi and Hussain 1974; Al-Hassan and Shamseldeen 1975). Fungicides are the only available measure to control the most serious date palm fungal diseases in Iraq; however, beneficial fungi such as *Trichoderma* have been used with positive results against date palm diseases.

4.2.4 Agroforestry Utilization and Potential

Renewal of standing date palm plantations is regularly reviewed by governmental officials and farmers alike. Replacement of overage trees is one of the future plans adopted by the Ministry of Agriculture. Thus, date palm tissue culture laboratories have been established across the date palm-growing regions in an attempt to maximize the number of in vitro offshoots besides the in vivo plantlets. Date palm trees have been incorporated into landscaping urban and suburban areas in Iraq (Ibrahim 2006, 2008). An ambitious plan is under way to combat desertification in date-growing governorates which are mainly subject to dust storms.

Although Iraq has the potential to restore date palm groves over the long term, some limitations currently represent obstacles for such restoration. One of these limitations is the meager contribution of the private sector in investing and introducing new technologies. This may be due to the current insecure political situation. The quantity of produced offshoots, whether sourced from already established groves or produced by tissue culture, is not able to meet the current demand. Dust storms have badly affected date production causing the spread of Old World date mite and consequently affecting the fruit quality. Accordingly, it is recommended that new technologies be introduced to improve the date production quantitatively and qualitatively. IPM is required to reduce the potential damage both in date palm groves and during storage. Possible improvement is necessary in sorting and packaging of dates to compete other producers and exporters.

4.3 Genetic Resources and Conservation

According to numerous references, including Biblical texts, the date palm is believed to be the oldest domesticated fruit tree. Representations of the tree appear in hieroglyphics from the Neolithic civilizations of Mesopotamia and ancient Egypt. The earliest cultivation was found around 4,000 B.C. in Eridu, what is now Tell Abu Shahrain, near the Euphrates River, Thi Qar governorate, and lower Mesopotamia in Iraq. In addition date palm is mentioned in Akkadian and Sumerian cuneiform sources dated as early as 2,500 B.C. Historically date palm covered areas extending from the Indus Valley (now Pakistan) to Mesopotamia in the Tigris/Euphrates valleys (Iraq) to the Nile Valley, Southern Persia (Iran), and the Eastern Mediterranean.

The center of date palm origin is still debated with claims that it originated from Babel in Iraq, Dareen or Hofuf in Saudi Arabia, or Harqan and other islands of the

Arabian Gulf in Bahrain. The spread of date palms from the center of origin near present-day southern Iraq into other areas of the Middle East, Northern Africa, and Eastern South Asia resulted in the establishment of new date palm oases in these areas. These human-created oases probably resulted from the introduction of a relatively small range of genotypes. Subsequent clonal propagation by the planting of offshoots of desirable types resulted in distinctive fruit types being associated with the various oases. Even in the center of origin, selection pressure resulting from clonal propagation of desired types over a long period of time resulted in a certain amount of genetic erosion and the same association of characteristic cultivars with specific oases.

4.3.1 Research in Genetics, Breeding, and Conservation

Date palm is one of the world's first cultivated fruit tree and one of the classical Old World fruits. It has 36 chromosomes ($n=18$; $2n=36$). Worldwide, about 3,000 date palm cultivars have been named. The accurate identification of cultivars for breeding purposes is one of the pillars of any successful improvement program, as well as assessing the current genetic variability within the germplasm and creating resources. In date palm, traditional breeding objectives always include yield enhancement, disease and pest resistance, and tolerance of saline water, saline soil, and drought.

Efforts have been made to assess the available variability within the local germplasm in Iraq. Al-Salih and Bader in 1983 began a date palm breeding research program of Barhi cultivar. The female palm was pollinated by Ghanami Ahmer cv. in which seeds were used as a starting material. After 6 months, the seedlings were transferred to the field and left until fruiting stage after 4–5 years. Approximately 48 % were female and 52 % male (Al-Saleh and Bader 2013).

Many crosses between selected male and female trees were carried out in 1990 and progeny seeds were collected. F2 seeds were sown and maintained until flowering. Several new clones of cv. Barhi were produced; one was superior compared with its parents. Al-Saleh (1988) described a rapid method for breeding and improvement of five Iraqi date palm cvs., namely, Sayer, Khsab, Lilwi, Khastawi, and Barhi. Seeds of these cultivars were grown under the same conditions until flowering. Selection for some distinguished male and female new offshoots was done. The research concluded that propagation of date palm by seed is a useful, simple, and cheap method for obtaining new cultivars.

Genetic invariability must be assured in offshoots. However, great confusion in the discrimination process may occur if the new cultivar is growing very near to its mother palm, which is called an *off-seed* tree. It is believed that off-seed trees of all cultivars originated either from seed fallen near the mother tree and germinated along with its true offshoots or from a vegetative bud mutation as in case of sectorial mutation.

4.3.2 Threats and Degradation

Iraq remains one of the leading date-producing countries in the world. Before 1991, it was the world's largest producer (FAOSTAT 2004) and had the most extensive date *forest* in the world (MacFarquhar 2003). However, a serious decline over the past 30 years occurred due to numerous military conflicts since the 1980s. During recent wars, large numbers of date palm trees were destroyed. Water quality became crucial due to the harmful effects of the wars which destroyed date palm orchards in various regions, but especially in Basrah governorate, Shatt Al-Arab region. The abandonment and neglect by many date palm growers of their orchards was another reason for the decline, caused by low returns from date sales. An additional reason is that farmers depend on the traditional irrigation methods; lack of efficient drainage systems leads to salt accumulation in the soil which affects the growth and yield of date palm orchards. The high cost of agricultural practices such as pollination, pruning, pest control, harvesting, and marketing, along with a deficiency in the number of skilled laborers, were also significant reasons for the deterioration.

4.3.3 Germplasm Banks and Genetic Conservation Efforts

Iraqi date palm resources have been exposed to dramatic deterioration especially of date palm populations during the last 30 years. However, a project to establish new mother date palm orchards and offshoot nurseries in the 14 governorates which grow date palm has been proposed by the General Board of Date Palm (GBDP) and approved by the Ministry of Agriculture to stop the deterioration in date palm sector and overcome the decline in date palm groves (Husien et al. 2009). Each governorate will collect the cultivars grown within their respective areas, while the governorates of Thi Qar, Al-Samawah, and Al-Najaf will collect both local and international cultivars for conservation at their date palm stations (Fig. 4.2). Date palm offshoots will be grown in orchards at a spacing 5×5 m. Each orchard will contain 80 % commercial cultivars, 16 % rare ones, and 4 % male trees, maintained under drip irrigation and modern agriculture practices. The project began in the spring 2004 in two stages: the first stage was finished in 2011, while the second will continue until 2021 to establish producing date palm orchards. Technical teams are instructed to investigate the offshoot sources within each governorate and maintain records at each date palm station with information on the cultivars grown and the region where the offshoot originated and other relevant data.

A project for mapping date palm cultivars will be executed soon at the genetic engineering labs of GBDP to classify and establish standard names of the cultivars. The initial plant spacing of 5×5 m after 8 years will be modified to 10×10 m by mechanical removal of planted offshoots which will be used to establish new mother

Fig. 4.2 Najaf date palm station established in 2008

date palms. The objectives of the project are to maintain local and international cultivars and to provide researchers known cultivar collections to work with, as well as for agricultural extension objectives. The total number of mother date palm orchards established so far is 30, covering 13 governorates in Iraq with a total area of 4,594 Iraqi dunam (1 dunam = 2,500 m^2). More than 497 cultivars had been collected up to January 2009 and efforts are continuing to collect all the cultivars grown in the various regions of Iraq (Table 4.4).

As a result of cooperation with the GBDP, the Date Palm Research Unit (DPRU), University of Baghdad, was recently established, using another type of conservation through extraction of the total genomic DNA of over 220 core Iraqi collection cultivars. Quantitative and qualitative tests were carried out and ultimately the DNA conserved by cryopreservation. This type of gene bank is now available for all researchers who are interested in conducting molecular analysis projects of Iraqi cultivars. Very limited work has been carried out on the cryopreservation of date palm embryogenic cultures and therefore the development of innovative procedures is needed for an efficient preservation of genetic resources and the management of commercial propagation.

Preliminary studies have revealed that embryogenic cultures provide prime plant material for in vitro mutagenesis experiments and selection of useful mutants, for the generation of protoplasts and somatic cell hybridization, and for genetic engineering. Genetic materials may be affected by diseases, climate, natural disasters, etc., so ex situ collections should be maintained even when there is no an immediate threat of habitat loss. In order to make germplasm more accessible for researchers, ex situ collections are subjected to characterization and evaluation of germplasm and utilization in breeding programs and other research activities. Maintenance of

Table 4.4 Date palm establishments in various Iraqi date palm stations

Station name	Governorate	Area (ha)	Establishment date
Rashdeia	Baghdad	100	2004
Rabee	Baghdad	15.5	2005
Zaafarania	Baghdad	8	2004
Latefia	Baghdad	25	2005
Madaen	Baghdad	34	2005
Fallujah	Anbar	29	2004
Dawar	Anbar	25	2005
Dawar (tissue culture)	Anbar	9	2006
Eshaqi	Salah Al-Deen	27.5	2004
Mandeli	Diyala	37.5	2004
Kut	Waset	32.5	2004
Azizia	Waset	13	2005
Amarah	Maysan	58.75	2004
Akad	Dhi Qar	99	2004
Nuria	Diwaniyah	37.5	2004
Diwaniyah 1	Diwaniyah	25	2004
Diwaniyah 2	Diwaniyah	5	2004
Samawah desert (tissue culture)	Muthanna	125	2008
Khedher	Muthanna	25	2004
Najaf	Najaf	61	2004
Najaf (tissue culture)	Najaf	63.75	2008
Husania	Karbala	16.75	2004
Razazza	Karbala	32	2004
Razazza (tissue culture)	Karbala	18	2006
Mahaweel	Babel	12.5	2005
Abu Sdera	Babel	25	2005
Basrah	Basrah	30.25	2004
Burjesia	Basrah	35	2004
Qurna	Basrah	53.25	2005
Faw	Basrah	69.75	2005

Source: Central Statistical Organization (CSO) (2012), Iraq

germplasm in a disease-free state is also desirable, and this is often possible only in ex situ collections.

Assessment of the genetic vulnerability of any crop requires knowledge of the extent and distribution of its genetic diversity. This is acquired by systematic sampling and mapping of the flora of the geographical areas in which the species in question is found, as well as an assessment of ex situ collections. Unfortunately, information on natural and seminatural germplasm is often limited at the international level. This is due to the remoteness of some of the material, a lack of resources devoted to assessing these areas, and political considerations. In some cases, information may be available at the local or national level, but not to the international community.

4.3.4 Current Status and Prospects of Genetic Resources

The date palm sector in Iraq needs urgent rehabilitation through (a) supporting and encouraging farmers and orchard owners to care for their trees and orchards and apply modern technologies, (b) pest control in palm groves especially against dubas bug and borers, (c) establishment of numerous field stations for growing mother trees from different parts of the country, (d) mass production of elite cultivars using plant tissue culture techniques, and (e) collection and characterization of all Iraqi cultivars using different genetic markers.

An inventory and assessment for all locally grown cultivars, individual seed trees, and rare clones to compile a database are vital steps which can be relied upon in formulating policies, economic productivity or marketing, and selection of the best cultivars in terms of production and quality, followed by traditional propagation and tissue culture means. The continuation of joint work between the GBDP in collecting date palm cultivars from farmers and DPRU for molecular characterization and genetic relationships determination will lead to achieving the ultimate goal of generating mapping of cultivars grown in Iraq.

One of the major limitations of date palm genetics and traditional breeding is the lengthy life cycle of this special crop. Some of these challenges are of worldwide concern, while others are regional and country specific. To respond to these constraints, the objectives of the breeding programs are set according to short- and long-term goals. The short-term objective is to replace and provide date palm orchards with offshoots in order to sustain the perennial aspect of the culture. This goes along with the diversification of the genetic basis that has been narrowed in recent years due to the expansion of monoculture of elite international cultivars.

As for the long-term objective, the development of new cultivars using conventional and/or nonconventional approaches would lead hopefully to resolve some of the date palm constraints, especially dubas bug and red palm weevil, and respond to greater opportunities and understanding of the evolution of monocotyledonous families. Moreover, more detailed characterization and evaluation data are needed to adequately assess the actual amount of genetic diversity present. These data should include both descriptive data and molecular-level genetic analysis of germplasm existing both in situ and ex situ, such as abnormal somatic embryo differentiation, endophytic bacteria proliferation, and somaclonal variations.

4.4 Plant Tissue Culture

4.4.1 Role and Importance

Date palm cultivation has been expanding in Iraq, as it has in most Arab countries. Dates are propagated traditionally by seeds or offshoots, but because of heterozygosity that progeny produces from seeds, the resulted trees are not identical, are poorer in quality than the mother plant, and are approximately 50 % males. Therefore, propagation by offshoots is better, but the number produced from a tree is limited, especially from superior and rare cultivars, so it does not satisfy the need when farmers consider establishing new orchards.

The use of plant tissue culture to supplement offshoot propagation is necessary. Before 1981, Iraq was one of the top producers of dates and had the largest date *forest* in the world. A sharp decline over the past 30 years has occurred due to numerous conflicts, long-term soil salinization, shortages of electricity, and official negligence. While date palm trees numbered 32 million in the mid-twentieth century, that figure dropped to 16 million by 2000. Thus, exploitation of plant tissue culture is crucial for micropropagating distinguished cultivars toward effective rehabilitation of Iraqi date palm orchards.

Accordingly, the government has adopted many programs aiming to increase the number of date palms to 40 million trees in the next 10 years. On the other hand, a few private laboratories for commercial date palm micropropagation which are located in Baghdad (Iraqi Center for Plant Tissue Culture Ltd.), Nasriyah (Uruk Center for Date Palm Tissue Culture), and Basrah (Technical Tissue Culture Lab. and Fadk Company Lab.) have ongoing projects for scale-up production of several cultivars which hopefully will succeed and thus meet the country's demand for the required number of offshoots.

4.4.2 Research Progress

In Iraq, early attempts at date palm micropropagation began in the 1980s. The first took place in Basrah by Mater (1983) involving callus induction and subsequent asexual embryo formation. Other attempts were made in Baghdad by some researchers in the Iraqi Atomic Energy Organization (Omar and Arif 1985) and the Scientific Research Council (Bader et al. 1986) using zygotic embryos for callus induction. These efforts led to the development of a complete protocol for in vitro propagation of date palm (Mater 1986; Omar 1988b, 1992).

During the 1990s, military conflicts restricted the research efforts of many institutes. Nevertheless, a successful protocol was accomplished in the beginning of 2000s for direct organogenesis including axillary branching and enhanced adventitious bud formation from shoot tips (Hameed 2001). Meanwhile, the College of Science and the Date Palm Research Unit, University of Basrah, had also participated in date palm micropropagation efforts. Many factors were studied influencing date palm somatic embryogenesis and plantlet regeneration of several cultivars cultivated in Basrah (Al-Kaabi 2010; Al-Mayahi 2010; Al-Meer and Al-Ibresam 2008; Al-Meer and Al-Ibresam 2010; Al-Musawi 2001; Jasim et al. 2009; Al'utbi and Al-Husaibi 2007; Jasim 1999, 2000; Jasim and Saad 2001, 2003; Jasim et al. 2008; Muhsen 2007; Saleh et al. 2006).

In addition, factors affecting vegetative bud formation via callus tissues were also investigated (Abdul-Sameed 2009; Al-Khalifa et al. 2008, 2009). Khierallah and Bader (2007) reported a complete protocol for date palm micropropagation by organogenesis using shoot tips of Maktoom cv.; immature inflorescence explants were also utilized for in vitro propagation by Khierallah (2007) at the Date Palm Research Unit, University of Baghdad. The study comprised of direct organogenesis, callus induction and subsequent shoot regeneration, rooting, acclimatization, and genetic fidelity detection using amplified fragment length polymorphism (AFLP) analysis for Barhi and Maktoom cvs.

Since plant tissue culture techniques involve the possibility of undesirable genetic variability in the derived plants, which is not apparent until the fruiting stage, other markers were employed to detect genetic stability in tissue culture-derived date palm plantlets for Barhi and Maktoom cvs. using random amplified polymorphic DNA (RAPD) markers (Ali et al. 2007; Khierallah et al. 2008) and for Bream cv. using RAPD markers (Khierallah and Husien 2013).

4.4.3 Micropropagation Strategies

Since the initial attempts at date palm propagation by tissue culture (Reuveni 1972; Schroeder CA 1970), two strategies of micropropagation have been achieved: organogenesis and somatic embryogenesis.

4.4.3.1 Offshoot-Derived Explants

Selected offshoots should have a weight of 10–15 kg, be 2–3 years old and 80–100 cm in height. Offshoots should be dissected acropetally until the shoot tips appear. The excised shoot tip ought to be about 3–4 cm in width and 6–8 cm in length. Explant region (apical meristem with soft inner leaves) of 2 cm diameter is excised and placed in antioxidant solution. The explant region is usually sterilized using commercial bleach (sodium hypochlorite) 20 %. Other effective disinfection protocols are available according to Hameed (2001), Jasim (1999), Khierallah and Bader (2007), and Omar et al. (1992).

4.4.3.2 Somatic Embryogenesis

The most commonly used and widespread method for date palm micropropagation is somatic embryogenesis. In Iraq, several protocols are being implemented and can be summarized as follows: transfer disinfected date palm shoot tip explants to petri dishes where all leaf primordia are removed except for two pairs surrounding the apical meristem, which can be divided longitudinally into four equal segments aseptically and cultured on callus initiation medium. Modified MS medium was used for this purpose. Jasim and Saad (2001) produced high callus fresh weight with certain combinations of plant hormones.

Recently, Ibrahim (2012) initiated callus and transferred it to embryogenic callus proliferation medium. The medium varies from one researcher to another, for example, it can consist of MS salts and vitamins without hormones (Omar et al. 1992) or the same supplements described for callus induction medium (Jasim and Saad 2001 and Ibrahim 2012). Embryogenic calli have been produced by various researchers (Jasim and Saad 2001; Omar et al. 1992).

Low rates of somatic embryogenesis and germination have prompted researchers to enhance the processes. It was found that supplementing the culture medium with apple seed powder (Saleh et al. 2006), corn seed powder (Jasim et al. 2008), vitamin E (Al-Meer and Yaseen 2010), polyethylene glycol (Al-Mayahi 2010), and coconut water or casein hydrolysate (Khierallah and Husien 2013) or exposure to a laser beam (Al-Kaabi 2010) increased somatic embryogenesis and germination rates for several Iraqi date palm cultivars.

Mature or germinating embryos initiated roots on MS medium supplemented with 0.1 mg/l NAA plus 0.01 mg/l BA (Omar et al. 1992). Jasim and Saad (2001) used half-strength MS salts for Barhi cv (Table 4.5). Maintenance of plantlets longer in the culture medium increased their survival rate in soil.

4.4.3.3 Adventitious Organogenesis

Direct regeneration of vegetative buds minimizes the risk of somaclonal variation among regenerants. Moreover, the duration of the culture period is limited by frequent renewal of the plant material. Disinfected date palm shoot tips can be dissected as cultured explants. It is necessary that axillary buds to be used are suitable explants. On average 16–20 explants may be obtained from each offshoot shoot tip.

In general, MS medium is used for the initiation stage. Supplementation of plant growth regulators depends on the culture stage (Table 4.6). Cultures preferably are incubated in darkness to reduce phenolic secretions (Fig. 4.3a). They require four subcultures at 4-week intervals until bud initiation (Fig. 4.3b).

The formed buds subcultured on a liquid agitated multiplication medium raised the number of buds up to an average of 12.6 (Khierallah and Bader 2007). The physical status of the medium may be evaluated when the buds transferred to a liquid medium contain the proper combination of plant growth regulators obtained from the multiplication stage. Agitating the medium is better as compared with stationary or solid medium to increase the number of buds (Hameed 2001; Khierallah and Bader 2007). Hameed (2001) reported that MS stationary liquid medium produced a high number of shoots (Fig. 4.3c). Khierallah and Bader (2007) used MS medium a supplement for rooting. According to Khierallah and Bader (2007), acclimatization was reported. More than 85 % survival was achieved at the end of acclimatization (Fig. 4.3d).

Table 4.5 Plant growth regulators added to MS medium during callus induction, embryonic callus proliferation, embryo germination, and rooting medium of date palm micropropagation (values in parentheses are in mg/l)

Cultivars	Explant source	Callus induction	Embryonic callus proliferation	Embryo germination and maturation	Rooting	Reference
Ashrasi, Barban, Bream, Khastawi, Khadrawy, Maktoom, Suada Zahidi	Shoot tips, lateral buds, root tips leaf primordia, mantel meristem, bundle sheath	2,4-D (100), 2iP (3)	2,4-D (100), 2iP (3)	None	NAA (0.1), BA (0.01)	Omar et al. (1992)
Zahidi, Tibarzal, Barhi, Maktoom, Khastawi, Ibrahimi	Shoot tips, lateral buds, root tips leaf primordia, mantel meristem, inflorescence	2,4-D (10), 2iP (3) or NAA (20), 2iP (3)	2,4-D (20), 2iP (3) or NAA (40), 2iP (3)	NAA (0.1), BA (0.05) liquid	NAA (0.5)	Hameed (2001)
Askar, Barhi, Guntar, Halawy, Khadrawy, Sayer, Zahidi	Apical buds	2,4-D (100), 2iP (3) or NAA (30), 2iP (3)	NAA (30), 2iP (3)	NAA (0.1)	NAA (0.1)	Jasim (1999); Jasim and Saad (2001)
Bream	Apical buds	Picloram (50), 2iP (3)	Picloram (50), 2iP (3)	NAA (0.1), BA (0.05)	NAA (0.5), BA (0.1)	Ibrahim (2012)

Table 4.6 Plant growth regulators added to MS medium during initiation and multiplication stages of date palm micropropagation (values in parentheses are in mg/l)

Cultivars	Explant source	Initiation	Growth responses	Multiplication	Elongation	Rooting	Reference
Zahidi, Tibarzal, Barhi, Maktoom, Khastawi, Ibrahimi	Shoot tips	2iP (5) BA (5)	Enhancement of axillary branching	2iP (4) BA (4) NAA (0.5) (liquid)	–	NAA (0.5)	Hameed (2001)
	Shoot tips	ZT (1) NAA (2)	Adventitious bud formation	Zeatin (4) NAA (2) (liquid)	GA_3 (0.3)	NAA (0.5)	Hameed (2001)
Maktoom	Shoot tips	2iP (2) BA (1) NAA (1) NOA (1)	Adventitious bud formation	2iP (4) BA (2) NAA (1) NOA (1) (liquid)	GA_3 (0.3), NAA (0.1)	NAA (1)	Khierallah and Bader (2007)
Barhi, Maktoom	Shoot tips	BA (1) 2iP (2) NAA (1) NOA (1)	Adventitious bud formation	BA (1) 2iP (4) NAA (1) NOA (1)	GA_3 (0.3), NAA (0.1)	NAA (1)	Bader and Khierallah (2007)
Barhi, Maktoom	Immature inflorescence	BA (2) NAA (1)	Vegetative buds	BA (1) 2iP (1.5) NAA (1)	GA_3 (0.3)	NAA (1)	Khierallah (2007)
Sayer	Shoot tips	2iP (10) NAA (3)	Lateral buds	2iP (10) NAA (3)	–	NAA (1)	Al-Khalifa et al. (2008)

Fig. 4.3 Date palm organogenesis stages. (**a**) Shoot tip explants of Maktoom cv. cultured in darkness, (**b**) adventitious bud initiation, (**c**) elongation, and (**d**) acclimatization (Hameed 2001)

4.4.3.4 Micropropagation Using Inflorescence Explants

Inflorescence explants have proved to be a promising alternative explant source for micropropagation of elite cultivars, particularly rare ones. Khierallah (2007) described a stepwise micropropagation protocol relying on direct and indirect organogenesis for the two cvs. Barhi and Maktoom in early spring. Explants excised from spathes for callus induction were cultured in MS modified medium containing 100 μM 2,4-D plus 15.0 μM 2iP (Fig. 4.4a). Adventitious shoots were obtained after callus transfer to MS liquid agitated medium supplemented with 10.0 μM 2iP plus 5.0 μM NAA.

For direct organogenesis, spathes with spikes were cultured in MS medium. Direct organogenesis was achieved after 24 weeks (Fig. 4.4b). Multiplied buds

Fig. 4.4 In vitro micropropagation stages of date palm using inflorescence explants. (**a**) Callus formation after 12 weeks. (**b**) Adventitious bud formation and development from callus after 12 weeks. (**c**) Bud multiplication directly from floral explants, elongation, and rooting R. (**d**) Plantlet acclimatization under greenhouse conditions

increased by increasing the glutamine concentration using liquid agitated medium.

Several factors affecting rooting and plantlet acclimatization were investigated. Acclimatization of plantlets reached 70 and 80 % for Barhi and Maktoom cvs., respectively, after transfer to the greenhouse (Fig. 4.4d).

4.4.4 Genetic Conformity of Tissue-Cultured-Derived Plants

Plant tissue culture is considered a means of vegetative propagation in which phenotypically and genetically identical clones are produced rapidly. However, somaclonal variation is known to result from changes in the nuclear, mitochondrial, or chloroplast genomes of regenerants. Accordingly, detection of genetic stability of date palm plantlets is necessary to confirm genetic fidelity.

Ali et al. (2007) used RAPD-PCR technique for detection of genetic stability in regenerated plantlets of Barhi cv. Reproducible RAPD patterns were obtained using 30 primers. Three (OPC.16, OPG.O8, and OPN.16) produced polymorphic

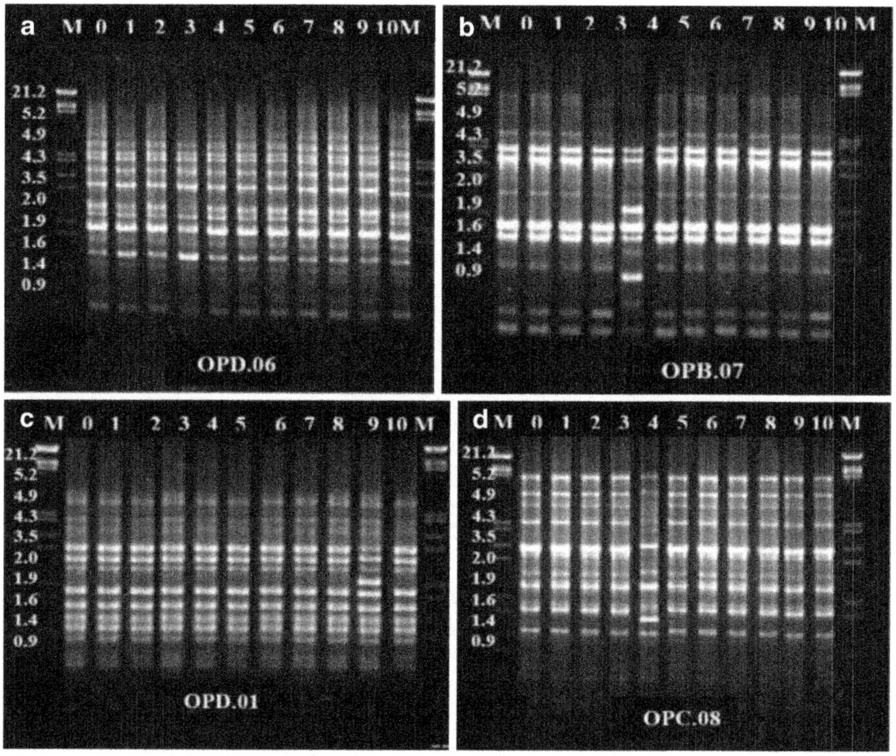

Fig. 4.5 (**a**) Monomorphic banding patterns of samples tested revealed by OPD.06 primer for Barhi cv., (**b–d**) polymorphic banding patterns using the primers (OPB.07, OPC.08 for Maktoom and OPD.01 for Barhi cv.). Numbers on the left indicate the fragment size of molecular weight markers (lane M) in kb. The lanes 0 are the banding pattern of the intact trees, while the lanes 1–10 are the banding pattern of the samples selected randomly from tissue culture-derived plantlets

bands in some of tested samples when compared with the DNA fingerprints of the mother offshoot suggesting the possibility of genetic variation among the resultant plants.

Khierallah et al. (2007) used RAPD markers and tested 25 universal primers performed on DNA extracted from fresh leaves of the mother tree and from samples randomly taken from plantlets derived from tissue culture for cvs. Barhi and Maktoom. Reproducible RAPD patterns were obtained with 20 primers; 17 primers showed completely monomorphic bands in all tested samples of the progeny (Fig. 4.5). Only three primers showed some polymorphic bands for both cultivars in some of the tested samples compared with the DNA banding pattern for the intact trees; these were OPD.01 primer for Barhi cv. and OPB.07 and OPC.08 for Maktoom. Meanwhile Khierallah et al. (2008) employed ALFP markers successfully to trace genetic fidelity of date palm in vitro plantlets cvs. Barhi and Maktoom derived from inflorescence explants (Fig. 4.6).

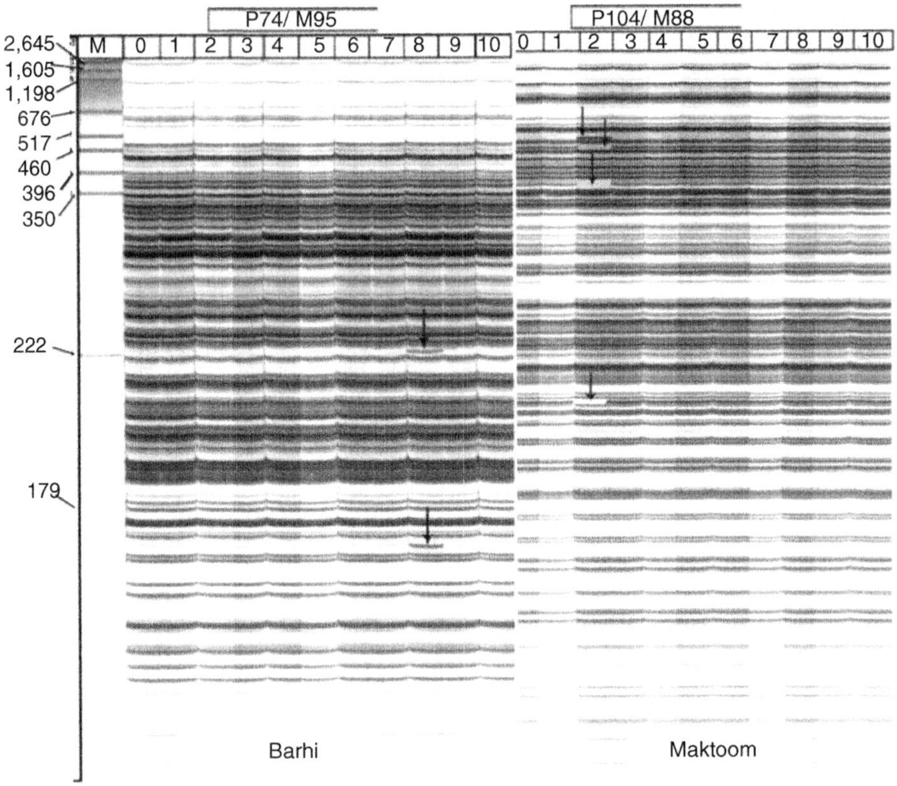

Fig. 4.6 AFLP banding pattern of the two date palm cvs. Barhi and Maktoom as revealed by primer combinations P74/M95 and P104/M88, respectively. Numbers on the left indicate the fragment size of molecular weight markers (lane M) in base pairs (bp). The lanes 0 are the banding pattern of the mother trees, while lanes 1–10 are the banding pattern of the samples selected randomly from direct organogenesis (Barhi) and indirect organogenesis (Maktoom)-derived plantlets

Recently, Khierallah and Husien (2013) generated RAPD pattern with 12 primers to identify genetic fidelity of date palm in vitro plantlets produced using picloram instead of 2,4-D for callus induction and some natural organic extracts (coconut water and casein hydrolysate) for somatic embryogenesis. Genetic variations may occur in plantlets derived from callus proliferated from shoot tips as well as in those derived from inflorescences. RAPD and AFLP appear to be an efficient technique and simple and fast DNA markers for the early detection of genetic variations in plants propagated by tissue culture.

4.4.5 Tissue Culture Limitations and Future Directions

Expansion of date palm cultivation is faced with propagation challenges of genetic improvement limitations. The heterozygous nature of this dioecious plant restricts the use of seeds which produce off-type seedlings and normally are not used to propagate elite cultivars. The limited availability of offshoots and the difficulties of establishing propagules from offshoots render this traditional propagation method unsuitable, particularly for large-scale propagation. Based on recent advances in plant tissue culture, micropropagation techniques have been developed for the rapid mass propagation of date palm.

Some limitations associated with genetic improvement have been circumvented by taking advantage of tissue culture and molecular methodologies. The scaling-up of tissue-cultured date palm in Iraq is still in flux for several reasons: the high cost of micropropagation technology; tissue culture of date palm, which is generally time-consuming; complicated biological events involved in each step of micropropagation; the lack of trained labor and management; poor planning; limited laboratory infrastructure; and mixing of research and production concepts.

Accordingly, investment should be made toward large-scale production of date palm in vitro. The government should offer facilities for investors in date palm production technology. On the other hand, further research is needed to reveal the most appropriate biochemical and molecular markers of embryogenesis in date palm since most of the molecular methods used to assess somaclonal variations have shown limitations. Indeed, cytogenetic analysis has not shown any alterations in genome structure, and isozyme markers are subjected to large ontogenic variation. Molecular markers are able to investigate only a small part of the genome and they are useless in the case of epigenetic changes. Very limited work has been carried out on the cryopreservation of date palm cultivars in Iraq with special reference to embryogenic cultures, and therefore the development of innovative procedures is needed for efficient preservation methods.

4.5 Cultivars Identification

Date palm is considered a crop of great socioeconomic importance in the Arabian region. The tree has been, and still is, at the center of the region's comprehensive agricultural development. It is believed that the date palm tree originated in Mesopotamia (Wrigley 1995). The numbers of known date palm cultivars that are distributed all over the world may be as high as 3,000, out of which about 600 are found in Iraq. Wars and economic sanctions imposed on Iraq have negatively affected both the production and natural genetic diversity of the crop and inhibited the much needed impetus to rebuild the date palm industry. The collection and genetic characterization of all Iraqi cultivars using different genetic markers is one of the important goals in date palm rehabilitation efforts. During the last few years, the General Board of Date Palm, Ministry of Agriculture, has collected over 510 date palm cultivars and planted them in 30 date palm stations dispersed in the south and middle parts of the country. Such a number of genotypes reflect the much-needed efforts to get powerful, accurate, and practical genetic markers that can be used in a high-resolution mapping for these cultivars. The success of any plant genetic conservation or breeding program depends on understanding the amount and distribution of the genetic variation present in the genetic pool. This understanding will be helpful in:

(a) Gathering date palm cultivars in newly established orchards using genetically diverse cultivars to avoid genetic vulnerability to various biotic and abiotic stresses
(b) Selection of the diverse parents in combination with the aim of segregating progenies with genetic variability
(c) Providing further gain for selection and molecular mapping

4.5.1 Research Progress in Morphological Descriptors

Morphological traits have been used to describe genetic variation in date palm cultivars, which are mainly related to the fruit, leaf, trunk, and other parts of the tree. An early survey of Iraqi cultivars using morphological description was done by Al-Baker (1962); he distinguished 531 cultivars and their origin, availability, and fruit quality. Another morphological identification was done on 110 Iraqi cultivars by Al-Jboory et al. (1971). In 1972, Al-Baker again described 627 Iraqi cultivars in his celebrated book in Arabic, *The Date Palm*, adding some fruit traits to the earlier description. Three decades later, Husien (2002) and Husien and Greab (2004) also gave a brief description of 36 and 50 cultivars, respectively. Al-Saleh and Al-Ansary (2005) adopted 12 phenotypic traits to describe 110 cultivars, documented with color photos. Genetic identification of date palm cultivars using morphological markers is usually not possible until fruits are produced and frequently requires a large set of phenotypic data that is often difficult to assess and is sometimes variable due to environmental influences.

4.5.2 Research in Molecular Descriptors

The development of molecular tools has changed the way in which individual cultivars can be identified and useful information concerning the genetic control of many agronomic characteristics can be analyzed. The ability to apply these molecular tools depends to some extent on the amount of other genomic information available for the specific plant species. Molecular marker technologies involve the use of isozymes followed by a series of DNA marker technologies and, most recently, by possibilities to compare complete genomes (Cullis 2011). Development of suitable DNA molecular markers for this crop may allow researchers to estimate genetic diversity, which will ultimately lead to the genetic conservation of date palm. The success of particular genetic conservation or breeding program is dependent on understanding the amount and distribution of the genetic variation already present in the genetic pool (Jubrael et al. 2005). In addition, biochemical studies including

Table 4.7 Names, gender and sources of collection of 30 date palm cultivars grown in Iraq

Cultivar	Gender	Cultivar	Gender	Cultivar	Gender
Collection at Al-Mahaweel date palm station					
Usta Umran	Female	Buliani	Female	Chipchab	Female
Tebarzal	Female	Leelwi	Female	Zahidi	Female
Um Al-Dihen	Female	Shwethi Ahmer	Female	Shwethi Asfar	Female
Guntar	Female	Jamal Al-Dean	Female	Khadrawy	Female
Khestawi	Female	Qul Husaini	Female	Baw Adem	Female
Bream	Female	Halawy	Female	Dayri	Female
Ashrasi	Female	Qitaz	Female	Barhi	Female
Maktoom Asfar	Female		Female		Female
Collection at Al-Zaafarania date palm station					
Meer Haj	Female	Ghanami Akhder	Male	Smeasmi	Male
Um Al-Blaliz	Female	Ghnami Ahmer	Male	Ghulami	Male
		Khekri	Male	Greatli	Male

isozyme and activity analyses of peroxidases have been used to characterize date palms in Morocco and Tunisia (Baaziz 1988; Baaziz and Saaidi 1988; Bendiab et al. 1998; Majourhat et al. 2002; Ould Salem Mohamed et al. 2001). As such, analysis does not reflect precisely the polymorphisms which have occurred (Al-Jibouri and Adham 1990).

DNA marker analysis in Iraqi date palm is still at the developmental stage. Employment of DNA marker to identify cultivars was started by Jubrael (2001) at the IPA Center for Agriculture Research, Baghdad. Random amplified polymorphic DNA (RAPD) markers were used to identify nine female cultivars, whereas (Al-Khateeb et al. 2001) used the same analysis to identify eight male cultivars. Amplified fragment length polymorphism (AFLP) markers were also been used for genetic fingerprinting of 18 Iraqi date palm cultivars (Jubrael et al. 2005). Microsatellite markers were also employed to assess genetic diversity in 30 Iraqi date palm cultivars (Khierallah et al. 2011b). As a member of the research team, Khierallah also used eight Iraqi date palm cultivars to evaluate the activity of over 1,000 simple sequence repeat (SSR) primer pairs developed by Hamwieh et al. (2010), mining genome sequencing data for this vital crop. This work was awarded the Khalifa International Date Palm Award for research in 2012. Recently, inter-simple sequence repeat (ISSR) markers were adopted to estimate genetic relationships among 17 Iraqi date palm cultivars (Khierallah et al. 2014).

4.5.3 Molecular Strategies for Genetic Identification

Leaves were collected from 30 well-defined reference Iraqi date palm cultivars grown in date palm stations of the Ministry of Agriculture (Table 4.7). A total of 22 female cultivar samples were collected from Al-Mahaweel Date Palm Station, Hilla governorate, 80 km south of Baghdad. The remaining eight cultivar samples were collected from Al-Zaafarania Station, Baghdad. Total genomic DNA was extracted from young and healthy leaves according to the procedure described by Benito et al. (1993) with minor modifications. After purification, the resultant DNA was quantified on 1 % agarose gel electrophoresis as described by Sambrook et al. (1989).

4.5.3.1 AFLP Analysis

According to Khierallah et al. (2011a), AFLP analysis was carried out for 18 of these 30 cultivars by following the method of Vos et al. (1995), with a few modifications. The total number of bands and those exhibited polymorphism were scored visually.

A total of 83 polymorphic AFLP fragments were detected with an average of 13.8 polymorphic fragments/primer combinations. Genetic distance was estimated using Jaccard's genetic similarity index and ranged from 0.07 to 0.75. Unweighted pair group method with an arithmetic mean UPGMA ordered date palm cultivars into two main clusters independent of their origin and sex. The first cluster consisted of three subclusters. The first one consisted of five female cultivars and one male, while the second subcluster consisted of five male cultivars. The third one consisted of five cultivars; four were females and one male. The second main cluster consisted of the remaining two female cultivars. Moreover, all primer combinations contributed to the discrimination of date palm cultivars, suggesting the efficiency of AFLP method in assessing genetic diversity in date palm. DNA fingerprinting techniques have an advantage in that the DNA content of a cell is independent of environmental conditions, organ specificity, and growth stage (Ainsworth et al. 1996). None of the diagnostic marker techniques so far applied to organisms has fulfilled all of the requirements in terms of cost and ease of use in cultivar identification. However, AFLPs satisfy more conditions than any other technique and are becoming the tool of choice for many applications, and the advantage of AFLP is that its utility can be assessed with a small number of primer pairs that can be extended for more studies.

AFLP is a powerful DNA fingerprinting technique that uses polymerase chain reaction (PCR) to amplify a limited set of DNA fragments from a specific DNA sample (Blears et al. 1998; Vos et al. 1995). The reliability of the restriction fragment length polymorphism (RFLP) technique is combined with the power of PCR. AFLP marker study showed that all primer combinations used in this study were effective in distinguishing date palm cultivars when used individually, revealing high level of polymorphism (Tables 4.8 and 4.9). Jaccard's genetic similarity index showed clusters consisted of five male cultivars. Two of them, Ghnami Ahmer and Ghanami Akhder, were very closely related with 85 % similarity reflecting high similarity in their morphological traits (Ibrahim 2008). Results also showed that there is large genetic diversity among the studied date germplasm. Jubrael et al. (2005) reported that high level of intervarietal polymorphism among another 18 Iraqi date palm cultivars could be partly due to the strong outcrossing mechanism in this species, which is likely to increase the degree of polymorphism.

Table 4.8 Number of fragments amplified, polymorphic bands, primer efficiency, and discrimination power of six primer combinations used for AFLP analysis

Primer combination	Total fragment (no.)	% Primer efficiency	No. polymorphic fragment	% Polymorphism	% Discrimination power
P11-aacg/ M88-tgc	63	25	12	19	14
P104-aagc/ M95-aaaa	33	13	7	21	8
P74-ggt/ M95-aaaa	46	18	26	57	31
P11-aacg/ M95-aaaa	31	12	5	16	6
P293-ggt/ M62-tgc	45	18	21	47	25
P101-aacg/ M95-aaaa	34	13	12	35	14
Total	252		83 (33 %)		

Source: Khierallah et al. (2011a)

Table 4.9 Major allele frequency, gene diversity, and polymorphism information content (PIC) estimated by AFLP markers in 18 date palm cultivars of Iraq

Primer combination	Major allele frequency	Range of gene diversity	Average of gene diversity	Range of PIC	Average of PIC
P11-aa/M88-tgc	0.75	0.10–0.48	0.29	0.10–0.36	0.23
P104-aagc/ M95-aaaa	0.78	0.10–0.48	0.29	0.10–0.36	0.23
P74-ggt/ M95-aaaa	0.72	0.10–0.50	0.30	0.10–0.38	0.24
P11-aa/M95-aaaa	0.75	0.10–0.49	0.30	0.10–0.37	0.24
P293-taca/ M62-ctt	0.72	0.10–0.50	0.30	0.10–0.38	0.24
P101-aacg/ M95-aaaa	0.70	0.20–0.50	0.35	0.18–0.38	0.28
Mean			0.31		0.25

Source: Khierallah et al. (2011a)

Although Iraqi date palm cultivars have been grown in Iraq for many years, Jaccard's similarity index and principal component analysis (PCA) revealed diverse relationships among them. In addition, the topology of the dendrogram and the distribution of cultivars by PCA analysis showed that a typically continuous genetic diversity characterizes Iraqi date palm germplasm (Fig. 4.7). In fact, the cultivars were clustered independently of their geographic origin in spite of their phenotypic distinctiveness.

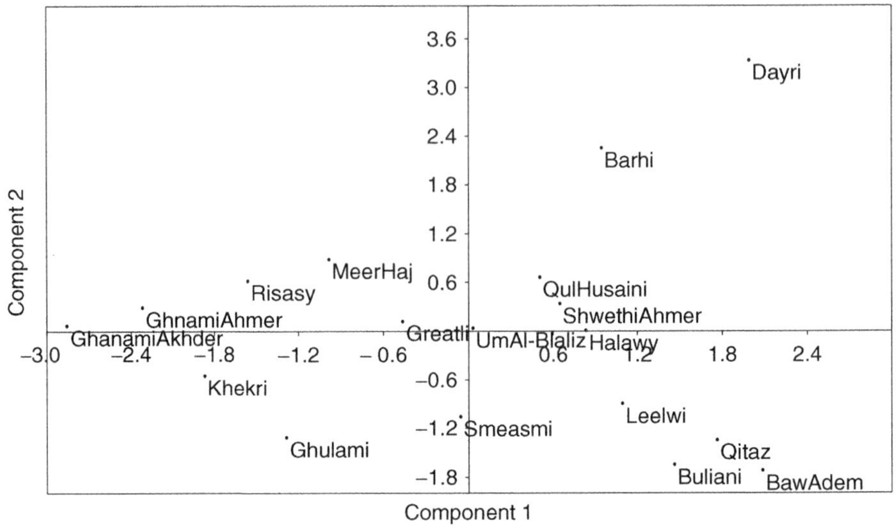

Fig. 4.7 Principle component analysis of the 18 date palm cultivars estimated by AFLP marker

4.5.3.2 Microsatellites Analysis

Microsatellites, SSRs, are ideal DNA markers for population studies and genetic mapping due to their abundance, high level of polymorphism, dispersion throughout diverse genomes, ease to assay by PCR, and ease to disseminate among laboratories (Udupa and Baum 2001). Microsatellite markers were used for investigating genetic diversity in date palm (Billotte et al. 2004). Zehdi et al. (2004) applied these markers to characterize Tunisian cultivars, while Al-Ruqaishi et al. (2008) utilized these primers to screen and analyze the genetic diversity among clonal genotypes of Omani cultivars. Recently, the technique spread to other date palm-producing countries. Elshibli and Korpelainen (2008) investigated genetic diversity in Sudan germplasm representing 37 female and 23 male accessions using 16 SSR primers. In Qatar, Ahmed and Al-Qaradawi (2009) employed 10 primers to analyze genetic diversity among 15 cultivars. Akkak et al. (2009) developed further 17 microsatellite markers, whereas Hamwieh et al. (2010) developed 1,000 SSR markers for date palm. A total of 33 date palm-specific primer pairs were tested, 16 of them were developed by Billotte et al. (2004) and 17 were developed by Akkak et al. (2009) as indicated in Table 4.10.

These microsatellite markers were employed to assess genetic diversity in 30 well-known Iraqi date palm cultivars (Khierallah et al. 2011b). Electropherogram data were analyzed by using Foundation Data Collection software (Genetic Analyzer Data Collection, Version 2.0, Applied Biosystems 3100). Allele size scoring was performed by GeneMapper software (GeneMapper® Software Version 3.7, Applied Biosystems, Carlsbad, California). The Jaccard similarity matrix (Jaccard 1908)

Table 4.10 Major allele frequency, number of genotypes that showed polymorphic bands, number of alleles generated, heterozygosity, gene diversity, and polymorphism information content (PIC) estimated by 22 SSR markers in 30 Iraqi date palm cultivars

Marker	Major allele frequency	Genotypes (no.)	Alleles (no.)	Heterozygosity	Gene diversity	PIC
mPdCIR010	0.204	19	14	0.852	0.890	0.880
mPdCIR016	0.500	8	5	0.529	0.654	0.602
mPdCIR025	0.224	15	8	0.690	0.822	0.798
mPdCIR032	0.333	13	5	0.815	0.772	0.736
mPdCIR035	0.481	10	7	0.519	0.695	0.659
mPdCIR050	0.370	17	12	0.926	0.813	0.797
mPdCIR057	0.909	3	3	0.136	0.168	0.160
PdCIR070	0.533	3	3	0.000	0.604	0.536
mPdCIR078	0.320	16	15	0.760	0.832	0.816
mPdCIR085	0.313	8	7	0.250	0.779	0.746
mPdCIR090	0.391	10	9	0.261	0.781	0.757
mPdCIR093	0.741	6	6	0.296	0.434	0.415
PDCAT4	0.407	13	7	0.556	0.738	0.701
PDCAT 5	0.276	13	7	0.862	0.816	0.791
PDCAT 6	0.231	21	21	0.654	0.903	0.897
PDCAT 11	0.600	7	7	0.100	0.598	0.566
PDCAT 12	0.850	5	4	0.100	0.269	0.256
PDCAT 14	0.220	19	10	0.960	0.865	0.850
PDCAT 15	0.440	8	6	0.480	0.732	0.699
PDCAT 17	0.385	11	9	0.500	0.774	0.746
PDCAT 18	0.140	19	19	0.560	0.923	0.918
PDCAT 21	0.731	6	4	0.269	0.431	0.394
Mean	0.436	11.364	8.545	0.503	0.695	0.669

Source: Khierallah et al. (2011b)

was used for cluster analysis using the unweighted pair group method arithmetic average (UPGMA) to study the genetic relationships among the cultivars. Jaccard similarity index, major allele frequency, heterozygosity, gene diversity, and polymorphism information content (PIC) estimation were done using a software package (PowerMarker Version 1.31) (Liu and Muse 2005). The phylogenetic diagram was drawn by PAST software Version 1.91 (Hammer et al. 2001) on the basis of the Hamming similarity index with 100 bootstrap (Fig. 4.8). PCA was performed according to Euclidean similarity index using the PAST software (Fig. 4.9). The 33 primer pairs of Billotte et al. (2004) and Akkak et al. (2009) were tested for their ability to generate expected SSR banding patterns in Iraqi date palms. A total of 22 primers successfully showed polymorphic bands among the 30 cultivars (24 female and 6 male) (Table 4.10). The genetic diversity was widely varied among cultivars ranging from 0.168 to 0.923 at loci mPdCIR057 and PDCAT 18 estimated number of 600 cultivars.

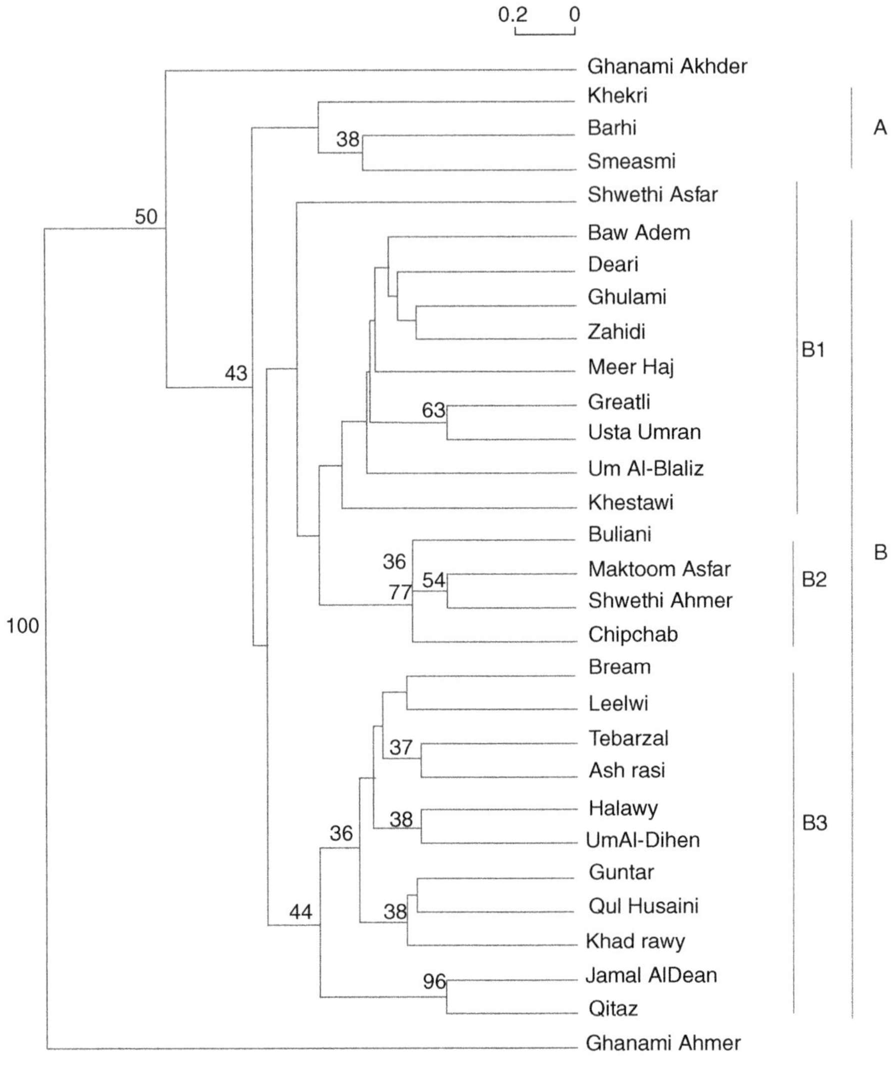

Fig. 4.8 Genetic relationships among 30 Iraqi date palm cultivars based on 22 microsatellite loci with 100 bootstrap (*Source*: Khierallah et al. (2011b))

Discrimination between cultivars is based on phenotypic differences, which are often influenced by the environment. In this study, microsatellite genotyping was used to display the genetic diversity and relationships among 30 widely grown cultivars in Iraq. Although the frequency of microsatellites varies among species, they are abundant and dispersed throughout the genome and show higher levels of polymorphism than other genetic markers. These features, coupled with their ease of detection, suggest their use as molecular markers. Their potential for automation and inheritance in a codominant manner are additional advantages when compared with other molecular tools (Holton 2001).

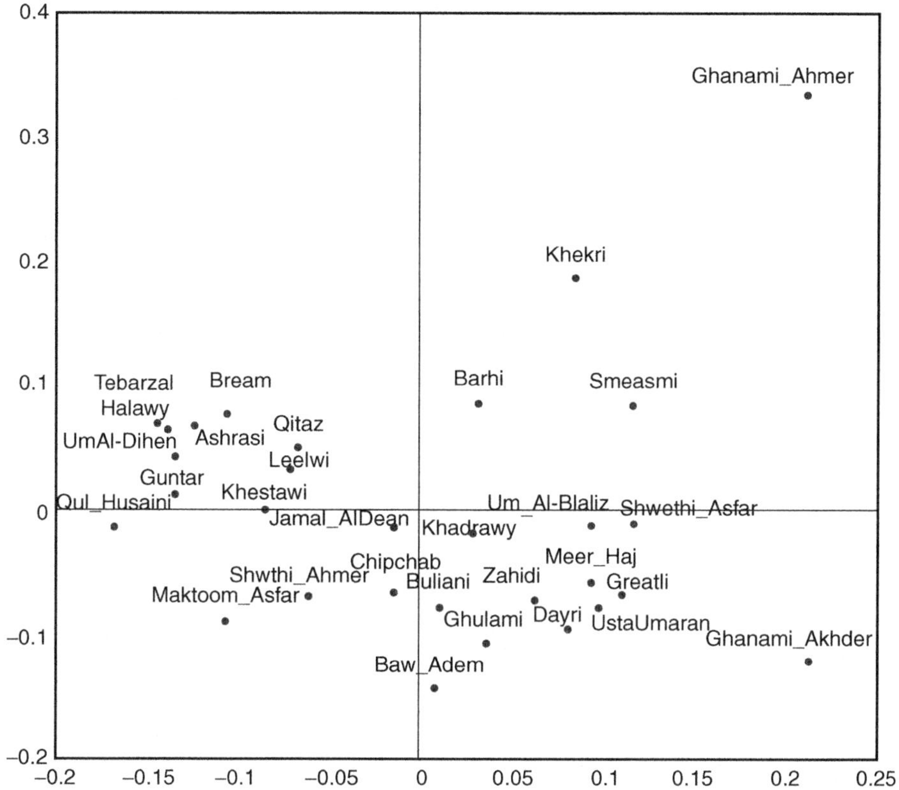

Fig. 4.9 Principal coordinate analysis (PCA) of 30 Iraqi date palm cultivars based on 22 microsatellite loci

The microsatellites examined in this study showed highly polymorphic patterns with a great number of alleles (188) distributed in 30 date palm cultivars. The number of detected alleles per locus in this study (8.54) was higher than 7.6 alleles per locus scored by Zehdi et al. (2004) in 46 date palm cultivars cultivated in Tunisia and those studied by Ahmed and Al-Qaradawi (2009) representing 15 cultivars grown in Qatar. The mean heterozygosity value detected in the Iraqi cultivars was 0.503, indicating the presence of high genetic diversity. Date palm growers believe that Ghanami Akhder and Ghnami Ahmer are two clones belonging to the same cultivar. However, the present investigation revealed high divergence between those two cultivars, suggesting they are independent cultivars and confirming the results of Al-Khateeb and Jubrael (2006) that used RAPD markers for the same purpose. Jubrael et al. (2005) and Khierallah et al. (2011b) suggested a common genetic basis among date palm genotypes in spite of the differences in fruit characters and tree morphology. Other authors reported similar results in Tunisian date palms (Sedra et al. 1998; Zehdi et al. 2004).

The long history of date palm domestication of an unknown origin (Wrigley 1995) and the nature of date palm culture may have played an important role in the composition of date palm genome (Elshibli and Korpelainen 2008). New cultivars may appear as a result of the continuous selection carried out by farmers following sexual reproduction. Exchange of propagules, which are a mixture of vegetative and seed-propagated materials, is conducted between farmers. All these processes together may result in a mixed genome within the same country (Elshibli and Korpelainen 2008). SSR analysis showed a high level of polymorphism among Iraqi cultivars. However, cultivar nomenclature and classification still remains based on fruit characters, including morphological, physical, and chemical traits.

4.5.4 Prospective of Germplasm Identification

Many Iraqi date palm cultivars collected from date palm stations are currently now at a juvenile growth stage. They will take some years to reach fruiting stage and be characterized by genetic markers. In addition, many farmers hesitate to give offshoots or even information about their cultivars. Undoubtedly, this will restrict cultivar identification efforts.

Iraqi date palm germplasm requires further investigation and adoption of proper markers to assist in identifying the economically and agronomically important cultivars. RAPD, AFLP, SSR, and ISSR markers are powerful methods to discriminate date palm genotypes and to assess genetic diversity in this fruit crop. Obviously, this would be enhanced by using more primer sets and/or a larger number of female and male cultivars. Work is currently in progress to find DNA markers linked to agronomic traits. Such markers would be important to assist in the selection and improved cultivation of this historic tree. The network (www.Iraqi-datepalms.net) was established in 2007, created and managed by Ibrahim Al-Jboory; it targets the collection of all scattered information on Iraqi date palm archives, past and present. The number of visitors to the website has exceeded one-half million. This website is a reference for researchers, farmers, and industry from different parts of the world, providing they are able to read Arabic.

4.6 Cultivars Description

4.6.1 Growth Requirements

The date palm tree is adapted to climates of long dry summers and mild winters. It has unique characteristics to thrive in deserts and oases where temperature are high but with underground water close to the surface (Al-Baker 1972). Under these circumstances the tree is described as having its feet in running water and its head in the fire of the sky.

Date fruit production and full ripening is dependent on the availability of certain heat requirements, according to cultivar. Most dry fruit cultivars are found in the dry areas, whereas soft and semidry ones are confined to more humid and semidry areas. During winter, temperatures may go below 0 °C. The low vegetative growth point of a date palm is 7 °C, and at above that level, growth is active and reaches an optimum at about 32 °C (Zabar and Borowy 2012). The growth rate remains continuous and stable until the temperature reaches 38–40 °C when it starts decreasing.

When the temperature drops for a certain period to below 0 °C, it causes metabolic disorders which lead to partial or total damage to leaves. At −6 °C, pinna margins turn yellow and dry out. Inflorescences are also heavily damaged by frost. When a frost period is forecast, inflorescences should be protected with craft paper bags, put in place immediately after pollination. At temperatures of −9 to −15 °C, leaves at the middle and outside of the tree canopy will be damaged and dry out (Zabar and Borowy 2012). If these low temperatures continue for a long period (12 h to 5 days), all leaves will exhibit frost damage and the tree will look as if it was burnt (Dowson 1982).

The date thrives in sandy, sandy loam, clay, and heavy soils but requires good soil drainage and aeration. The tree is remarkably tolerant of alkali soils. A moderate degree of salinity is not harmful but excessive salt stunts growth and lowers the quality of the fruit (Al-Baker 1972).

4.6.2 Cultivars Distribution and Production Statistics

About 600 date cultivars were grown in Iraq before 1980; however, the number has been reduced to about 500 cultivars now. Cultivars can be identified by many characteristics such as fruit appearance, texture, size, and shape. Fruits are of three general types: soft, semidry, or dry. The type of fruit depends on the glucose, fructose, and sucrose content. The classification is based on the texture or consistency of fruit under normal conditions of ripening. Soft dates are distinguished by soft flesh, high moisture, and high sugar content. Semidry dates feature a firm flesh, fairly low moisture, and high sugar content. Dry dates are characterized by high sugar content and low moisture.

Many Iraqi cultivars are considered to be commercial, such as Zahidi, Halawy, Sayer, Khadrawy, Khastawi, and Dayri. The distribution of cultivars across the date-producing provinces is detailed in Table 4.11 and illustrated in Fig. 4.10. Percentages of the leading cultivar production are shown in Table 4.12. It is clear that Zahidi occupies first place with 57.4 %, while other cultivar production is much lower. The distribution of cultivars over Iraqi date-producing provinces is shown in Table 4.13. Babylon governorate comes in first in production; others fluctuate in their yield and vary in the type of cultivar.

Table 4.11 Iraqi dates production by cultivars and provinces for the year 2012

Governorate	Grand total	Production (mt)						
		Zahidi	Dayri	Halawy	Khadrawy	Sayer	Khastawi	Others
Kirkuk	115	48	2	1	16	–	3	45
Diyala	83,792	47,327	4,986	256	2,048	3,499	10,511	15,165
Al-Anbar	43,196	34,248	14	11	164	31	7,568	1,160
Baghdad	93,782	56,963	323	274	1,717	248	11,467	22,790
Babylon	100,343	76,081	1,829	219	1,286	593	12,710	7,625
Kerbela	72,217	56,140	320	61	518	98	9,534	5,546
Wasit	43,649	24,063	370	565	1,736	285	6,588	10,042
Salah Al-Deen	19,734	10,796	67	504	368	188	2,756	5,055
Al-Najaf	32,304	23,063	246	424	1,466	791	3,724	2,590
Al-Qadisiyah	31,436	24,227	293	380	916	689	1,357	3,574
Al-Muthanna	25,275	10,078	1,739	968	2,027	1,366	1,508	7,589
Thi Qar	40,086	7,800	4,541	582	10,193	5,954	1,421	9,595
Maysan	8,967	1,733	612	135	2,320	1,902	656	1,609
Basrah	60,554	3,589	4,487	19,803	4,366	9,008	1,684	17,617
Total	655,450	376,156	19,829	24,183	29,141	24,652	71,487	110,002

Source: Central Statistical Organization (CSO) (2012), Iraq

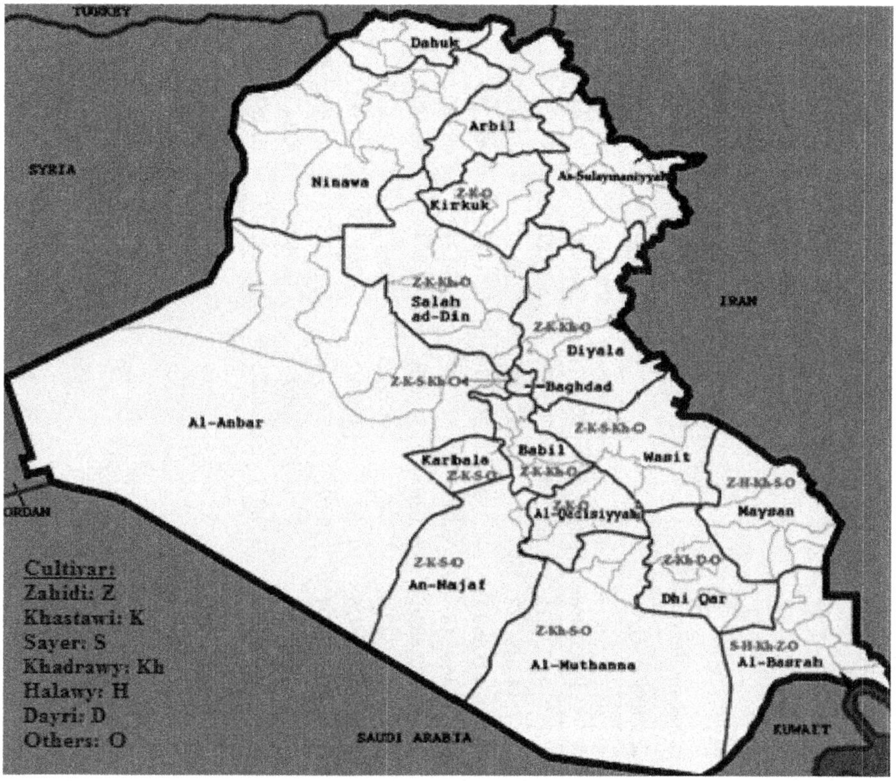

Fig. 4.10 Distribution of date palm cultivars in Iraq

4.6.3 Nutritional Aspects

The high nutritional value of dates is well documented; it has been said that dates can be a full meal since they contain reasonable quantities of macro- and microelements essential for human health. Minor variations in sugar content occur among cultivars, with a few exceptions (Table 4.14). Zahidi cultivar contains almost three times the amount of sucrose as compared with others, while Sayer contains approximately four times the amount of ascorbic acid than other studied cultivars. Other components exhibit similar trends. Generally, dates represent a rich source for primary and secondary metabolites especially with vitamins (Table 4.15).

4.6.4 Cultivars Description

Cultivars can be classified by characteristics of fruit appearance and texture and fall into three types:

(a) Soft dates: These are distinguished by a soft flesh, high moisture content (>30 %), and high sugar content; main cultivars belonging to this type are Maktoom, Barhi, Khyara, Shwethi, and Leelwi.

Table 4.12 Date palms, average yield, and production by cultivars grown in Iraq for the year 2012

Cultivars	%	Production (mt)	Average yield kg/palm		Palms in production stage			Planted in 2012	Young palms	Total palms
			In stage of production	Productive	Total	Productive	Not productive			
Zahidi	57.4	376,156	73.3	77.2	5,134,681	4,872,271	262,410	206,471	625,271	5,966,423
Khastawi	10.9	71,487	67.7	67.7	1,055,577	1,055,577	–	95,211	210,232	1,361,020
Sayer	3.8	24,652	55.3	60.8	446,028	405,669	40,359	247,276	210,431	903,735
Khadrawy	4.4	29,141	58.1	62.2	501,861	468,848	33,013	51,774	130,242	683,877
Halawy	3.7	24,183	58.6	63.2	412,903	382,883	30,020	58,527	98,444	569,874
Dayri	3.0	19,829	64.2	64.9	308,898	305,320	3,578	407,815	259,822	976,535
Others	16.8	110,002	63.4	63.7	1,735,623	1,725,926	9,697	1,366,373	1,201,136	4,303,132
Total	100.0	655,450	68.3	71.1	9,595,571	9,216,494	379,077	2,433,447	2,735,578	14,764,596
Males		–	–	–	481,736	481,736	–	9,109	32,969	523,814
Total		–	–	–	10,077,307	9,698,230	379,077	2,442,556	2,768,547	15,288,410

Source: Central Statistical Organization (CSO) (2012), Iraq

Table 4.13 Date palms, average production allocated by cultivars and provinces for the year 2012

		Average yield						
	Grand total	Zahidi	Dayri	Halawy	Khadrawy	Sayer	Khastawi	Others
Kirkuk	65.6	79.3	60.6	71.4	71.1	–	49.2	55.3
Diyala	63.6	62.5	62.0	68.0	69.0	57.0	62.5	70.0
Al-Anbar	82.1	85.3	69.7	74.8	67.8	77.1	72.0	72.0
Baghdad	67.5	78.5	61.3	62.7	58.0	57.1	60.5	53.0
Babylon	78.6	83.9	60.5	73.9	63.0	70.0	64.5	69.0
Kerbela	58.5	56.3	47.0	56.6	55.0	52.2	70.0	67.0
Wasit	87.4	90.7	64.0	72.0	77.0	86.1	99.5	78.0
Salah Al-Deen	90.6	95.2	82.6	88.0	89.9	91.0	94.0	81.0
Al-Najaf	68.0	68.5	58.0	69.0	65.0	64.0	67.0	69.0
Al-Qadisiyah	75.1	83.2	58.1	55.1	52.0	50.0	55.0	60.0
Al-Muthanna	57.9	69.6	48.6	58.0	45.2	51.0	56.0	54.0
Thi Qar	60.8	69.1	66.1	68.5	56.3	59.7	63.1	58.0
Maysan	63.9	75.2	53.2	59.3	62.9	62.5	60.8	63.2
Basrah	60.4	61.5	83.0	57.2	54.2	49.8	72.0	68.0
Total	68.3	73.3	64.2	58.6	58.1	55.3	67.7	63.4

Source: Central Statistical Organization (CSO) (2012), Iraq

Table 4.14 Nutritional value of four Iraqi date palm cultivars

Components	Halawy	Sayer	Khadrawy	Zahidi	Mean
Element (mg.kg^{-1})					
P	160	130	150	140	145.0
K	8,540	8,330	8,940	8,870	8,670.0
S	100	200	140	210	162.5
Ca	1,840	2,030	1,330	2,070	1,357.5
Mg	560	580	600	590	582.5
Cl	2,600	3,120	2,660	3,420	2,950.0
Fe	52.6	32.1	45.0	103.7	58.4
Mn	58.6	52.5	51.4	51.6	53.5
Cu	27.7	28.9	25.4	27.5	27.4
Zn	13.9	18.2	12.9	7.4	13.1
Co	7.6	9.6	9.6	9.5	9.1
Fe	2.0	1.2	1.4	1.2	1.5
Sugar (% d.w.)					
Total sugar	87.9	86.1	87.7	86.8	87.1
Reducing sugar	82.7	82.6	81.9	73.4	80.2
Sucrose	4.8	3.5	4.5	12.7	6.4
Glucose	43.7	44.8	44.8	32.8	41.5
Fructose	37.2	38.0	38.0	39.2	38.23
Vitamins (mg. 100 g^{-1} d.w.)					
Thiamine	99.0	130.0	94.0	80.0	100.8
Riboflavin	173.0	135.0	149.0	167.0	156.0
Biotin	4.6	4.7	4.1	5.7	4.8
Folic acid	57.0	70.0	43.0	63.0	58.2
Ascorbic acid	3.6	17.5	3.2	2.4	6.7
Other components					
Moisture (% f.w.)	7.3	7.5	9.5	8.3	8.1
Total soluble solid (% f.w.)	84.2	81.3	80.8	82.1	82.1
Total insoluble solid (% f.w.)	17.9	10.0	9.5	9.2	11.7
Protein (% d.w.)	2.3	2.8	2.4	2.2	2.4
Fat (% d.w.)	0.5	0.3	0.5	0.4	0.4
Ash (% d.w.)	1.9	1.8	2.1	1.8	1.9
Crude fiber (% d.w.)	1.8	1.7	2.3	2.5	2.1

Source: Yousif et al. (1982)

(b) Semidry dates: These have a firm flesh, fairly low moisture content (20–30 %), and high sugar content; this type includes Zahidi, Dayri, and Khadrawy.

(c) Dry dates: These have a high sugar and low moisture (<20 %) content; their flesh is dry and hard; this type includes Banafsha and Bedray.

Detailed descriptions of the most important cultivars grown in Iraq are presented in Table 4.16 and Fig. 4.11, showing the variability in fruit morphology of the best-known Iraqi cultivars.

Table 4.15 Chemical composition of dates in 100 g fresh weight

Component	Content (%)	Minerals	Content (mg)
Moisture	18	Ca	167
Total sugar	80	P	13.8
Sucrose	5.9	K	798
Glucose	39	S	14.7
Fructose	35	Na	10.1
Total soluble solid	82	Cl	2,71
Total insoluble solid	12	Mg	53.3
pH	6.0	Fe	5.3
Protein	2.2	Mn	4.1
Fat	0.37	Cu	2.4
Ash	1.7	Zn	1.2
Fiber	1.9	Co	0.9
		Fe	0.13
Vitamins	Content (μg)		
Thiamine	93		
Riboflavin	144		
Biotin	4.4		
Folic acid	5.3		
Niacin	5.3		
Ascorbic acid	61,000		

Source: Yousif et al. 1982

Table 4.16 Characteristic of the most important date palm cultivars grown in Iraq (Al-Baker 1972)

Cultivar	Description
Zahidi	Semidry date, native to Iraq. Medium size, cylindrical, light golden brown, very sugary; sold as soft, medium-hard, and hard. Distinguished by its large seed in proportion to the fruit itself. This date lends itself well to processing and softening by steam hydration. This date is known for its high invert sugar level and is widely used to make diced dates and date sugar products. It features a crunchy and fibrous flesh. Industrial uses
Sayer	The most prevalent cultivar in Basrah, dry, dark orange-brown, medium size, soft, and syrupy; oval curved to one side; industrial uses. Originates in Iraq
Halawy	Soft. Extremely sweet, small to medium in size. Thick flesh, caramel taste, is somewhat wrinkled in appearance, with a yellow color ripening to a light amber and then to a golden brown. Originally from Iraq. Fresh or industrial markets
Khastawi	Leading fresh date in Iraq; it is syrupy and small in size, prized for dessert. For fresh market
Khadrawy	Semidry, very dark date. Originally from Iraq, it has many desirable qualities. It cures well; it ripens to amber and is then cured to reddish brown, with a caramellike texture and a sweet flavor. Industrial uses for export and fresh consumption locally
Barhi	Nearly cylindrical, light amber to dark brown when ripe; soft, with thick flesh and rich flavor; of superb quality. For shipment needs refrigeration as soon as picked, then curing and special packing. Native to Iraq

Fig. 4.11 Variations in fruit morphology of common cultivars of date palm in Iraq (*Source*: www.Iraqi-datepalms.net)

4.7 Dates Production and Marketing

4.7.1 Practical Production Approaches

4.7.1.1 Pruning

To optimize date yield, several cultivation practices are followed in the groves. Pruning is an important agricultural practice to remove dry leaves and leaf bases, as well as fiber, spines, and high offshoots. Pruning reduces insects and the spread of diseases. It also facilitates workers when they maneuver to perform other basic agricultural practices such as pollination, thinning, pulling down bunches, and bagging. Moreover, it enhances lighting thus decreasing the humidity around the bunches. Pruning is usually carried out once a year after harvesting, along with pollination or when the bunches are pulled down. This operation can be done manually or mechanically using a hydraulic lift and/or ladder.

4.7.1.2 Bunch Management

For effective date palm management, three practices should be carried out on bunches: thinning, bending, and pulling down bunches and bagging.

Thinning is the first practice and it provides more nutrients to a fewer number of fruits. It increases the fruit size, enhances the quality, prevents delay in ripening, and reduces the weight and compactness of the fruit bunch. Thinning also decreases fungal infection and mechanical damages. This will benefit both harvesting and packing operations. There are three different methods of thinning: the first reduces the number of strands from the central part of each bunch, the second reduces the number of bunches per palm, and the third reduces the number of fruits per strand. Thinning can be also carried out either by certain chemicals, by using diluted pollen grains, or by mechanical means (Soliman et al. 2010).

Bending and pulling down bunches is the second important management practice. Most Iraqi date palm orchards carry out this process (Fig. 4.12). The bunch is bent over the leaves and the fruit stalk is tied to the midrib of one of the lower leaves. Pulling down bunches to release them from among the leaves should be done carefully to avoid breaking the fruit stalk. Bunches should not be pulled down until the fruit stalk is thick and sufficiently long. This process facilitates harvesting and bagging.

Bagging is the third practice relying on covering of fruit bunches with a certain type of paper or cloth bag. These bags protect fruit from dust and sunburn, while also decreasing the loss or damage from insects or birds (Al-Obeed and Harhash 2010; Kassem et al. 2010).

Fig. 4.12 Bunches bent down in date palm

4.7.2 Date Palm Fruit Storage and Packing

Date fruits are an important and nutritional food as they contain high percentages of sugars, minerals, and vitamins. Date fruits can spoil easily, in particular the softer cultivars; care in handling must be taken during harvesting, gathering, and grading according to the cultivar. Most Iraqi date cultivars are harvested by cutting the entire bunch when most of the fruits are ripe. Skilled palm grove workers cut the fruit bunches, lower them to the ground, and place them on mats. Dates are then removed from their bunch and carefully packed. Some growers fumigate fruits while they are in the field to protect them from insects and various diseases. Storage is considered an important factor for marketing to increase sales and income. Date fruits are normally stored on the farm in centrally located, cooled storage areas and/or in regular storage areas of a packing house. Usually, as part of the preparation process, date fruits are fumigated to control insect populations. Packaging is a crucial process in date fruit marketing locally and abroad and must be done according to international standards to make the product suitable for consumers and more profitable to the producer.

4.7.3 Date Palm Fruit Processing and Pressing

In Iraq, processing of date fruit into various products is still at an early stage of development. Although fruit processing can be very profitable, Iraqis are still using primitive implements and methodology. *Dibis* production, alcohol, vinegar, liquid sugar, bread yeast, and citric acid are a few examples of products derived from processed dates. Previously, Iraq had the largest packing and pressing factories for date fruits and its various products. Different types of packaging were used, including plastic, cardboard boxes, and wooden crates. In the 1990s, the number of pressing factories grew to 250; however, since 2003, this sector has suffered from severe neglect leading to the closing most of the pressing facilities. This sector, therefore, should be given high priority to reactivate this important division of date palm fruit processing (Zabar and Borowy 2012).

4.7.4 Marketing Status and Research

Date fruit marketing represents an obstacle for Iraqi producers because of current regulations as well as the additional reasons, including the following: (a) The wholesale price of dates is rather low compared with the high cost of production. This has led to the neglect of trees or poor management, and this reflects negatively on fruit quantity and quality. (b) As the standard of living in Iraq has risen dramatically during the last 5 years in Iraq, wages have increased and consequently the costs to maintain date palm groves have increased. (c) The shortage of modern machinery has negatively affected the cost input, including tree management and fruit packaging.

Despite all of that, the Iraqi government has intervened and raised the price of Zahidi cultivar fruit from IQD 190 per kg (USD 0.15) up to IQD 450 (USD 0.675), which is considered to be a promotional amount; however, it still is undervalued. Nevertheless, more attention has been paid to date growers, as reflected in an increase in production over the last 2 years. A study in 2008 by FAO, using a matrix analysis, revealed that the profitability of date growers has increased several times. Research is now underway by different specialized institutions on handling, harvesting, storage, processing, and marketing of dates. This will lead to improve date production and enable competing on the international market with better fruit quality. The private sector should invest strongly in this vital economic aspect. Government loans are available for the private sector which may encourage date processing industry in the near future.

4.7.5 Current Import and Export

Iraqi date production was low before 1993 (Fig. 4.13); however, exports represent a high proportion of the dates produced. When Iraq was under economic sanctions, most of dates were consumed locally. A general increase trend was obvious in 2002–2004, accompanied with a slight improvement in export. As stated earlier, poor fruit quality, lack of infrastructure, inferior packaging, and the slow and inefficient routines of export facilities are one of the main causes for the deterioration of date trading. Recently the government has made facilities available for use by the private sector with the hope of stimulating exports.

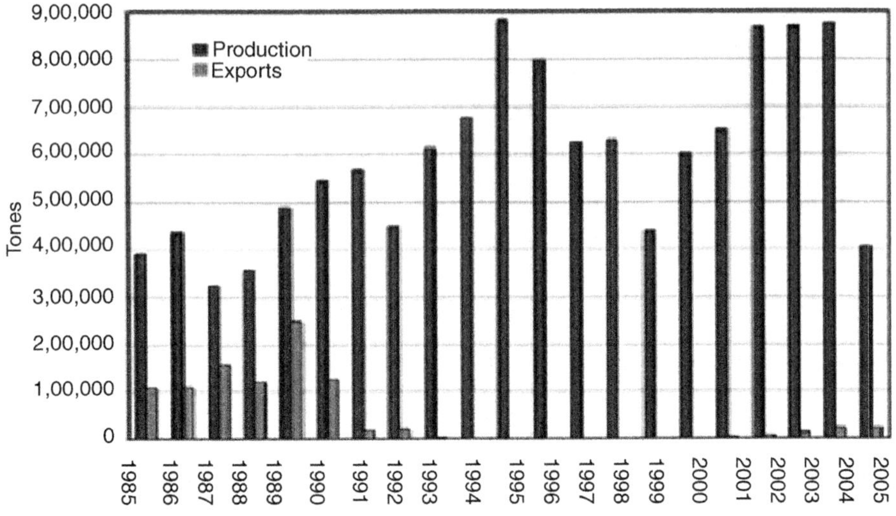

Fig. 4.13 Date production and exports (mt) in Iraq 1985–2005 (*Source*: FAOSTAT 2009)

4.8 Processing and Novel Products

4.8.1 Industrial Processing Activities

Although commercial date palm populations have declined significantly in Iraq over the last 30 years, some industrial activities are still going on. Government authorities are encouraging date growers by offering incentives, such as increasing the price of dates purchased by the government date processing factories. Zahidi is the most commonly grown cultivar across in Iraq. Its fruits are used in several manufacturing activities. *Dibis* processing factories are located in several Iraqi governorates. Some of these are run by the General Establishment of Dates, of the Ministry of Agriculture. Others are run by the private sector. The *dibis* product is consumed domestically and any surplus exported. Traditional date and tree products are fully utilized in date-producing areas especially in the southern governorates. These activities make use of all parts of the date palm tree producing furniture, folk tools, and simple raw materials for building houses.

4.8.2 Commercial Dates Processers

No current accurate survey is available on date fruit processors in Iraq, although several are distributed across the country. Date processing is the only manufacturing activity supervised by the Ministry of Agriculture and has a wide spectrum of activities including *dibis* production and fresh date packaging. Tens of factories belonging to the private sector are located in the date palm-growing regions, but most are not registered with the Ministry of Planning.

4.8.3 Secondary Metabolites

Date fruits are practically the richest in calories and nutrients among the various types of fruits. Ibrahim (2008) stated that 100 g of dates contains 274 cal, 72.9 g sugars, 2.3 g fibers, 59 mg Ca, 63 mg P, 3.0 mg Fe, and 648 mg K, although those are cultivar dependent. In addition to vitamins, dates are also rich in secondary metabolites including pectins, tannins, green pigments, carotenoids, anthocyanins, organic acids, and volatile compounds. Research in Iraq has focused on full utilization of the date palm tree, and numerous papers have been published on the subject by Iraqi researchers with special reference to extracting valuable materials for industrial, nutritional, and pharmaceutical purposes. Such scattered research work is rather difficult to summarize; however, it can be said to document the potentially wide applications in food and feed alike (Ibrahim 1998).

4.8.4 Bioenergy

Date palm is a principal agricultural crop in the Middle East and North Africa. Date palms are abundant yielding about 3,316,500 mt of palm secondary products per year including palm midribs, leaves, stems, and fronds (Taha et al. 2001). In Iraq, 16 million trees are grown in the middle and southern portions of the country, resulting there in a surplus production of dates and other secondary biomass such as leaves and sheaths.

Dates have high sugar content (55–70 %/wt.). Traditionally they are used for food or to produce sweets, *dibis*, vinegar, and alcohol products. In Iraq, a portion of the biomass waste of all these industries is used as animal feed. However, in the rural areas where local primitive *dibis* factories are located, biomass waste is usually dumped without treatment, while it could be used as a biofuel for domestic use especially for operating generators. Power generators could serve thousands of local inhabitants, save fossil fuel, create new jobs, and sequester CO_2 to keep a cleaner environment. Date waste biomass mainly consists of cellulosic compounds with some sugar, fats, and minerals. That makes it a high potential resource for biofuel production through fermentation. This potential has been recently considered in the Middle East and Arabic regions. The technology has been used in many agricultural countries such as India, China, Thailand, and Brazil since at least 2006. In Iraq the trend toward bioenergy is still in the primary stages. To the best knowledge of the authors, only a small amount of research work has actually been done on the date palm biomass as a biofuel resource thus far. In some seasons biomass waste is burned in the open or left to decay. Bioenergy may be generated from pruning remains of date palm trees. These include dry leaves and fruit pits that could be utilized to produce local thermal and electrical power in the rural areas without the need for long distribution grids and large power stations. Research work should be undertaken to investigate date palm fresh fruits, pits, and leaves for composition and biomass value assessment.

Jaafar and Baldwin (2010) studied Zahidi cv. as a source for biomass since it is the most abundant in Iraq, representing nearly 60 % of the country's production. The cellulosic, hemicellulosic structural sugars and lignin contents of Iraqi Zahidi cultivar date pits were investigated, including extractable components for functional food and nutrition. Structural sugars were measured to evaluate the date pits as a biomass resource for food or bioenergy. Total lignin was 15.32 %, mainly acid-soluble lignin, and structural sugars 59.87 %, mainly arabinose (51.6 %). This was the first study documenting such a high content of this sugar in date pits. The results reveal the high potential of date pits and leaves for biofuel production. Nevertheless, further investigations are still needed to achieve an adequate economic capacity of such potential renewable energy resources in Iraq.

The economic and environmental impacts of bioenergy production can be summarized as follows: (a) it supports local and national economies; (b) it is carbon neutral; (c) it is a cheap and renewable resource; (d) it is relatively friendly to the environment; and (e) energy can be produced locally instead of by expensive power-generating units.

4.9 Conclusions and Recommendations

Iraq is the oldest historic domestication center of date palm. Whereas the country had over 32 million date palms in the mid-twentieth century, this figure dropped dramatically to a low of approximately 12 million by 2000. The decline has come as a result of numerous conflicts in the date-producing regions since 1980. Thus, the government had adopted many programs aimed at increasing the numbers up to 40 million trees in the next 10 years. Plant tissue culture to support propagation by offshoots is essential to increase date palm numbers of desired cultivars as well as for cultivar identification using DNA marker technology. Date palm pests and diseases—dubas bug, lesser date moth (Humera), and Ghobar mites—are also responsible for the decline in a number of palms and deterioration in production of up to 50 % in some seasons. The Ministry of Agriculture, scientific institutes, and universities have conducted extensive investigations, together with FAO experts, to control these pests in Iraqi date groves. Accordingly, it is recommended that new technologies be introduced in order to improve the date production quantitatively and qualitatively. Integrated pest management is vital to reduce the potential damage in both date palm groves or during storage. Possible improvement is necessary in sorting and packaging of dates to compete with other producers and exporters to expand global date marketing.

References

Abbas EH, Abood HM (1996) Isolation and identification of date palm neck bending disease. Al-Mustansiriyah J Sci 6(1):14–16 (in Arabic)

Abbas EH, Muhee MN (1991) Relation of *Chalaropsis radicicola* and *Oryctes elegans* with date palm neck bending disease. Iraqi J Microbiol 3(1):219–221

Abdullah SK, Lopez Lorca LV, Jansson HB (2010) Diseases of date palms (*Phoenix dactylifera* L.). Basrah J Date Palm Res 9(2):15–22

Abdul-Sameed AR (2009) Effect of age of offshots and type, supplement time of explants on formation of lateral buds of date palm cv. Sayer *in vitro*, Basrah J Date Palm Res 8(1)100–110 (in Arabic)

Ahmed T, Al-Qaradawi A (2009) Molecular phylogeny of Qatari date palm genotypes using simple sequence repeat marker. Biotechnol 8:126–131

Ainsworth C, Beynon J, Buchanan-Wollaston V (1996) Techniques in plant biology: the practical manual. Wye College, University of London, London

Akkak A, Scariot V, Dorello T et al (2009) Development and evaluation of microsatellite markers in (*Phoenix dactylifera* L.) and their transferability to other *Phoenix* species. Biol Plant 53: 164–166

Al'utbi SD, Al-Husaibi FM (2007) The effect of IAA and other auxins on the growth of date palm (*Phoenix dactylifera* L.) shoot tips and callus. Basrah J Date Palm Res 6(1):1–14

Al-Abbassi SH (1987) Jumping mechanism of dubas bug *Ommatissus binotatus* de Berg. (Homoptera: Tropiduchidae). J Agric Water Res 6(1):29–38

Al-Abbassi SH (1988) Biology of *Ommatissus binotatus* de Berg. (Homoptera: Tropiduchidae) under laboratory conditions. Date Palm J 62:412–423

Al-Ali HA, Ismail SI (1987) Morphology of the palm stem borer *Jebusaea hammerschmidti* Reiche (Cerambycidae: Coleoptera). Date Palm J 5(1):38–65

Al-Ani HY, El-Behadli AH, Majeed HA, Majeed MM (1971) The control of inflorescence rot. Phytopath Mediter 10:82–85

Al-Asady RM (2003) Susceptibility of different date palm cultivars to the terminal rot caused by *Thielaviopsis paradoxa* (Deseyn) Hohn. MSc thesis, University of Basrah (in Arabic)

Al-Badran BM (2008) Study the inflorescence rot caused by *Mauginiella scaettae* and *Fusarium* sp and the possibility of chemical and biological control of this disease. MSc thesis, University of Basrah (in Arabic)

Al-Bahili AZ (2004) Biological and chemical control of longhorn date palm stem borer, *Jebusea hammerschmidti*. MSc thesis, University of Basrah (in Arabic)

Al-Baker AJ (1962) The Iraqi dates varieties. The Government Press, Bagdad, Iraq (in Arabic)

Al-Baker AJ (1972) The date palm: a review of its past and present status, and the recent advances in its culture, industry and trade. Al-Ani Press, Bagdad, Iraq (In Arabic)

Al-Beldawi AS, Hussain FM (1974) Studies on date palm inflorescence rot and its control in Iraq. Plant Protection Yearbook No 1:207–212

Al-Delamy KA (2004) Ecological and economic studies on lesser date moth (*Batrachedra amydraula* Meyrick) (Cosmopterygidae: Lepidoptera) in the middle of Iraq. MSc thesis, College of Agriculture, University of Baghdad (in Arabic)

Al-Dhamin ASA (2002) Field efficacy of *Melia azedarach* L. fruit extracts on the survival of *Ommatissus binotatus lybicus* deBerg. (Homoptera: Tropiduchidae). MSc thesis, College of Science, University of Baghdad

Al-Dhamin ASA (2008) Evaluation of the field efficacy of crude extracts of (*Albizia lebbeck* L.) Benth. and Actara insecticide on biological performance of *Ommatissus lybicus* Bergevin (Homoptera: Tropiduchidae). PhD thesis, College of Science, University of Baghdad

Al-Fahdawi TM (1988) Eradication effect and permethrin residue and the effect of temperature on its efficacy to control lesser date moth, *Batrachedra amydraula* and fig moth, *Ephestia cautella*. MSc thesis, College of Agriculture, University of Baghdad (in Arabic)

Al-Hamdany MA, Jaber HY, Kadhem AH, Sabar JA (2011) Role of *Chalara radicicola* in date palm trees decline. Arab J Plant Prot 29:118–121 (in Arabic)
Al-Hassan KK, Abbas GY (1987) Outbreak of terminal bud rot of date palm caused by *Thielaviopsis paradoxa*. Date Palm J 5(1):117–119
Al-Hassan KK, Shamseldeen SA (1975) Survey of inflorescence rot in Iraq during 1974. In: The third international conference of dates and date palms, Baghdad, 30 Nov–4 Dec 1975
Ali TA, Jubrail JM, Jassim AM (2007) The use of RAPDs technique for the detection of genetic stability of the regenerated plantlets (Barhi cv.) in Iraq. Acta Hort 736:127–134
Al-Jboory IJ (1999) Date palm Old World date mite (Ghobar mite) *Oligonychus afrasiaticus* (McGregor), vol 9, Tech Broch. Ministry of Agriculture, Bagdad, Iraq (in Arabic)
Al-Jboory IJ (2001) New record of entomopathogenic nematode for Iraq. Arab Near East Plant Prot Newsl 32:6 (in Arabic)
Al-Jboory IJ, Saleh SJ (2001a) Survey and identification of mites inhabiting date palm trees in Iraq with some observations on parasitism efficacy of Diplogynid mites on date palm borers. Basrah J Date Palm Res 1(2):1–5 (in Arabic)
Al-Jboory IJ, Saleh SJ (2001b) New record of entomopathogenic nematode isolated from longhorn stem borers and stalk borers in Iraq. Basrah J Date Palm Res 1(1):38–45 (in Arabic)
Al-Jboory IJ, Saleh HM (2002a) New record of pathogenic virus isolated from longhorn date palm stem borer *Jebusaea hammerschmidti*. Basrah J Date Palm Res 1(1):38–45 (in Arabic)
Al-Jboory IJ, Saleh SJ (2002b) New records of dipteran parasite *Megaselia* sp. from the adult of longhorn date palm stem borers. Iraqi J Agric Res 33:4–12 (in Arabic)
Al-Jboory NM, Khalaf Z, Numan K (1971) A description of 110 Iraqi dates palm cultivars, Bull 38. General Horticulture Directorate, Ministry of Agriculture, Bagdad, Iraq (in Arabic)
Al-Jboory IJ, Tewag NS, Al-Jumaili SA, Alwan MM (2002) Production of bio pesticide from the vaccine of the fungus *Beauveria bassiana* isolated from longhorn date palm stem borers. Basrah J Date Palm Res 2(1,2):7–13 (in Arabic)
Al-Jboory IJ, Al-Zubae IA, Al-Dahwe SS (2006) Evaluation of two isolates of *Beauveria bassiana* in the controlling of some insects and mite pests and testing some production cultures. Univ Aden J Nat Appl Sci 10:1–10 (in Arabic)
Al-Jibouri AAM, Adham KM (1990) Biochemical classification of date palm male cultivars. J Hort Sci 65:725–729
Al-Jorany RS, Al-Delamy KA (2010a) Economic losses of lesser date moth (*Batrachedra amydraula* Meyrick) (Cosmopterygidae: Lepidoptera) on two cultivars, Khastawi and Zahdi of date palm in the middle of Iraq. Al-Anbar J Agric Sci 8(2):256–265 (in Arabic)
Al-Jorany RS, Al-Delamy KA (2010b) Model design to estimate the economic threshold level of lesser date moth *Batrachedra amydraula* Meyrick in the center of Iraq. Iraqi J Agric Sci 43(special issue 1):85–91 (in Arabic)
Al-Kaabi AM (2010) Effect of laser radiation on propagation of date palm *Phoenix dactylifera* L. cv. Barhi *in vitro*. Basrah J Date Palm Res 9(1):49–66 (in Arabic)
Al-Khalifa AA, Jasim AM, Abbas MF (2008) Effect of plant growth regulators on adventitious buds formation from date palm callus (*Phoenix dactylifera* L.) cv. Barhee in vitro. Basrah J Date Palm Res 7(2):32–50 (in Arabic)
Al-Khalifa AA, Al-Meer UN, Al-Jabary KM (2009) Effect of MS salts and sucrose on formation of lateral buds of date palm cv. Sayer *in vitro*. Basrah J Date Palm Res 8(1):39–50
Al-Khateeb AA, Abdalla GR, Dinar HM et al (2001) Auxin:cytokinin reactions in the in vitro micropropagation of date palm (*Phoenix dactylifera* L.). Egypt J Appl Sci 17(10)409–415
Al-Khateeb TA, Jubrael JM (2006) The use of RAPD markers for sex and male cultivars identification in (*Phoenix dactylifera* L.) in Iraq. Acta Hort 736:162–170
Al-Mayahi AMW (2010) Effect of NAA, 2iP inorganic sulphates, PEG and adenine sulphate on maturation of the somatic embryos in date palm *Phoenix dactylifera* L. cv. Quntar propagated by in vitro culture. Basrah J Date Palm Res 9(1):30–44 (in Arabic)
Al-Musawi AH (2001) Effect of auxin, cytokinine, and activated charcoal on embryo development and germination of date palm (*Phoenix dactylifera* L.) cultured in vitro. Basrah Date Palm Res J (1):8–18

Al-Meer UN, Al-Ibresam OT (2008) Effect of ammonium and potassium nitrate in some characteristics of callus and somatic embryos of date palm (*Phoenix dactylifera* L.) *in vitro*. Basrah J Date Palm Res 7(2):16–32 (in Arabic)
Al-Meer UN, Al-Ibresam OT (2010) Effect of vitamin E on some callus and embryos characteristics of date palm cv. Barhee propagated in vitro. Basrah J Date Palm Res 9(2):1–12 (in Arabic)
Al-Obeed RS, Harhash MM (2010) Effect of bunch bugging colour on 'Succary' and 'Khalas' date palm cultivars: fruit chemical characteristics. Acta Hort 882:1213–1217
Al-Rubaee RJ (2003) Testing the efficiency of hydraulic apparatus designed locally for insecticide injection to control date palm pests. MSc thesis, College of Agriculture, University of Baghdad (in Arabic)
Al-Rubeai HF, Al-Dhamin AS, Al-Rawy MA (2010) Field efficacy of *Melia azedarach* L. fruit extracts on the survival of *Ommatissus lybicus* Ashe and Wilson (Homoptera: Tropiduchidae). In: Date palm production and processing technology. Proceeding of the international Conference on Date Palm Production and Processing Technology, Sultan Qaboos University, Muscat, Oman, pp 221–226 (in Arabic)
Al-Ruqaishi IA, Davey M, Alderson P, Mayes S (2008) Genetic relationships and genotype tracing in date palm (*Phoenix dactylifera* L.) in Oman based on microsatellite markers. Plant Genet Resources 61:70–72
Al-Saleh AA (1988) A rapid method for date palm breeding and improvement. Arab Organization for Agriculture Development. In: Date palm propagation and management in Arab Homeland Symposium, Al-Ain, 5–10 Sept 1988 (in Arabic)
Al-Saleh AA, Al-Ansary NA (2005) A pictorial for Iraqi date palm cultivars. Part 1. Ministry of Agriculture, Al-Ezza Press, Bagdad, Iraq (in Arabic)
Al-Saleh AA, Bader SM (2013) Personal communication (unpublished data)
Al-Sewidi TM (2003) Heat accumulation and life table of Old World date mite *Oligonychus afrasiaticus*. MSc thesis, College of Agriculture, University of Baghdad
Al-Sewidi TM, Al-Jboory IJ (2006) Fecundity of Old World date mite *Oligonychus afrasiaticus* (McGregor) on date palm. Arab J Plant Prot 24(1):23–30
Anonymous (1912) A danger on Iraqi date palms. Arab Lang J
Aziz FM (1990) Sensitivity of some date varieties to infestation by *Batrachedra* sp. (Lepidoptera: Cosmopterygidae). MSc thesis, College of Science, University of Baghdad (in Arabic)
Aziz FM (2005) Biological and ecological studies on the lesser date moth *Batrachedra* spp. (Lepidoptera: Cosmopterygidae) and predicting of its appearance infesting the date palm early in the spring. PhD thesis, College of Science, University of Baghdad (in Arabic)
Baaziz M (1988) The activity and preliminary characterisation of peroxidases in leaves of cultivars of date palm (*Phoenix dactylifera* L.). New Phytol 111:403–411
Baaziz M, Saaidi M (1988) Preliminary identification of date palm cultivars by esterase isoenzymes and peroxidase activities. Can J Bot 66:89–93
Bader SM, Jarrah AZ, Naji MT (1986) Effect of nutrient medium on growth and development of *Phoenix dactylifera* L. embryo culture in vitro. J Agric Water Res 4(4):51–60 (in Arabic)
Bader SM, Baum M, Khierallah HSM, Choumane W (2007) The use of RAPDs technique for the detection of genetic stability of date palm plantlets derived from *in vitro* culture of inflorescence. J Edu Sci 20(3):149–159
Bendiab K, Baaziz M, Majourhat K (1998) Preliminary date palm cultivar composition of moroccan palm groves as revealed by leaf isozyme phenotypes. Biochem Syst Ecol 26:71–82
Benito C, Figueiras AM, Zaragozal C et al (1993) Rapid identification of Triticeae genotypes from single seed using the polymerase chain reaction. J Plant Mol Biol 21:181–183
Billotte N, Marseillac N, Brottier P et al (2004) Nuclear microsatellite markers for the date palm (*Phoenix dactylifera* L.): characterization and utility across the genus *Phoenix* and in other palm genera. Mol Ecol Notes 4:256–258
Blears M, De Grandis S, Lee H, Trevors J (1998) Amplified fragment length polymorphism (AFLP): a review of the procedure and its applications. J Ind Microbiol Biotechnol 21:99–114
Buxton RA (1920) Insect pests of date palm in Mesopotamia and elsewhere. Bull Ent Res 11:287–303

Central Statistical Organization (CSO) (2012) The annual report of agricultural statistics. Ministry of Planning, Bagdad, Iraq

Cullis C (2011) Molecular markers in date palm. In: Jain SM et al (eds) Date palm biotechnology. Springer, Dordrecht/New York, pp 361–370

Dowson VHW (1936) A serous pest of date palm *Ommatissus binotatus* Fieb., (Homoptera: Tropiduchidae). Trop Agr Trin 13:180–191

Dowson VHW (1982) Date production and protection, vol 35, Plant Production and Protection Paper. FAO, Rome

Dutt A (1922) Supplementary note on the pests of the date palm in Iraq. Mesopotamia Dept Agr Mem 6:13–21

El-Haidari HS, Al-Hafidh EM (1986) Palm and date arthropod pests in the Near East and North Africa. Regional Project for Palm and Dates Research Center. Al-Watan Press, Baghdad (in Arabic)

Elshibli S, Korpelainen H (2008) Microsatellite markers reveal high genetic diversity in date palm (*Phoenix dactylifera* L.) germplasm from Sudan. Genetica 134:251–260

FAOSTAT (2004) Crop Production, Statistics Division, Food and Agriculture Organization of the United Nations, Rome. http://faostat.fao.org/

FAOSTAT (2009) Crop production 2008. Statistics Division, Food and Agriculture Organization of the United Nations, Rome, Italy. http://faostat.fao.org/

Ghali FS (2001) Date palms decline disease caused by the fungus *Chalara paradoxa*, disease conditions and its control. PhD thesis, College of Agriculture, University of Baghdad (in Arabic)

Hama NH, Abdul Razaq AS, Alsharnmary NS (2010) An alternative approach to control date palm dubas bug *(Ommatissus lybicus* De Berge) (Tropiduchidae: Homoptera). Date palm production and processing technology. Proceeding of the international Conference on Date Palm Production and Processing Technology, Sultan Qaboos University, Muscat, Oman, pp 205–212 (in Arabic)

Hamad BS (2005) Ecological and biological studies on *Chrysoperla mutata* (MacLachlan) (Neuroptera: Chrysopidae) natural enemy of dubas bug *Ommatissus lybicus* DeBerg. (Homoptera: Tropiduchidae). PhD thesis, College of Science, University of Baghdad

Hameed MK (2001) Propagation of some date palm (*Phoenix dactylifera* L.) cultivars through tissue culture technique. PhD dissertation, College of Agriculture, University of Baghdad (in Arabic)

Hammer Ø, Harper DAT, Ryan PD (2001) PAST: palaeontological statistics software package for education and data analysis. Palaeont Electr 4:1–9

Hamwieh A, Farah J, Moussally S et al (2010) Development of 1000 microsatellite markers across the date palm (*Phoenix dactylifera* L.) genome. Acta Hort 882:269–277

Hasan BH, Al-Jboory I, Al-Rubeai H, Viggiani G (2003) *Pseudoligosita babylonica* n. sp. (Hymenoptera: Trichogrammatidae), egg parasitoid of *Ommatissus lybicus* Berg. (Homoptera: Tropiduchidae) in Iraq. Boll Lab Ent Agr Filip Silv 59:75–78

Hasoon AH (1988) Biological and ecological study on Old World date bug, under laboratory condition: *Ommatissus binotatus* de Berg. (Homoptera: Tropiduchidae). MSc thesis, College of Agriculture, University of Basrah

Holton TA (2001) Plant genotyping by analysis of microsatellites. In: Henry RJ (ed) Plant genotyping: the DNA fingerprinting of plants. CABI, Cambridge, UK, pp 15–27

Husien FA (2002) A description of some Iraqi dates palm cultivars. Ministry of Agriculture, Bagdad, Iraq (in Arabic)

Husien FA, Greab SH (2004) A description of 50 Iraqi dates palm cultivars. Ministry of Agriculture, Bagdad, Iraq (in Arabic)

Husien FA, Raheef AH, Esmaeel RM, Hraib SH (2009) A project of establishing new mother date palm orchards and offshoots nurseries in Iraq. A study submitted to Khalifa International award. The first session 2009, Bagdad, Iraq (in Arabic)

Hussain FA (2002) Description of some Iraqi date palm varieties. Part 1. Ministry of Agriculture, Bagdad, Iraq (in Arabic)
Hussein AA (1974) Dates and date palm pests and their control in Iraq. Aledara Al-Mahalea Press, Baghdad
Ibrahim AA (1979) Study of the yearly levels of N-P-K in leaves, fruits and soil in some commercial date palm varieties. MSc thesis, College of Agriculture, University of Baghdad (in Arabic)
Ibrahim AA (1995) The physiological relationship between growth regulators and date palm fruits var. Hellawi. PhD dissertation, College of Agriculture, University of Basrah (in Arabic)
Ibrahim AA (1998) From the history of Saydat Al-Shajar, date palm. In: Scientific symposium on date palm, Yemen, 27–29 June
Ibrahim KM (2006) The role of date palm tree in landscaping. In: International congress on date palm, production and processing, Sultan Qaboos University Oman, Muscat, June 2006
Ibrahim AO (2008) Date palm the tree of life. Arab Center for the Study of Arid Zones and Dry Lands (ACSAD), Damascus
Ibrahim KM (2010) The role of date palm tree in improvement of environment. In: The fourth conference on date palm. Abu-Dhabi, pp 15–17
Ibrahim NHH (2012) Some factors affecting embryogenic callus initiation of date palm (*Phoenix dactylifera* L.) cv. Bream *in vitro*. MSc thesis, College of Agriculture, University of Baghdad
Jaafar KA, Baldwin R (2010) Characterization of Iraqi Zahdi dates pits. Blessed Tree 2:2–5
Jaccard P (1908) Nouvel le srescherchessur la distribution florale. Bulletin de la Socieìteì Vaudoise des Sciences Naturelles 44:223–270
Jasim AM (1999) Response of different date palm cultivars (*Phoenix dactylifera* L.) to in vitro culture. Basrah J Agric Sci 12(2):9–16
Jasim AM (2000) Production of somatic embryos of date palm in vitro by liquid media culture. J Basrah Res 24(1):1–6
Jasim AM, Saad AA (2001) Effect of some media component on growth and somatic embryos formation and germination of date palm (*Phoenix dactylifera* L.) cultured in vitro. Basrah J Date Palm Res 1:1–7 (in Arabic)
Jasim AM, Saad AA (2003) Effect of date of explants supplementation and pH on propagation of date palm (Phoenix dactylifera L.) in vitro. Abstract. In: Date palm international meeting, KSA, April 2003 (in Arabic)
Jasim AM, Abass KI, Abdul Samed AR (2008) Effect of some plant substances on contamination and growth somatic embryos of date palm (*Phoenix dactylifera* L.) cv. Ashkar cultured in vitro. Basrah J Date Palm Res 7(2):1–16 (in Arabic)
Jasim AM, Al-Mayahi AM, Attaha AH (2009) Propagation of four rare cultivars of date palm (*Phoenix dactylifera* L.) by tissue culture technique. Basrah J Date Palm Res 8(1):72–99 (in Arabic)
Jassim HK (2007) Studies on the biology of palm dubas bug *Ommatissus lybicus* (Debergevin.) Asche and Wilson. (Homoptera: Tropiduchidae) and its biocontrol by some isolates of entomopathogenic fungi *Beauveria bassiana* (Balsamo) Vuill. and *Lecanicillium* (=*Verticillium*) *lecanii* (Zimm.) Zare and Oami. PhD thesis, College of Agriculture, University of Baghdad, Bagdad, Iraq (in Arabic)
Jubrael JMS (2001) Genetic characterization for some date palm cultivars in Iraq using RAPD markers. *IPA* Journal for Agricultural Researches 11(1):138–148 (in Arabic)
Juber KS, Al-Mohamadawi AZ (2010) Determination of some pathogens of date palm offshoots death phenomenon and the effect of salinity as a predisposition factor for disease. Iraqi J Agric Sci 41(2):105–116 (in Arabic)
Jubrael JM, Udupa SM, Baum M (2005) Assessment of AFLP-based genetic relationships among date palm (*Phoenix dactylifera* L.) varieties of Iraq. J Am Soc Hort Sci 130(3):442–447
Kassem HA, Ezz TM, Marzouk HA (2010) Effect of bunch bagging on productivity, ripening speed and postharvest fruit quality of 'Zaghloul' dates. Acta Hort 882:132–141

Khalaf AN (2002) The role of plant hormones in growth and maturity of date palm seedlings and parthenocarpy var. Barhi. PhD, College of Agriculture, University of Basrah, Basrah, Iraq (in Arabic)
Khalaf MZ, Al-Taweel AA (2014) Palm borers in the Iraqi environment: species, damage and control. Ministry of Science and Technology, Agricultural Research Directorate, Bagdad, Iraq, p 31
Khierallah HSM (2007) Micropropagation of two date palm (*Phoenix dactylifera* L.) cultivars using inflorescences and study the genetic stability using AFLP-PCR markers. PhD dissertation, College of Agriculture, University of Baghdad, Bagdad, Iraq
Khierallah HSM, Bader SM (2007) Micropropagation of date palm (*Phoenix dactylifera* L.) var. Maktoom through direct organogenesis. Acta Hort 736:213–224
Khierallah HSM, Bader SM, Baum M, Al-Khfaji MA (2007) Assessment of AFLP variations in date palm in vitro plantlets derived from inflorescence explant. The 4th Symp, date palm, KSA, 2008 Proc. The 4th Symp, date palm, KSA, p552–564
Khierallah HSM, Husien NH (2013) The role of coconut water and casein hydrolysate in somatic embryogenesis of date palm and genetic stability detection using RAPD markers. Res Biotechnol 4(3):20–28
Khierallah HSM, Bader SM, Baum M, Choumane W (2008) Assessment of AFLP variations in date palm in vitro plantlets derived from inflorescence explant. In: Proceeding of the 4th symposium date palm, 2008. KSA, pp 552–564
Khierallah HSM, Bader SM, Baum M, Hamwieh A (2011a) Assessment of genetic diversity for some Iraqi date palms (*Phoenix dactylifera* L.) using AFLP markers. Afr J Biotech 10(47): 9670–9676
Khierallah HSM, Bader SM, Baum M, Hamwieh A (2011b) Genetic diversity of Iraqi date palms revealed by microsatellite polymorphism. J Am Soc Hort Sci 136(4):282–287
Khierallah HSM, Al-Sammarraie SKI, Mohammed HI (2014) Molecular characterization of some Iraqi date palm cultivars using RAPD and ISSR markers. J Asia Sci Res 4(9):490–503
Liu K, Muse SV (2005) PowerMarker: integrated analysis environment for genetic marker data. Bioinformatics 21:2128–2129
MacFarquhar N (2003) Forbidden fruit: Iraq dates hit by war and sanctions. 2010. http://www.iht.com/articles/83194.html. Food and Agriculture Organization of the United Nations. Accessed 1 July 2008
Majourhat K, Bendiab K, Medraoui L, Baaziz M (2002) Diversity of leaf peroxidases in date palm (*Phoenix dactylifera* L.) as in an example of marginal (seedling derived) palm groves. Sci Hort 95:31–38
Mater AA (1983) Plant regeneration from callus culture of *Phoenix dactylifera* L. Date Palm J 2(1):57–77 (in Arabic)
Mater AA (1986) In vitro propagation of (*Phoenix dactylifera* L.). Date Palm J 4(2):137–152 (in Arabic)
Memarian A (1947) Date palm dubas bug in Basrah. MOA, Baghdad (unpublished report)
Ministry of Agriculture (1992) Date palm disorder phenomena in Iraq. Technical Report. Ministry of Agriculture, Iraq, pp 28 (in Arabic)
Mohamed HE, Al-Jboory IJ (2001) New record of *oryctes like virus* from date palm fruit stalk borer *Oryctes elegans* from Iraq. Basra Date Palm Research Journal 1(2):7–9 (in Arabic)
Muhsen KA (2007) Regeneration of date palm (*Phoenix dactylifera* L.) cv. Sherafy from different apical explants *in vitro*. Basrah J Date Palm Res 6(1):64–80 (in Arabic)
Omar MS (1988a) Callus initiation, asexual embryogenesis and plant regeneration in (*Phoenix dactylifera* L.). Date Palm J 6:265–275
Omar MS (1988b) In vitro response of various date palm explants. Date Palm J 6(2):1–8
Omar MS, Arif MB (1985) An investigation of the fat of (*Phoenix dactylifera* L.) Carpels cultured in vitro. Date Palm J 4(1):15–24 (in Arabic)
Omar MS, Hameed MK, Al-Rawi MS (1992) Micropropagation of date palm (*Phoenix dactylifera* L.). In: Bajaj YPS (ed) Biotechnology In agriculture and forestry, vol 18, High-Tech and micropropagation II. Springer, Heidelberg, pp 471–492

Ould Salem Mohamed A, Trifi M, Sakka H et al (2001) Genetic inheritance analysis of four enzymes in date palm (*Phoenix dactylifera* L.). Genet Res Crop Evol 48:361–368
Rao YR (1921) A preliminary list of insect pests of Iraq, vol 71, Dept Agric Iraq, Mem. Time Press, Bombay
Rao YR, Dutt A (1922) The pests of the date palm in the Iraq, vol 6, Dept Agric Mesopotamia, Memoir. Times Press, Bombay
Reuveni O, Adato Y, Kipnis HL (1972) A study of new and rapid methods for the vegetative propagation of date palms. Date Grow Inst Rep 49:17–24
Saaidi M (1992) A study of new date palm disorders in Iraq. Consultancy Mission Report FAO TCP/IRQ/2255 (E)
Sahi AA (1986) The scientific book of dates technology. University of Basrah (in Arabic)
Saleh AM, Waheed AM, Kadhim MJ, Husien AS (2006) Effect of some fruits seed powders addition on the growth and development of embryogenic callus of date palm *Phoenix dactylifera* L. cultivar Barhi. Basrah J Date Palm Res 5(2):127–136
Sambrook J, Fritsch EF, Maniatis T (1989) Molecular cloning: a laboratory manual, 2nd edn. Cold Spring Harbor Laboratory Press, Cold Spring Harbor
Schroeder CA (1970) Tissue culture of date shoots and seedlings. Date Grow Inst Rep 47:25–47
Sedra MH, Lashermes P, Trouslot M, Hamon S (1998) Identification and genetic diversity analysis of date palm (*Phoenix dactylifera* L.) varieties of Morocco using RAPD markers. Euphytica 103:75–82
Shalash JS, Hamood HH (1989) Flower and fruit abscission in Zahdi and Khustawi date palm varieties. J Agric Water Res 8(1):33–41 (in Arabic)
Soliman SS, Al-Obeed RS, Harhash MM (2010) The effect of bunch thinning on yield of fruit quality of 'Khalas' date palm. Acta Hort 882:726–731
Taha HS, Bekheet SA, Saker MM (2001) Factors affecting in vitro multiplication of date palm. Biol Plant 44(3):431–433
Udupa SM, Baum M (2001) High mutation rate and mutational bias at (TAA)n microsatellite loci in chickpea (*Cicer arietinum* L.). Mol Genet Genom 265:1097–1103
Vos P, Hogers R, Bleeker M et al (1995) AFLP a new technique for DNA fingerprinting. Nucl Acid Res 23:4407–4414
Wrigley G (1995) Date palm. In: Smartt J, Simmonds NW (eds) Evolution of crop plants, 2nd edn. Longman Group, Essex, pp 399–403
Yousif AK, Ben Jamin ND, Kado A et al (1982) Chemical composition of four Iraqi date cultivars. Date Palm J 1(2):285–294
Zabar AF, Borowy A (2012) Cultivation of date palm in Iraq. Ann. Univ. Mariae Curie Sklodowska, Lublin 32(1):39–54
Zehdi S, Trifi M, Billotte N, Marakchi M, Pintaud JC (2004) Genetic diversity of Tunisian date palms (*Phoenix dactylifera* L.) revealed by nuclear microsatellite polymorphism. Hereditas 141:278–287

Further Reading

Abass MH, Hameed MA, Alsadoon AH (2007) Survey of fungal leaf spot diseases of date palm (*Phoenix dactylifera* L.) in Shaat Al-Arab orchards in Basra and evaluation of some fungicides. Basrah J Date Palm Res 6(1):1–21 (in Arabic)
Al-Ali AS (1977) Phytophagous and entomophagous insects and mites of Iraq, Publication, No.33 Ministry of Higher Education, Baghdad, Iraq
Al-Ameri ANA (2009) Study of effect of some environmental factors on date palms offshoots decline and death disease caused by *Chalaropsis radicicola* (Bliss) C. Moreau and its integrated control in Basrah. MSc thesis, University of Basrah
Al-Azawi AF (1980) General and applied entomology. Al-Zahraa Press, Baghdad (in Arabic)

Al-Beldawi AS (2002) Date palm diseases in the United Arab Emirates. Ministry of Agriculture and Fisheries Wealth UAE, Abu Dhabi, UAE (in Arabic)
Al-Beldawi AS, Sadik H, Shamiseldeen SA (2001) Isolation and identification of neck bending, growth point and trunk rot of date palm. Emir J Agric Sci 12:33–44
Al-Dabagh J, Al-Hakak TH (2011) Date marketing and processing in Iraq. Ministry of Agriculture, Bagdad, Iraq
Al-Jboory IJ, Al-Suaide TM (2006) Effect of different temperatures on the biology of Old World date mite, *Oligonychus afrasiaticus* (McGregor). In: Paper presented at 12th International Conference of Acarology, Amsterdam, 21–26 Aug
Al-Safi GS, Theab EM, Swear EA (1975) Control of lesser date moth by aerial application. In: The third international conference of dates and date palms, Baghdad, 30 Nov–4 Dec 1975
Al-Shamsy BH (2003) Biological performance of dubas bug *Ommatissus lybicus* under field conditions and the predication of dubas bug appearance by using degree day's model. MSc thesis, College of Agriculture, University of Baghdad (in Arabic)
Al-Shamsy BH, Al-Jboory IJ, Al-Rubae HF (2002) Observation on the dubas bug parasitoid *Oligosita* sp. Basrah J Date Palm Res 2(1,2):34–37 (in Arabic)
Al-Sosey AJ (1967) The scale insect *Parlatoria* on date palm in Iraq, vol 66, Ext Broch. General Directorate of Research of Agricultural and Projects, MOA, Baghdad
Al-Yaseri II, Ismail MZ, Al-Fadhal F (2011) Efficacy of some fungicides to controlling date palm inflorescence rot caused by *Mauginiella scaettae* Cav. Alkufa J Agr Sci p 203–207 (special issue) (in Arabic)
Al-Zubaidi AA (2005) Studies on the leaflet spot diseases and its chemical control in Basrah province. College of Agriculture, University of Basrah, Basrah, Iraq (in Arabic)
Bader SM, Khierallah HS (2007) The role of silver thiosulphate and glutamine on direct organogenesis of two date palm (*Phoenix dactylifera* L.) cultivars. J Biotech Res Centre Al-Nahrain Univ 32:58–71
Djerbi M (1983) Diseases of the date palm. FAO Regional Project for Palm and Dates Research Center, Baghdad
El-Behadli AH, Alrubai JT, Jasim HM (1989) Studies on date palm decline phenomena. In: Proceeding of the fifth conference of Scientific Research Council, Baghdad, 7–11 Oct 1989, 1(6):81–86
El-Haidari HS (1966) Use of aircraft for dubas bug control. Ministry of Agriculture, Ext Newsl, p 17
Food and Agriculture Organization of the United Nations Statistics - FAOSTAT (2011) http://faostat.fao.org/site/339/default.aspx. Accessed Feb 12 2011
Hamee MH (2005) Susceptibility of different cultivars of date palm (*Phoenix dactylifera* L.) to *Mauginiella scaettae* the causal agent of inflorescence rots. Basrah J Date Palm Res 4:1–2 (in Arabic)
Hussain AA (1963) Biology and control of dubas bug *Ommatissus binotatus* var. *lybicus* de-Berg. (Homoptera: Tropiduchidae) infesting date palm in Iraq. Bull Entomol Res 53: 737–745
Hussein AA (1985) Dates and date palm and their pests. College of Agriculture, Basrah University, Basrah
Jaafar KA (2010) Biogas production by anaerobic digestion of date palm pulp waste. Al-Khawarzmi Eng J 6:3–14
Khalaf MZ, Naher FH, Ali AS (2010) Population density of *Oryctes elegans* Prell. (Coleoptera: Scarabaeidae) on some date palm varieties in south Baghdad orchards. Agric Biol J North Am 1(3):238–242
Kinawy MM (2012) Date palm pests in Oman. Royal Court Affairs, Royal Gardens and Farm, Sultanate of Oman
Murashige T, Skoog FA (1962) Revised medium for rapid growth and bioassays with tobacco tissue culture. Physio Plant 15:473–497
Salih MHA, Al-Jboory I (2001) Isolation and identification of Oryctes like virus from stalk borers *Oryctes elegans* for the first time in Iraq. Basrah J Date Palm Res 1(2):1–12 (in Arabic)

Chapter 5
Date Palm Status and Perspective in Pakistan

Adel A. Abul-Soad, Shaimaa M. Mahdi, and Ghulam S. Markhand

Abstract Date palm (*Phoenix dactylifera* L.) is the third most important fruit crop after citrus and mango in Pakistan. This crop is found in all four provinces of Pakistan on 90,000 ha with a production of around 600,000 mt yr^{-1}. Pakistan's position is always among the seven largest producers and exporters of dates in the world. Sindh and Balochistan provinces contribute more than 90 % of production and crop area in Pakistan. In Sindh, Khairpur is the biodiversity center having more than 300 date palm cultivars. Date palm has a very strong effect on socioeconomic activities of the area. The top Pakistani cultivars are Aseel, Dhakki, Begum Jangi, Rabai, and Muzawati. Pakistani date fruit is being exported as fresh, dry, or *chohara*. Most date-processing factories in Pakistan are located in Khairpur. This chapter provides basic information describing date palm cultivation in Pakistan, major flood-affected date palms and fruit, protection from monsoon rains, the first commercial trial for date palm micropropagation using inflorescence explants at the Date Palm Research Institute (DPRI) at Khairpur, and major date palm problems generally in Pakistan, including diseases and pests, with emphasis on the Khairpur and Balochistan areas, and date fruit drying through the use of solar dryers.

Keywords Chloroplast DNA • Cultivars • Date palm diseases • Floods • Monsoon rains • Solar dryer • Molecular markers • Tissue culture

---

A.A. Abul-Soad (✉)
Botany Department, Horticulture Research Institute,
Agricultural Research Center, 9 Cairo University St., Orman, Giza 12619, Egypt
e-mail: adelaboelsoaud@gmail.com

S.M. Mahdi
Tropical Fruit Department, Central Agricultural Pesticides Laboratory (CAPL),
Agricultural Research Center, Al Giza, Egypt
e-mail: shiminour@gmail.com

G.S. Markhand
Bioassay Department, Date Palm Research Institute (DPRI),
Shah Abdul Latif University, Khairpur, Sindh, Pakistan
e-mail: gmarkhand@gmail.com

© Springer Science+Business Media Dordrecht 2015

J.M. Al-Khayri et al. (eds.), *Date Palm Genetic Resources and Utilization: Volume 2: Asia and Europe*, DOI 10.1007/978-94-017-9707-8_5

5.1 Introduction

Date palm (*Phoenix dactylifera* L.), or *khajoor* in the local language, is called the tree *of divine providence* and provides livelihood to many people. The presence of date seeds in the excavation of Mohenjo Daro (an archaeological site at Larkana, Sindh province) storage godown indicates that there was date palm cultivation in Sindh as early as 5,000 years old (Ahmad and Smead 2005; Kumar 2009; Marshal 1931). Some researchers believe that date palm was brought to the Indian subcontinent by Alexander the Great (Nixon 1951; Pasha et al. 1972). Another school of thought believes that date palm was probably found before Alexander the Great as the soldiers of his army fed on harvested dates from Kech valley of Balochistan during traveling along the Makran coast in the fourth century BC (Qasim and Naqvi 2012). Then dates were spread in Sindh by Mohammed Bin Qasim in 712 AD along with the Islamic religion. Arab soldiers discarded date seeds at camp sites during wars, from which date palms grew and flourished in the Indus Valley (Ahmad and Tahir 2005; Dhillon et al. 2005; Jatoi et al. 2010). Since the dawn of ancient history, the date palm has been an important source of food for the Arabian Peninsula (Beech 2003; Popenoe 1913).

In the last century and during the colonial period of the Indian subcontinent, a number of Arabian date palm cultivar offshoots were imported in 1910–1912 from Basra (Iraq) by the British Indian Government and planted in Multan and Muzaffar Garh (Punjab). These cultivars were Halawy, Khadrawy, Sayer, Zahidi, and Dayri (Milne 1918).

The date palm tree provides communities in rural areas of Pakistan with many benefits. People eat dates as high-energy food. During Ramadan, Muslims break their fast each day by eating dates. Wood from the tree is used to build homes and to make artifacts and is chopped and burned for fuel. Palm fronds are used to make handicrafts and leaflets to weave fruit baskets and mats.

Pakistan is the seventh largest producer of dates with a total production of 557,279 mt in 2011 and is rated among the largest producers of date palm in the world from different cultivars, grown over 93,088 ha. On average the yield of dates is 59,866 hg ha^{-1} (hectogram per hectare) in 2011 which is very low as compared to Egypt, the largest producer in the world, 329,773 hg ha^{-1} (FAOSTAT 2013). It is worth mentioning that Pakistan also is the largest exporter in the world. The major countries importing both dried and fresh dates from Pakistan are India, USA, UK, Canada, Germany, Denmark, Malaysia, and Indonesia (FAOSTAT 2013). The major cultivars are Begum Jangi of Balochistan, Aseel of Sindh, and Dhakki of Dera Ismail Khan.

Dates are grown in all four provinces of Pakistan. The main date-producing areas are Kech (the administrative center is Turbat) and Panjgur (Balochistan), Khairpur and Sukkur (Sindh), Jhang, Dera Ghazi Khan (Dera Ismail Khan), Muzaffar Garh, Multan and Bahawalpur (Punjab), and Dera Ismail Khan (Khyber Pakhtunkhwa). Table 5.1 shows comparative date production, production area, and yield of dates in Pakistan (Pakistan Statistical Year Book 2012).

Table 5.1 Area, production, yield, and ranking of dates in Pakistan (2010–2011)

Area	Area (ha)	Production (mt)	Yield per ha (kg)	% area	% production
Balochistan	50,100	204,300	4,078	55.6	38.98
Sindh	32,700	268,600	8,214	36.92	51.25
Punjab	5,800	42,500	7,328	6.43	8.11
Khyber Pakhtunkhwa	1,500	8,600	5,733	1.66	1.64
Pakistan total	90,100	524,000	5,816		

It is clear that Balochistan and Sindh provinces account for about 90 % of the total Pakistan date production and the harvested area. However, dispersed date palm cultivation of less than 10 % of the total harvested area is carried out in a few places in the provinces of Punjab and Khyber Pakhtunkhwa (KPK), which are located in northern Pakistan where the temperatures are low; annual average temperature is 25 °C.

5.1.1 Current Date Palm Problems

There are a number of problems and limitations for the development of the date palm industry, which include all activities related to date palm, e.g., production, protection, and processing. The problems vary among the four provinces of Pakistan. Monsoon rains during the ripening of date fruit are always the most serious risk and the largest challenge facing Pakistani growers. Irregular rainfall from the end of June to August can destroy the entire crop whether on the tree or date fruit curing on mats. Use of bunch covers is not cost effective, especially with the lower-quality cultivars and high relative humidity. The unavailability of tissue-cultured palms of local elite cultivars is hindering the establishment of new plantations and spreading the problems associated with conventional methods using small offshoots. Modern equipment and mechanical lines of factories, limited number of high-quality local cultivars, hygiene, and lack of cold storage facilities are major problems facing the date-processing sector in Pakistan.

In the Sindh date palm area, Khairpur district, growers have been suffering from an emerging pest for the last one to two decades, which is destroying many date palms. Red palm weevil (RPW) causes losses of date palms particularly in abandoned orchards. A local treatment and early diagnoses of symptoms are used to control RPW in Pakistan. Termites, date palm lesser moth, and stem borers are among the major pest problems affecting date palm not only in the Sindh but all over Pakistan.

In Balochistan province, limited water resources and dependence on conserved water in the aquifers to irrigate date palm orchards necessitate installation of new irrigation systems such as drip irrigation to save water and avoid yield reduction during drought periods. Dubas bugs, which suck the sap of leaflets of date palm, became a major threat in Kech district, particularly with the high usage of chemical treatment to control the pest infestation levels during the last couple of decades. Rodents, i.e., rats, attack the trunk of date palms and destroy newly planted offshoots in the northwest of the province.

In Punjab and KPK, in addition to monsoon rains, the humid climate of above 60 % relative humidity and low temperature, below 25 °C during fruit development and ripening stages, restrict the type of cultivars to the soft or semidry dates and reduce fruit quality (Abul-Soad 2013b).

In this chapter, these problems are addressed in detail, and the efforts made by Pakistani scientists and others to solve these problems are discussed (Table 5.3).

5.2 Cultivation Practices

5.2.1 Description of Current Cultivation Practices

Date palm is an important crop of Sindh province, mainly grown in Khairpur and Sukkur districts. Almost 85 % of the Sindh dates are produced in Khairpur alone. Date palm production is a major source of income for local growers. In Khairpur, maximum benefit is gained through intercropping with other plant species under the canopy of date palms (average temperature in summer is 30–35 °C). Date palms allow other species such as maize, banana, wheat, rice, alfalfa, and vegetables to thrive in terms of growth. Some of these plant species need full sun and frequent irrigation such as rice and wheat. Subsequently the distance between adult date palms is sometimes increased to reach 10 m to allow light penetration to cereal crops (Fig. 5.1). The relative productivity of adult date palm trees has decreased.

Fig. 5.1 Intercropping of rice with date palm trees at Khairpur

Newly established date palms, 3–4 years old, also have been affected. Flood irrigation has occasionally caused water to enter the young central fronds of the trees and cause fungal infection from *Diplodia* disease (Abul-Soad 2011a).

In principle, there is no contradiction between the date palm and other intercropped species in terms of fertilizer application. Nevertheless, growers formerly ignored specific application recommendations for date palms while they applied fertilizer for the intercrops. Date palm responds positively to the lateral addition of organic and inorganic fertilizers, but the focus is on nitrogen fertilizer sources required for wheat and rice. Daily flood irrigation might be causing problems for the date palm roots especially if there is present the soilborne fungus which causes palm wilt (*Fusarium solani*). Furthermore, a suspicion exists that short intervals of frequent irrigation, leading to waterlogging, could be one of the reasons for date palm decline in Khairpur where the intercropping is a common practice. Intercropping with fruit crops, such as banana and citrus, is preferred over cereals and maize crops which require frequent water supplement.

It is worth mentioning that during the date harvest season in Khairpur, a new *settlement* is created associated with different activities of the date palm for a period of about 2 months (July–August). Many people earn their livelihood from the date palm. It is like a large festival which can be seen everywhere in the area. The growers are picking the rutab fruit of different cultivars, harvesting the khalal fruit to make *chohara,* the local name given to the boiled date fruit at khalal stage (before ripening). Harvested fruits are boiled for 15–30 min and then spread on mats, made from date palm fronds, and dried in the sun for 5–6 days to obtain *chohara*. This is practiced in order to harvest early to avoid the monsoon rain damage threat in July and August (Abul-Soad 2012b). After drying, the fruits were transported by tractor trolleys within the area (Fig. 5.2). The settlements are made up of temporary immigrants from surrounded areas that have come for work and are a common sight in Khairpur. Also, the odor of *chohara* is in the air during these months at Khairpur. The *city of dates* has become the cynosure of date exporters to look for a business opportunity to ship Khairpur dates abroad while international demand is increasing.

Fig. 5.2 Curing boiled date fruit at khalal stage on mats under solar radiation and open air for 5–6 days

In Khairpur, growers are accustomed to manually pick the early-ripening fruit of Aseel and other cultivars, by climbing the palm and shaking the bunch a few times before finally cutting the fruit bunch. The fallen fruit are collected in baskets or shallow trays and lowered down by a rope to another person receiving fruit on the ground.

The conventional practice to place the fruit bunches on the ground contaminates the fruit with dust and reduces fruit quality. In reality, the old practice should be updated by the international better practice to use plastic sheets beneath the date palm tree during harvest process to collect the harvested fruit bunches and keep fallen individual fruit free of dust.

Fruits harvested at early rutab stage are often spread on mats made from the leaflets of date palm fronds. After 5–6 days of sun curing, partially dried dates are usually collected in wooden boxes. Ruptured skin (pericarp), reshaped, mashed, and rotten fruit cured on mats placed in a dusty place without any conscious effort to avoid contamination create technical problems related to cleaning the fruit in the processing factories. The majority of dates are usually harvested during khalal stage when the fruit is not yet edible, to avoid damage from the monsoon rains.

Pakistan occupies a primary position in terms of date fruit exports with 121,681 mt (FAOSTAT 2013) due to *chohara* exports to India and Afghanistan. India was the second largest date importer in the world in 2010 (FAOSTAT 2013). *Chohara* is mainly used by Hindus for ritual celebration purposes not necessarily for eating. Without a doubt, this type of boiled date fruit would not be acceptable for human consumption in most of the date-importing countries due to its tannin taste, somewhat bitter, fibrous, and very dry. Nevertheless, in 2013 China placed an order for a container of *chohara*; hygienic analysis by the Chinese revealed 45 % contained dust particles although the fruits were washed before export. The requirements of the Chinese for *chohara* called for it to be packaged in transparent plastic bags of 500 g each, be labeled, be free from dust particles, and have an accreditation certificate. Hygienic handling through sun curing, sorting, transportation, and washing of *chohara* fruit with certified FAO detergents should resolve the problem.

At Pir or Bethak, locations where trucks carrying date fruit bunches are unloaded by laborers, the fruit is manually detached from the fruit stalks. This is the conventional practice before the drying process in a solar dryer or curing on mats in the sun or making *chohara*. Fruit detachment process is usually carried out by pulling the fruit bunch over a wooden comb fixed in the ground (Fig. 5.3), a tedious and laborious process. The use of mechanical methods may be an alternative solution for the process that would help date palm growers save time and effort and reduce the end product cost.

Near the end of the festival, another activity of date processing begins. It is packing the cured *chohara* date fruits, mostly in wooden boxes, handmade baskets, or jute sacks.

Packed fruits are commonly transported by trucks to the Khajoor date market, which is called *mandi* (Sindh language) and *bazaar* (Urdu) at Khairpur or Sukkur. Dates usually come from everywhere in the country to be sold during August and early September to the processing factories as whole date fruit or for export as *chohara*.

Fig. 5.3 Pulling fruit bunches over a wooden comb to detach the fruit before they are boiled to make *chohara*

5.2.2 Practical Approach to Establish a New Date Palm Farm

Establishing date palm nurseries is not a common practice in Pakistan, to grow out offshoots for field cultivation with a low mortality rate. Date palm growers customarily obtain small offshoots by separating them from the parent plant then cultivating them directly in the field; however, this carries a mortality rate which may reach 80 %. For transporting offshoots to distant areas, the outer fronds are trimmed off and offshoots of different sizes shipped by truck. Challenging the conventional wisdom in the date palm world, most of the growers prefer to establish new plantings during June and July, in the middle of summer. It is recommended that the spring or autumn months are the best time for offshoot cultivation to avoid high summer temperature and cold stress in winter. This is a reasonable practice under the current situation in Pakistan although there is a high infestation rate of RPW in winter. Typical mortality rates after offshoot cultivation are about 20–40 %. Moreover, in the Thar Desert in India, near the border of Sindh province, transplanting offshoots in the month of September showed higher survival (90 %) as against 65 % when transplanted in the months of March–April (Mertia et al. 2010).

Growers in Pakistan commonly cultivate the detached offshoots within the same orchard, dispersing them among the adult date palms, which reduces the distance between trees and increases relative humidity (Fig. 5.4). This is the reason Graphiola leaf spot disease is widely spread in date palm orchards in Khairpur. The regular planting distance between palms is 5–6 m. However, with the addition of offshoots, the distance is reduced to 3–4 m in many cases in Sindh. Growers establishing new date palm cultivation should specify a suitable piece of land in the orchard for a small nursery and properly manage it. The distance between offshoots in the nursery could be reduced to 1–2 m in this case.

After 1–2 years of proper management, i.e., fertilizer application, irrigation, and pest and disease management, the grown offshoots can be transplanted to the permanent site with an expected survival of close to 100 %. The common and

Fig. 5.4 Offshoots were commonly transplanted among adult trees for new cultivations in Sindh

Fig. 5.5 Offshoots transplanted to their permanent sites to establish a new date palm farm in Sindh

inappropriate (for date palms) practice of intercropping vegetables associated with frequent flood irrigation has resulted in 20–40 % of the established date palms being devastated by *Diplodia phoenica* infection, a fungus causing root rot.

Only rarely are offshoots transplanted to a permanent location to establish a new farm such as the Jatoi Farm in Moro district, Sindh (Fig. 5.5). In such a case, the planted offshoots may take longer time to begin the fruiting stage (2–3 years) compared to the nursery offshoots (1–2 years). In addition to that, the date palm nursery assures reduced mortality, homogeneity in fruiting time, and more efficient use of land and provides a higher number of palms under better management.

5.2.3 *Fertilizer*

Date palms need fertilizer like other cultivated crops. Intercropping with vegetables and other crops in Sindh, the poorly managed date palm plantations in Balochistan make chemical fertilization financially unviable. The only fertilizer use is through the addition of farm manure in winter, a practice found in Sindh. Nevertheless, there is an indirect application of chemical fertilizers through the amounts given to the intercropped species with date palm. It is important to mention that the soil of Sindh has Mg and P deficiency which requires exogenous application with a source for these major elements. A fertilizer program should be tailored to the date palms, especially after the establishment phase for rapid and optimum growth. On the other hand, most date palm plantations are not regularly fertilized in other provinces. Moreover, many of the date palm plantations in Punjab are scattered and abandoned.

Date palm can be grown in a wide range of soil types. Deep sandy soils with a good moisture supply are the best. Good drainage and aeration are the main soil requirements for ideal production. Date palm tree will grow in heavier soils, but care must be taken not to waterlog the soils. Date palms can grow in alkali and saline soils, but growth and productivity will be affected (Abul-Soad et al. 2008).

A fertilizer application program should begin prior to cultivation in the permanent site, during the land preparation phase, and continue until the date palms reach adult status, at the optimum production phase (Table 5.2). Trade names in the recommended fertilizer programs represent examples from the Pakistani fertilizer market. These additives are:

(a) NPK: 17–17–17 or 23–8–17 of nitrogen, phosphorus, and potassium fertilizer
(b) DAP: diammonium phosphate, 23 % P_2O_5+8 N/ 46 % P_2O_5+8 % K_2O
(c) SOP: K_2O, 50 %
(d) FM: farm manure in a raw form
(e) HA: humic acid (Biofertilizer, Biopower, Bluetus)

Salt accumulates on the soil surface during January when irrigation is routinely stopped, and there is no water in the main irrigation canals of the Indus River. Salt accumulation is increasing in lower level lands. Heavy irrigation 1–2 times is required to leach the salt of such lands in the Sindh Valley.

Due to the high pH of soil in Pakistan (alkali) which may reach 7–8.5, the preferred formulas of chemical fertilizers could be nitrogen in the form of ammonium sulfate (21 % N), phosphorus in the form of calcium superphosphate (45–47 % P_2O_5), potassium in the form of potassium sulfate (50 % K_2O), and magnesium in the form of magnesium sulfate (27–33 % MgO). Unfortunately, fertilizer sources of MgO are virtually unavailable in Pakistan.

5.2.4 Irrigation

Flood irrigation is the common practice when vegetables and other crops are intercropped with date palm, a practice which may be increasing the decline of disease infections associated with root rot. The common irrigation system for date palm plantation in Pakistan can be divided into two types:

5.2.4.1 Flood Irrigation

Flood irrigation is practiced in Sindh, Punjab, and KPK provinces through smaller irrigation canals branching from the Indus River. This irrigation method is the oldest method known and widely applied in date palm cultivation. It has advantages of low running costs and easy water application, and initial costs are low if the area is fairly

Table 5.2 The annual fertilization program (in g palm^{-1} yr^{-1}) for an adult productive date palm tree

Fertilizers	Jan.	Feb.	Mar.	Apr.	May	June	July	Aug.	Sept.	Oct.	Nov.	Dec.	Total
NPK	–	–	300	300	500	300	300	300	300	–	–	–	2,300
DAP	500	–	–	500	–	–	–	500	–	–	–	–	1,500
SOP	–	–	–	–	300	600	300	–	–	–	–	–	1,200
FM	50,000	–	–	–	–	–	–	–	–	–	–	–	50,000
HA	–	–	300	–	–	–	–	–	–	–	–	–	300

flat. Disadvantages are low efficiency (percolation, time, and wasted area among trees), labor intensive, and does not work for sandy soils. Growers are of the opinion that this irrigation system may be responsible for date palm decline and mango degradation in Sindh.

5.2.4.2 Groundwater Irrigation

This source is mainly used in Balochistan province where the desert environment is prevalent. Furthermore, the groundwater depth (availability) has a major impact on the uneven productivity of date palm in the desert areas of Balochistan. A drop of water level to 8–10 m has caused drought impact on the date palm forcing the growers to pump deeper groundwater for irrigation of the trees. The annual rainfall varies between 50 and 250 mm which recharges the aquifers (Abul-Soad et al. 2009).

The daily net irrigation requirement (NIR) of date palm in such areas is varied from 97 L $tree^{-1}$ d^{-1} in December to 854 L $tree^{-1}$ d^{-1} in June according to a study undertaken by the Kuwait University at similar areas (Abdul-Salam and Al-Mazrooei 2007). In Saudi Arabia, the average daily water use of a mature date palm tree was estimated at 184.4 L $tree^{-1}$ d^{-1} (Alamoud et al. 2012). The NIR of 1-year tissue-cultured palms under Kuwait's environment ranged between 23 and 27 m^3 ha^{-1} yr^{-1} (Bhat et al. 2012). In fact, the water requirement is the same for the irrigated as well as nonirrigated plantations irrelevant of the water source.

The minimum water requirement for one tree per year is estimated at 25 m^3. That quantity of water is enough for minimal vegetative growth and yield. The minimum productivity of one date palm tree of the dominant Rabai cv. in northwestern Balochistan was estimated at 10 kg yr^{-1}. The low productivity was after 3–4 years of drought (no rain) at those areas.

Newly planted offshoots in these areas are, however, manually irrigated for a period of 2–3 years, especially during the first 6 months, until the young trees are able to extend their roots to reach the groundwater. The extent of date palm plantations is dispersed to isolated locations due to the availability of the groundwater in northwestern Balochistan. Thus, the plantations can be divided into two categories (Abul-Soad et al. 2009):

(a) Irrigated plantations (furrow and basin irrigation): where the plantations are irrigated, the date palm tree will not extend its roots more than 2 m below the surface. If there is a deficiency of irrigation water, the tree will extend its roots to access the water in the range of its root system. In these areas where the plantations are irrigated, the groundwater is at a depth of over 17 m. Mature trees rely exclusively on pumped groundwater to provide the irrigation water, with the root systems adapted to the depth at which the water is available. The groundwater is deeper, and its access appears to be supported by the availability of cheap electricity and diesel fuel. Pumps powered by diesel engines are used, while some of the pumps are electric, sometimes the latter available from Iran. Water is delivered to the date palm plantations through unlined surface channels connected to a holding basin or a small reservoir which is fed by a pump. The increasing cost of energy is likely to make irrigated agriculture uneconomic. In irrigated areas, plantations are irrigated approximately twice a week in the summer and once a week in the winter.

(b) Nonirrigated plantations: the depth of groundwater typically is less than 4 m in these areas where the mature trees depend on the shallow groundwater and are not irrigated by the growers. Date palms can easily extend their roots to access the groundwater. Introduction of improved irrigation systems for the date palm plantations is required whether in the Indus River Basin or desert areas in Balochistan using drip, sprinkler, or micro irrigation. However, water management must take into account the level of skills available and will require training and education of the growers. Farmer organizations could be organized for the collective benefit of the growers, and the district authorities may have to be involved to organize the training, collecting of production, and marketing and to settle irrigation conflicts among date growers.

5.2.5 Flood-Affected Date Palms

Excessive flooding occurred in Pakistan in August 2010. A torrent of water threatened date palms, mostly at Khairpur. Floodwater rose 1–3 m up the palm trunks in some areas of Khairpur. The impact on the tree and fruit of date palm was recorded (Abul-Soad 2010, 2011a). It was observed that some of the young trees (offshoots) of new plantations died, whereas the offshoots that still attached to parent trees were more tolerant. However, adult date palm trees were able to continue growth and development although the lower part of the trunk was entirely immersed in floodwater for a period of time (Fig. 5.6). Large air pockets in the root tissue apparently play a role in the respiratory system of the date palm tree (Barreveld 1993).

Fig. 5.6 Flooded date palm orchards at Khairpur during monsoon rain in August 2010

Date palm is the cornerstone for the livelihood of people at Khairpur, and flooding hinders all activities in the area during and after the monsoon rain. Water covered the date palms and houses in the vicinity, as well. Moreover, the fruit crop from young trees bearing fruit bunches was covered with floodwater and completely destroyed. In addition, flooding stopped other date palm activities such as harvesting, processing, and curing in the sun.

Some areas of Khairpur were protected by barriers made from soil mixed with stones, to avoid the water overflow during the flood. At the time of the flooding, a significant number of people who were living in unprotected areas were resettled by the government and other relief agencies in the date palm orchards. Date trees provided them shelter. Some of those people were the seasonal laborers who work in the date palm industry at Khairpur during the harvest season. Rain also had a deleterious impact on boiled fruit on mats which were nearly dried *chohara*, but fermented due to the rainfall. These fruits were collected and used to feed animals (Fig. 5.7).

Date palms were subject to heavy rainfall for many days during the flooding. A positive result was that trees appeared flush with green afterward. Fruit production increased relatively in the following year through photosynthesis which expectedly has been enhanced by rainwater. In addition, soil conditions were, as expected, improved after floodwaters receded through the leaching of salts. Furthermore, river silt deposited after the flooding contributed to enhancing soil fertility as it is a good source of nutrients, particularly trace elements.

No exceptional disease symptoms emerged, particularly of the wilt disease (or as known in Pakistan, sudden decline disease of date palm). About 1–2 months after floodwaters receded, regular agricultural activities resumed with intercropping under the date palms.

Date fruits take approximately 180–210 days from fruit set until maturity. The stage sensitive to rainfall starts from full maturity (khalal stage) up to ripening (rutab and tamar stages). Summer monsoon rains always cause a problem for the date crop, particularly during July and August, whether fruits are still on the tree or being

Fig. 5.7 Deleterious impact of rain on cured dates on mats which were used as animal feed

cured in the sun on mats. An alternative way is to cure date fruit artificially in controlled champers (dehydrators) or solar dryers. This process is necessary in order to dehydrate rutab fruit to reach the tamar stage (kharak in Urdu) where the fruit moisture content is less than 24 %. Nevertheless, the available dehydrator units are insufficient to handle the considerable amounts of dates during the harvest season.

The harvesting process before the time of the monsoon rains is a critical period. Huge amounts of date fruit have to be harvested before the rains. Most of date palm areas in Pakistan experience monsoon rains; consequently, a plan should be prepared to minimize crop losses. A suggested plan may include:

(a) Establishment of early cultivar plantations which can be harvested earlier and avoid the impact of the monsoon rains, e.g., Gajjar and Kasho Wari cvs.
(b) Rain-resistant cultivars may be introduced such as Orabi cv. from Damietta Governorate, Egypt. It is a soft, late-season cultivar suited to coastal areas.
(c) Using cost-effective covering material to protect the fruit bunches.
(d) Using solar dryers to avoid curing dates in the open air on mats.

5.2.6 Fruit Bunch Covering

Date palm is usually grown in dry and arid zones where a climate of hot dry summer and little rainy winter prevails. However, a few countries are faced with the monsoon rains during the fruit-ripening season in July, August, and September, such as Pakistan, Oman, Sahara Desert (Sahel) in Africa, and USA (southern California and Arizona). Rainfall can damage the date fruit within only 1–2 days. Rainfall for few intermittent hours can cause cracks in the epicarp (the outermost layer of the fruit) at late khalal stage of Otakin cv. in Khairpur (Fig. 5.8), whereas other cultivars at early green khalal stage are less damaged.

Fig. 5.8 Cracks of date fruit epicarp caused by intermittent rainfall of a few hours accompanied by ambient humid conditions

There is a beneficial effect of a little rain by washing away dust and sand particles from the fruits. Zaid and de Wet (2002) mention that the amount of any particular rain is of less importance than the conditions under which it occurs. A light shower followed by a prolonged period of high relative humidity and clouds may cause more damage than heavy rain followed by clear weather and dry winds. However, the case in Sindh, KPK, and Punjab provinces is often different as the relative air humidity range is 50–60 % in July through September (SLUBGH 2013).

Rain during fruit ripening causes the fruit to crack and ferment and then mold can develop. In order to protect the fruit bunches from monsoon rains, several types of covering material have been experimented with, such as bituminized paper, nylon bags, and Tyvek bags. No significant difference in chemical composition was recorded between covered and uncovered fruit bunches with bituminized paper (Abul-Soad et al. 2010).

Bunch covering is, however, not only practiced against rain damage, but traditionally is also used to protect maturing fruit from birds and late season pests and prevent early ripening fruit from falling to the ground. Covered fruit bunches of Aseel cv. ripen a week earlier as compared to uncovered bunches in Khairpur (Abul-Soad et al. 2010). A bunch cover may capture heat inside the covers which accelerates the fruit ripening. To the contrary, it was reported that rainfall near harvesting lowers the temperature and delays ripening (Barreveld 1993).

In Balochistan, covering baskets made from the date palm leaflets (baat) have been used to cover the date fruit bunches and protect them from hot, dry winds bearing sand particles, catching falling fruit, or to prevent over drying. Mesh bags and waterproof bags may be employed together to protect fruit bunches against rainfall and to realize other benefits.

A unique type of bag (Tyvek®) recently appeared on the Pakistani market which has been used on other fruit crops for similar purposes (Fig. 5.9). Tyvek® is part of

Fig. 5.9 Protection of the date fruit bunches from monsoon rains by Tyvek bags in Pakistan

a family of tough, durable sheet products used in a variety of applications. It is a product of DuPont Company which uses rolls of Tyvek paper to manufacture the covers. It is made from very fine high-density polyethylene fiber which makes Tyvek® fabric lightweight yet strong, vapor permeable, yet water, chemical, puncture, tear, and abrasion resistant.

Subsidized Tyvek® bags were distributed among date palm growers all over Pakistan in 2010, but the trial failed. Growers stated that this type of bunch cover captured moisture within the bag and rapidly deteriorated the fruit. The number of spoiled fruit and fruit drop increased after bunches were covered with Tyvek bags (30–40 %). Such results turned the overwhelming majority of growers against their use.

In fact, this type of bags may be feasible to cover date fruit bunches. The optimal use is to protect date fruits against monsoon rains; the appropriate time to cover fruit bunches of different cultivars at different places in Khairpur was investigated by DPRI, Shah Abdul Latif University, Khairpur, with aid from DuPont Pakistan and the Agribusiness Support Fund (ASF) (Abul-Soad et al. 2010).

In reality, the size of the Tyvek bags might need to be modified to fit the variable sizes of fruit bunches. Bags should not be tied at the base, leaving the lower portion of the fruit bunch exposed to the air for more ventilation. Fruit bunches can be safely covered 1–3 weeks before harvest in July, at the site where the experiment has been conducted.

The climate varies within Khairpur area itself which may affect covered fruit bunches. Covered fruit bunches near the mountains mature earlier than other areas due to the dry weather and high temperatures. Early-ripening cultivars of dates always have the advantage to gain higher revenue since they reach the retail market first. It was noted that the Fasli/Toto cultivar was very sensitive to fruit drop if the harvest is delayed. The usual harvesting time of a cultivar must be taken into consideration when a grower needs to harvest the covered fruit bunches. The covered fruit bunches of Fasli cv. must be harvested 1 week earlier (last week of June) than the traditional harvesting time of this cultivar (first week of July). Otherwise, moisture and mold are subsequently going to increase inside the bag, and the spoilage rate will subsequently increase, and eventually 90 % of fruits will fall. Due to climate change, harvest time may be delayed 1–2 weeks after a prolonged winter season.

Covering fruit bunch was found to be a time-consuming process; a single date palm tree needs the efforts of one or two laborers for 30–45 min at least to cover all bunches with Tyvek bags. Fears about moisture accumulation inside the bag remain the main concern of growers.

In general, covering fruit bunches is a good practice to protect them from birds, pests, dust, as well as monsoon rains. The collected fallen fruit inside the net cover also helps avoid pest and disease problems in the following season. Using covers should be carried out with a complete package of proper instructions according to the needs of local cultivars, prevailing conditions at the cultivation area and time of covering.

5.2.7 Pruning and Dethorning of Fronds and Pollination Approaches

Pakistani growers mostly carry out frond pruning of adult trees after harvest in August and before spathe emergence in February. The third occasion is during pollination in March and April. The lower 2–3 whorls of fronds are cut and 15 cm of the base left in place to be used as steps to climb the tree for the common practice of manual pruning. Simultaneously 2–3 leaf whorls are retained on the tree beneath where the spathes are developing. Thus those fronds can help support the heavy fruit bunches to avoid bunch stalk break, in most of the cultivars. During the pruning operation, unwanted offshoots are removed to encourage growth of the parent palm and remaining offshoots and to allow better access to the bearing palm.

Removal of spines is called dethorning. With a pruning knife, date spines are usually removed from the new growth of fronds in the crown of the palm just before the pollination season to allow easy access to the date spathes as they bloom. Such an operation will ensure a safe approach to the spathes for their pollination and also avoid any risk of injury to laborers during other technical practices (e.g., tying fruit bunches down, covering of bunches, and harvesting).

In most of date palm countries, including Pakistan, the traditional pollination technique is to place a few male spikelets in the center of the opened female spathe. Artificial pollination can be done according to the traditional method or by using a mechanized device. The number of spikelets depends on the cultivar and climatic conditions. For instance, 4–5 male spikelets are used for pollination of Aseel cv. to achieve good fruit quality. On the other hand, date palms in mountain areas, where the climate is dry and hot, some growers use only 2–3 spikelets. It would be a good idea to introduce powerful and famous males to Pakistan to improve the fruit quality such as Khikri Adi and Ghannami Akhdar (Abdul Wahid et al. 2010).

Fruit quality depends mainly on the pollen grain source. Growers typically obtain male spathes at the beginning of the season from the market with wide variation in quality (Fig. 5.10). Using stored pollen grains for pollination is not a practice

Fig. 5.10 Fresh spathe market at Khairpur as a source of pollen grains used for pollination

Fig. 5.11 Mechanical pollinator donated by FAO to DPRI and date palm growers in 2010

followed in Pakistan. Nevertheless, a study to evaluate the pollen viability of different date palm cultivars growing in the Punjab showed that the stored pollen of Khadrawi cv. at–20 °C had higher germination ability (71 %) after 12 months of storage compared with Hillawi cv. (34 %). This study also found that the pollen grains needed 24 h after thawing to germinate on an in vitro medium at 30 °C compared to lower temperatures (20 and 25 °C). Air-dried pollen is enclosed in sealed capsules before storage (Maryam et al. 2013).

Mechanical pollination was developed mostly in the USA where labor is expensive and not always available. It has been one of the most important alternatives to reduce associated labor costs by 50–70 % (Galeb et al. 1987; Nixon and Carpenter 1978). The FAO office at Islamabad, in 2010, donated a few pollinator machines which were made in UAE to progressive growers in Balochistan and Khairpur and also to DPRI (Fig. 5.11). But most of the growers avoided using them for reasons such as the high price of fuel, tall trees, dense plantations, intercropping, lack of technical training of usage, small property, and the large quantity of pollen grains needed to fill the machine tank.

5.2.8 *Pest and Disease Control*

Date palm trees in Sindh, particularly Khairpur, suffer from a lethal disease called sudden decline syndrome (Abul-Soad et al. 2011; Maitlo et al. 2013). Symptoms resemble those of similar wilt diseases, such as palm lethal yellowing caused by *Phytoplasma* which is a fatal disease of coconut and also infectious to date palm (Al Awadhi et al. 2002; Ammar et al. 2005; Thomas 1974). The drying of fronds exhibited in date palm decline disease is also similar to symptoms of bayoud disease caused by *Fusarium oxysporum* Schlecht. f. sp.*albedinis* in Morocco and Algeria (Djerbi 1982).

Fig. 5.12 (**a**) Healthy date palm tree, (**b**) tree infected with sudden decline disease in Khairpur

The common symptoms of sudden decline syndrome in Khairpur start with an orange-yellowish rachis followed by gradual drying of the pinnae. Drying starts from old lower fronds and moves toward younger central fronds and finally with a complete dried crown followed by death within 6 months (Fig. 5.12). The disease may afflict adult date palm trees in winter preventing spathe emergence in February, or infection may occur during fruit development in summer causing immature and unripe fruit formation.

Age is not a factor for a date palm tree to be infected with the disease in view of the fact that all ages showed common symptoms; however, males were found to be less infected than productive females. Among the females, unhealthy trees and those growing under adverse conditions, particularly waterlog stress, were more vulnerable. There is a suspicion that the presence of an infection source (*Fusarium solani*) associated with an adverse condition of waterlogged soils may be responsible for the date palm decline disease in Khairpur. All the infected trees showed root rot symptoms with vascular root tissues turning brown (2 cm in width) at 60 cm depth. Therefore, the number of roots decreased, and the normal roots were white and newly emerged (Fig. 5.13). It appears that soil conditions become unsuitable for adult root growth, and the tree attempts to bring forth new adventitious roots to survive. Intercropping with date palm and associated frequent flood irrigation leading to waterlogging may be the cause. At present it is a critical period for this disease. The number of infected trees is increasing day by day in Khairpur and surrounding districts without a cost-effective and viable solution.

A method was developed to treat the diseased date palms consisting of a four-tiered procedure. This procedure included cutting off infected fronds, foliar application of copper oxychloride (0.5 % solution), soil application of Topsin M (0.3 % solution), and chemical fertilizer application (Maitlo et al. 2013). Taking into consideration the adverse impact of waterlogging, measures have to be made to improve soil drainage.

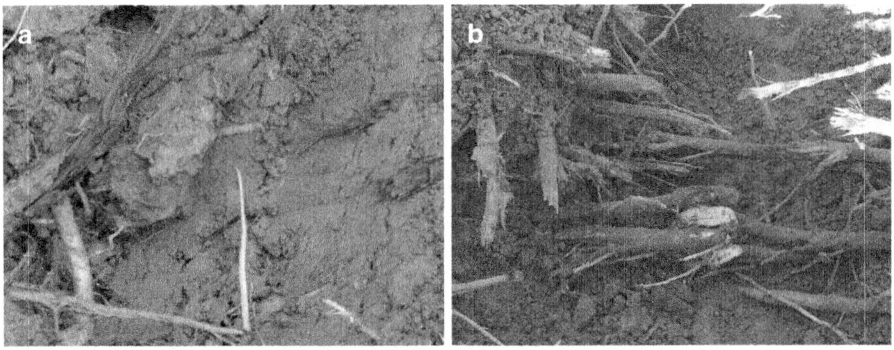

Fig. 5.13 Root rot associated with the sudden decline disease. (**a**) The vascular colorization of infected roots, (**b**) healthy roots of date palm at 60 cm depth in clay soil

On the other hand, concerted efforts with sanitary procedures should be taken to prevent the transmission of the disease to other producing areas. Also, cost-effective and innovative solutions should be created to control such a major disease in Khairpur.

An extensive survey was made by DPRI, Shah Abdul Latif University, Khairpur, during 2009 on the tissue boring pest of date palm, red palm weevil (*Rhynchophorus ferrugineus* Oliver). The survey covered 46 orchards of 8 different locations at Khairpur area. Results of the survey indicated that 219 date palms (5.81 %) were infested with RPW. The pest incidence revealed that the maximum infestation percentage was 9.29 % for trees 8–10 years old, followed by 7.96 % for trees in the 11–15-year-old age group. However, trees 35–40 years of age had a very low infestation incidence (0.02 %), (Jatoi et al. 2010).

In Khairpur, the most tolerant cultivars with a lower infestation percentage against RPW were Muzawati (8 %) followed by Pathri (12 %), Thothar (15 %), and Khar and Shakkri (16 %). Those cultivars which suffered a high infestation rate were Karbaline, Dedhi, Fasli, Aseel, and Eidan Shah (Shar et al. 2012). The clear-cut symptom of RPW infestation is brown sap oozing from the trunk associated with a bad smell or strong evidence of holes in the trunk where the larvae make tunnels to enter the tree. Offshoots are particularly vulnerable to RPW mostly at the point in where they are joined to the mother palm (Fig. 5.14).

The conventional method to deal with an infested tree is to keep monitoring date palm trees in the orchard to discover any early infestations by RPW. The control treatment starts once the grower sees brown sap oozing from the trunk or the surrounded offshoots (Fig. 5.14a). If it was an early infestation, the larvae are removed from the tunnels and the wound covered with clay. But if the infestation is heavy, the superficial tissue around the trunk base is removed with a metal tool and any larvae found collected and destroyed. Tablets of Phostoxin or/and pieces of natural stone of calcium carbide are inserted into this tissue by hand. The base of the trunk is covered by a mud mix made of a pesticide solution (malathion at 0.3 %) clay and wheat or rice straw (Fig. 5.14b). A few tubes may be inserted above the covered area of the trunk and frequently filled with any suitable pesticide. In addition, fertilizer treatment

Fig. 5.14 Infestation and control of red palm weevil. (**a**) Symptoms of infection in adult trees, brown sap with bad smell, (**b**) controlled date palm tree after remedy application

with N-P-K chemical fertilizer was applied (500 g NPK Zarkhaze 17–17–17) every 15 days before irrigation.

Pheromone traps are used to survey and control any RPW infestation, but it is not a common practice in Pakistan. In a study conducted in the Therhi area, Khairpur district, 18–21 adults of RPW were captured during the first, second, and third week of July (Shar et al. 2012).

Heavy fruit drop in Dhakki cv. has become a serious problem for the growers at Dera Ismail Khan. More than 70 % of the date fruits dropped before maturity compared to 10 % for other cultivars. About 15 % regular fruit drop every month of Dhakki cv. was recorded (Marwat et al. 2012). Also, the irregular fruit drop of Dhakki cv. has been noted at other locations in Sindh (Abul-Soad 2011a). In reality, it might be a genetic character for that cultivar. The dried fruit beetle (*Carpophilus hemipterus* L.) is a cosmopolitan pest that attacks a large number of agricultural commodities both before and after harvest (Bartelt et al. 1990). Host species include onion and wheat which are widely cultivated with the date palm in all four provinces. Dried fruit beetle could be the main pest causing the late date fruit drop in Pakistan. Dried fruit beetle was detected to vastly infest fruit at late stage during harvesting seasons in Khairpur and Dera Ismail Khan (Fig. 5.15a). The adult female, 4 mm in length, can lay 500–1,000 eggs, and within a single month these will reach adult stage to attack other fruits. It attacks fruit of soft, semidry, and dry dates and also can attack stored dates. Measures that could be used to control such pest are:

(a) Fruit bunches should be covered with nylon or net covers once fruits start taking color, to prevent adult attack. Collecting the infested fruit after harvest and burning it is recommended.
(b) Trap baits of 1:1:1 mixture of acetaldehyde, ethyl acetate, and ethanol (Smilanick et al. 1978).
(c) Using pheromones mixed with dates.

Fig. 5.15 Symptoms of some major problems affecting date palms in Pakistan. (**a**) Dried fruit beetle infestation during fruit ripening; (**b**) frond bases infested with termites; (**c**) early fruitlets dropped 2 months after fruit set showing the healthy, damaged, and infested fruit; (**d**) dwarf date palms growing in saline soil and low soil percolation; (**e**) symptoms of early infection of Graphiola leaf spot disease as yellow sori; (**f**) brown spots of dried leaflets of infected leaflets with Graphiola leaf spot disease; (**g**) fruit blemishes caused by high air humidity during khalal stage of Hillawi cv. in Jhang, Punjab province; (**h**) furrow caused by rat attacks on small offshoots in northern Balochistan

Termites infest an overwhelming majority of date palm trees in Khairpur, especially near the desert. Adult termites make tunnels in the fruit bunches and frond bases on the trunk (Fig. 5.15b). No attention has been paid to the control of this pest by either the Agriculture Department of Khairpur district or the growers, although agricultural practices and chemical treatment can reduce the damage.

Early fruit dropping caused mainly by the date palm lesser moth (*Batrachedra amydraula* Meyr.) was detected in Khairpur, particularly in the Therhi area, and causes severe fruit loss 1–2 months after fruit set each year with more than 30 % loss. Symptoms can be recognized by the red, damaged small fruit on the ground (Fig. 5.15c). Also, most of the fallen fruits exhibit black spots near the cap (site of the infestation). Frequent chemical treatment every year with cypermethrin exacerbated the problem and produced resistant strains as the spray was not effective after a few years of application. The recommendation to control this problem is illustrated in Table 5.3.

Elevating the soil level by 0.5–1.0 m is suggested to provide a new medium for the roots away from stagnant subsoil water, particularly in the saline soils of

Table 5.3 Major problems facing date palm production and protection in Pakistan (Abul-Soad 2013b)

Province	Problem	Description	Recommendations
Sindh	Monsoon rains	Sudden rainfall during fruit ripening and harvesting season destroys fruit bunches, drying fruit on mats and small offshoots	Planting local early cultivars, e.g., Kasho Wari, Gajjar
			Fruit bunch covering
			Solar dryers
	Sudden decline disease	Yellowing then drying out of the mature fronds from the outside toward the center. Root number is decreased and root rotting is occurred followed by palm death within few months	Avoid new plantations with polluted offshoots. Avoid waterlogging and excessive irrigation
			Stop intercropping with rice, wheat, banana, and alfalfa which need regular flood irrigation
			Avoid cultivation on water canals
	Red palm weevil	Brown sap with a bad smell mostly on the lower part of the trunk of adult trees or small offshoots. Small furrows and holes on the lower part of the trunk which may be followed by tree decline. As there is no management to control RPW, symptoms are spread over all cultivation area particularly in the abandoned orchards	Regular monitoring of date palm trees to control early infestation by Phostoxin tablets
			Pesticide treatment of cut surfaces and wounded tissues after pruning and offshoot detachment
			Offshoot detachment
			Chopping, burning, and burying heavily infested trees

(continued)

Table 5.3 (continued)

Province	Problem	Description	Recommendations
	Termites	Small tunnels on different parts of the remaining part of fronds after pruning, fruit bunch stalks, and frond midribs. Soil canals on the base of the plants	Spray with a termite killer
			Removing and burning destroyed offshoots
			Soil plowing with light irrigation
	Date palm lesser moth	Small fruit dropping after 1–2 months of pollination from 30 to 50 %. Black spots near to the fruit cap and red deteriorated fruit on ground	Palm maintenance and cleaning
			Bunch covering and burning fallen fruit of last season
			A couple of spray with appropriate pesticide before flowering and after fruit set with 15 days
Balochistan	Monsoon rains	As aforementioned in Sindh	As aforementioned in Sindh
	Dubas bugs	Dubas excrete heavy honeydew and dust sticks to it. Dripping of honeydew may damage the crops growing under the trees. High infestation causes 50 % fruit yield reduction	Windbreaks help to reduce injury and cultural practices
			Using natural enemies
			Systemic insecticides at the appropriate time are useful in the high levels of infestation
	Groundwater deficiency	Irrigated date palm orchards watered by pumped water from the aquifers and nonirrigated date palm orchards that depend on shallow water conserved after monsoon rains resulted in yield fluctuations	Using drip irrigation systems
			Padding the irrigation canals to reduce water leakage
			Flood irrigation system is prohibited
Punjab and Khyber Pakhtunkhwa	Monsoon rains	As aforementioned in Sindh	As aforementioned in Sindh
	High humidity	Brown scars on the fruit at khalal stage	Balanced irrigation

Khairpur and similar areas. It was observed that the date palms grown in soil covered with salty sewage and drainage water were dwarfed; the number of fronds dramatically decreased and fruiting ceased (Fig. 5.15d). River silt after floodwaters receded can be a good source of material for soil elevation. On the other hand, this process is necessary for the intercropped species and has been followed in northern Egypt as a method for reclamation of saline soils in coastal areas of the

Mediterranean Sea. Mango, apple, and vegetables have been successfully intercropped with date palm at Damietta Governorate in Egypt by annual or biannual soil elevation.

Virtually all date palms seen in Khairpur were suffering from the Graphiola leaf spot disease, or false smut, caused by *Graphiola phoenicis*. This problem mainly appears because of intercropping which increases the relative humidity to above 50 %. Severe attacks of the fungi occur in October after the monsoon rains. Symptoms begin with yellow spores called *sori* (Fig. 5.15e). The fungi dry out leaflet tissue where it occurs and then all the leaflets become dried (Fig. 5.15f). Infections are transmitted from infected to uninfected tree by air and the touching fronds of very closely planted palms. Maintaining proper planting distance and a prohibition of intercropping are the first step to control such a problem. Other means are listed in Table 5.3.

In Punjab and KPK districts, the presence of the clouds and monsoon rains during summer causes fruit blemishes on the fruit during khalal stage. Most of the harvested fruit of the prominent Hillawi cv. in Punjab showed brown blemishes (Fig. 5.15g).

A date palm sap-sucking insect known as the dubas bug (*Ommatissus lybicus*), locally known as *sherrgo*, is spreading in Balochistan, particularly in Panjgur district in the center of the province, and affecting overall date production. The dubas bug was first detected in 1999 in Panjgur district, and it is one of the major reasons for the low date production in Balochistan. Chemical control using different pesticides has increased the problem by creating resistant strains. Both the nymphs and adults suck the sap from the fruit stalk and fronds and may cause about 50 % of economic losses. Biological control is one of the sustainable methods to curb the spread of this pest (Shah et al. 2012).

In northern Balochistan, villagers in Gualishtop near Nok-Kundi claim that within the last decade a rodent resembling a large rat has become a threat to the date palm plantations of the area. Abul-Soad et al. (2009) conducted a study at these areas in 2009 and recorded that the rodent moves in tunnels 30–60 cm in diameter burrowed by them, extending to more than a hundred meters. The rodents dig their tunnels in sandy soil approximately 0.3 m below the surface and reach the pith of the date palm. The rodent damages the fronds at the base of the younger date palms (10–15 years) which leads to a fungal infection and consequently devastates the tree (Fig. 5.15h). In mature trees the rodent is said to eat the sweet and soft apical meristem, resulting in death of the tree.

Rodent attacks on date palms have been recorded in Nok-Kundi, Mashkel, and Dalbandin causing complete destruction of both adult palms and small offshoots. This exotic pest mostly became a major threat due to some change in the equilibrium of natural forces, particularly because of a prolonged drought period of 4–5 years in this arid region of the province.

The Vertebrate Pest Control Institute, Karachi Univ. Campus, has made a study covered the area of Nok-Kundi and Dalbandin. Their salient findings indicated that mole rat (*Nesokia* spp.) has a narrow feeding niche and in noncrop lands of Balochistan is largely herbivorous. Food habit studies showed that rodents throughout the season feed principally upon the date fruits, stem, grasses, and roots, but date palm pith was the main component of the diet.

Rodenticide bait and trapping were used to eliminate the rodents. This attempt to control the rodents was not successful hitherto. However, the rodenticide baiting used was unsafe for grazing ruminants and livestock. Recognizing the damage caused and the risk of the threat of spreading to Rajai, Wadian, and other areas, further studies are needed to understand the nature of the problem and to develop control techniques such as launching an augmentation program for its natural predator the barn owl. This is necessary in order to free date palm plantations in Balochistan from this menace (Abul-Soad et al. 2009).

On an international scale, the only control measure for rodents is poison, composed of a mixture of zinc phosphate at 30–50 g, 1 kg of millet flour, and 3 % cooking oil. The paste is placed around the palms at the entry to the galleries. A chemical product Finale gave excellent results at the Eersbegin project in Namibia. It is a highly active anticoagulant bait at 0.025 g/kg as an active ingredient. The rodents die in 4–12 days. The chemical was used (July and August 1997) in both the Eersbegin and Naute date plantations in Namibia with a good success rate against *Mus musculus* (Zaid and de Wet 2002). Other problems are briefly discussed in Table 5.3.

5.2.9 *Major Problems Facing Date Palm Cultivation*

In Punjab, awareness about cultivars, irrigation practices, fertilizer application, offshoot transplantation, and insect/pest of date palm are the major factors hindering better yield and ultimately the profit for growers. There is the need to train growers about irrigation and fertilizer application, disease and pest management, offshoot transplanting, and processing of dates (Ata et al. 2012). The major problems facing date palm cultivation, production, and protection in the four Pakistani provinces are discussed according to their potential in Table 5.3.

A study was carried out in 2007 in northwestern Pakistan to investigate the causes of late adoption and slow dissemination of date palm cultivation and to examine the role of education. The study showed that growers with little or no formal education were late adopters of date palm cultivation. In order to increase date palm production, the growers' educational level, both formal and agricultural, should be enhanced with specific focus on adoption of new agricultural practices reflecting the latest research (Khan et al. 2009).

5.3 Genetic Resources and Conservation

5.3.1 *Threats and Degradation of Date Palm in Balochistan*

Balochistan has dry hot weather during the date-growing season. Annual rainfall is below 100 mm, and the wind speed is mostly high and charged with dust. Date palm plantations are situated in areas where the groundwater is shallow (5 m), mainly

Fig. 5.16 Date palm plantations are threatened by sand dunes in the desert area of Hamun-e-Mashkel in Balochistan

around the Hamun-e-Mashkel (a salt lake in the middle of the desert), and a belt along the Iranian border where the Tahlab river basin joins the Hamun-e-Mashkel. Date palm populations in this area may reach a half million trees. About 20 % of these trees are dispersed in scattered villages. Livelihood in the areas is principally dependent on date palm.

Sand dunes, drought periods, poor management, and shortage of skilled labor are the current problems in such areas. Sand dunes can cover date palm plantations (Fig. 5.16). Many date palms have disappeared under mountainous sand dunes built up by steady winds during the date season.

Growers utilize *baat* or *sondh* (Fig. 5.17), i.e., baskets made from date palm leaflets to protect the developing fruit from strong winds carrying sand particles. Every 1–3 fruit bunches were covered with a single *baat*. At harvest time, the practice is to cut the *baat* along with upper frond to which it is tied.

Harvesting begins in July in the Mashkel area south of the Hamun-e-Mashkel, and the harvest time varies slightly from area to area due to climatic conditions. Harvest in areas located toward the north of the Hamun-e-Mashkel, such as Gualishtop, Rajai, and Wadian, begins around the end of the first week of August. In southern Balochistan at Kech and Panjgur, fruit harvest is usually initiated from the first week of July and extends to mid-September.

The date palm is drought tolerant as compared to other fruit species and is adapted to water stress and low water availability found in the desert and oasis conditions. However, a prolonged period of drought in the southwest of Balochistan has led to a reduction in the rate of formation of new fronds and yield. The distribution of roots in the irrigated and nonirrigated date palm plantations varies; root numbers are higher for regularly irrigated date palms.

For appropriate growth and development, however, extremes of water level, neither waterlogged conditions where the roots cannot breathe nor water levels deeper

Fig. 5.17 Date palm plantations in northern Balochistan. (**a**) Nonirrigated date palm trees, (**b**) covering bag, *baat*, made from date palm leaflets placed over the fruit bunches

than 5–6 m, are undesirable. The trees extend their roots to adapt to a drop in the level of groundwater. Yield was measured of about 50 % of normal annual production at a groundwater depth 5–6 m. The impact of lowering the groundwater level depends mainly on the availability of supplementary water from rainfall. The date palm tree can conserve the water for months to prolong its life.

It is difficult to measure the maximum depth to which roots can reach. However, during a drought period, the date palm trees near the dry waterway of the Tahlab and Washab rivers were able to extend horizontal roots up to 20 m in length to reach the waterway.

Long periods of drought and deep groundwater levels, which measured 17 m in the irrigated date palm plantations of Yak Mach village, forced growers to rely on pumped water to establish new date palm plantings in the area. Wide variation in water quality was detected between the irrigated and nonirrigated areas. The EC of groundwater at Gualishtop area was 10,760 $\mu S\ cm^{-1}$ compared to 4,140 $\mu S\ cm^{-1}$ at the irrigated date palm plantations of Yak Mach village. Higher salt concentration in the irrigation water could be a contributing factor to the lower productivity at Gualishtop (on average of 10–30 kg $palm^{-1}\ yr^{-1}$) as compared to 30–50 kg at Yak Mach (Abul-Soad et al. 2009).

Similar figures were reported for other areas in the world. A drop of groundwater level to 20 m at the Moroccan oasis of Jorf caused drought impact on the date palms, even though the annual rainfall varied between 50 and 250 mm (El Baali et al. 2002), forcing the growers to initiate irrigation.

Due to the poor management in such Pakistani areas, potentially useful cultivars of very small populations (10–15 trees) are at risk of degradation and extinction. Exceptional among the landrace cultivars is the Shandishkand which currently exists in Gualishtop village near Hamun-e-Mashkel Lake. The crispy dried fruit of this fabulous cultivar is in high demand by the local growers, and the challenge is to

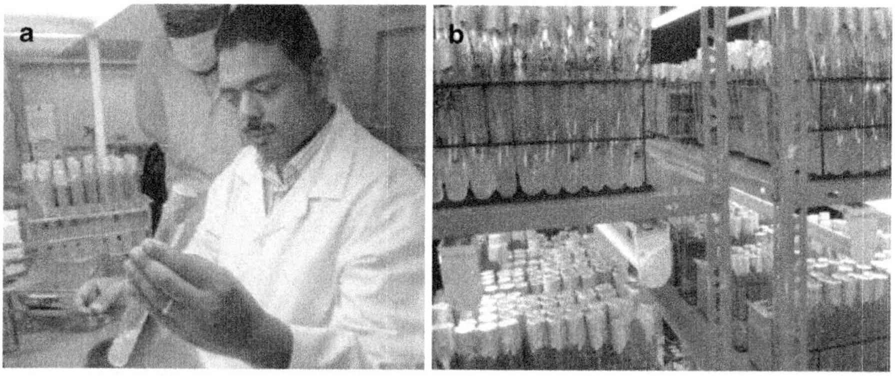

Fig. 5.18 Date Palm Laboratory of DPRI, SAL University, Khairpur; (**a**) culture room, (**b**) in vitro cultures of different cultivars using shoot tip and inflorescence explants

propagate it. Tissue culture is the only solution to conserve such cultivars and other similar landraces that are well adapted to the local environment.

Unfortunately, poor horticultural management when new plantations are established, along with stress brought on by pest attacks, particularly rodents and mites, has restricted new plantation expansion and reduced the productivity of these dry land areas.

There are no genebanks for date palms in Pakistan. Nevertheless, in Khairpur, Faisalabad, Bahawalpur, and other date palm locations within the country, collection farms have been established to collect and conserve different local and prominent cultivars.

5.4 Plant Tissue Culture

The development of tissue culture techniques for mass propagation of date palm plants has revolutionized the world date palm industry. In Pakistan, tissue culture of date palm began two decades ago. A few plants were produced through somatic embryogenesis and transplanted to open fields for evaluation, but the fruits were inferior, which discouraged growers (Hussain et al. 1995; Rashid and Quraishi 1994). A very few research laboratories are trying to develop a successful protocol for commercial propagation of date palm, but without notable results (Khan and Bibi 2012). However, on both commercial and research levels and without dedicated funding, the Biotechnology Laboratory of DPRI, SAL University, Khairpur, succeeded in producing, in 2010, a few thousand plantlets of six Pakistani and three foreign cultivars. The local cultivars were Dhakki, Gulistan, Dedhi, Kasho Wari, Gajjar, and Kurh, along with the foreign Barhee, Zaghloul, and Partamoda cvs. (Abul-Soad 2011b, 2012a; Abul-Soad and Mahdi 2010, 2012; Abul-Soad et al. 2007). The production of tissue-cultured palms took 4–6 years and involved three sequential phases:

(a) Laboratory-based plant production (in vitro plants) for 2–3 years (Fig. 5.18).

Fig. 5.19 The ex vitro date palm plants 2 years of age in the greenhouse

Fig. 5.20 In vivo date palm plants 1 year of age planted in open field of SAL Univ., Khairpur

(b) Plants establishment in a greenhouse (ex vitro plants) for 1–2 years (Fig. 5.19).
(c) A growing-on phase in a shaded nursery or net house before field cultivation for 1 year, then plants were transplanted to the open field in 2012 (Fig. 5.20).

After the full cycle of tissue culture production period, plants were cultivated in the open field for evaluation and fruiting was expected within 1–2 years. There is demand for tissue-cultured plants by date growers in Pakistan, but the high price of cultivars from outside Pakistan is delaying their dissemination. The DPRI team used the knowledge gained by a visiting Egyptian scientist who arrived in 2006 to use inflorescence explants for date palm micropropagation to satisfy the local demand for Pakistani and elite world cultivars.

5.4.1 Micropropagation Using Inflorescence Explants

Inflorescence-based micropropagation holds great potential for the multiplication of recalcitrant individual male and female date palms and cultivars of commercial interest with limited populations. A female palm from seed and bearing a superior crop is great repository for new cultivars. This can be accomplished in a short time with minimal efforts as compared to the traditional practice of using shoot tip explants (Abul-Soad 2011b; Abul-Soad et al. 2006). Pilot production trials in DPRI revealed that 1–2 immature spathes were enough to produce thousands of the plants, while about 100 offshoots may be required for producing a similar number of plants using the traditional explant, the shoot tip (Abul-Soad and Mahdi 2010). Using inflorescence explants may be the dream of any laboratory in the world to micropropagate date palm. The achievement of DPRI could be a breakthrough in the date palm growing world.

Details concerning the major research laboratories working on date palm micropropagation in Pakistan are (a) PRI, SAL University, Khairpur, Sindh; (b) the Biotechnology Laboratory in HEJ Research Institute of Chemistry, Karachi University, Sindh; and (c) Plant Biotechnology Program, National Agriculture Research Centre (NARC), Islamabad.

5.5 Cultivar Identification

Morphological data along with the DNA-based techniques and sequencing of different gene fragments from the chloroplast genome were used as a means to reliably identify different date palm cultivars in Pakistan.

5.5.1 Morphological Characterization

Commercially important cultivars from Bahawalpur and Jhang have been morphologically characterized. The traits studied were frond length, frond width, frond base width, pinnae number, spine number, midrib length, midrib length with spines, midrib length with pinnae, length of the top pinnae, fruit weight, fruit volume, fruit length, fruit diameter, and seed weight. The data were subjected to principal component analysis which determined that frond length, midrib length, fruit weight, fruit volume, fruit length, and seed weight together contributed to 34.1 % variability of the data set. Variability of 16.9 % was caused by spine number and spine area length, while the fruit weight and fruit volume together contributed to 15.8 % variation. Variation among different cultivars in terms of seed weight was limited (8 % in the data set). Ten traits including frond length, midrib length, spine area length, spine number, fruit weight, fruit volume, fruit length, and seed weight contributed more than other traits to the observed variability among cultivars (Nadia et al. 2013).

Twenty-one cultivars were used to determine physical fruit characteristics. Results indicated that among all cultivars, Dhakki recorded the maximum fruit weight (13.89 g $fruit^{-1}$), flesh weight (12.89 g $fruit^{-1}$), fruit length (4.56 cm $fruit^{-1}$), and volume (11.94 cm $fruit^{-1}$). Edible/nonedible ratio of different cultivars ranged from 1.94 for Desi simple cv. to 14.50 for Aseel cv. (Nadeem et al. 2011b).

The anatomical features of root, frond, and rachis of ten Pakistani cultivars were described. Date palm has some structural variation of these tissues to cope with an

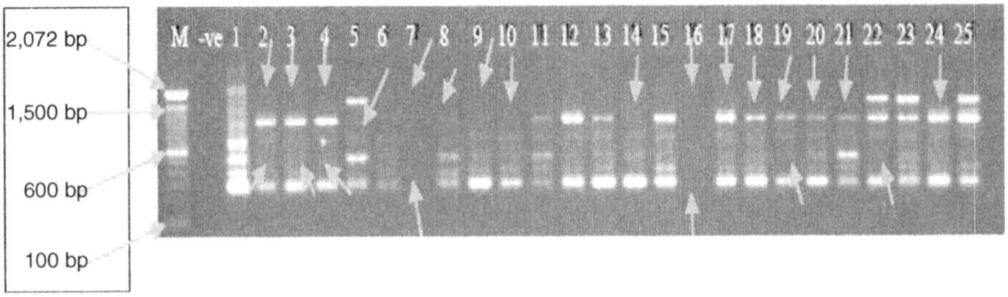

Fig. 5.21 Polymorphism level among 25 major Pakistani date palm cultivars with GLG-16 RAPD marker

adverse environment. The presence of stomata on one or both sides of the frond and variation in the stomata complex were highly dependent on the environment. Mesophyll in the frond was highly compact with large cells closely associated with vascular bundles in the laminar portion. The roots among cultivars varied in epidermal size, inner and outer cortex thickness, presence of sclerenchyma layers and sclerenchyma bundle, thickness of endodermis and pericycle, number of metaxylem vessels, and phloem area in the vascular region. Cultivars from saline soil and low relative humidity conditions possessed a thick sclerenchyma region, while cultivars growing under high relative humidity developed a larger area of cortex tissue. The date palm rachis has axial vascular bundles which were constant in size and the center of the rachis. Among these ten cultivars, Chuwara, Zahidi, and Makran showed discrete morphological characteristics and high structural variation (Ahmad et al. 2013).

5.5.2 Molecular Characterization

Genetic diversity and the phylogenetic relationships among 25 Pakistani date palm cultivars were analyzed. The study was carried out by using six universal random amplified polymorphic DNA (RAPD) primers. The RAPD primers showed polymorphism among all date palm cultivars (Fig. 5.21). Based on the pairwise comparison of amplification products, the genetic relationship was estimated. All date palm cultivars showed variation at the DNA level. The average of genetic diversity among the date palm cultivars ranged from 0.64 to 0.95. A dendrogram was constructed using NTSYSpc program. On the basis of this analysis, the populations were clustered into two main clusters and subclusters (Mirbahar et al. 2014).

For example, the RAPD primer GLG-16 produced maximum 8 DNA bands with Aseel cv., and on the other hand no band was generated by Seedless and Kasho Wari cvs. The most common and dominant number of 05 bands was observed with cvs. Asul Khurmo, Asul Kurh, Karbaline, Khar, Kupro, Noori, Otakin, Dhakki, Hillawi, Begum Jangi, Muzawati, and Basra.

On the average 4 bands were generated by this primer with all 25 cultivars. Shakkar and Rabai cvs. produced 07 DNA bands and Sher Shahi Dokka,

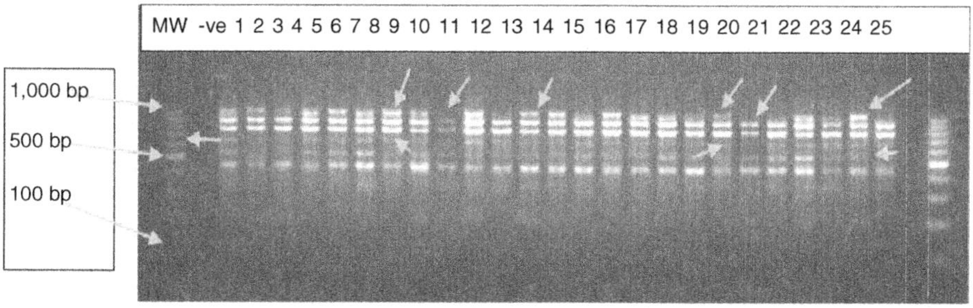

Fig. 5.22 Polymorphism level with ISSR marker HB-15 among the major 25 Pakistani date palm cultivars. *MW* Molecular weight, *1* Aseel, *2* Asul Khurmo, *3* Asul Kurh, *4* Dedhi, *5* Gajjar, *6* Karbaline, *7* Kasho Wari, *8* Khar, *9* Kupro, *10* Nar Aseel, *11* Noori, *12* Otakin, *13* Dhakki, *14* Gulistan, *15* Hillawi, *16* Seedless, *17* Sher Shahi Dokka, *18* Begum Jangi, *19* Muzawati, *20* Shakkri, *21* Ab-e-Dandan, *22* Shakkar, *23* Husaini, *24* Basra, *25* Rabai

Ab-e-Dandan, and Husaini produced 06 bands, respectively, with GLG-16 RAPD primer. This RAPD marker produced total 121 bands and an average number of 4.84 DNA bands produced with the 25 cultivars. GLG-16 marker produced 08 polymorphic bands and without monomorphic. This marker showed 100 % polymorphism with 20 cultivars (Fig. 5.21).

The study was carried out using seven anchored inter simple sequence repeat (ISSR) markers. The ISSR primers showed high polymorphism (84 %) among the 25 cultivars. The amplicon obtained was successfully used to differentiate the cultivars. The 25 cultivars showed variations at the DNA level. NTSYSpc software was used to check similarity matrices and phylogenetic relationship among the cultivars. The pairwise similarity among the cultivars ranged from 0.608 to 0.980. Therefore, ISSR markers have divided the 25 cultivars into two main clusters and subclusters (Mirbahar et al. 2013). For example, the primer HB-15 generated maximum 06 bands with DNA templates of Aseel, Asul Kurh, Dedhi, Gajjar, Karbaline, Khar, Nar Aseel, Hillawi, Seedless, Sher Shahi Dokka, Muzawati, Shakkar, Husaini and Basra and minimum 04 DNA bands with DNA templates of Kupro, Noori, Otakin, Dhakki, Shakkri, and Rabai cvs. Five cultivars, Asul Khurmo, Kasho Wari, Gulistan, Begum Jangi, and Ab-e-Dandan, produced 05 bands with this primer. The primer HB-15 produced total 133 DNA bands with all 25 cultivars. The average number of 5.32 bands was generated by each cultivar with HB-15 primer. This primer produced 3 monomorphic bands and 3 polymorphic bands which showed 50 % monomorphism and 50 % polymorphism, respectively. The bands were of 400–1,000 bp (Fig. 5.22).

The complete date palm chloroplast genome (cpDNA) from Aseel cv. was sequenced using a combination of Sanger-based and next-generation sequencing technologies. The size of the genome was 158,458 bp with a pair of inverted repeat (IR) regions of 27,276 bp that were separated by a large single-copy (LSC) region of 86,195 bp and a small single-copy (SSC) region of 17,711 bp. Genome annotation demonstrated a total of 138 genes, of which 89 were protein coding, 39 were tRNA, and eight were rRNA genes (Khan et al. 2012).

Similarly, a sequence from the Saudi Arabian date palm cultivar Khalas has been made (Yang et al. 2010). Comparison of cpDNA sequences of cvs. Aseel and Khalas showed following intervarietal variations in the LSC region. These were the two SNPs in intergenic spacers and one SNP in the rpoc1 gene and polymorphism in two mononucleotide simple sequences repeats (SSR). Also, a 4-bp indel in the accD-psaI intergenic spacer was detected between the two cultivars. The SSC region has a polymorphic site in the mononucleotide SSR located at position 120,710. Aseel cv. cpDNA sequence with partial *Phoenix dactylifera* cpDNA sequence entries deposited in Genbank was also compared and identified a number of potentially useful polymorphisms in this species. Analysis of date palm cpDNA sequences revealed a close relationship with the common bulrush, *Typha latifolia*. Occurrence of small numbers of forward and inverted repeats in date palm cpDNA indicates conserved genome arrangement (Khan et al. 2012).

Eleven cultivars were selected for molecular-based varietal identification, and 16SrRNA, ribulose bisphosphate carboxylase large subunit (RBLC) gene, and maturase K (MATK) gene from date palm chloroplast were sequenced. The studied cultivars showed highly similar/identical sequences for these genes which are also identical to already sequenced genome of date palm (Nadia et al. 2013).

5.5.3 Research Activities of the Date Palm Research Institute

Date palm is of economic importance in Pakistan. Nevertheless, there are only a few small research stations specialized in date palm belonging to the agriculture departments in Sindh and Balochistan provinces. The Date Palm Research Institute (DPRI) was established in 2006 at Shah Abdul Latif University, Khairpur. Its main purpose is to conduct research on various aspects of date palm. Hence, the Institute has engaged in various research activities since its establishment and has achieved remarkable recognition within country in a short period spell of time. Adel A. Abul-Soad is a date palm expert from Egypt and one of the lead persons in this Institute who has worked as a foreign professor/consultant since 2006. A team of 12 researchers is engaged in research on various aspects of date palm, in addition to other crop species such as mango, banana, and orchids, some for their graduate degrees. DPRI has four departments: Biotechnology and Tissue Culture, Crop Production and Plant Protection, Postharvest and Processing, and Germplasm Collection Farm/Nursery.

The micropropagation of elite cultivars of Pakistan and abroad is a mandate of the Institute. Efforts are being made to modify the cultivar structure of dates in the country to save the crop from the monsoon rains which are synchronized with the ripening of crop. The Institute is conducting research to enhance the quantity and quality of dates and addressing current disease and pest problems to generate more revenue and alleviate poverty. Despite Pakistan being one of the top date-producing countries in the world, century-old pre- and postharvest techniques and a dearth of advanced processing units, standardized packaging and poor marketing tools are major problems being dealt with by the Institute (Abul-Soad 2011a, 2013b). Continuous efforts have been made to improve the existing date-drying facilities at Khairpur and to work for their optimization.

The Institute is engaged in conducting studies, training courses, seminars, and workshops for growers on different aspects of date palm. The annual Dates Festival and Seminar at Khairpur is a mandate of DPRI. A number of local and international NGOs are collaborating with DPRI, such as the Pakistan Agriculture Cold Chain Development (PACCD); WINROCK International, USA; DuPont Company; Pakistan Horticulture Development & Export Company (PHDEC); and USAID's Agribusiness Fund and Firms project (Abul-Soad 2013a). DPRI was assigned to conduct few projects on different aspects of date palm, such as a case study on the use of Tyvek bags to cover fruit bunches (Abul-Soad et al. 2010). Also, a study was carried out to survey the elite cultivars found in the different provinces of Pakistan (Markhand et al. 2010). Nevertheless there are only a few dispersed date palm stations and projects in Pakistan. DPRI is recognized as a Center of Excellence for date palm in Pakistan.

5.6 Cultivar Description

5.6.1 Date Palm Distribution in Pakistan

Date palm cultivars are distributed in all over Pakistan; however, about 90 % are found in the Kech, Panjgur, Dera Ismail Khan, and Khairpur districts (Table 5.1). The predominant Pakistani cultivars are Aseel in Khairpur and Sukkur districts and Dhakki in Dera Ismail Khan which is now being dispersed to other parts of Punjab and Sindh provinces. Begum Jangi is the common cultivar in Balochistan province especially in Kech district, and Rabai and Muzawati are found in the desert areas of Hamun-e-Mashkel salt lake and Panjgur (Fig. 5.23).

More than 300 cultivars and landraces of dates have been indentified in Khairpur, which is the biodiversity center of dates. Elite provincial date cultivars of Sindh include: Aseel, Karbaline, Fasli, Dedhi, Kupro, Gajjar, Kasho Wari, Began, Mithri, Bhedir, Khar, Autaqin, and Asul Khurmo (Mahar 2007). Only Aseel and Karbaline are of sufficiently high quality to be promoted.

In another study, fruits of 85 cultivars of Pakistani dates were collected through the harvest season. An appropriate format was designed to record the scientific and commonly used characteristics. These characters included fruit color at khalal stage; shape, size (length and diameter), color, and height of perianth; fruit group (soft, semidry, dry); edible stage; and the micropyle position and split (wide, narrow, deep). Furthermore, photographs of the entire fruit, longitudinal and transverse sections and dorsal and ventral sides of the seed, were taken. The objective is to evaluate the quality of different Pakistani dates. For example, Aseel cv. is elliptical in shape like most of the Pakistani dates (Fig. 5.24). Measured production of fruit production per tree is 152 kg (Markhand et al. 2008). However, the actual productivity figure is 80–90 kg tree^{-1} under current cultivation practices.

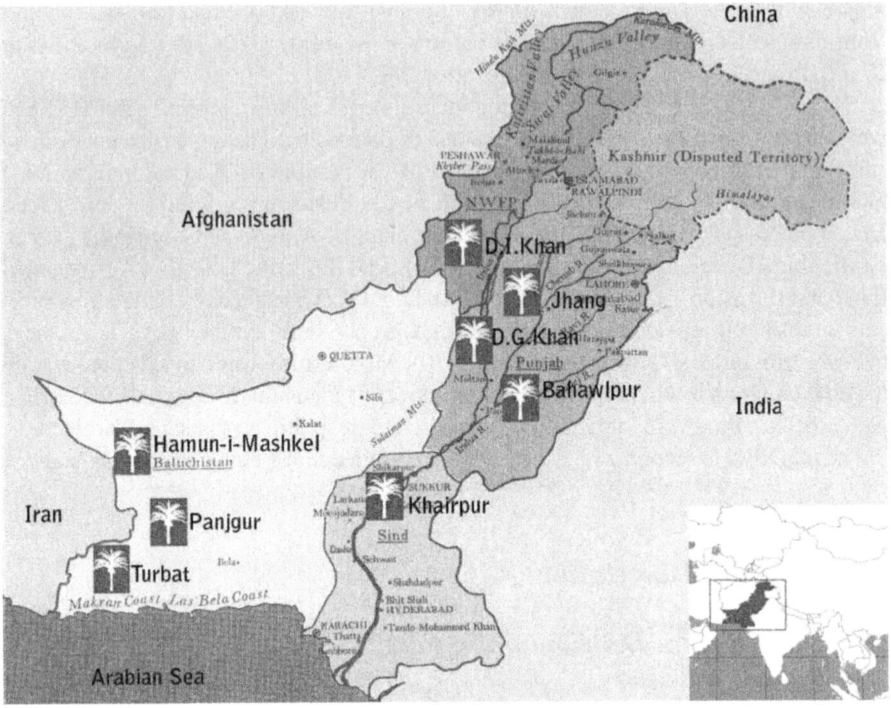

Fig. 5.23 Date palm distribution in the four provinces of Pakistan

A few of the evaluated cultivars show good properties. The majority of Pakistani dates fall in the semidry group. Most of the cultivars, if not all, are sensitive to monsoon rains, which occur during with ripening/harvest season (Markhand et al. 2010).

Very early cultivars could be harvested before monsoon rainfall, i.e., the end of July/ early August, to avoid rain damage. The earlier cultivars are Kasho Wari and Gajjar in Khairpur and Hillawi in Jhang, Punjab province and Rabai in Balochistan (Fig. 5.24).

In the Punjab, date palm cultivation is concentrated at Dera Ghazi Khan, Muzaffar Garh, Jhang, and Bahawalpur districts. Date palm cultivation is limited at Jhang; Halawy or Hillawi is the common cultivar of the area (Fig. 5.24).

There is a seedless cultivar named Dora Kaheri, also called Dora Sanawan, cultivated in Sanawan village near Kot Adu, Punjab. There are few thousand trees scattered at the area. The general practice is that growers do not carry out pollination. Hence, fruits ripen in September. It is desirable to get late fruit after the monsoon rains in July–August. Sometimes, the first early whorl of spathes is pollinated because of the availability of pollen. The khalal red fruit is harvested and mixed with vinegar and let stand overnight to develop into a ripe soft date. These dates are dark brown in color and seedless (unpollinated fruit). This cultivar is famously known in the area as being seedless. If it is pollinated it will ripen in late July and have seeds. It is consumed as a fresh date prepared with vinegar or sun dried on mats.

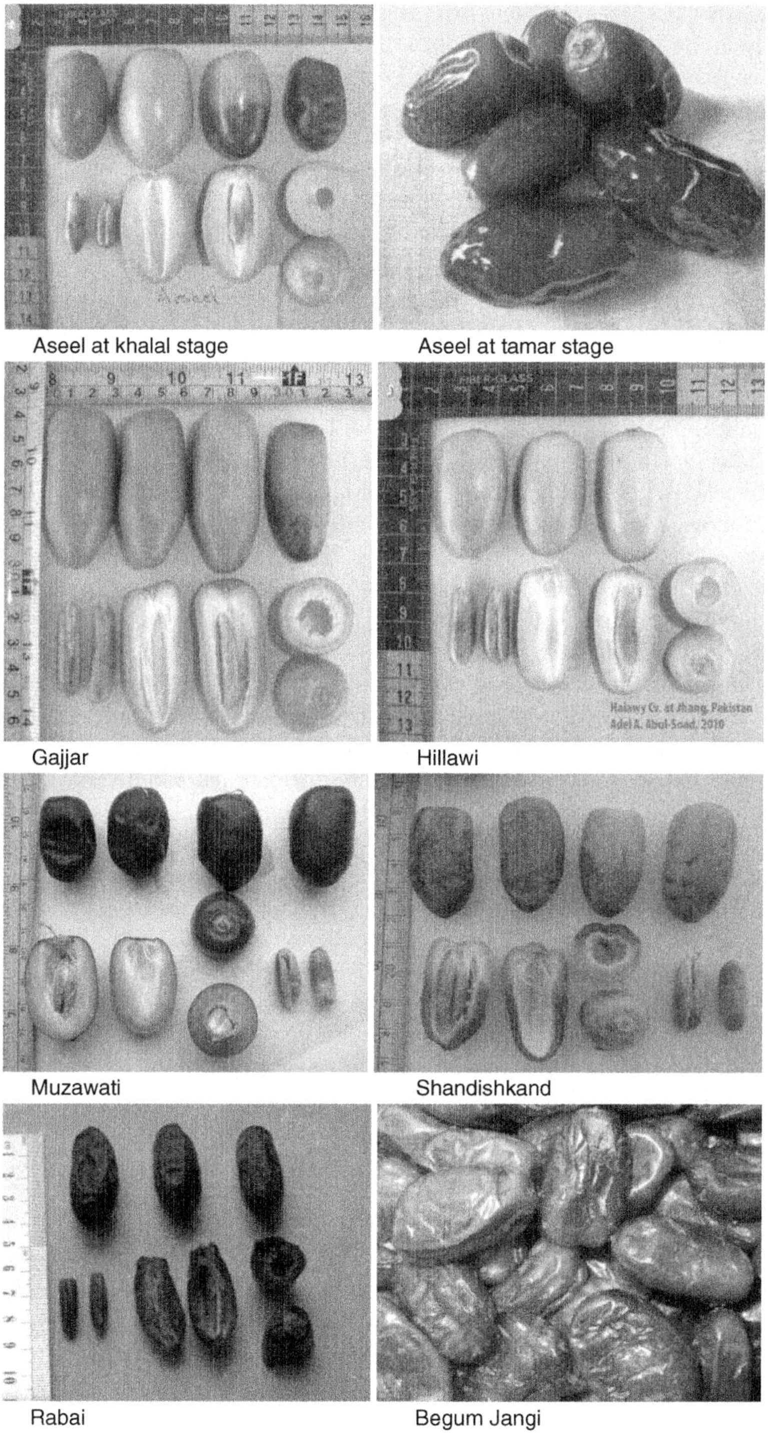

Fig. 5.24 Important Pakistani date palm cultivars

In Khyber Pakhtunkhwa, date palm is cultivated at Dera Ismail Khan district particularly in the Paniala, Paharpur, Chawdwan, and Dhakki villages. The prominent cultivars grown are Dhakki, Gulistan, Shakkri, Zaidi, Hillawi, Basra, Azadi, and Khudarawi. Dhakki dates are famous among all other cultivars, and there is rapidly growing demand for its offshoots. Total production is 11,000 mt dates, 75 % of the total production of dates in the area from Dhakki cv., going mainly for *chohara* preparation. Cured date production ranges from 800 to 1,000 mt yr^{-1} (Marwat et al. 2012).

Dhakki fruit has been physically and chemically characterized (Imranullah and Abdul Qayum 2013). Findings revealed that the average fruit length at khalal stage was 5 cm, fruit diameter 3.2 cm, fruit weight 26.5 g/fruit, and pulp weight 25.1 g/fruit. Astringency in terms of citric acid reaches a maximum at rutab stage, 1.5 %. The total yield per tree was 96 kg. However, common observation for the production in Dhakki village at Dera Ismail Khan indicated that harvested fresh fruit at khalal stage ranged from 50 to 60 kg $tree^{-1}$ of large-sized fruit.

In Balochistan, the famous cultivars are Begum Jangi, Muzafati or Muzawati, Kehraba, and Rabai, together accounting for more than 90 % of the date palm population, followed by the 10 % representing minor cultivars including Carba, Drage Trownge, Shandishkand, and Ringno (Fig. 5.24). The most outstanding is cv. Shandishkand with its crisp and tasty fruit; however, the number of palms is very small, and they produce no offshoots. It is landrace found nowhere else (Fig. 5.24).

In the Kech district of Balochistan, minor cultivars make up about 10 % of the total date palms, including Roghni, Hussaini, Goknah, Pashbugh, and Hallini or Haleene. The renowned cultivar in Panjgur district is Kehraba followed by a few other cultivars such as Muzawati, Rabai, Jawan, and Sour. Unfortunately, there are no reliable statistics about the precise number of date palm cultivars in Pakistan. The average fruit production of a palm of Begum Jangi cv. is about 100 kg $tree^{-1}$ at khalal stage giving about 1–1.5 *mond* (40–60 kg) after drying. *Mond* is the local weight measure used in the date market.

The weight of one bunch of, for example, Rabai cv. ranges from 3 to 6 kg. The average annual production of the preferred cultivars is 50–70 kg $tree^{-1}$. The Makran division (including Kech and Panjgur districts) produces about 90 % of all Balochistan dates. Only a few thousand tons are marketed.

Urgent and rapidly expanding demand exists for elite cultivars. It is unknown if the demand can be met through offshoot micropropagation of the evaluated and selected cultivars or by importing offshoots (Markhand et al. 2010).

5.6.2 Nutritional Aspects of Pakistani Cultivars

Twenty-one Pakistani date palm cultivars were subjected to antinutritional factor determination (phytate, oxalate, and tannins) at tamar stage based on dry fruit weight. Tannin content range was 0.22–0.87 % of dry fruit weight. The highest mean value was recorded in Desi cv. (0.87 %), and the lowest tannin content was recorded in Aseel cv. (0.22 %), Dhakki cv. (0.25 %), and Hillawi cv. (0.28 %).

Maximum phytate content was found in Dora Desi cv. (0.77 %), and minimum value was detected in Dora cv. (0.30 %). Oxalate content ranged from 3.63 to 6.49 % among date cultivars. Maximum mean value for oxalate content was found in Simple basray cv. (6.49 %) and the lowest value found in Dora cv. (3.63 %). The study concluded that the cultivars studied are suitable for consumption and processing due to their low level of antinutrients (Nadeem et al. 2011a).

Some 21 cultivars were used to determine the textural profile and total phenolic content. Maximum water activity was found in Desi Basray cv. (0.482) and minimum in Karbaline (0.323). The maximum hardness from a textural profile analysis was recorded in Dora cv. (59.64 g), and the maximum value of fruit color was observed in Desi red small cv. (149.10 CTn). Phenolic content ranged from 296.67 to 140.67 mg (GAE) 100 g^{-1} of fresh fruit weight. The study revealed that fruit of Dhakki, Aseel, and Hillawi cvs. might be preferred for table purpose because of their good physical characteristics such as large size, high fruit and flesh weight, edible/nonedible ratio, firm texture, and nutritional properties. Other date cultivars like Karbaline, Zaidi, and Dora are suitable for processing due to low moisture content and higher sugar content (Nadeem et al. 2011b).

5.7 Date Production and Marketing

5.7.1 Fruit Thinning

The impact of manual fruit thinning on the quantitative and qualitative characteristics of Aseel cv. has been reported. Sixteen different combinations of possible manual thinning methods were performed. Number of bunches, number of spikes, and spike lengths were reduced manually after 6 weeks of pollination. The average fruit yield was 152 kg $tree^{-1}$ when the number of bunches was reduced to 19 bunches with removal of the central forth spikes. The average fruit weight increased from 10.9 g $fruit^{-1}$ in control treatment (no spikes removed) to 13.1 g $fruit^{-1}$. Also, fruit thinning significantly increased fruit diameter and flesh weight but did not affect total sugar, reducing and nonreducing, or the pH of fruits (Markhand et al. 2008).

5.7.2 Solar Dryer Technology

Solar dehydration is an ancient process for food preservation. The purpose of dehydrating fully mature date fruit is to reduce the water content below 24 % of fresh weight. Date fruits naturally dry on the tree reducing water content to 20 % of the fresh weight. The shelf life of such fruit can reach up to a year. Fruit of the soft and semidry types require dehydration if harvested at khalal stage due to particular reasons during the ripening period. These reasons could be monsoon rains coincident with ripening such as in Jordan, Palestine, Morocco, Tunisia, USA, and Pakistan.

Fig. 5.25 The solar dryer (solar tunnel) unit. (**a**) The transparent plastic tunnel, (**b**) thirty stands inside the tunnel used for drying the unripe date fruit

Early harvesting because of the higher relative humidity during ripening can cause fruit drop before full ripening in Arabian Gulf countries or delay ripening in Pakistan. Also, early harvesting could occur because of marketing demand or late season pest attack on fruit bunches.

In Pakistan, monsoon rains mostly coincide with the date ripening months of June to August. Monsoon rains falling continuously for a couple of hours can damage the date fruit bunches at khalal stage from fungi and mold growth and skin cracks (Abul-Soad 2010). Many trial experiments have been done by the local growers using electric dehydrators and units made from the wood and baked bricks, to accelerate fruit ripening after harvesting of khalal fruit (Abul-Soad 2013b). Finally in 2012, the trails were concluded with successful trials using a plastic tunnel for drying unripe date fruit (Abul-Soad 2013a). Also, the tunnels can be used during the off-season for drying other fruits, vegetables, and medicinal plants.

The solar dryer or solar tunnel is a unit comprised of a metal frame covered with transparent plastic sheeting within which unripe date fruit continues ripening to reach tamar stage within a period of 3 days (Fig. 5.25). By comparison, spreading unripe fruits on mats for sun curing takes 7 days, and fruits are exposed to changeable climatic conditions and dust. The solar dryer is provided with doors at each end to allow room for the drying stands and the access of laborers. On one end, two windows are covered with a metal mesh, and on the other end two exhaust fans draw the moisture out of the tunnel.

Fans in the current model in Pakistan are controlled manually and provided with an external source of electricity. It could be controlled automatically by a thermostat to keep the temperature inside the tunnel in the targeted range of 60–65 °C. At times, the temperature reaches 70 °C because of electricity fluctuations, which is a major problem in Pakistan. Increasing the temperature to 70 °C causes an adverse impact on fruit drying. It is worth mentioning that in the older models of solar dryers at Khairpur, a solar cell was fixed beside the tunnel to provide electricity to run the fans, but it is rather expensive for the growers. Adoption of solar cells for power has been hindered by the attitude of the growers that solar cells should be made available free or at a subsidized cost.

The range of the temperature inside the tunnel was measured from 50–65 °C from 12 a.m. to near sunset when the temperature outside the tunnel was 35–40 °C. Relative humidity measured inside the tunnel was 40–50 %, while outside the tunnel in the open air, it was 60–65 % on 10–12/08/2012 (Abul-Soad 2013b).

The capacity of a single dryer is 2.5 mt approximately, loaded on 30 stands arranged in 3 rows. Each stand consists of 7 shelves bearing about 84 kg of fresh fruit. Each shelf is composed of a couple of perforated stainless steel trays. The minimum production of a single solar dryer, by contrast, under the usage at Khairpur by the growers, revealed that the output of dried fruit was about 1 mt after 2.5–3 days (Abul-Soad 2013b). The stepwise protocol of using the solar dryer can be summarized as follows:

(a) Appropriate fruit stage: date fruit is preferably harvested at early rutab stage, as compared to khalal, to avoid failure of fruit ripening. Early harvested fruit at khalal stage could take more time to dry inside the solar dryer and may fail to reach the desired fruit quality. The fruit at early rutab stage is changing to a yellow color (Aseel, Dhakki, and Begum Jangi cvs.) or red (Muzawati cv.) with the fruit tip turning brown.
(b) Harvesting: fruit skin of Aseel cv. is very delicate at rutab stage. Fruit must be handled carefully during harvest to avoid dropping early-ripened fruit on the ground and crushing them. The ground under the date palm is covered with a plastic sheet to collect the harvested fruit bunch and any individual fruit which falls during the harvest. This practice has recently been adopted by progressive growers; however, the conventional harvesting practice is still commonly being used (see Sect. 5.2.3).
(c) Containers: harvested fruit should be placed in plastic crates for transfer to the solar dryer by small hand carts or by trucks if the fields are distant.
(d) Sorting: once date fruit reach the location of the solar dryer, they should be sorted to discard any crushed, deteriorated, mashed, infected, ruptured, and immature fruit types before the drying process. Sorting includes separating the fruit into three categories according to size: small, medium, and jumbo. It is worth mentioning that jumbo fruits are commanding high prices in the market and have export potential. There is need to establish quality standards for the commercial Pakistani date cultivars.
(e) Fruit detachment: at Pir or Bethak, manual fruit detachment from the fruit stalks is the conventional practice before drying in the solar dryer or sun curing on mats (Fig 5.3).
(f) Washing: loose fruit cannot be washed before drying in solar dryers under current practices. However, fruit loaded on the perforated metal trays can be washed under running tap water for 1–2 min and then placed directly into the solar dryer at the highest shelf in the stand. Fruit is not to be loaded on the lower shelves or left for some time after washing. This is because of the high sugar content of the date fruit which encourages fermentation and deterioration of the fruit.

Fig. 5.26 The quality of dried date fruit of Dhakki cv. in the upper trays loaded with 3 fruit layers after 3 days inside the solar dryer at Therhi, Khairpur

(g) Number of fruit layers on the tray: fruit loaded on the upper shelf is exposed to higher temperature, which expands the fruit skin, than those on the lower shelves. Increasing the number of fruit layers from 1 to 3 layers in the trays of the upper shelf lessens the impact of high temperature and reduces the number of swollen fruit (Fig. 5.26). Swollen fruit exhibit a separation of skin from the underlying flesh which reduces quality. It appears that the moisture produced by three fruit layers adds moisture to the ambient air thereby reducing the negative impact of heat on the date fruit. Also, a few swollen fruits are observed in the lower shelves which are loaded with a single fruit layer. Loading lower shelves with more than one layer increases the number of rotten fruit due to moisture accumulation.

(h) Packaging and storage: fruit should not reach full dryness; moisture content should be at about 24 %. Fruit is expected to lose moisture during subsequent steps from dryer through cold storage to the retail market. Dried fruits are loaded into plastic crates and kept for 12–24 h under shade in a protected place or room temperature and then transferred to the market or cold storage.

It is important to protect mature dates from any new infestations in the late season and from storage pests after drying. Heat during curing or drying can be considered a sterilization treatment. In a study on cvs. Khalas, Rizez, and Shiashy in Saudi Arabia, to evaluate the effect of solar radiation as a postharvest control treatment against insect pests for mature date fruit, direct sunlight for 2 days was found to be an excellent substitute for chemical fumigation, the use of which creates safety

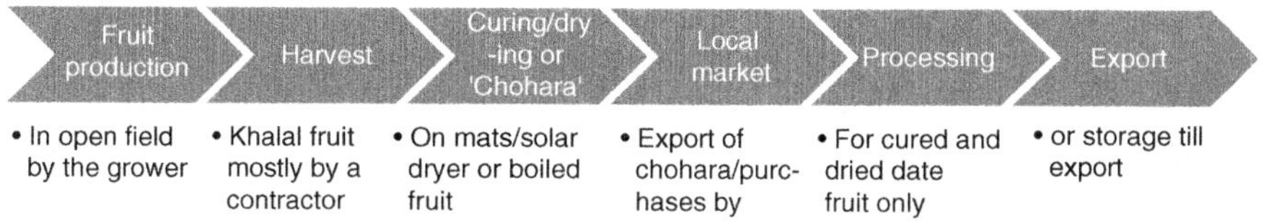

Fig. 5.27 Date fruit value chain in Pakistan

concerns, to maintain quality characteristics for human consumption. The sunlight treatment significantly increased total phenolics level to more than 175 % and anti-radical activity in all cultivars (Saleh 2012).

5.7.3 Date Market in Pakistan

5.7.3.1 Pakistani Market Structure

In Pakistan, at least 80–90 % of the harvested fruit goes to *chohara* preparation, and the remaining amount is cured for direct consumption as dates; no specialized date fruit warehouses exist. The cured khalal and rutab fruit are packed in wooden boxes or handmade baskets to go directly to cold storage or local wholesale markets. Cold storage facility used for vegetables and other fruits are used to store dates as well. Recently, three cold storage facilities have been built with a subsidy and given to three processing factories in Khairpur, through financial assistance from USAID Pakistan to support development in the date sector. Once cured dates reach local markets, agents of the processing plants will be competing through auctions to buy the good quality cured dates. Procured dates from the wholesale market are stored under ambient conditions at processing plants in *godam*. The processing plants will start their work to prepare the dates for the export outside Pakistan. The fruit for processing is mostly from Aseel, Begum Jangi, Dhakki, Muzawati, and Rabai cvs. and also a mixture of minor cultivars commonly called *fruit mix*. Imported dates of exotic cultivars are mostly offered for domestic consumption as fresh dates, especially during Ramadan, and to be used also in the processing plants to cover the shortage of local dates. The value chain of dates in Pakistan is illustrated as shown in Fig. 5.27.

The second type of marketed date fruit, *chohara*, is mostly collected in fiber sacks after sun curing, then transported to the local market at Khairpur or Sukkur for auction by brokers and exported, mainly to India. The annual export of *chohara* typically exceeds 100,000 mt annum^{-1} and accounts for why Pakistan is always among the top three world exporters of dates. Nevertheless, the amount and export value of Pakistani dates is relatively very low.

The usual measurement unit for dates in Pakistan is 40 kg (mond) which is approximately equal in value to PKR 1,500–7,000 (USD 15–70). Production of a single tree of Aseel cv., for example, is 100–120 kg of fresh fruit but can yield

40–50 kg of *chohara* (Abul-Soad 2011a, 2012b). This is the local market price, and the date grower receives less because the contractors are mostly carrying out the harvest and bearing transportation costs to the local market.

There is a smaller wholesale market for dates in Karachi as compared to Khairpur and Sukkur. Most Balochistan dates and imported dates from Iran are received in the Karachi market, mostly packed in containers. The price of Aseel cv. reaches PKR 120 kg^{-1} at the maximum in Sukkur Market in December.

Only sweaty dates of soft cultivars or semidry with a low amount of tannin at khalal stage are suitable for consumption such as Dedhi and Mithri cvs. at Khairpur, Gulistan cv. at KPK, and Hillawi cv. at Punjab. Prices of early season dates are always higher than the mid-season cultivars. In general, when dates reach khalal stage, they are regarded as ready for trading as fresh fruit. Dates in khalal stage are the first in the harvesting season and therefore have a ready market. Transportation by refrigerated trucks delays early harvested rutab of Begum Jangi sales from Balochistan and distant areas to reach retail markets in Karachi or wholesale markets.

An economic zone is under construction by the Pakistan government on the main national road at Khairpur to replace the current wholesale market. It will be equipped with digital mega balances, a cold storage facility, and additional features. This project is expected to facilitate and accurately track the date market.

5.7.3.2 Current Pakistan Date Imports and Exports

Date fruit trading is an expanding business both locally and globally. Pakistan was the sixth largest date-producing country of the world in 2011. Export quantity data shows that Pakistan's date exports are very small compared to production. There is a significant gap between date production and export figures. Pakistan on average exports 20 % of production, and 80 % of the crop is either consumed locally or wasted. Pakistan is the second most important world exporter, following UAE, and in 2011 recorded 113,358 mt were shipped to 26 different destinations all over the world (FAOSTAT 2013). This suggests outlets for Pakistani dates in the international markets.

India in 2011 was the largest importer from Pakistan with an 86 % share and Afghanistan with 6 % (FAOSTAT 2013), these representing mainly *chohara* dates. The USA, UK, and Germany are the major countries importing processed dried dates with 5, 2, and 1 % share, respectively (Fig. 5.28). It is necessary to strengthen Pakistan's position in the existing markets, especially USA, UK, and Germany, through the support of the entire chain value of dates coupled with strong market promotion and also to identify new potential markets such as Denmark, Japan, and Australia which imported about 650 mt each in 2011 (FAOSTAT 2013).

Also, there is a need to export Pakistani dates to nondate-producing countries near Pakistan, such as Indonesia and Malaysia which are among the top ten date-importing countries of the world. Nevertheless, Pakistan thus far has failed to export any dates to those countries. There is growing demand in such Muslim countries for dates especially in the holy month of Ramadan.

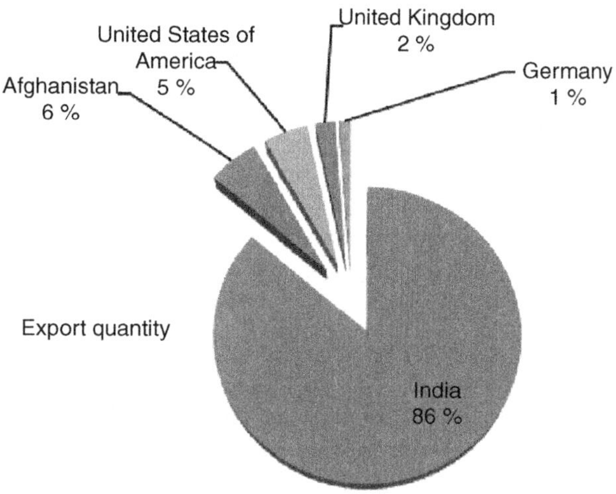

Fig. 5.28 Major countries of exported date fruit of 113,358 mt into Pakistan in 2010

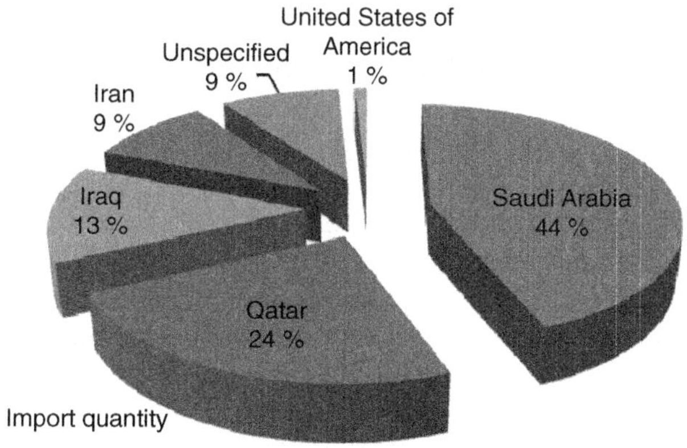

Fig. 5.29 Major countries of imported date fruit of 13,205 mt into Pakistan in 2010

The export value of Pakistani dates is very low. Pakistan is not among the top 20 countries in terms of value of exports, although it is the second largest exporter of the world in terms of the export quantity, with 121,681 mt in 2010, which decreased to 113,358 mt in 2011 (FAOSTAT 2013). This production fluctuation could be because of losses due to monsoon rains. Date exports can be increased through improved agricultural practices and postharvest management.

A rough estimate of the export of processed dates is about 11,000–13,000 mt yr^{-1}. Therefore, processing plants mostly need high-quality imported dates fit for processing (Abul-Soad 2011a, 2012b). Keeping in mind that the shortage of dates occurs because a large quantity of stock is lost to spoilage because of limited and far-flung processing facilities, substantial revenues could be achieved through proper fumigation and storing. Also, the yearly demand for dates is about 10,000 mt, but in Ramadan it jumps to 40,000 mt in Karachi alone (SMEDA 2009). To fulfill these requirements, Pakistan imported 13,205 mt in 2010 (FAOSTAT 2013). The major sources are Saudi Arabia with 44 %, Qatar 24 %, Iraq 13 %, and Iran with 9 % shares of the total imported dates (Fig. 5.29).

A promotion plan for various date markets should be the mandate of the key associations, especially the Pakistan Horticulture Development Export Company (PHDEC). Export limitations and challenges are as below (PHDEC 2008):

(a) Date procurement from wholesale market
(b) Lack of hygienic processed and value-added products
(c) Lack of market promotion outside Pakistan
(d) Limited attempts for market expansion – access to new potential markets
(e) Transportation by nonrefrigerated modes
(f) Lack of R&D support

Moreover, the situation in Dera Ismail Khan is about to worsen regarding the lack of information about customers, marketing, harvesting, processing, packaging, and infrastructure to produce good quality fresh dates. If production and processing constraints are properly handled, this may have a significant positive socioeconomic impact on local inhabitants (Marwat et al. 2012).

5.8 Processing and Novel Products

5.8.1 Date Processing

Dates are processed to deliver products of uniform size, which are healthy and neat in appearance. Processing in Pakistan includes storage, fumigation, sorting, washing, drying, grading, and packaging. In common practice, once cured dates arrive at date factories in Khairpur from the wholesale market (Fig. 5.30), fruits are stored for few weeks or months under ambient temperature. Store rooms, *godam*, are provided with fans and windows covered with wire net for ventilation purpose.

Stored dates under such conditions are exposed to pest infestation whether existing in the fruit or from a new infestation. Thus fumigation is carried out to kill the different pests of all ages inside the fruit and make them fit for human consumption. Fumigation consists of piles of dates covered with polythene plastic sheets and

Fig. 5.30 Date market of cured dates packed in wooden boxes and jute sacks, *chohara* at Khairpur

exposed to a noxious gas within the sheets for 72 h. Tablets of hydrogen phosphide, trade name Phostoxin, have been widely used by technical personnel. After fumigation, rooms are carefully ventilated and the powder residue of the tablets removed. The process of dumping/fumigation keeps on going on the one hand, while on the other hand the fumigated dates are then passed to the another section for initial sorting or storage. An initial sorting is performed to discard the rotten fruit.

For individual sale orders, date fruits are weighted and sent to villagers' houses on small donkey carts for pitting and decapping, washing, and drying in the sun. Sometimes processing is extended to include chopping as well. Fruits returned to the factory are subjected to a manual sorting and grading process by young girls seated on wooden tables (Fig. 5.31). According to current situation, dates are classified into A, B, C, and industrial grades according to size and blemishes they have. The sorting process could be done before sending the dates to the households; however, grading is usually carried out before packaging and sometimes according to the specific requirements of the buyer.

Dates are graded usually by size, shape, color, and skin. Fruit properties and taste vary among the different cultivars. A survey in the leading processing plants in Khairpur was performed to collect fruit samples and set quality standards for the local commercial cultivars. The preliminary results of this study, done by DPRI members for Aseel cv., indicated that the category of jumbo fruit, i.e., grade A, consisted of 55–60 fruit 500 g^{-1} and grade B contained 61–74 fruit 500^{-1}, while grade C composed of 75–115 fruit 500 g^{-1} package according to the general concept of fruit quality in Pakistan date factories. The lowest-quality dates (grade C) are of sufficient quality to be consumed as fresh whole table fruit; however, the inferior fruits were classified as industrial grade in this study (Zangejo et al. 2013). Most date factories are processing certain Pakistani cultivars such as Aseel, Begum Jangi Dhakki, Rabai, and Muzawati or imported cultivars from Iran and Iraq such as Zahdi, Sayer, and Rabai.

Before the sealing of packaged date fruit, cartons are exposed to a metal detector. Nevertheless, tiny metal particles are a big challenge for export. It needs x-ray detection which somewhat could be unable to detect the very tiny metals within the

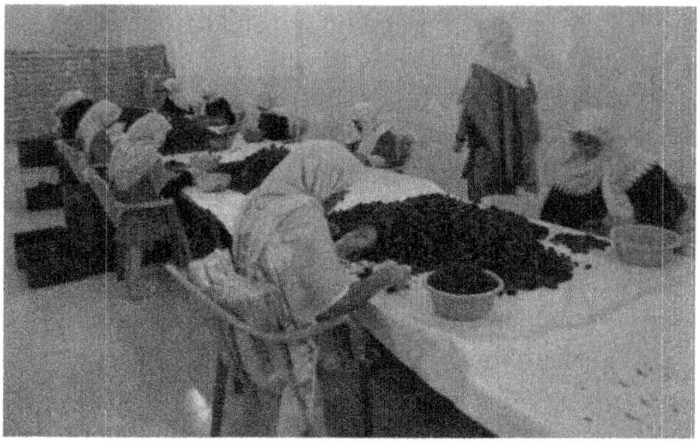

Fig. 5.31 Sorting the Rabai cv. dates into three categories according to size and freedom from blemishes

fruit. The processed dates are packed by a wrapping machine with polythene plastic bags which preserve the fruit moisture. The boxes can be of different design and weight (0.5, 1, 2, and 10 kg) based on requirements of the purchaser (Fig. 5.31). Packaged dates are stored in cold storage at 0–5 °C for about 10 months without change in color or freshness. Stored dates in the cold storage at 5–18 °C resulted in color and freshness change of 10 % approximately which reduces the quality.

There are about 10–12 date-processing plants in the *date city* of Khairpur. The practices used for processing are noticeably varied and need to be standardized. This represents one of the challenges facing the industry to increase the value of Pakistani dates and compete in the international market.

5.8.2 Limitations and Challenges Facing Processing

(a) Product development: the key problem in Pakistani date processing is the unhygienic manner of harvesting and postharvest handling of the fruit. The presence of dates contaminated with mud, sometimes found on packed Pakistani dates, is due to the poor handling of fruit during harvest rather than from processing. Harvest practices affect the later steps during processing. Thus it is suggested that the fruit be harvested directly into plastic crates rather than being transported in bulk to the packinghouse. This would prevent dust and dirt contamination of the fruit and ensure that they arrive in good condition and not crushed. It is worth mentioning that the NGOs working in Pakistan, such as USAID-funded companies, provide support to the growers of Khairpur with a sufficient number of plastic crates and other materials to develop the date sector (Abul-Soad 2013a).

(b) Cold chain system: when transporting the fruit, one must also take into account its sensitivity. It is a large problem to transfer the fruit, particularly at rutab stage, from distant locations to the wholesale market at Khairpur or Karachi. It requires a trip of 7–14 h to transport Dhakki cv. dates from Dera Ismail Khan or Begum Jangi or from Turbat, Balochistan. The fruit is soft and contains moisture, and high temperatures during transportation should be alleviated with refrigerated compartments. The scenario now is that cured dates are transported, but the early-harvested rutab, which command higher prices, is not.

(c) Introduction of new cultivars: although refrigerated transport is expensive, introducing elite international cultivars such as Medjool would justify such a cost and broaden export opportunities. In Karachi supermarkets, the price of a 400 g box of imported Medjool is PKR 350, while 1 kg of Aseel cv. of cured dates reaches PKR 80–120 in the month of October.

(d) Farm management: from the above discussion, it is obvious that date palms in Pakistan do not receive special care as far as fertilizer application and other agricultural practices are concerned. This results in slow growth rates for the palms and low production. A national program for date palm maintenance and improvement is recommended in order to increase the productivity and fruit quality and consequently increase the export value of Pakistani dates.

(e) Quality standard creation: it is worth noting that the international markets are mainly looking for semidry dates, tamar, with specific quality parameters, while the Khairpur region is mainly producing dry dates, *chohara*, with limited export market expect to India. Less than 10 % of exported Kharak are table fruit, while about 90 % were for the industrial purposes such as diced fruit and syrup production. Hygienic and big size fruits of Begum Jangi and Aseel cvs. are the main two requirements for the Pakistani dates to compete with other countries' cultivars in the international market. Therefore, the tissue-cultured plants of Medjool, Deglet Noor, Ajwa, and Barhee cvs. need to be introduced to Pakistan as they are on the top in the international market.
(f) Regulations and legislations: inspection and accreditation by local and international entities is very important and recently was required by the Chinese to import *chohara* from Pakistan. Also, there is a need for more exports to other international markets, particularly Europe and the USA.
(g) Product promotion: through participation in international expositions, international media, pamphlets and leaflets, workshops, and training for key stakeholder (e.g., growers, laborers, contractors, transporters, merchants, processors).
(h) Research and development support: lack of funding is the main hindrance facing the research institutes to continue in this sector and conduct the required investigations on either the output of current factories or to contribute in the establishment of a world-standard processing plant.

5.9 Conclusions and Recommendations

Date palm is an essential Pakistani crop, ranked in importance after mango and citrus. Despite the position of Pakistan among the top ten producing countries, and one of the three largest exporters, the actual cash value of date fruits is very low. A major reason behind this fact is the lack of knowledge about Pakistani cultivars in the international market. Efforts are needed to promote cultivars such as Dhakki, Aseel, Muzawati, and Begum Jangi. In reality, more proficiency and efficiency in the date industry will lead to increased prices for Pakistani dates in the local and international markets. There are a few trials for the commercial micropropagation of date palms, but, under the current situation, tissue-cultured plants of the international elite cultivars such as Medjool, Barhee, Ajwa, and Safawi should be introduced with subsidized prices to growers to change the cultivar structure in Pakistan and to increase date export opportunities.

As a rule of thumb, increasing tree productivity and enhancing fruit quality can be achieved through elevating awareness as well as the benefits of improved agricultural practices in the field. Using solar dryer technology is an emerging practice in Pakistan to protect the harvested date fruit from monsoon rains. According to conventional wisdom, attention should be paid to providing growers offshoots of early local cultivars such as Gajjar and Kasho Wari in Sindh province and Hillawi in Punjab. Unfortunately, the use of various materials as bunch covers has failed as a means to protect the fruit during the ripening period, and growers prefer to harvest the crop at khalal stage to avoid fruit deterioration. Research must be supported to evaluate other alternative materials to cover fruit bunches during monsoon rains. Using plastic nets to cover the fruit bunch at khalal stage also could reduce early and late fruit drop. About 80 % of harvested fruits are processed and boiled to prepare *chohara*, mainly exported to India, with the remaining 20 % spread on mats for sun curing and/or recently dried using solar dryers.

Sudden decline disease, associated with root rot, is currently a major threat to date palm cultivation in Sindh. The fungal disease hits all ages and devastates the palm within 6 months without any effective control measure at the moment. Dubas bug pest reduces the productivity of the date palm trees in Kech, Balochistan province, and chemical applications have failed to control it. The date palm lesser moth and late season pests cause considerable fruit drop in Sindh (Therhi area of Khairpur) and in Dera Ismail Khan (Khyber Pakhtunkhwa province). Three things are required in order to increase the export window of Pakistani dates: hygienic and bigger size fruits of Begum Jangi and Aseel cvs. and production of the world top cultivars Medjool, Deglet Noor, and Barhee that have already market with higher prices. Decision-makers and NGOs in Pakistan are exhibiting considerable interest in the date-processing sector. Nevertheless, only 12–13 factories of medium capacity currently exist in Khairpur where about 50 % of the cured dates are processed. Work inside these factories is laborious and machines rarely used. The quantity of processed dates is 15,000 mt yr^{-1}. There is a need for transportation of fresh fruit under refrigeration and cold storage to extend the shelf life of processed dates and renovation of the machinery lines with pitting and cleaning machines. In reality, Pakistan is a promising country for a date palm industry but needs technology and information transfer to achieve its potential.

Acknowledgment The authors would like to express their deep thanks to the DPRI, SAL Univ., Khairpur, Sindh, Pakistan, and the whole work team for their sincere efforts and kind contribution during the writing of this manuscript.

References

Abdul Wahid LAH, Abbas MF, Abbas KA et al (2010) Effect of pollen parent on protein pattern during growth and ripening of date palm fruit Hellawi cv. Egypt J Hortic 37(2):133–138

Abdul-Salam M, Al-Mazrooei S (2007) Crop water and irrigation water requirements of date palm (*Phoenix dactylifera*) in the loamy sands of Kuwait. Acta Hortic 736:309–315

Abul-Soad AA (2010) Impact of rain fall and floods on the fruit and trees of date palm in Pakistan. Bless Tree 2(4):56–59

Abul-Soad AA (2011a) Date palm in Pakistan, current status and prospective. Report USAID Pakistan

Abul-Soad AA (2011b) Micropropagation of date palm using inflorescence explants. In: Jain SM, Al-Khayri JM, Johnson DV (eds) Date palm biotechnology. Springer, Dordrecht, pp 91–118

Abul-Soad AA (2012a) Influence of inflorescence explant age and 2,4-D incubation period on somatic embryogenesis of date palm. Emir J Food Agric 24(5):434–443

Abul-Soad AA (2012b) The mechanisms of dates market in Pakistan. Bless Tree 4(1):60–69
Abul-Soad AA (2013a) Standard operation protocol (SOP) of the solar dryer. Report, USAID Firms Project, Karachi
Abul-Soad AA (2013b) Current status and perspective of date palm in Pakistan. International conference on date palm: present status & future prospects. Islamia University Bahawalpur, Bahawalpur, 2–3 Sept 2013, p 1
Abul-Soad AA, Mahdi SM (2010) Commercial production of tissue culture date palm (*Phoenix dactylifera* L.) by inflorescence technique. J Genet Eng Biotechnol 8(2):39–44
Abul-Soad AA, Mahdi SM (2012) Micropropagation of date palm males and females. Bless Tree 4(2):80–86
Abul-Soad AA, Zaid ZE, Sidky R (2006) Improved method for the micropropagation of date palm (*Phoenix dactylifera* L.) through elongation and rooting stages. Bull Fac Agric Cairo Univ 57:791–801
Abul-Soad AA, El-Sherbeny NR, Baker SI (2007) Effect of basal salts and sucrose concentrations on morphogenesis in test tubes of female inflorescence of date palm (*Phoenix dactylifera* L.) cv. Zaghloul. Egypt J Agric Res 85(1B):385–394
Abul-Soad AA, Mounir A, Amar M et al (2008) Date palm: field operations and pests and diseases. ISBN: 977-302-198-X. http://www.caae-eg.com
Abul-Soad AA, Markhand GS, Shah SA (2009) Study on potential impact of lowering of groundwater on date palm plantations, Final report, Hagler Bailly Pakistan (Pvt.) Ltd. 39, Street 3, E7, Islamabad 44000, Pakistan
Abul-Soad AA, Markhand GS, Memon S (2010) DuPont Tyvek® bags impact on dates. Final report. DuPont Pakistan Operations (Pvt.) Ltd., 2G-4, Johar Town, Canal Bank Road, Lahor-54790, Pakistan
Abul-Soad AA, Maitlo WA, Markhand GS et al (2011) Date palm wilt disease (sudden decline syndrome) in Pakistan, symptoms and remedy. Bless Tree 3(4):38–43
Ahmad S, Smead PF (2005) Dates in Pakistan. In: Ahmad S (ed) Date palm. Horticulture Foundation of Pakistan, Islamabad, and Pakistan Science Foundation, Islamabad, pp 71–94
Ahmad S, Tahir A (2005) Dates culture. In: Ahmad S (ed) Date palm. Horticulture Foundation of Pakistan, Islamabad, and Pakistan Science Foundation, Islamabad, pp 41–45
Ahmad F, Hameed M, Abbas S et al (2013) Comparative anatomical studies in some date palm (*Phoenix dactylifera* L.) cultivars from diverse origin. International conference on date palm: present status & future prospects, Islamia University of Bahawalpur, Bahawalpur, 2–3 Sept 2013, p 38
Al Awadhi HA, Hanif A, Suleman P et al (2002) Molecular and microscopical detection of Phytoplasma associated with yellowing disease of date palm *Phoenix dactylifera* L. in Kuwait. Kuwait J Sci Eng 29:87–109
Alamoud AI, Mhohammad FS, Al-Hamed SA et al (2012) Reference evapotranspiration and date palm water use in the Kingdom of Saudi Arabia. Int Res J Agric Sci Soil Sci 2(4):155–169
Ammar MI, Amer MA, Rashed MF (2005) Detection of phytoplasma associated with yellow streak disease of date palms (*Phoenix dactylifera* L.) in Egypt. Egypt J Virol 2:74–86
Ata S, Shahbaz B, Ahmad M et al (2012) Factors hampering date palm production in the Punjab: a case study of D.G. Khan District. Pak J Agric Sci 49(2):217–220
Barreveld WH (1993) Date palm products, vol 101, FAO agricultural services bulletin. Food and Agriculture Organization of the United Nations, Rome
Bartelt RJ, Dowd PF, Plattner RD et al (1990) Aggregation pheromone of dried fruit beetle, *Carpophilus hemipterus*. J Chem Ecol 16(4):1015–1039
Beech M (2003) Archaeological evidence for early date consumption in the Arabian Gulf. In: The date palm: from traditional resource to green wealth. Emirates Centre for Strategic Studies and Research, Abu Dhabi, pp 11–33
Bhat NR, Bhat VS, Lekha MK et al (2012) Estimation of water requirements for young date palms under arid climatic conditions of Kuwait. World J Agric Sci 8(5):448–452
Dhillon BS, Tyaagi RK, Saxena S et al (2005) Plant genetic resources: horticultural crops. Narosa Publishing House, New Delhi, pp 174–176

Djerbi M (1982) Bayoud disease in North Africa, history, distribution, diagnosis and control. Date Palm J 1:153–197
El Baali E, Azouggagh M, Ahl Rchid O (2002) Water pumping for irrigation in a southern Moroccan oasis. International research on food security, natural resource management and rural development, 9–11 Oct, Proceeding Deutscher Tropentag, Kassel-Witzenhausen, p 1–8
Galeb HH, Mawlood EA, Abbass MJ et al (1987) Effect of different pollinators on fruit set and yield of Sayer and Hallawy date palm cultivars under Basrah conditions. Date Palm J 5(2):155–173
Hussain I, Rashid H, Muhammad A et al (1995) In vitro multiplication of date palm. Pak J Bot 27(1):101–104
Imranullah, Abdul Qayum (2013) Physico-chemical characterization of fruit of six date palm cultivars during various maturity stages. International conference on date palm: present status & future prospects, The Islamia University of Bahawalpur, Bahawalpur, 2–3 Sept, p 14
Jatoi MA, Solangi N, Markhand Z (2010) Dates in Sindh: facts and figures. In: Markhand GS, Abul-Soad AA (eds) Proceedings international dates seminar, 28 July 2009, Khairpur, p 59–71
Khan S, Bibi T (2012) Direct shoot regeneration system for date palm (*Phoenix dactylifera* l.) Cv. Dhakki as a means of micropropagation. Pak J Bot 44(6):1965–1971
Khan NM, Akram M, Amanullah et al (2009) Impact of education on diffusion of dates palm orchards in Northwest Pakistan. Sarhad J Agric 25(3):495–500
Khan A, Khan IA, Azim MK (2012) The chloroplast Genome sequence of date palm (*Phoenix dactylifera* L. cv. 'Aseel'). Plant Mol Biol Rep 30:666–678
Kumar RJ (2009) New interpretations on Indus Valley Civilization. All Right Publishers, Chennai
Mahar AQ (2007) Post-harvest studies of different cultivars of date palm (*Phoenix dactylifera* L.) fruits, their protection, identification, processing and preservation at district Khairpur, Sindh. PhD thesis, Date Palm Research Institute, Shah Abdul Latif University, Khairpur, Sindh
Maitlo WA, Markhand GS, Abul-Soad AA et al (2013) Chemical control of sudden decline disease of date palm (*Phoenix dactylifera* l.) in Sindh, Pakistan. Pak J Bot 45(S1):7–11
Markhand GS, Abdul RM, Abul-Soad AA et al (2008) Effect of fruit thinning on the quantitative and qualitative characteristics of date palm (*Phoenix dactylifera* L.) cv. Aseel. Hortic Sci 43(4):1142
Markhand GS, Abul-Soad AA, Mirbahar AA et al (2010) Fruit characterization of Pakistani dates. Pak J Bot 42(6):3715–3722
Marshal J (1931) Mohenjo-Daro and the Indus civilization. Reprint 2004, AES Publications Pvt. Ltd, New Delhi
Marwat SK, Usman K, Khan EA et al (2012) Ethnobotanical studies on dwarf palm (*Nannorhops ritchieana* (Griff.) Aitchison) and date palm (*Phoenix dactylifera* L.) in Dera Ismail Khan, KPK, Pakistan. Am J Plant Sci 3:1162–1168
Maryam, Jaskani MJ, Fatima B et al (2013) Evaluation of pollen viability in date palm cultivars under different storage temperatures. International conference on date palm: present status & future prospects, The Islamia University of Bahawalpur, Bahawalpur, 2–3 Sept 2013, p 9
Mertia RS, Birbal, Kumawat RN (2010) Performance and field management of six prominent cultivars of date palm (*Phoenix dactylifera* L.) in extreme arid regions of Thar Desert, India. Acta Hortic 882:65–68
Milne D (1918) The date palm (*Phoenix dactylifera*) and its cultivation in the Punjab. Thacker, Spink and Co., Calcutta
Mirbahar AA, Markhand GS, Khan S et al (2013) Genetic diversity analysis among important date palm (*Phoenix dactylifera* L.) cultivars from Pakistan using ISSR markers. International conference on date palm: present status & future prospects, The Islamia University of Bahawalpur, Bahawalpur, 2–3 Sept, p 26
Mirbahar AA, Markhand GS, Khan S et al (2014) Molecular characterization of Pakistani date palm (*Phoenix dactylifera* L.) cultivars. Pak J Bot 46(2):619–625
Nadeem M, Salim-ur-Rehman, Anjum FM et al (2011a) Anti-nutritional factors in some date palm (*Phoenix dactylifera* L.) varieties grown in Pakistan. Int J Food Safety 13:386–390
Nadeem M, Salim-ur-Rehman, Anjum FM et al (2011b) Textural profile analysis and phenolic content of some date palm varieties. J Agric Res 49(4):525–539

Nadia, Aish M, Muhammad ZH (2013) Morphological and genetic diversity analysis of date palm cultivars of Pakistan. International conference on date palm: present status & future prospects, The Islamia University of Bahawalpur, Bahawalpur, 2–3 Sept, pp 27–28

Nixon RW (1951) Fruit thinning experiments with the Medjhool and Barhee varieties of dates. Date Grs Inst Ann Rep 28:14–17

Nixon RW, Carpenter JB (1978) Growing dates in the United States. Agriculture information bulletin no. 207. USDA, Washington, DC, Technical Document 63

Pakistan Statistical Year Book (2012) Statistics Division, Government of Pakistan, Agricultural Department, Islamabad, Pakistan

Pasha SA, Hussain A, Gajani IB (1972) Date palm of Sindh. Punjab Fruit J 33(4):9–14

PHDEC (2008) Dates marketing strategy. Pakistan Horticulture Development Export Company, Ministry of Commerce, Government of Pakistan. www.phdeb.org.pk

Popenoe PB (1913) Date growing in the old world and the new. West India Gardens, Altadena

Qasim M, Naqvi SA (2012) A fruit from heaven. In: Manickavasagan A, Essa MM, Sukumar E (eds) Dates: production, processing, food, and medicinal values, CRC Press, Taylor & Francis Group, 6000 Broken Sound Parkway NW, Suite 300, Boca Raton, pp 341–349

Rashid H, Quraishi A (1994) Micropropagation of date palm (*Phoenix dactylifera* L.) cv. Dhakki through tissue culture. Pak J Agric 15(1):1–7

Saleh A (2012) Effect of sunning as post harvest treatment for insect pests on antioxidants and physicochemical properties of date fruit. Am J Food Technol 7:715–725

Shah A, Ata-ul-Mohsin, Naeem M et al (2012) Biology of Dubas Bug, *Ommatissus lybicus* (Homoptera: Tropiduchidae), a pest on date palm during spring and summer seasons in Panjgur, Pakistan. Pak J Zool 44(6):1603–1611

Shar MU, Rustmani MA, Nizamani SM et al (2012) Evaluation of different date palm varieties and pheromone traps against red palm weevil (*Rhynchophorus ferrugineus*) in Sindh. J Basic Appl Sci 8:1–5

SLUBGH (2013) Temperature and humidity for the month of August 2010–13. E-Talk Newsletter, Published by SAL University, Khairpur, Sindh 6(8):11

SMEDA (2009) Pre-feasibility study, dates processing plant. Small Medium Enterprise Development Authority, Government of Pakistan, 6th Floor, LDA Plaza, Egerton Road, Lahore, Pakistan

Smilanick JM, Ehler LE, Birch ME (1978) Attraction of *Carpophilus* sp. to volatile compounds of figs. J Chem Ecol 4:700–701

FAO Statistics Division (2013) Production, harvested area, import and export of dates in Pakistan in 2011. http://faostat.fao.org

Thomas DL (1974) Possible link between declining palm species and lethal yellowing of coconut palms. Proc Fla State Hortic Soc 87:502–504

Yang M, Zhang X, Liu G et al (2010) The complete chloroplast genome sequence of date palm (*Phoenix dactylifera* L.). PLoS One 5:e12762

Zaid A, de Wet PF (2002) Pollination and bunch management. Chapter 8 In: Zaid A (ed) Date palm cultivation. FAO, plant production and protection, vol 156, Rome

Zangejo KM, Abul-Soad AA, Markhand GS (2013) Setting standards for processed dates in Sindh. MSc thesis, Botany Department, Faculty of Sciences, Shah Abdul Latif University, Khairpur, Sindh

Chapter 6
Date Palm Status and Perspective in Oman

Rashid Al-Yahyai and M. Mumtaz Khan

Abstract Date palm is the primary agricultural crop in Oman, and it constitutes 80 % of all fruit crops produced and 50 % of the total agricultural area in the country. Oman is the eighth largest producer of dates in the world with an average annual production of 260,000 mt per annum. There are approximately more than over seven million date palms and 250 cultivars in cultivation, primarily in the northern governorates of the sultanate. However, around 70 % of the total date production is harvested from only 10 cultivars, and a small fraction (2.6 %) of the total date production is exported. Only half of the dates produced are used for human consumption, with the other half being utilized primarily for animal feed or considered surplus and wasted. Dates are mainly harvested for fresh fruit consumption; however, alternative uses such as date syrup, date sugar, and other by-products can also be found in the local market. Dubas bug and red palm weevil are the dominant biotic factors that affect date quality and yield in Oman. Traditional methods of cultivation, small farm size, enough labor available, and poor postharvest handling and marketing are the main issues that face date palm production in Oman. New plantlets are produced from tissue culture with a primary focus on superior cultivars that are kept, among other cultivars, in the only date palm ex situ gene bank in the interior of Oman. Enhancing fruit quality by optimizing fruit size and nutritional content and rapid cultivar selectivity based on molecular techniques for better or improved commercial cultivars will increase the marketability of Omani dates. Furthermore, employing modern orchard layouts and mechanization of the labor-dependent cultural practices such as irrigation, pruning, pollination, and harvesting is vital for the sustainable and profitable production of dates in Oman.

Keywords Dates • Oman • *Phoenix dactylifera* • Subtropics • Horticulture • Postharvest • Food

---

R. Al-Yahyai (✉) • M.M. Khan
Department of Crop Sciences, College of Agricultural and Marine Sciences,
Sultan Qaboos University, 34, Al-Khod 123, Oman
e-mail: alyahyai@squ.edu.om; mumtaz@squ.edu.om

© Springer Science+Business Media Dordrecht 2015

J.M. Al-Khayri et al. (eds.), *Date Palm Genetic Resources and Utilization: Volume 2: Asia and Europe*, DOI 10.1007/978-94-017-9707-8_6

6.1 Introduction

The date palm (*Phoenix dactylifera* L.) is an ancient crop plant that has been cultivated for centuries in many arid parts of the world, particularly in West Asia and North Africa. In Oman, date palm is the most important agricultural crop that continues to occupy the largest cultivated area today. The nutritious year-round fruits from hundreds of cultivars, and the multiple benefits of the palm by-products to the growers and the community, have given date palms special status among other crops with significant economic, cultural, and religious importance in the country. The large number of cultivars also contributes to the permanence of date palm cultivation because of the extensive range of climatic adaptation, given that Oman has a diverse topography and a range of climatic zones (Al-Yahyai and Al-Khanjari 2008). Because of its extensive cultivation in Oman and the multiple uses of the date palm in a variety of products, date palm has a special social status among Omanis. Consequently, date palm is the main plant for home garden and landscaping, and it is widely grown even in areas where the climate does not allow for high-quality fruits or yield, such as in Salalah in southern Oman or the Hajar mountains. Nonetheless, in regions suitable for date palm cultivation, dates constitute the major source of household agricultural income for many farmers.

This chapter presents the status and prospects of date palm cultivation in Oman. Date palm cultivation in terms of its history, current practices, and the social and economic aspects of date production as well as ongoing research activities is discussed. The current challenges and future prospects of date production in Oman are also addressed.

6.1.1 Historical and Current Agricultural Aspects

Throughout history, growers in Oman have mastered date palm cultivation and the cultural practices involved during the pre-petroleum era. Many travelers to Oman reported date palm cultivation. Popenoe (1913) wrote, "The Fardh date growers of Oman, who are the cleverest Arab cultivators I have seen."

This is not surprising since date palm has a long association with religion and civilizations (Al-Yahyai and Manickavasagan 2012). The date palm tree has been in cultivation since 2,400 BC. Drawings and manuscript writings have indicated the importance of date palm to various civilizations in and around the Arabian Peninsula. Date palm reinforced its importance and status following the rise of Islam in Arabia. Due to its importance to this region in particular, 21 references to date palm are cited in the Holy Quran and 300 in the writings of the Hadith. These emphasize the religious and cultural significance, in addition to the nutritious and economic importance, of date palm in the sultanate.

Given its adaptation to the harsh conditions that dominate the Arabian Peninsula, such as high temperature and long drought periods, the date palm is the most appropriate crop plant suitable for cultivation in Oman. This is particularly important

Table 6.1 Regional date palm distribution in the Sultanate of Oman

Governorate	No. of date palms	Percentage of total	Production (mt)	Percentage of total
Al-Batinah (North and South)	3,263,862	41.87	105,929.14	45.85
Al-Sharqia (North and South)	1,502,525	19.27	37,294.65	16.14
Al-Dhahira (and Buraimi)	1,333,898	17.11	29,515.50	12.78
Al-Dakhlia	1,112,959	14.28	44,006.17	19.05
Muscat	322,222	4.13	10,282.98	4.45
Musandam	234,453	3.01	3,874.09	1.68
Dhofar	23,679	0.30	132.38	0.06
al Wusta	2,188	0.03	–	–
Total	7,795,786		231,034.91	

Source: MAF (2005)

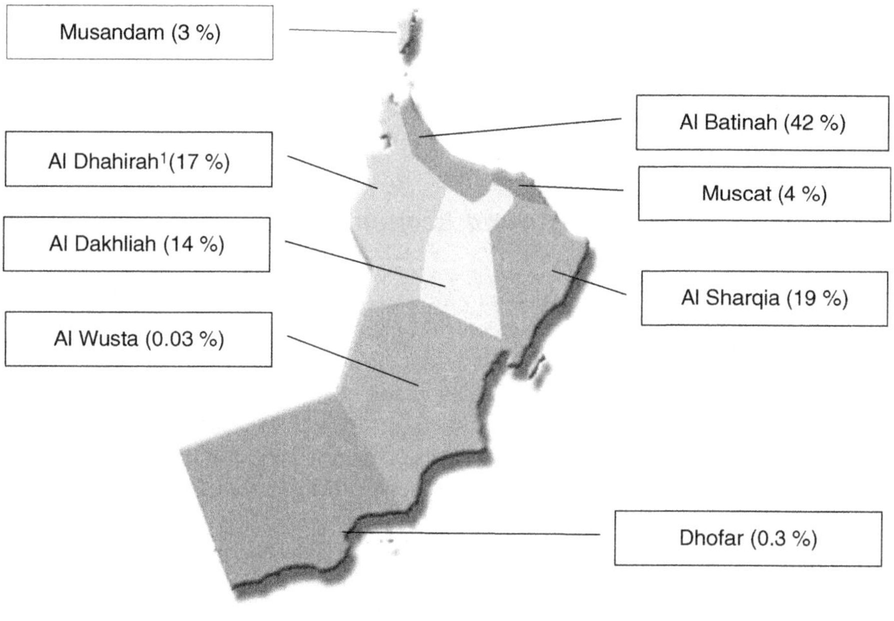

Fig. 6.1 Regional distribution and percentage (from a total number of 7,795,786) of date palm trees in each governorate of the Sultanate of Oman ([1]Includes the Buraimi governorate. *Source*: Al-Yahyai and Al-Khanjari (2008))

in this country that is characterized by low rainfall (an average 100 mm per year), whereas the rate of evapotranspiration is tenfold. Date palm is grown primarily in the northern part of Oman, where agroclimatic requirements meet the needs of date production (Al-Yahyai and Al-Khanjari 2008). Date palms occupy half of the cultivated area and 82 % of all fruit crops grown in Oman (Al-Yahyai and Al-Khanjari 2008). Date palm is cultivated in various governorates of Oman as illustrated in Table 6.1 and Fig. 6.1.

6.1.2 Importance to Omani Agriculture

Oman exports mainly only two fruit crops, dates and bananas. Date palm is by far the most significant export commodity, mainly to the United Arab Emirates (UAE), Saudi Arabia, and India. However, date exports remain very small compared to what the country produces. In 2011, Oman exported 7,171 mt of dates, which is only 2.6 % of the total production for that year. The low level of exports was attributed to the poor handling of dates, poor fruit quality, and the numerous pests and diseases that affect postharvested dates in Oman (Al-Marshudi 2002; Al-Yahyai 2007; Manickavasagan et al. 2013). Compared to other countries, Omani dates fetched a lower price in the international market (USD 102 per mt), while countries of limited production such as Tunisia attained four times the price (USD 418 per mt) of dates (Mbaga et al. 2011). Improvement of fruit quality may lead to greater exports of dates to various parts of the world where demand for dates is increasing such as in Oceana, the Americas, and Europe (Al-Yahyai and Manickavasagan 2012). In the local market, price per kg for Fardh, the main processed cultivar, ranges from USD 1.5 to 2.5 compared to USD 4–6 for dates imported from Iran, Saudi Arabia, and UAE for comparable cultivars.

6.1.3 Production Statistics and Economics

Oman is ranked eighth in the world in date production (Table 6.2). The cultivated date area in 2011 was 31,348 ha which produced 268,011 mt of fruit (FAOSTAT 2013). Since 1980, there has been an increase in production reaching nearly 300,000 mt; however, that was followed by a rapid decline in the following years for unknown reason. The production has picked up since 2004 but continues to fluctuate from year to year (Fig. 6.2). Little of the Oman's vast production is exported, averaging 2.6 % in the past 10 years. However, the import of dates has rapidly

Table 6.2 Top 10 date-producing countries of the world

Country	Date production (mt)
Egypt	1,373,570
Saudi Arabia	1,122,822
Iran	1,016,608
Algeria	724,894
Iraq	619,182
Pakistan	557,279
Sudan (North and South)	432,100
Oman	268,011
United Arab Emirates	239,164
Tunisia	180,000

Source: FAOSTAT (2013)

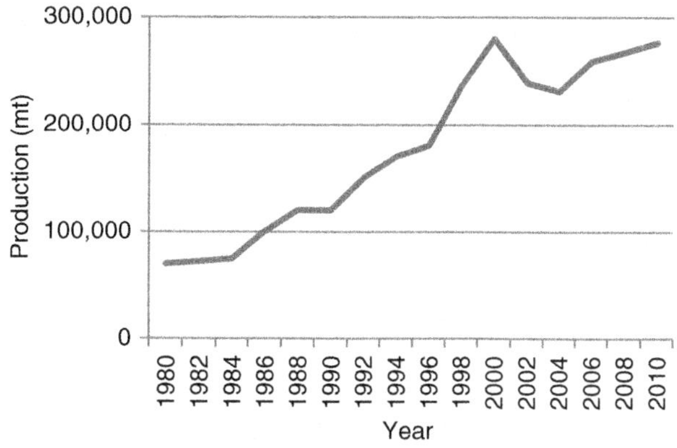

Fig. 6.2 Date production in the Sultanate of Oman (*Source*: FAOSTAT 2013)

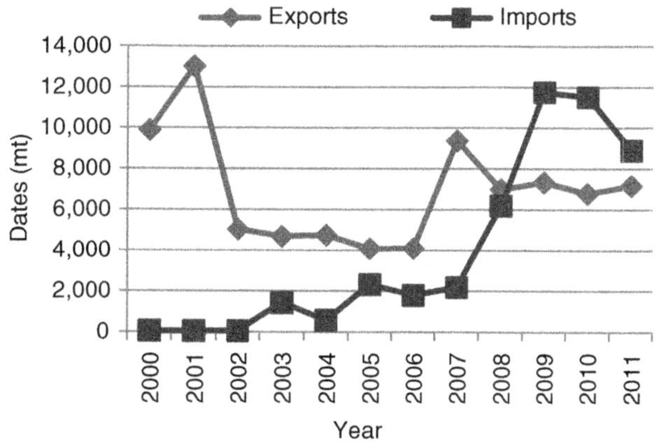

Fig. 6.3 Oman's date exports and imports (*Source*: FAOSTAT 2013)

increased from 2,000 mt in 2007 to 12,000 in 2009 (Fig. 6.3). By 2011, imports have increased by 74 %, whereas exports only increased by 43 %. This indicates that a portion of the imported dates, mostly from neighboring countries, is reexported. However, no data is available on the volume of reexports, as the figure 7,171 mt of dates exported in 2011 includes all types of exports (fresh dates, dry dates, and processed).

6.1.4 Current Agricultural Problems

Date production in Oman remains traditional (Al-Marshudi 2002; Al-Yahyai 2007). Due to the subsistence nature of date production in Oman, farmers face many issues, namely, the traditional irrigation and fertilizer application methods (Al-Yahyai 2007). Sufficient irrigation water quantity and quality have also been a major concern to date palm growers, limiting field expansion because date-growing regions are fully dependent on groundwater extraction for irrigation. Biotic issues are also of great concern to growers, particularly pests and diseases. Although presently under chemical control, pests such as the dubas bug (*Ommatissus lybicus* Bergevin, Homoptera: Tropiduchidae) and red palm weevil (*Rhynchophorus ferrugineus* Olivier, Coleoptera: Curculionidae) are a threat to date palm production in the sultanate. Logistical problems include insufficient number of skilled laborers and underdeveloped transport and storage facilities, market outlets, and large-scale processing factories. In recent years, extreme weather conditions such as cyclones, Gonu in 2007 and Phet in 2010, and flooding have severely affected date production; events associated with climate change may be the next challenge to date palm cultivation in Oman. The government has subsidized the replacement of date palms destroyed by cyclones and storms. The issue of climate change is also being studied for better preparedness in regions that may potentially be impacted by the adverse effects of climatic catastrophes.

6.2 Cultivation Practices

6.2.1 Research and Development

Published research work is lacking on various aspects of cultivation practices in Oman. Earlier research carried out at Sultan Qaboos University (SQU) focused on pollination, fertilizer application, irrigation, and utilization of date palm as the most significant issues (El Mardi 1995; El Mardi et al. 1998). Recent work has focused on yield and fruit quality improvement and postharvest technology (Al-Yahyai and Al-Kharusi 2012a; El Mardi et al. 2007; Manickavasagan and Al-Yahyai 2012; Pillay et al. 2003; Williams et al. 2005).

6.2.2 Description of Current Cultivation Practices

6.2.2.1 Pollination

Pollination of date palm is carried out by hand in almost all date palm groves in Oman. There are several male palm cultivars (called *fahl*) that are used for pollination, most notably Khori and Bahlani. Hand pollination is done by placing strands of the male cultivar in the middle of the female inflorescence (i.e., spadix) (Fig. 6.4). The number of strands varies from 3 to 12 depending on both the male and female cultivars. Newly introduced mechanical pollen dusters have not gained acceptance and trust among date palm growers. El Mardi et al. (2002) pollinated Fardh date palm by hand, hand duster, and motorized duster with no effect on fruit yield, despite the larger fruit volumes when dusters were used. They also reported that a pollen/flour (1:5) ratio for mechanical pollination produced a lower sucrose and dry matter and higher yield.

Fig. 6.4 Hand-pollinated date palm female inflorescence with male strands placed in the middle

6.2.2.2 Irrigation

Irrigation water is traditionally delivered to date palm groves through open canals. The water source is mainly underground aquifers and from wells or via a system of underground tunnels called *falaj*. *Falaj* (aka *qanat*) is an ancient system of delivering water to farms in Oman. There are three types of falaj depending on the source of water: (1) *ghaili falaj*, where water is diverted to canals from a flowing wadi; (2) *aini falaj*, spring water; and (3) *daudi falaj*, where water is sourced from an underground aquifer (Al-Ghafri 2006). Water is distributed to date palm groves in units of time. Other sources of water have also been explored, such as the utilization of treated sewage water (El Mardi et al. 1995, 1998), but with limited use until now. The timing and frequency of irrigation is largely dependent on the allocated shares of water for each grove and is not based on empirical methods. Adoption of new methods of irrigation (such as bubbler irrigation, which is a localized, low pressure, solid permanent installation drip irrigation system), particularly in well-irrigated groves, is slowly gaining momentum as the government subsidizes installation of modern irrigation systems.

Al-Yahyai and Al-Kharusi (2012b) reported that chemical quality attributes of date palm (cv. Khalas) grown in northern Oman varied in response to decreased frequency of irrigation water applied during fruit development. Palms irrigated daily had higher fruit water content, juice volume, and titratable acidity (TA), whereas total and reducing sugar content, total soluble solid content (TSS), pectin, and dry matter were high under deficit irrigation. They concluded that fruit quality may be enhanced using various deficit irrigation regimes.

6.2.2.3 Fertilizer Application

Date growers apply fertilizers manually using animal manure twice a year (Al-Yahyai 2010). Chemical fertilizers are not commonly used in date palm groves in Oman. However, most of date palm trees are intercropped with legumes used as fodder, thus perhaps making sufficient nitrogen available to date palm. Al-Kharusi et al. (2009) evaluated the effects of mineral and organic fertilizers with and without the supplementation of micronutrients on the chemical characteristics and quality of date fruits from two cultivars of date palm (Khalas and Khasab). They found that dry matter content was highest when mineral fertilizers (NPK, in 4 applications) were supplemented with organic peat and micronutrients. They also reported that organic peat application over a 3-year period increased the tannin contents, whereas the mineral fertilizers reduced it. Dry matter, tannins, and titratable acidity were affected by cultivar and fertilizer treatment. Overall, the study found that mineral fertilizers (i.e., NPK) have a significant impact on date fruit quality.

6.2.2.4 Pruning and Thinning

Pruning of date palm involves the removal of old, dead, or damaged leaves. The leaves are used as by-products for making various handicrafts including crates, ropes, and baskets, as well as a source of fuel for heating and cooking, and for house construction. Using two types of tools (i.e., a sharp sickle called *makhlab* and a harvest knife called *majaz*), pruning is carried out annually following harvest and to clear up remaining fruit bunches, older leaves, and fibers. Although it increases fruit size and quality, bunch and fruit thinning is not commonly practiced, except in rare cases where the palm has over 20 bunches. Studies by Al-Yahyai (unpublished) showed that thinning to 10 bunches per Khalas cv. gave the best fruit yield and quality.

6.2.2.5 Harvest and Postharvest Handling

Dates are consumed at three stages in Oman: at the khalal stage (via cooking, called *tabseel*), fresh at the rutab stage, and as dry fruits harvested at the tamar stage. Harvesting of date palm in Oman is done following traditional methods of cutting the date bunch then dropping it or lowering it by rope to the ground (Al-Yahyai 2007). This leads to losses in dates due to fruit separation from the bunch, fruit

Fig. 6.5 Traditional sun drying of dates

bruising and damage, and exposure to insect infestation and pathogen infection. After harvest, dry fruits are sun dried in the open air (Fig. 6.5), which also exposes dates to insect damage from wasps and ants. Dried dates are stored in plastic or wood containers. Due to poor conditions, dates are also infested during storage (Manickavasagan et al. 2013). A recent study showed that dates can be frozen at the khalal stage for up to a year, with little effect on the chemical composition of the fruit (Al-Yahyai and Al-Kharusi 2012a). During fruit formation and development of date palm, khalal stage occurs between 19 and 25 weeks after fruit set. At this stage sucrose contents are increased, while fruit weight increment, acidity, and moisture contents are reduced. In some cultivars at this phase, the fruit becomes palatable and is considered of commercial maturity. Microwave energy and computer vision technology have shown promising results in identification and elimination of many of the defective fruits during factory processing (Manickavasagan and Al-Yahyai 2012; Manickavasagan et al. 2013).

6.2.2.6 Pest and Disease Control

Date palms in Oman are infested with two major pests: the dubas bug (*Ommatissus lybicus* Bergevin, Homoptera: Tropiduchidae) and red palm weevil (*Rhynchophorus ferrugineus* Olivier, Coleoptera: Curculionidae). Chemical control is the method utilized by the government to combat dubas bug. Several insecticides have been evaluated for dubas bug control in Oman with SUMI-ALPHA® 5 EC being effective as a ground spray and KARATE® 2 ULV, TREBON® 30 ULV, and SUMICOMBI® 50 ULV achieving some measure of success as aerial sprays (MAF 2006a). Strict quarantine measures and eradication of suspected palms infected by red palm weevil is the common method of controlling the spread of this pest. Al-Sadi et al. (2012) investigated fungal and oomycete pathogens associated with root diseases of date palms in Oman. Isolated fungi and oomycetes showed that they belong to 34 species. They found *Fusarium solani* (27 %), *Ceratocystis radicicola* (25 %), and *Lasiodiplodia theobromae* (19 %) to be the most common pathogens associated with root diseases of date palms. Among the 21 pathogenic species, 13 are reported for the first time in the study as new root pathogens of date palm on a global basis. These include the following species:

Ceratocystis omanensis	*Fusarium acuminatum*	*Pythium indigoferae*
Cochliobolus hawaiiensis	*Fusarium redolens*	*Pythium spinosum*
Exserohilum rostratum	*Fusarium thapsinum*	*Pythium ultimum* var. *ultimum*
Corynascus kuwaitiensis	*Nigrospora sphaerica*	
Fusarium brachygibbosum	*Phoma multirostrata*	

Al-Sadi et al. (2012) also reported 22 other new fungal and oomycete species that occur in Oman. However, many of these fungal (as well as viral and bacterial) infections do not constitute a direct threat to date palm cultivation in Oman, and little is done to control them.

6.2.3 Agroforestry Utilization and Potential

Agroecological research suggests that date palm is a suitable plant for a range of uses in Oman, particularly in saline and marginal soil areas. Date palm plantings help combat desertification, if well managed. However, no studies have been conducted to determine the extent of date palm agroforestry utilization and potential in Oman.

There are several issues related to date palm cultivation practices in Oman including traditional methods of management and cultural practices, fruit handling, enhanced fruit quality, and marketing. Application of mineral fertilizers and introduction of modern irrigation systems and mechanical pollination will greatly make date production more economical and increase grower income. Al-Yahyai (2007) outlined the areas where date palm cultivation needs further improvement. Other areas include date processing and by-product utilization, value addition of dates, and enhancement of marketing and exports and processing in the country.

6.3 Genetic Resources and Conservation

6.3.1 Research in Genetics, Breeding, and Conservation

The Sultanate of Oman is located in the southeastern part of the Arabian Peninsula. The average annual rainfall in the country is less than 100 mm, which results in making 75 % of the total area a desert or desert-like with little or no vegetation. The remaining area varies greatly in topographic and climatic characteristics; this allows the cultivation of various types of field and fruit crops. Date palm is the major crop in Oman, as it covers about half of the cultivated area in the country. Due to the diverse date plan gene pool and variability in the topographic/climatic conditions in Oman, the date palm production season expands from May to

November. This long season of date production makes the country distinct from all other major producing countries. Variability in climatic conditions, topography, and cultivation practices have also resulted in selections of numerous male and female cultivars adapted to different regions of the sultanate. Currently there are about 250 cultivated varieties of date palm in Oman. Thirty out of the 250 varieties are cultivated on large areas because of their high quality and consumer demand (Al-Yahyai and Al-Khanjari 2008).

The main objective of conserving date palm genetic material is to protect this precious heritage from the threat of extinction, which may be caused by deforestation, environmental pollution, and natural calamities such as cyclones, floods, drought, or fire. Another prime goal of keeping large date palm genetic resources is for crop breeding and improvement programs.

To aid in the preservation of the extensive genetic diversity of date palm in Oman, the Ministry of Agriculture and Fisheries Wealth (MAF) has carried out various trait analyses, for example, morphological and molecular characterization of various date palm cultivars. Various molecular techniques are currently being utilized to determine the genetic diversity of date palm cultivars in Oman. A study on DNA fingerprinting was carried out on six Omani cvs. (Bunaringa, Hilali, Khalas, Khalas Oman, Khenazi, and Zabad) and three foreign cvs. (Barhi, Deglet Noor, and Medjool). These cultivars were mapped because of their usage for large-scale propagation through tissue culture. The Random Amplified Polymorphic DNA (RAPD) analysis showed the similarities and the differences between the genotypes tested (Fig. 6.6) (MAF 2006a). A total of 78 alleles were counted from which 11 were shared in entire cultivars. The number of alleles differed according to the genotype/primer combination. The similarities between cultivars were calculated based on the analysis, which showed the close relationship between Omani cultivars compared to the foreign date palm genotypes as illustrated in Fig. 6.7 (MAF 2006a). A microsatellite marker (SSRs) tool was used to identify the genetic diversity in date palm

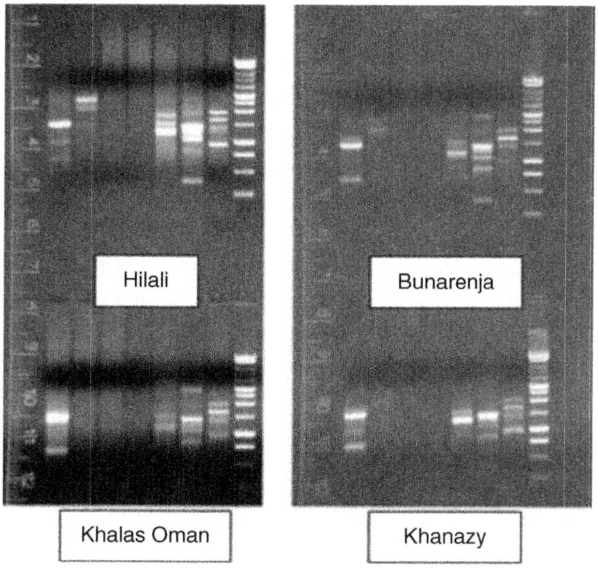

Fig. 6.6 Fingerprinting of four Omani date palm cultivars using RAPD technique (*Source*: MAF 2006a)

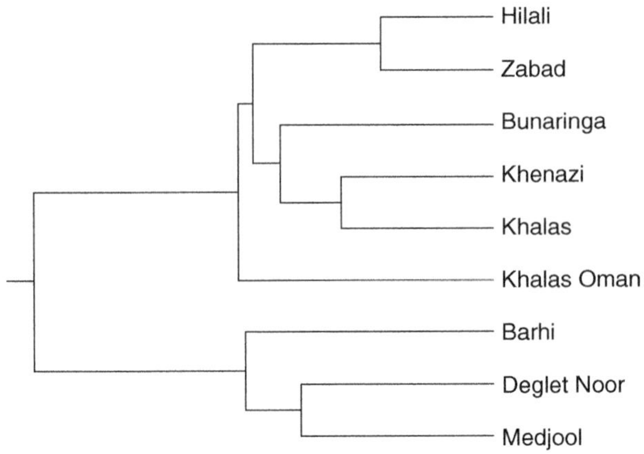

Fig. 6.7 The relationship among different cultivars based on RAPD analysis (*Source*: MAF 2006a)

(*Phoenix dactylifera* L.) clonal genotypes, which were derived through somatic embryogenesis in Oman. Twenty-one palm genotypes representing different countries (14 Omani, 5 Bahraini, 1 Iraqi, and 1 Moroccan) were screened with ten microsatellite markers. All primer pairs generated an amplification product and a high level of polymorphism among all the analyzed samples. Results revealed that the genotypes from Bahrain and Iraq have a close relationship with accessions already grown in Oman. The Moroccan genotype (Medjool) performed distinct from the rest of the date palm genotypes (Al-Ruqaishi et al. 2008). The value of this work was to describe and preserve the date palm genetic resources of Oman.

The government of Oman recognizes the significance of plant genetic resource conservation and people's involvement in such a principled initiative. Therefore, the activities are shared and complimented with public awareness campaigns to highlight the importance of preserving biodiversity of date palm cultivars in the country.

6.3.2 *Preservation of Date Palm Biodiversity*

There has been a steady decline in the production of dates in Oman over the past decade. This decline is mainly attributed to a combination of various factors including increased soil salinity in major date palm-growing regions (Al-Batinah and Al-Sharqia), desertification in areas adjacent to the desert in central Oman, heavy insect pest infestation such as dubas bug and red palm weevil, and urbanization of rural areas. These biotic and abiotic factors have led to a decrease in area of plantations and depletion of the gene pool of some existing cultivars.

Keeping plant biodiversity intact through conservation of the existing plant material, in particular, endangered old cultivars and introduction of new gene pool, is a prerequisite for sustainability of any crop. However, since the sustainability of in situ gene banks of perennial fruit crops, such as date palm, has been questioned, newer approaches for preserving local varieties should be investigated apart from the use of gene banks and tissue culture laboratories. These approaches may include in vitro preservation using cryopreservation (Bagniol et al. 1992; Engelmann et al. 1995; Finkle et al. 1979; Mater 1987; Ulrich et al. 1979) and ex situ gene banks based on farmer involvement (Aaouine 2000; Arunachalam 2000; Jarvis et al 2004; Sawadogo et al. 2005). This can be achieved through certain economic incentives to farming communities for preserving traditional agricultural practices (Siebert 2004).

At present, to assist in the preservation of the extensive genetic diversity of date palm in Oman, detailed studies on morphological and phenological description, chemical /biochemical analyses, and molecular identification and characterization of various cultivars are under way.

At the Date Palm Horticulture Research Laboratory, molecular markers have been established to map date palm genetic diversity (MAF 2006b). In 1996, conventional breeding in date palm was initiated where KL96-13 accession was used as male parent in the two different crosses (El-Kharbotly et al. 1998). This male was used to develop a backcross population (BC1) suitable for the construction of a genetic map by crossing it to its mother (Khalas). Khalas is known to produce high-quality date fruits. More offsprings of the male were developed by crossing the same male with Um-Sella cv. to produce an F1 population. Amplified fragment length polymorphism (AFLP) technique was used to analyze BC1 population (57 palms) from the Khalas cv. and of their parents. In collaboration with the College of Medicine at SQU, the polymorphism studies were conducted on a DNA analyzer. The primary analysis showed a genetic linkage between two molecular markers and the maleness character on a distance between 9 and 29 map units. The research is in progress to increase the number of reactions and the analysis of genetic relationships. The most common cultivar in the Al-Batinah region is Um-Sella but it bears low-quality fruit. This cultivar was also incorporated in the program of genetic improvement of date palm. It was crossed with the male KL96-13 (originated from cv. Khalas) to produce an F1 population. The female from this cross is being evaluated for the fruit quality. The results showed improvement in shape, color, and size of the produced fruits (MAF 2006b).

To study the effect of pollen grains' source on Khalas fruit quality, fruit set, and ripening, research was conducted in two locations: Al-Barka Nursery Farm (Al-Batinah) and Nizwa Al-Rafa'Farm (interior) in the 2007 season. Seven male cvs. (Khori, Alnaseeb, Medgahdel, Abosaaba, Bahlani, Ghareef, and Naghayli) were used to pollinate Khalas A'Dhahirah date palm trees. There were significant differences in fruit set and fruit characteristics in both locations (MAF 2007). A similar study was conducted on Zabad cv. to overcome the low percentage of fruit set and yield. Pollen of five male cvs. was used to pollinate the flowers of Zabad cv. The results showed that there were no significant effects of pollen grains on the fruit set and the quality of Zabad cv. (MAF 2007).

In 2009, microsatellite or simple sequence repeat (SSR) markers were used to fingerprint some Omani date palms cvs. (Bahlani, Buhabisha, Bunaringa, Khinizi, Thameed, and Zabad female; Bahlani and Khori male). All samples were analyzed using 20 SSR primer pairs which were designed for date palm genetic mapping (Fig. 6.8). To identify genetic linkage between Khalas cv. from Gulf Coast countries, 15 different microsatellite primers were used (Fig. 6.9) (MAF 2009). Date palm genetic diversity was further explained on basis of physical attributes. The physical phenotypic diversity index results

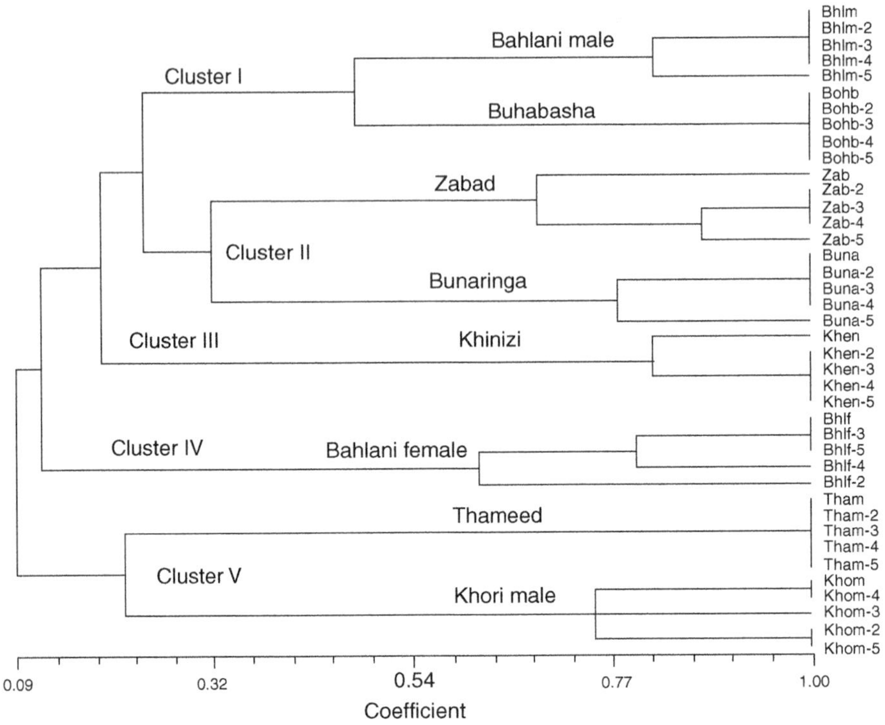

Fig. 6.8 UPGMA cluster analysis of SSR data generated by 20 primer combinations for 40 of Omani date palm cultivars showing patterns of genetic similarity using Jaccard's coefficient (*Source*: MAF 2009). Abbreviations are samples of the corresponding cultivars

showed large biodiversity among the selected date palm cultivars. Likely, the similarity matrix also exhibited high similarity among date palm cultivars ranging from 74 to 90 %. Another study by Al-Ruqaishi et al. (2008) used microsatellite markers (SSRs) to screen and analyze the genetic diversity among clonal genotypes of date palm derived by somatic embryogenesis. The study detected high levels of polymorphism among 21 cultivars, 14 of which were from Oman. These studies emphasized that further studies are needed on chemical and molecular evaluations to discover the genetic linkages among cultivars (Al-Yahyai and Al-Khanjari 2008).

6.3.3 Quarantine Regulations

Quarantine regulations have been established and implemented by the MAF in Oman. Keeping in view the trans-boundary or inland invasion of insect pests and diseases, unwanted plant material movement is strictly prohibited. Therefore, stringent plant material quarantine rules are in place which ensures a safeguard against any plant-related catastrophic breakout. Plant quarantine principles are followed strictly, and if any threatening disease is discovered, the site is monitored and

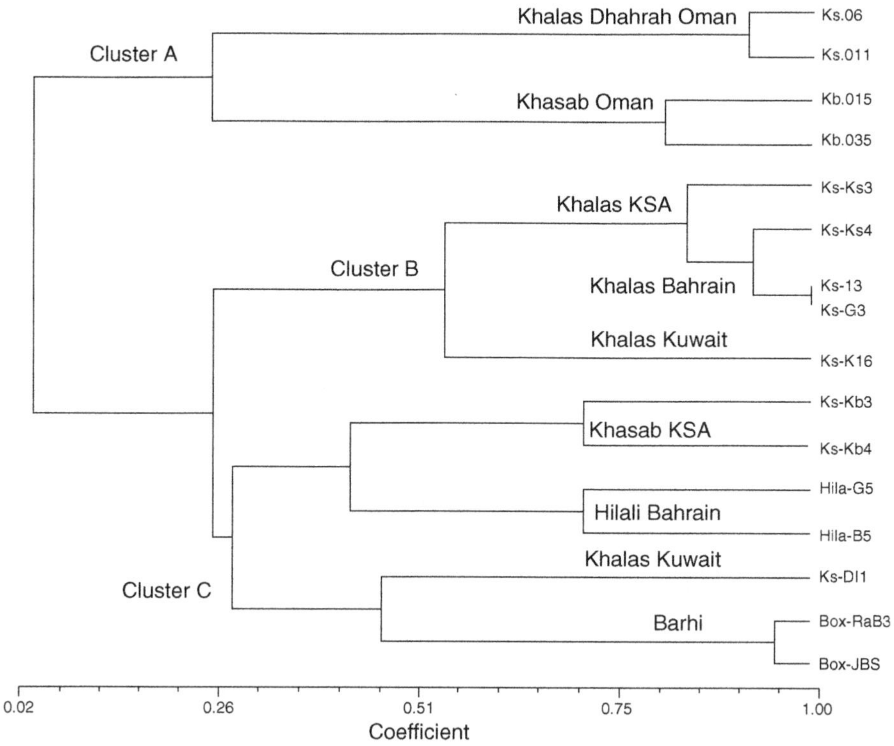

Fig. 6.9 Dendrogram of similarity J. coefficients based on UPGMA cluster analysis in 16 genotypes from Oman, Saudi Arabia, Bahrain, and Kuwait using 16 microsatellite primers (*Source*: MAF 2009). Abbreviations are samples of the corresponding cultivars

movement of plant material is restricted until the problem has been resolved. Importation of all palm types is prohibited without appropriate phytosanitary certification.

6.4 Plant Tissue Culture

In Oman, the first tissue culture and biotechnology laboratory for date palm mass propagation and research was established at Jimmah Agricultural Research Station. Since its inception in 1992, it has contributing significantly to date palm tissue culture mass production. This laboratory has its long-term strategy to replace 3.1 million date palm trees with newly produced true-to-type plantlets. The plan to produce 500,000 Fardh cv. seedlings in 5 years is a noteworthy ambition of this laboratory. More than 250,000 plantlets have been distributed to growers and significant numbers of genotypes (>70) have been tested/produced (Fig. 6.10). At the Jimmah facility, research studies are also being conducted on cryopreservation and improvement of existing methods to manipulate and harvest plants derived from somatic embryos as shown in Fig. 6.11 (MAF 2009).

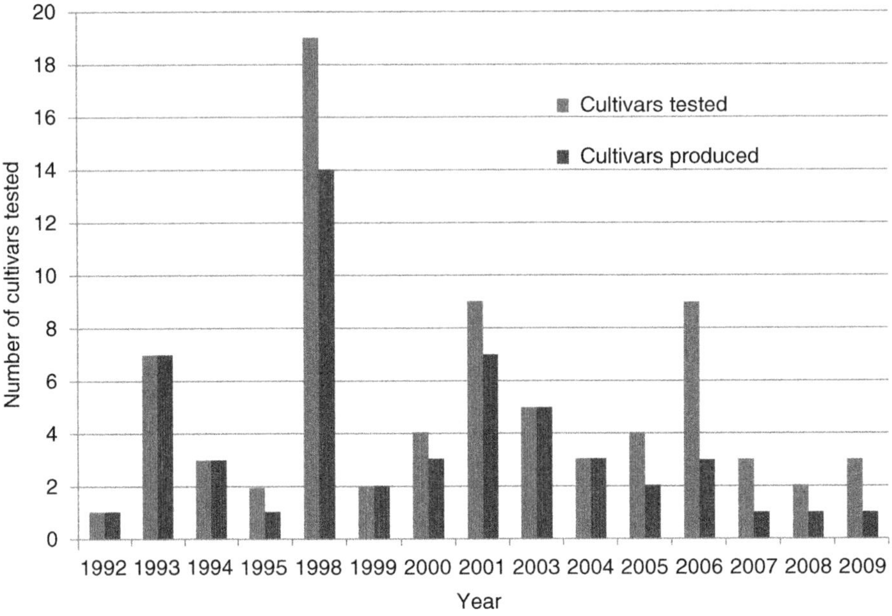

Fig. 6.10 Number of cultivars tested/produced at Jimmah Tissue Culture Laboratory, 1992–2009 (*Source*: MAF 2009)

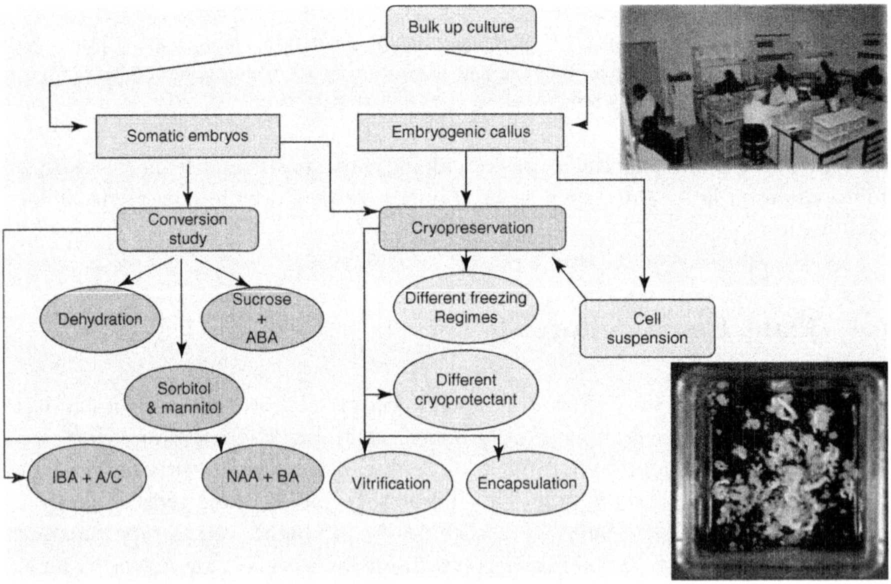

Fig. 6.11 Schematic illustration of date palm tissue culture (somatic embryogenesis) followed at Jimmah Tissue Culture Laboratory, Oman

Fig. 6.12 Stages of date palm plantlet acclimatization in (**a**) growth room, (**b**) greenhouse, and (**c**) shadehouse conditions at Jimmah, Oman, prior to distribution for cultivation

In the Jimmah tissue culture and biotechnology laboratory, a series of experiments have been conducted on standardizing micro-propagation techniques for date palms; most of the developed techniques can be widely applied to many cultivars with necessary modifications. In this institute, two types of procedures have been successfully employed in tissue culture of date palms (Fig. 6.11). The most common method is somatic embryogenesis, which has many advantages and disadvantages. With this method, mass cloning of date palm plantlets through repetitive somatic embryogenesis has existed since 1996, and a good number of propagules have been distributed to date farmers (Figs. 6.12 and 6.13). However, it involves the possibility of some undesirable genetic variability in tissue

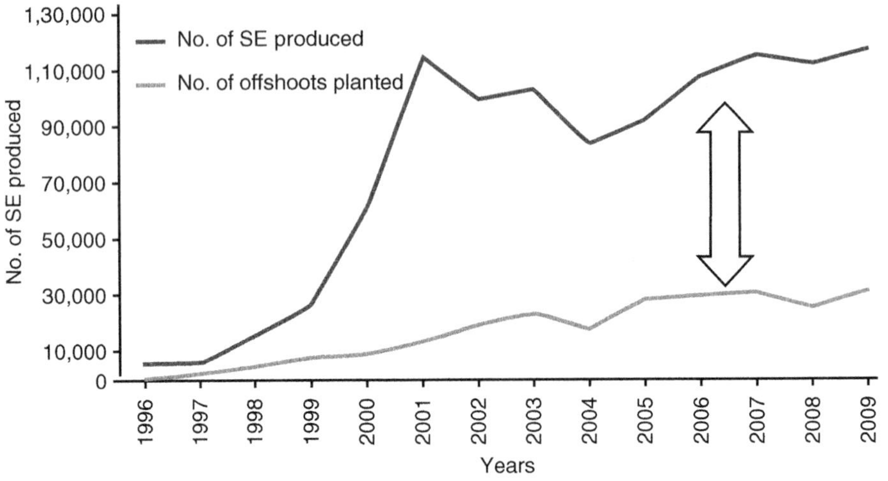

Fig. 6.13 Production of somatic embryos (SEs) of date palm at Jimmah Tissue Culture Laboratory and the total number of saplings distributed to farmers since 1996 (*Source*: MAF 2009)

culture-derived plants, resulting in some abnormal vegetative traits as well as unusual flowering and fruiting habits, which are not usually apparent until the fruiting stage. Genetic variations were recorded in 6- and 12-month-old cultures. It was observed also that all morphologically abnormal shoots showed genetic variations at the molecular level (Saker et al. 2000). One of the alternative methods used now is direct organogenesis. The successful development of this new technique is expected to reduce the number of steps in culture, thereby shortening the duration of culture with higher concentrations of auxins which may lead to reduced somaclonal variations.

A comparable plant tissue culture laboratory at SQU was also created with the main objectives to train technicians in the field of plant tissue culture and to conduct other related research and development studies. However, at Jimmah, commonly practiced methods are adopted with little modifications according to local cultivar requirements.

6.5 Cultivar Identification

6.5.1 Role and Importance

The diverse topographical and climatic ecoregions in Oman have allowed for the cultivation of various types of date palm cultivars (Al-Yahyai and Al-Khanjari 2008). The documented number of cultivars is 180 female and 48 male of the 7.8 million trees of date palm grown throughout the country. However, the majority of the total date production, in each region and in Oman in general, comes only from a

Table 6.3 Major date palm cultivars in various governorates of the Sultanate of Oman in a descending order of production per cultivar

Northern and Southern Al-Batinah	Al-Dakhliah	Al-Dhahira and Buraimi	Northern and Southern Al-Sharqia	Dhofar	Musandam	Muscat	Oman (overall)
Um Sella	Naghal	Fardh	Mabsli	Sarna	Shahl	Um Salla	Um Sella
Mabslui	Khasab	Naghal	Madloki	Madloki	Qash	Khunaizi	Mabsli
Khasab	Fardh	Khasab	Barni	Qash	Qash Habash	Khalas	Khasab
Shahl	Mabsli	Khunaizi	Khasab	Hajar	Lulu	Naghal	Naghal
Naghal	Khalas	Khalas	Naghal	Faradhi	Khunaizi	Khamri	Fardh
Mesli	Khunaizi	Bu Maan	Bu Narenja	Qash Jabrin	Qash Ahmar	Qadmi	Shahl
Khalas	Handhal	Suwaih	Bu Daan	Khunaizi	Naghal	Khasab	Khunaizi
Khunaizi	Barshi		Fardh	Naghal	Bu Alathoq	Barni	Khalas
Munawmah	Barni		Khunaizi	Hassassi	Baql	Qash	Madloki
Salani	Zabad			Basri	Hilali	Hassassi	Barni

Source: Al-Yahyai (2010)

few commercial cultivars (Table 6.3). These cultivars are dominant because of their marketable high fruit quality or early and late season production (Al-Yahyai 2010). The date palm gene bank of the MAF, established at Wadi Quriat in 1988, contains 166 female and 21 male cultivars, of which 81 produce yellow fruits, 24 produce red, and the remaining produce various other fruit colors (Al-Yahyai and Al-Khanjari 2008; Al-Zidjali 1996). This is the largest field gene bank in the country along with a sizeable collection at the SQU agriculture experiment facility.

6.5.2 Research in Date Palm Descriptors

Date palm descriptors, both morphological and molecular, are lacking for cultivars grown in Oman. Information gathered for record keeping that include date of flowering time, yield, and fruit quality parameters from the only gene bank in Oman has not been published yet. Two publications describing the physical and chemical characteristics of fruits were published as illustrated books in Arabic by the Diwan of Royal Court Affairs (Macki et al. 1998). Al-Yahyai and Al-Khanjari (2008) used that information to assess the biodiversity of date palm in Oman. Jaradat and Zaid (2004) also used fruit quality traits to assess the origin and diversity of date palm in southern Arabian Peninsula, including Oman.

Date palm molecular descriptor development for cultivars in Oman now is in progress. The MAF in Oman is currently working on developing DNA markers and molecular descriptors for date palm cultivars.

No commercial laboratories specializing in identification and characterization of date palm cultivars exist in Oman. The MAF administers one tissue culture laboratory that is also carrying out molecular identification of cultivars for quality monitoring purposes.

6.6 Cultivar Descriptions

6.6.1 Growth Requirements

Date palm is chiefly grown in the northern part of Oman. Al-Yahyai and Al-Khanjari (2008) identified two main agroclimatic zones where date palm is grown: (a) the coastal plains that include the governorates of Batinah (north and south), Sharqia (north and south), and Musandam and (b) the interior that includes Dakhlia, Dhahira, and Buraimi (Fig. 6.1). Both agroclimatic zones are characterized by hot summers and cool winters, with the coastal plains being more humid, particularly during the summer. The central governorate, Wusta, is largely dominated by desert and nomadic lifestyle, thus the number of date palms is minimal. The southern governorate of Dhofar is not suitable for date palm due to monsoon rains during the ripening months of dates (May–August); however, farther inland, date palms are grown in oases bordering the Empty Quarter desert. Different cultivars are grown in these various regions due to a range of agroclimatic conditions (Table 6.5), depending primarily on the cultivar performance under the specific climatic conditions of that region.

6.6.2 Cultivar Production Statistics and Economics

Despite the large number of cultivars, as mentioned previously, the production of dates is largely dominated by only a few cultivars (Table 6.4). It can also been noted from the table that the largest cultivar in production, Um Sella, is not

Table 6.4 Major date palm varieties and its total production in the Sultanate of Oman

Cultivars	2004 yield (mt)	% of total	Cumulative %
Um Sella	32,696.48	14.15	14.15
Mabsli	30,583.24	13.24	27.39
Khasab	26,678.61	11.55	38.94
Naghal	24,423.38	10.57	49.51
Fardh	18,051.93	7.81	57.33
Shahl	11,435.75	4.95	62.28
Khunaizi	11,340.99	4.91	67.18
Khalas	11,139.04	4.82	72.01
Madloki	5,423.58	2.35	74.35
Barni	4,966.3	2.15	78.65
Total	231,034.91		

destined for human consumption and used mainly as livestock feed. The most popular cultivars for human consumption and for processing, Khalas and Fardh, respectively, only contribute around 12 % of the total date produced in Oman. Earlier cvs., such as Naghal which reaches rutab stage in May, and late cvs. such as Khasab, harvested until late October, are consumed fresh and not commonly stored for consumption during the winter months. Mabsli is a cultivar that constitute a large proportion of the export market as boiled dates, most of which (70 %) is destined for India. However, efforts are under way to expand the list of cultivars that are suitable for boiling to expand the exports of the important niche commodity.

6.6.3 Nutritional Aspects

The date fruit nutritional composition of sugar, acidity, fiber, minerals, and other components have been studied under various treatments, in the field (Al-Kharusi et al. 2009; Al-Yahyai and Al-Kharusi 2012b; Al-Yahyai 2010; El Mardi et al. 2002), during storage (Al-Yahyai and Al-Kharusi 2012a), at the consumption stage of the dates (Al-Farsi et al. 2005a, b, 2007), and its by-products (Al-Farsi et al. 2007). This information indicated that cultural practices and postharvest handling have a great impact on date nutritional quality. Various cultivars have varying quantities of nutrient composition, particularly minerals and antioxidants.

6.6.4 Morphological Description

The only publication on the date palm cultivars of Oman is by Macki et al. (1998) in Arabic. However, this catalog details the general physical and chemical attributes of the reproductive part of the date palm rather than a comprehensive descriptive analysis of the whole plant. The most popular cultivars for fresh consumption are Khalas, Barni, and Khunaizi (Fig. 6.14), whereas Fardh (Fig. 6.15) is commonly used for industrial packing and processing. The physical and chemical characteristics of Khalas, Fardh, Khunaizi, and Barni are shown in Table 6.5.

6.7 Date Production and Marketing

The date palm tree has the ability to survive under relatively harsh climatic and soil conditions while many other crops cannot thrive under such adverse conditions. Therefore, it is an irreplaceable tree in irrigable desert lands that provide protection to ground-level crops from heat, wind, and other threats and is an excellent plant species in terms of combating desertification. Due to these characteristics, date palm tree has been grown in Oman and the Arabian Peninsula for centuries and it is

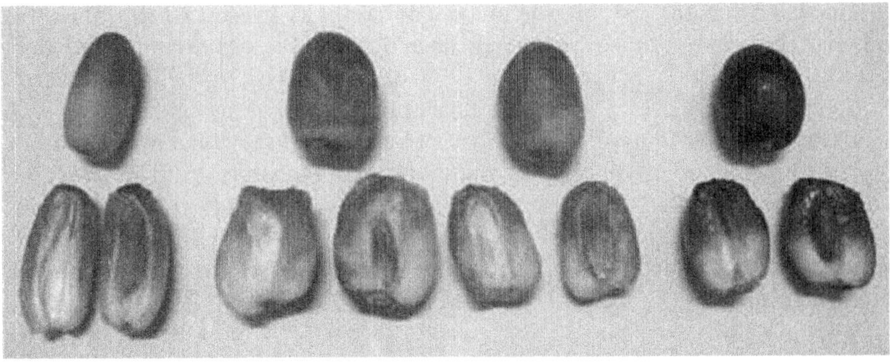

Fig. 6.14 Dates of Barni, Khalas, and Khunaizi cultivars grown in northern Oman

Fig. 6.15 Fardh dates at the transitional stage from khalal to rutab stage, from northern Oman

considered to be the oldest fruit tree. Dates have high nutritive value and are the main food source for Omanis since they began to sail the seas of this region, which can be traced back to the seventeenth century (El Mardi 1995). Date cultivation and use of the fruit as a basic food commodity is believed to be an integral part of the Omani national heritage and social life.

Date palm cultivation is one of the most essential agricultural activities in Oman. This country has diverse topographical and climatic ecoregions that allow for cultivation of various date palm cultivars, particularly in the northern coastal and the interior regions (Al-Marshudi 2002). Date production is sufficient to meet the domestic demand and a significant surplus is also exported. The majority of palm growers in Oman still use traditional methods for growing this crop from planting until it is marketed. They are using the farming techniques which require only limited inputs of capital and cause minimal disturbance to the environment. Hence, the patterns of production seem truly sustainable and skills are being

Table 6.5 Fruit physical and chemical characteristics of four date cultivars from northern Oman

Parameter	Fardh	Khalas	Khunaizi	Barni
Fruit physical characteristic				
Color: khalal	Orange	Yellow	Scarlet red	Yellowish green
Color: rutab	Reddish brown	Dark yellow	Dark red	Yellow
Color: tamar	Dark brown	Reddish yellow	Black	Brown
Fruit weight (g)	15.4	9.6	10.4	9.8
Seed weight (g)	0.6	0.9	0.5	1
Fruit size	Large	Medium	Medium	Medium
Fruit length (cm)	3.9	3.4	3.3	4
Fruit diameter (cm)	2.1	2.1	2.2	1.7
Sphericity (length/diameter)	2.1	1.6	1.5	2.7
Chemical characteristics of pitted dry dates				
Ash (%)	1.19	1.16	1.14	1.25
Moisture content (%)	39.69	55.17	48.87	42.96
Fiber (%)	1.17	0.74	1.59	2.37
Total carbs (per 100 g) (%)	95.00	89.20	94.30	91.30
Reducing sugars (%)	88.17	80.40	86.60	84.35
Nonreducing sugars (%)	5.33	6.75	5.60	5.00
Starch (%)	1.50	2.05	2.43	1.88
Total protein (%)	3.69	3.57	2.72	3.80
Total lipids (%)	0.39	0.42	0.47	0.56
Pectin (%)	2.75	1.76	3.00	1.52
TSS (%)	60.30	44.85	50.87	56.96
Mineral content (mg per 100 g) on dry weight basis				
Potassium	762.00	800.50	786.33	790.50
Phosphorus	48.33	42.60	53.23	57.53
Magnesium	35.83	35.45	34.90	36.45
Iron	1.18	1.34	1.19	1.19
Zinc	0.35	1.78	0.51	0.57
Copper	0.56	0.72	1.43	0.86
Manganese	1.91	1.70	1.73	2.12
Sodium	4.93	5.15	4.90	5.28

Source: Adapted with permission from Macki et al. (1998)

passed from generation to generation (Al-Marshudi 2002). Most of the traditional date palm growers are considered subsistence producers with a main objective to only supply enough food for the family needs (Al-Yahyai 2007). Nonetheless, tremendous development has occurred in the production and distribution of dates during the last two decades. Now in many areas of the sultanate, date palm trees are very well developed in terms of cultivar selection, planting, harvesting, marketing, and storage.

6.7.1 Optimization of Yield

Significant advances have been made in traditional cultivation methods to allow the farmers to increase yield under all kinds of harsh environments.

The MAF and the College of Agricultural and Marine Sciences (CAMS) at SQU are doing a commendable job to increase date palm production and to improve fruit processing in Oman. The MAF has also developed tissue culture laboratory to mass propagate selected high-quality date palm trees for distribution to growers. Several MAF agricultural research stations conduct applied research and provide extension services to date palm growers throughout the country. The MAF has also designed a strategic plan for improving date palm production. The MAF plans to increase investments in the date sector that will help in production of high-quality dates as well as expansion of the national and international markets for dates throughout the year. The researchers in CAMS have also done considerable work for the optimization of date palm yield in the country. They have introduced labor-saving methods in cultivation, developed modern irrigation systems, and improved packaging of the final products. All these efforts have resulted in substantial increases in date palm production in Oman. It has been reported that in the last decade the area under cultivation of date remained almost the same, but the production has been almost doubled (Al-Marshudi 2002). Much work is yet needed to improve fruit quality. At present, Oman's current net production value of dates is USD 132,533,000 (Fig. 6.16).

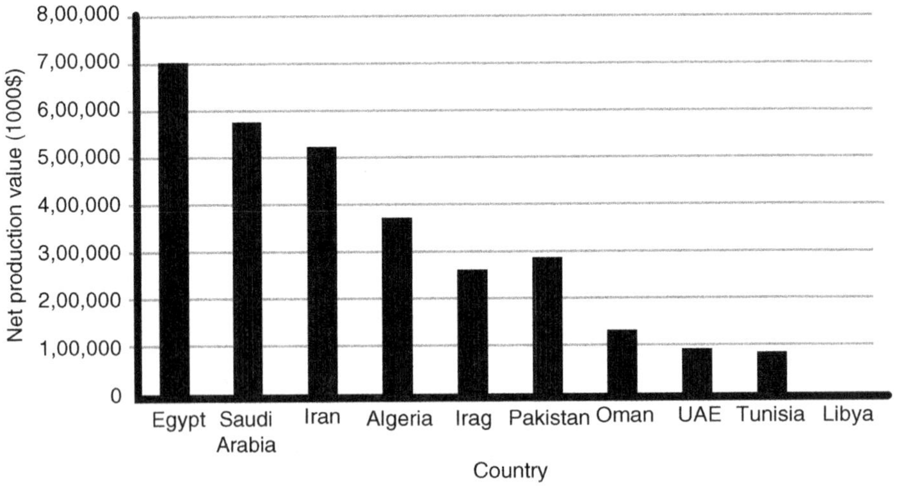

Fig. 6.16 The net production value in the top 10 date-producing countries in the world in 2011 (*Source*: FAOSTAT 2011)

Fig. 6.17 Loaded U-Shaped harvesting platform (*Source*: Courtesy of H. Jayasuriya)

6.7.2 *Harvest Mechanization*

Harvesting of the date palm fruit is very critical in terms of determining the quality of the produce and hence the ultimate price. Date fruit is harvested either manually or with some degree of mechanization, especially on large farms. The harvesting machinery needs to be efficient and clean since it will affect the rest of the downstream steps such as processing, packing, and marketing.

With the subsequent development of date palm industry in Oman, harvesting strategies have also undergone advancement, and now mechanized harvesting (Fig. 6.17) is becoming popular. Usually the mechanized date-harvesting systems consist of vehicles equipped with a number of long arms with baskets at the end, where a worker can stand in the basket to pick the fruit (Mazloumzadeha et al. 2008). Research labs at SQU are working to develop new harvesting machines adapted to the Omani date farms (Fig. 6.18).

6.7.3 *Postharvest Operations*

The date harvest in Oman is long, starting from May and extending until November, about 6 months, because of the large diverse gene pool (>250 biotypes) which are distributed throughout the northern parts of Oman (Al-Yahyai and Al-Khanjari 2008). Some date cultivars, for example, Mabsli Abou-Narenjah and Madloky, need special heat treatment processing, before drying, called *tabseel*. Dates are cooked in

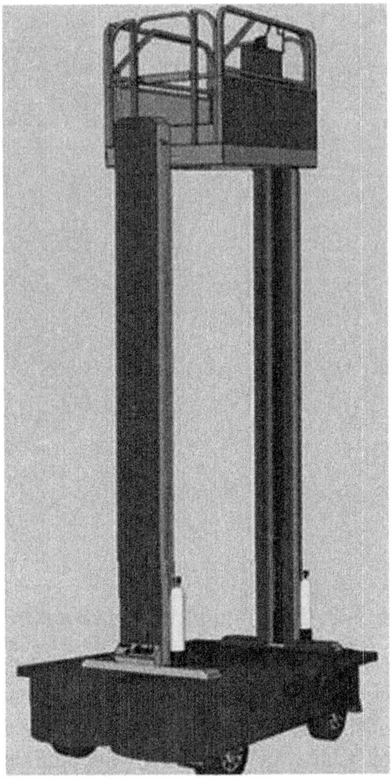

Fig. 6.18 Harvesting platform lift model (*Source*: Courtesy of H. Jayasuriya)

large quantities in a big boiling pan nearly for 45 min, until the dates are shiny and ready to be scooped out of the pan. Drying surfaces used are the ground, simple platforms, flat rooftops, and open containers and dates remain there for a week. Dry dates are packed in 50 kg jute sacks and kept in a dry storage area. The Ministry of Commerce and Industry purchases directly the dates farmers produce. The main consumer market is India; therefore, large proportion of the product (*tabseel*) is exported to India and little to other countries (Fig. 6.14). Date palm postharvest handling of *tabseel* season is not just an annual farming activity; it is the time when family, relatives, friends, and neighbors join hands to harvest, process, and work together in the village.

The date palm harvesting season is considered to entail the hardest agricultural work of the year as it involves most of the people who own the date palm groves. Many factors must be considered, for example, environmental hazards especially at the time of actual harvesting.

Before final packing, dates are graded and deformed or spoiled fruits removed, leaving the best for marketing. The Ministry of Commerce and Industry (MOCI) makes an announcement to the farmers about the purchase of their production from a collection point, for overseas export. The ministry rechecks the quality, presence of any pests or diseases, and uniformity and makes the payments according to their product quality. India is the major importer of the product, which has medicinal uses and is combined with chocolate and also has a role in festival celebrations. Dates can be eaten fresh, dried, or in different processed forms, for example, in cookies, cakes, and bread. Honey and vinegar are also made from the date fruits. The total dry date palm *tabseel* export in 2012 from Oman was 6,767 mt (MOCI 2012). Apart from traditional processing, modern factories have been established in Oman to process dates from conditioning to packing for domestic consumption and export.

6.7.4 Survey of Commercial Producers and Major Farms

Date palm cultivation in Oman is scattered throughout the country. However, large commercial farms are located in Dhakhlia region (Samail), Al-Dahira region (Ibri), and Al-Batinah and Al-Sharqia regions. Al-Sharqia regions because of the presence of wadis with flowing water, good soil, and favorable climatic conditions. Date plantations have been adversely affected due to the increasing groundwater salinity in the last few years caused by seawater incursion in Al-Batinah region. For the most part, date farms in date-growing areas do not have formally planned orchards, although new groves with the help of the agriculture department are being established scientifically following the prerequisites of good agricultural practices for sustainable crop production.

6.7.5 Marketing Status and Prospect

Dates are sold either fresh or dry. Sometimes the fresh dates are auctioned when they are still on the tree or brought to the local market. But generally these are sold or exported to various countries in dried form after some traditional postharvest treatment, such as boiling (Fig. 6.19). Typically, the marketing channel is directly from the farm to the local consumer in villages; those remote areas where transport and logistics are inefficient pose difficulties. But in the developed areas having good infrastructure, there are many alternatives for marketing. For example, these could be sold on-farm, to the retailers, in local markets, or to the date factories where they are processed for export. Whatever the selling channel, the success of marketing depends upon the way the dates are produced, harvested, sorted, graded, processed, packaged, and transported (Mbaga et al. 2011, Mbaga 2012). The government agencies as well as private organizations involved in date industry are doing their best to increase the marketability of Omani dates. The government is trying to improve the infrastructure by building new roads especially in the rural areas and constructing new local markets (Al-Marshudi 2002). The government is also supporting research programs to improve the quality of dates, particularly, postharvest handling and processing to improve the country's date export in coming years.

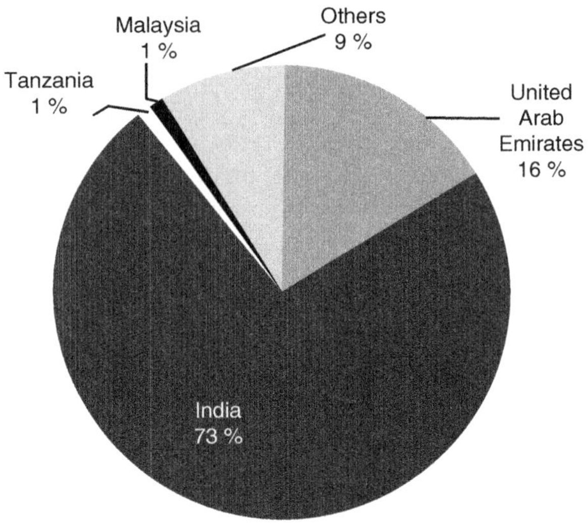

Fig. 6.19 Countries importing boiled dry dates (*tabseel*) from Oman (*Source*: MOCI 2012)

6.7.6 Current Import and Export

According to recent data from FAOSTAT (2013), Omani date exports have decreased from 2001, but import quantities have increased (Fig. 6.3). There could be various factors responsible for this trend, for example, pests and diseases, increasing soil salinity, decreasing water availability, degradation of soil and water quality, availability of skilled labor to carry out field operations, and harvest and postharvest losses. The government of Oman is providing incentives and technical support to growers and industry for quality crop production and to decrease postharvest losses.

6.8 Processing and Novel Products

6.8.1 Industrial Processing Activities

Dates are a major fruit crop grown in Oman, and growers use various methods and equipment to harvest, process, and preserve the picked fruit. In general, farmers use plastic containers and date leaves for postharvest operations; date fruit is sun dried in the open air or at room temperature and later transferred to cool storage places. There are some date cultivars which are heat treated before drying, such as cultivars used to make *tabseel*. The date palm processing factories follow standard operational protocols, for example, cleaning, sorting, fumigation, heat treatment, drying, grading, conditioning, packaging, and storing. The factories which were established in 1976 still function with their older technology and machines (Amprawum 1998). However, at present, with the advent of science and technology, date palm processing units are operating modern systems, which are much more efficient.

6.8.2 Survey of Commercial Dates Processers

Most commercial date palm farms are located in Dakhlia Governorate (Samail), Al-Dhahira Governorate (Ibri), and Al-Batina and Al-Sharqia Governorates. Processing units are located near the larger producing areas. The functional units of date palm processing are based in Nizwa, Nakhal, and Al-Rusail industrial area. The government of Oman has developed better road networks and date growers now have easy access to take their products for sale at the factory gate throughout the country. Keeping in view the market demand, the processing units are making varied products, for example, dried and conditioned dates, date syrup, confectionary, pits for *coffee*, and animal feed from fruit remains.

6.8.3 Secondary Metabolites

Higher plants, including date palm and many others, store a variety of chemical compounds in their plant parts. These compounds are commonly divided into two groups: the products of primary plant metabolism, for example, proteins, lipids, and carbohydrates, which play primary roles in plant growth and development, and secondary plant metabolism products, for example, phenolics. These phenolic products are broadly scattered in the plant body and have no direct relationship with primary metabolism; however, these compounds have a significant role in combating biotic/abiotic stresses. This group includes lignins and other phenolic substances, which gives mechanical support to the plant cell wall, while tannins, flavonoids, and a few other simple phenolics provide protection against biotic and abiotic stress. Certain phenolics show an allelopathic effect on adjacent plants. Other metabolites, for example, phytosterols, and alpha-tocopherols also have substantial roles in plant growth and development mechanisms. These metabolites have been recognized for their nutritional and health value in the human diet. The secondary metabolites in date palm have potential functions in human health and nutrition. An experiment was conducted at SQU on date palm secondary metabolites on three cultivars. Date palm cultivars and time of harvest behaved independently as to their bioactive ingredient contents. However, it was postulated that the dates are good source of antioxidants (Singh et al. 2012). Another study was undertaken to convert surplus date fruits into value-added products. It was demonstrated that date fruits have a high content of natural dietary fiber and antioxidants and could be explored further for nutraceutical prospects (MAF 2006b).

A similar study was carried out to extract the phenolics and dietary fiber from date palm seeds (cv. Mabseli). The results showed that date seed concentrates have a potential source of natural dietary fiber and antioxidant content. It could be possibly used as a bioactive food ingredient (MAF 2008). A study on citric acid production from Omani date cv. Um Sella concluded that conditioning experiments are needed for higher citric acid yield (MAF 2006b). Another similar experiment on citric acid production through a submergence technique showed good-quality citric acid recovery from date cv. Fardh. Extracted citric acid from date palm has significant industrial value (MAF 2008).

6.8.4 Indigenous Date Products

Dates can be eaten fresh, dried, or in different processed forms made from the fruits. The date fruit is processed into diverse products, which have significant commercial value at local and global markets. The indigenous products are mainly timber wood, dry dates, date bars, date syrup, date juice concentrates, date jam, date butter, date candy, date chutney, date biscuits, date pickles, etc. Date seeds (pits) have their own value for making date *coffee* and for vegetable oil.

6.8.5 Bioenergy

Entrepreneurs in Sohar, Oman, have initiated an ambitious plan to transform a petroleum-producing country into a biofuel producer. Oman will be the first Arab country to produce biofuel from date palm fruits at a large scale. The private Oman Green Energy Company is looking to invest a large amount of capital in new date plantations and biofuel refinery construction in Sohar. The company is targeting the production of green biofuel energy in particular for automobiles; it claims that they have been successful in producing and testing ethanol from biomass attained from the abundant date palms in Oman (Biopact 2007). Date palm sap carries a high concentration of sugars which makes it good substrate for fermenting microorganisms; it also has significant potential to be converted into ethanol and biofuel production (Gupta and Kushwaha 2011).

6.9 Conclusion and Recommendations

The date palm (*Phoenix dactylifera* L.) is the oldest cultivated crop in Oman and the Arabian Peninsula. Date palm is the main crop in Oman and occupies half of the agricultural cultivated land. It represents 80 % of the total fruit crop production in the sultanate with an average annual production of about 260,000 mt. To ensure a continuous supply of dates throughout the year, many cultivars have been grown for centuries in various agroclimatic zones of Oman. Furthermore, the date palms supply many household needs over time and their by-products have been used in ways ranging from feeding animals to providing construction materials. Date palm has also social, economic, cultural, and religious significance in the country. However, there has been a slow decline in production and export in recent years. Various reasons could be attributed to the decline in the production of dates in Oman, which include increasing soil salinity, decreasing water availability, labor shortage, and labor costs, pest and diseases, postharvest losses, inadequate processing facilities, etc. Research on date palm is developing slowly as there is no independent, public or commercial, research facility dedicated to the date palm in Oman.

Concerted effort is required to improve the date palm industry in Oman. Extensive research work should be conducted to increase the yield of this crop, particularly when it undergoes water or salt stress and a scenario of climate change. An integrated pest management approach should be used to control insect pests, especially in areas where pesticide spraying is ineffective. Maximum emphasis should be given to the awareness of local farmers about recent techniques in cultivation, postharvest handling, and processing of dates. The successful accomplishment of these tasks will ultimately provide new avenues for improving date palm cultivation and increasing production for domestic consumption and export-quality dates in Oman.

There has been marginal activity in processing and producing novel products from date palm to the present. However, experimentation on date palm products and their transformation into novel products has shown promise and potential, for example, biofuel products, bioactive ingredients, and nutraceutical products. It is clear that there is surplus production which needs more attention for conversion into value-added products.

Traditionally date palm is propagated through offshoots, but due to limited supply, the pace of further generation perpetuation is slow. The increased biotic and abiotic stresses have increased the vulnerability of ex situ-conserved date palm genetic resources. The short-term conservation of genetic material is difficult because of certain viral, bacterial, and fungal pathogen contaminations. The preserved genetic material in the form of embryonic cultures may lose its regenerative capacity gradually if it is not preserved at ultra-low temperature. At present callus-induced organogenesis has been commonly used for date palm micro-propagation and genetic conservation. Breeding for desired characters is a long-cycle activity and may show genetic variability or new mutations at the later stage of development. In this way unwanted characters may appear at later stages of growth.

To keep genetic diversity of date palm alive, it is important to preserve the existing gene pool with full support and interest of academics. A realistic number of people need to be trained in the field of plant genetic resource conservation and utilization. Such programs require massive investments which should be backed by the public sector. Live date palm gene banks must be maintained accordingly; however, in vitro date palm gene banks carry an advantage for keeping tissue or DNA in controlled environment (ultra-low temperature) for a longer term. Tissue culture is an ideal plant propagation method and therefore skilled human resources and all necessary facilities are prerequisites before starting such endeavor.

Lack of expertise and facilities in the area of morphological and molecular descriptors represents a great challenge toward methodological identification of Omani date palm cultivars. Greater attention is needed to conserve the large number of neglected (i.e., poor quality) cultivars, as they constitute an important germplasm resource for further potential cultivar development and improvement. Breeding and cultivar development is also lacking in the country as little work has gone into developing new commercial cultivars.

Overall, great steps have been taken to preserve date palm cultivation and utilization in Oman. The country intends to plant an additional one million date palm trees in the coming few years. However, besides increased production, further efforts are needed to enhance fruit quality by adhering to proper management practices including selection of male pollinizers, fruit thinning, and following proper harvesting and postharvest handling and processing methods. This will ultimately increase exports, which currently stands at around 2.6 % of the total production and to a limited number of importing countries.

References

Aaouine M (2000) Production of date palm vitro plants: the Moroccan experience. In: Proceedings of the date palm international symposium, Windhoek, 22–25 Feb 2000, pp 46–52

Al-Farsi M, Alasalvar C, Morris A et al (2005a) Compositional and sensory characteristics of three native sun-dried date (*Phoenix dactylifera* L.) varieties grown in Oman. J Agric Food Chem 53:7586–7591

Al-Farsi M, Alasalvar C, Morris A et al (2005b) Comparison of antioxidant activity, anthocyanins, carotenoids, and phenolics of three native fresh and sun-dried date (*Phoenix dactylifera* L.) varieties grown in Oman. J Agric Food Chem 53:7592–7599

Al-Farsi M, Alasalvar C, Al-Abid M et al (2007) Compositional and functional characteristics of dates, syrups, and their by-products. Food Chem 104(3):943–947

Al-Ghafri A (2006) Aflaj irrigation D/S ratio. In: Proceedings of the international conference on economics incentives and water demand management, Muscat, 18–22 Mar 2006

Al-Kharusi LM, El Mardi MO, Ali A et al (2009) Effect of mineral and organic fertilizers on the chemical characteristics and quality of date fruits. Int J Agric Biol 11:290–296

Al-Marshudi AS (2002) Oman traditional date palms: production and improvement of date palms in Oman. Tropiculture 20(4):203–209

Al-Ruqaishi IA, Davey M, Alderson P, Mayes S (2008) Genetic relationships and genotype tracing in date palms (*Phoenix dactylifera* L.) in Oman based on microsatellite markers. Plant Gen Res Char Util 6(1):70–72

Al-Sadi AM, Al-Jabri AH, Al-Mazroui SS, Al-Mahmooli IH (2012) Characterization and pathogenicity of fungi and oomycetes associated with root diseases of date palms in Oman. Crop Prot 37:1–6

Al-Yahyai R (2007) Improvement of date palm production in the Sultanate of Oman. Acta Hort 736:337–343

Al-Yahyai R (2010) Current status of date palm in the Sultanate of Oman. In: Proceedings of the international conference on date palm production and processing technology, Sultan Qaboos University, Muscat, 9–11 May 2006, pp 1–6

Al-Yahyai R, Al-Khanjari S (2008) Biodiversity of date palm in the Sultanate of Oman. Afr J Agric Res 3(6):389–395

Al-Yahyai R, Al-Kharusi L (2012a) Physical and chemical quality attributes of freeze-stored dates. Int J Agric Biol 14:97–100

Al-Yahyai R, Al-Kharusi L (2012b) Sub-optimal irrigation affects chemical quality attributes of dates during fruit development. Afr J Agric Res 7(10):1498–1503

Al-Yahyai R, Manickavasagan A (2012) An overview of date palm production. In: Manickavasagan A, Essa M, Sukumar E (eds) Dates: production, processing, food, and medicinal values. CRC Press, Boca Raton, Florida, USA, pp 3–12

Al-Zidjali TM (1996) Oman: country report to the FAO international technical conference on plant genetic resources. FAO international technical conference on plant genetic resources, Leipzig

Amprawum DB (1998) Post-harvest processing and technologies used by Oman date farmers and factories. Agric Mech Asia Afr Lat Am 29(2):61–66

Arunachalam V (2000) Participatory conservation: a means of encouraging community biodiversity. PGR Newsl 122:1–6

Bagniol S, Engelmann F, Michaux-Ferrière N (1992) Histo-cytological study of apices from *in vitro* plantlets of date palm (*Phoenix dactylifera* L.) during a cryopreservation process. Cryo Lett 13:405–412

Biopact (2007) http://news.mongabay.com/bioenergy/2007/06/oman-green-energy-company-makes-ethanol.html. Accessed 15 Sept 2013

El Mardi MO (1995) Traditional date culture. College of Agriculture, Sultan Qaboos University Press, Sultanate of Oman

El Mardi MO, Salama SB, Consolacion E, Al-Shabibi MS (1995) Effect of treated sewage water on vegetative and reproductive growth of date palm. Commun Soil Sci Plant Anal 26(11):1895–1904

El Mardi MO, Salama SB, Consolacion EC, Al-Solomi M (1998) Effect of treated sewage water on the concentration of certain nutrient elements in date palm leaves and fruits. Commun Soil Sci Plant Anal 29(5):763–776

El Mardi MO, Esechie H, AI-Kharousi LM, Abdelbasit KM (2002) Effect of pollination method on changes in physical and chemical characteristics of date fruit during development. Agric Sci 7(1):21–27

El Mardi MO, Al-Said FA, Bakheit SC et al (2007) Effect of pollination method, fertilizer and mulch treatments on the physical and chemical characteristics of date palm (*Phoenix dactylifera*) fruit I: physical characteristics. Acta Hort 736:317–328

El-Kharbotly A, El-Mardi O, Al-Saidi N, Al-Mahruki Y (1998) Towards the construction of genetic map of date palms using AFLP technique. In: Proceedings of the first international conference on date palm, United Arab Emirates University, Al-Ain, 8–10 March 1998, pp 194–207

Engelmann F, Assy-Bah B, Bagniol S et al (1995) Cryopreservation of date palm, oil palm, and coconut. In: Bajaj YPS (ed) Biotechnology in agriculture and forestry, Cryopreservation of plant germplasm I. Springer, Berlin, pp 148–169

FAOSTAT (2011) Crop Production, Statistics Division. Food and Agriculture Organization of the United Nations, Rome. http://www.faostat.fao.org

FAOSTAT (2013) Food and Agriculture Organization of the United Nation. http://www.faostat.fao.org. Accessed 2nd Sept 2013

Finkle BJ, Ulrich JM, Rains DW et al (1979) Survival of alfalfa, *Medicago sativa*, rice *Oryza sativa* and date palm *Phoenix dactylifera* callus after liquid nitrogen freezing. Cryobiology 16:583

Gupta N, Kushwaha H (2011) Date palm as a source of bioethanol producing microorganisms. In: Jain SM, Al-Khayri JM, Johnson DV (eds) Date palm biotechnology. Springer, Dordrecht, pp 711–727

Jaradat AA, Zaid A (2004) Quality traits of date palm fruits in a center of origin and center of diversity. Food Agric Environ 2(1):208–217

Jarvis D, Zoes V, Nares D, Hodgkin T (2004) On-farm management of crop genetic diversity and the convention on biological diversity programme of work on agricultural biodiversity. PGR Newsl 138:5–17

Macki MA, Hamouda AM, Al-Abri A (1998) Date palm cultivars of Oman, vol. II, The date palm. Diwan of Royal Court, Sultanate of Oman

MAF (2005) Oman date production (2003–04). Ministry of Agriculture, Sultanate of Oman

MAF (2006a) Biennial Agricultural Research Report 2005/2006. Ministry of Agriculture, Sultanate of Oman

MAF (2006b) Annual report. Ministry of Agriculture, Sultanate of Oman

MAF (2007) Oman date production (2005–06). Ministry of Agriculture, Sultanate of Oman

MAF (2008) Annual report. Ministry of Agriculture Oman, Sultanate of Oman
MAF (2009) Annual report. Ministry of Agriculture, Sultanate of Oman
Manickavasagan A, Al-Yahyai R (2012) Quality assessment of dates by computer vision technology. In: Manickavasagan A, Essa M, Sukumar E (eds) Dates: production, processing, food, and medicinal values. CRC Press, Boca Raton, Florida, USA, pp 217–226
Manickavasagan A, Alahakoon PMK, Al-Busaidi TK et al (2013) Disinfestation of stored dates using microwave energy. J Stored Prod Res 55:1–5
Mater AA (1987) Production and cryogenic freezing of date palm germplasm and regeneration of plantlets from frozen material. Iraqi J Agric Sci 5S:35–49
Mazloumzadeha SM, Shamsa M, Nezamabadi-Pourb H (2008) Evaluation of general-purpose lifters for the date harvest industry-based on a fuzzy inference system. Comp Elect Agric 60:60–66
Mbaga MD (2012) Date marketing. In: Manickavasagan A, Essa M, Sukumar E (eds) Dates: production, processing, food, and medicinal values. CRC Press, Boca Raton, Florida, USA, pp 155–172
Mbaga M, Al-Shabibi MS, Boughanmi H, Zekri S (2011) A comparative study of dates export supply chain performance: the case of Oman and Tunisia. Benchmark Int J 18(3):386–408
Ministry of Commerce and Industry (MOCI) (2012) Report of the Statistical Department. Ministry of Commerce and Industry, Sultanate of Oman
Pillay AE, Williams JR, El Mardi MO et al (2003) Risk assessment of chromium and arsenic in date palm leaves used as livestock feed. Environ Int 29(5):541–545
Popenoe PB (1913) Date growing in the old and new worlds. West India Gardens, Altadena, California, USA, p 64
Saker MM, Bekheet SA, Taha HS et al (2000) Detection of somaclonal variations in tissue culture-derived date palm plants using isoenzyme analysis and RAPD fingerprints. Biol Planta 43:347–351
Sawadogo M, Ouedraogo J, Belem M et al (2005) Influence of ecosystem components on cultural practices affecting the *in situ* conservation of agricultural biodiversity. PGR Newsl 141:1925
Siebert SF (2004) Traditional agriculture and the conservation of biological diversity in Crete, Greece. Int J Agric Sust 2:109–117
Singh V, Guizani N, Essa MM et al (2012) Comparative analysis of total phenolics, flavonoid content and antioxidant profile of different date varieties (*Phoenix dactylifera* L.) from Sultanate of Oman. Int Food Res J 19(3):1063–1070
Ulrich JM, Finkle BJ, Moore PH, Ginoza H (1979) Effect of a mixture of cryoprotectants in attaining liquid nitrogen survival of cells. Fiziol Rast 15:749–756
Williams JR, Pillay AE, El Mardi MO et al (2005) Levels of selected metals in the Fard cultivar (date palm). J Arid Environ 60(2):211–225

Chapter 7
Date Palm Status and Perspective in Yemen

Saeed Ba-Angood

Abstract In Yemen, there are more than four million date palms (*Phoenix dactylifera* L.) occupying an estimated area of 14,464–14,955 ha during the period 2008–2012 and producing 55,181–57,849 mt of dates in the same years. More than 321 date palm cultivars are recorded in Yemen, 42 are considered excellent. Most of the date palm plantings in Yemen are irregular, mainly concentrated in the governorates of Hadramaut and Hodeidah. Cultivars of Hamra, Mijraf, Sokotri, Serfaneh, Manasif, Tha'al, and Bqal are dominant. The date palm in Yemen is suffering from several insect and mite pests, namely, the red palm weevil which appeared in 2013; the Dubas bug first recorded in 2000; the stalk and stem borers, *Oryctes* spp.; the lesser date moth, termites, and white scale; storage pests; and the dust mite. Chemical insecticides are used for the control of most of them. Birds, bats, and rodents are among the serious pests. In addition, the diseases soft rot, black scorch, Graphiola leaf spot, and inflorescence rot are recorded. Sanitation is the current means for controlling these diseases. Two factories designed mainly for packing dates have been established, but they are facing problems to work efficiently. Date fruits are marketed locally. Utilization of date palm parts has been known for a long time in Yemen. Date palm growers face problems of poor marketing of their products; lack of handling procedures (packing and processing); high production cost (labor); social and human obstacles such as land fragmentation, poor infrastructure, urbanization, and difficulty of access to soft loans; as well as unavailability of specialized associations and alliances. Because the government of Yemen considers the date palm as one of the most important economic and strategic crop, a national program for promoting the date palm sector is being implemented targeting the following objectives: rehabilitation of old date palm trees, introducing new systems for crop management, and applying integrated pest and disease programs for major insects, mite pests, and diseases. Two tissue culture laboratories have been established.

Keywords Cultivars • Cultivation • Diseases • Genetic resources • Harvest • Historical • Manufacturing • Marketing • Pests • Processing • Propagation

---

S. Ba-Angood
Department of Plant Protection,
University of Aden, 6172, Khormaksar, Aden, Yemen
e-mail: baangood@yemen.net.ye; sabaangood1945@yahoo.com

© Springer Science+Business Media Dordrecht 2015
J.M. Al-Khayri et al. (eds.), *Date Palm Genetic Resources and Utilization: Volume 2: Asia and Europe*, DOI 10.1007/978-94-017-9707-8_7

7.1 Introduction

Since ancient times, date palm (*Phoenix dactylifera* L.) has been considered an important economic and food security crop in Yemen. It was the only food source in arid areas where poverty and starvation periods were recorded in the past. It plays an important role in food production, animal feed, income and employment generation, and foreign exchange earnings, particularly for those growing high-value introduced tissue culture cultivars such as Barhi date by-products are used in some raw materials for local industries. The prevalent climatic and geographic conditions make Yemen suitable for planting and producing date palms. Date palm ranks first among the fruit crops; in cultivated areas, date palm constitutes about 25 % of the total fruit area.

As shown in Table 7.1, the estimated area of date palms ranged from 14,464 to 14,955 ha during the period 2008–2012, producing 55,181–57,849 mt of dates, in the same time period (MoAI 2013a). There are more than four million date palms in Yemen; bearing palms constitute 67.4 % and the males 3 %. Date palm cultivation is found in most governorates of Yemen (Table 7.1 and Fig. 7.1). However, it is more concentrated in Hadramaut and Hodeidah governorates. Hadramaut ranks first in both area and production; the number of date palms there constitutes about 47 % of the Yemeni total. Hamra, Mijraf, Sokotri, Serfaneh, Manasif, Tha'al, and Bqal cvs. are dominant. The majority of date palm trees are old and in traditional plantations where both fruit quality and productivity are low. Local production and supply is insufficient to meet the demands of people. Yemen annually imports about 24,000 mt of dates, at a cost of more than USD 100,000 (MoAI 2013b).

Date palm farmers face a number of challenges, including weak extension services, the spread of pests and diseases, low-quality fruit production, low prices, lack of marketing facilities, and insufficient storage and packing facilities. Date palm farmers in Hadramaut and Al-Mahrah governorates have also experienced a large flood disaster in November 2008, which uprooted about 500,000 date palms.

In addition, date palm farmers also face poor marketing practices in their production, lack of efficient handling procedures (packing and processing), high production cost (labor), social and human obstacles such as land fragmentation, poor infrastructure, urbanization, difficulty of access to soft loans, and unavailability of specialized associations and alliances.

Because the government considers the date palm one of the most important strategic crops, a national program to promote the date palm sector in Yemen is being implemented which includes rehabilitation of old date palms, improving cultural practices and introducing new systems for crop management, and applying integrated pest and disease programs for major insects and diseases. The government has established ten pilot demonstration plots for farmers and nurseries to provide improved date palm cultivars.

Table 7.1 Date palm cultivation area (ha) and production (mt) in Yemen (2008–2012)

		Year				
Governorate	Area and production	2008	2009	2010	2011	2012
Hadramaut	Area (ha)	5,660	5,773	5,833	5,841	5,845
	Prod (mt)	23,762	24,475	24,719	24,833	24,856
Hodeidah	Area	4,934	5,033	5,088	4,986	4,963
	Prod	18,125	18,669	19,224	17,334	16.812
Al-Mahrah	Area	998	1,018	1,032	1,033	1,033
	Prod	3,066	3,158	3,196	3,203	3,219
Taiz	Area	787	803	813	781	733
	Prod	2,480	2,530	2,585	2,459	2,329
Shabwah	Area	555	572	581	586	586
	Prod	2,050	2,071	2,103	2,166	2,149
Lahej	Area	506	516	537	533	546
	Prod	1,918	1,956	2,035	2,008	2,011
Mareb	Area	425	438	454	458	461
	Prod	1,454	1,498	1,552	1,491	1,451
Al-Jawf	Area	352	363	374	374	381
	Prod	1,576	1,623	1,672	1,633	1,703
Sadah	Area	134	134	132	121	119
	Prod	423	427	412	381	359
Hajjah	Area	37	37	40	35	36
	Prod	115	115	132	119	115
Abyan	Area	27	27	31	36	36
	Prod	101	103	114	132	113
Sana'a	Area	21	21	20	15	13
	Prod	56	57	49	35	31
Amran	Area	7	7	6	3	3
	Prod	19	19	15	11	11
Al Bayda	Area	5	5	4	3	3
	Prod	13	13	10	8	8
Raimah	Area	7	7	6	4	4
	Prod	23	23	19	15	14
Al-Daleh	Area	4	4	3	0	0
	Prod	9	9	7	0	0
Sana'a City	Area	4	4	0	0	0
	Prod	8	8	0	0	0
Aden	Area	2	2	2	0	0
	Prod	6	6	5	0	0
Al-Mahweet	Area	0	0	0	0	0
	Prod	0	0	0	0	0
Dhamar	Area	0	0	0	0	0
	Prod	0	0	0	0	0
Ibb	Area	0	0	0	0	0
	Prod	0	0	0	0	0
Total	Area	14,464	14,764	14,955	14,809	14,762
	Prod	55,204	56,760	57,849	55,828	55,181

Source: MoAI (2013a), Agricultural Year Book 2012

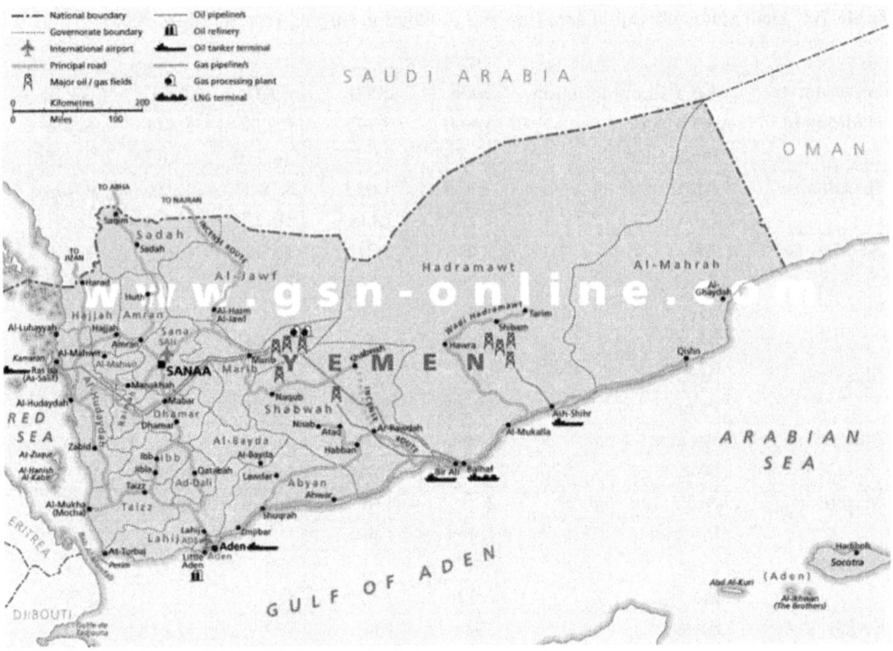

Fig. 7.1 Map showing date palm cultivation in the governorates (see Table 7.1) (*Source*: www.gsn-online.com)

7.2 Cultivation Practices

7.2.1 Chronological Account of Research and Development

There are more than 321 date palm cultivars recorded in Yemen; 42 are considered excellent; 65 produce good quality dates and these together constitute one-third of the total cultivars; the remaining two-thirds are considered acceptable or poor. The mean yield varies from 27 to 63 kg per tree, which is low compared with cultivars in neighboring countries.

The government is promoting the date palm as one of the strategic crops (date palm, cotton, coffee, mango, and guava). The emphasis is being given to the rehabilitation of old palm orchards, improving irrigation methods, and upgrading storage and packing facilities and marketing services. Modern date palm farms have been established through a partnership between government and owners with high-yielding tissue culture cultivars and a modern irrigation system.

In November 2008, date palm plantations experienced a flood disaster which uprooted almost 500,000 trees (Fig. 7.2). The Public Corporation for Agricultural Services (PCAS) contracted with the Flood Reconstruction Fund to provide 250,000 tissue culture date palm plantlets during 2011–2012 as assistance to farmers (Ba-Asher and Kasim 2011).

Fig. 7.2 Effect of floods on date palms in Sah (Wadi Hadramaut)

Recently, the Ministry of Agriculture and Irrigation (MoAI) established a tissue culture laboratory at the Seiyun Agricultural Research Center in Hadramaut Governorate for production of well-known date palm cultivars to help meet growing demands for date palms in Wadi Hadramaut. Unfortunately, the laboratory is encountering many problems in carrying out its task.

Fig. 7.3 Date palms grown in Yemen. (**a**) Date palms in the Tihama area, Al Hudaidah Governorate; (**b**) date palms in the Habban area, Shabwah Governorate; (**c**) date palm offshoots and parents; (**d**) date palms with old dry unpruned leaves

A program on the improvement of date palm production in Wadi Hadramaut was implemented by the Agricultural Research and Extension Authority (AREA) in cooperation with the Food and Agriculture Organization (FAO) and Ministry of Agriculture and Irrigation. AREA also organized training courses for farmers, technicians, extension agents, and researchers; prepared studies on the control of palm pests and diseases, fertilization, morphological features, productivity of date palm cultivars, and others; organized and participated in relevant local, regional, conferences, seminars, and workshops; and prepared booklets on main date palm varieties (AREA 2013).

7.2.2 Description of Current Cultivation Practices

Date palm culture in Yemen largely remains traditional. Local propagation of date palm is mostly by offshoots. Most of date palm plantings in Yemen are irregular; they are grown either at the sides of irrigation canals (Fig. 7.3a) or scattered in the orchard (Fig. 7.3b). Most of them (more than 95 %) depend on open well irrigation, others on springs and running water. Very few depend on floods that occur every 2–3 years. Old dry leaves are seldom removed, hanging down and impeding the operations of ascending the palms. Climbing trees is usually done by ropes. Older farmers are the only

Fig. 7.4 Tall trees in an old field of date palm in Doan area, Wadi Hadramaut

Fig. 7.5 New established date farm in Wadi Hadramaut

individuals who climb palms; they are always complaining that their sons and the youth never want to climb trees. Normal practice allows offshoots to develop on the parent tree and as a result of not removing them (Fig. 7.3c) for several years, along with the close distance between trees, clumps of unpruned trees are formed with crowded leaves competing with one another for light, water, and nutrients (Fig. 7.3d).

Tall nonproductive palms are also often found on old farms (Fig. 7.4). These random cultivations have led to poor productivity and low fruit quality. Surface irrigation predominates, where underground water is a source of irrigation for 61 % of the date palms, followed by springs 32 % and streams 7 %. General agricultural practices including fertilization, thinning fruits, and crop care are ignored by most farmers and is evidenced by the low productivity (29 kg/tree/year) comparing with neighboring countries.

However, in recently established farms, the date palms are laid out in regular lines facilitating service and harvest operations. Spacing is usually 8×8 m. Bunch management techniques of fruit thinning, bending, and covering are being advocated. Irrigation practices are mostly basin, but drip and bubbler water saving systems are being introduced (Fig. 7.5).

Recent field trials on fertilization have shown that treatments of 460 g N (urea fertilizer)+460 g phosphorous and 950 g N +460 P gave good yield when tested for Hamra cultivar (Al Bar 2003)

Pollination is done manually by tying a few strands of a male spathe and inserting them among the strands of a female spadix as soon as they burst open. Sometimes,

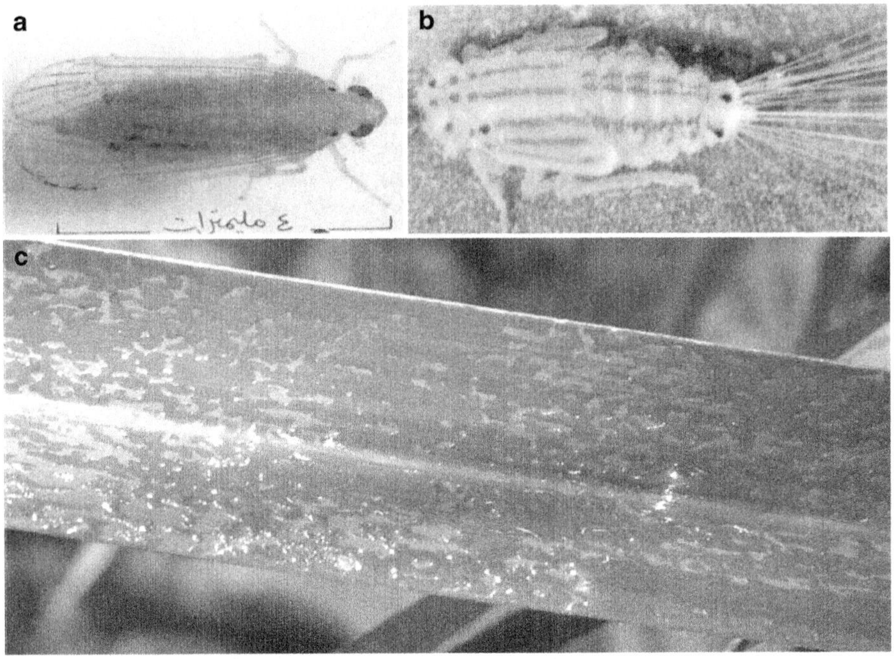

Fig. 7.6 Adult (**a**) and nymph (**b**) of the Dubas bug, (**c**) damage symptoms by Dubas bug

several female spathes open at the same time necessitating four or five visits from a pollinator. Dates in Yemen are harvested by workers climbing to the tops of date palms by stepping on the stubs of leaf bases or using ropes. Ladders are seldom used. The entire fruit bunch is cut when ripe. It is hoped that the use of machines may be coming soon to modernize the operation.

7.2.3 Pest and Disease Management

7.2.3.1 Insects

Dubas Bug The Dubas bug (*Ommatissus lybicus*) has emerged as a key pest of date palm since 2000 (Fig. 7.6a, b); it is said to have entered the country from Oman. It was recorded first in Mahrah Governorate and then spread west until it reached Hadramaut and other eastern and southern governorates of the country. Tihama, the second-ranked date-palm-producing area in Yemen, currently is free of the pest. The Dubas bug causes two kinds of damage (Fig. 7.6c) on date palm. First, direct feeding of nymphs and adults on the sap of the leaves and to some extent on fruit bunches. Second, there is indirect damage on leaves and fruit caused by excessive accumulation of honeydew excreted by the insects, particularly during their nymph develop-

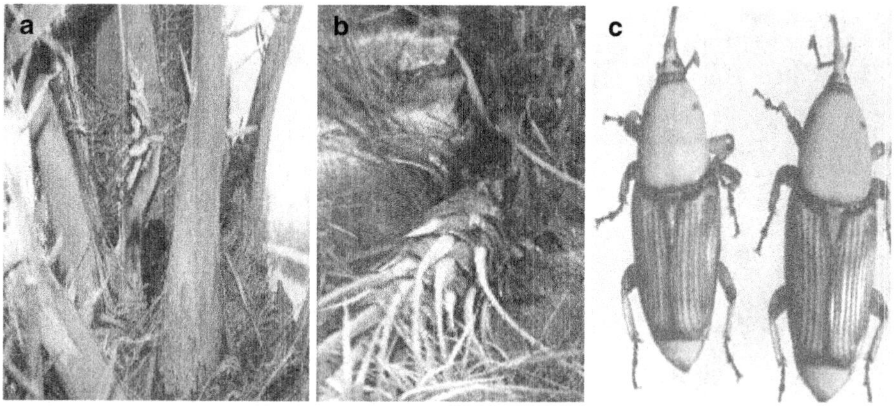

Fig. 7.7 Damage caused by the red palm weevil (**a**) and (**b**) adult weevil (**c**)

ment stages, which interferes with vital biological processes such as photosynthesis, transpiration, and crop harvest operations. Damage due to honeydew deposits also affects intercalary crops within date palm plantations. The pest is controlled by an annual national chemical control campaign using deltamethrin insecticide (Ba-Angood et al. 2009). However, there is a promising predator in the common green lacewing insect (*Chrysoperla carnea*) and a parasite (*Pseudoligosita babylonica*) contributing to the management of the pest in some areas of Hadramaut Governorate.

Red Palm Weevil Until May 26, 2013, Yemen was free of the devastating red palm weevil (*Rhynchophorus ferrugineus* Oliv.) (Fig. 7.7a–c). Strict quarantine measures were being taken to keep the country free from this destructive pest. However, suddenly the pest was first recorded in a farm in Qatn area in Wadi Hadramaut, (Al-Habshi 2014); the pest is said to have come into Yemen with tissue culture offshoots from Saudi Arabia. The MoAI and Seiyun Agricultural Research Centre are doing their best to stop the spread of the attack. An FAO TCP project was expected to help in the management of the pest; but unfortunately, the FAO offices in Sana'a and Cairo offices have been slow in providing assistance and the pest is spreading (Ba-Angood 2014). For the management of this pest, pheromone traps are used for monitoring and mass trapping. Once infected palms are recorded, the entire field is sprayed with fenitrothion insecticide. Some of the infected palms are removed.

Fruit Stalk and Stem Borer A number of date palm fruit and stalk borers were recorded in Yemen; the most important are *Oryctes elegans*, *O. agamemnon*, *O. rhinoceros* (Fig. 7.8a), and *O. boas*. This pest causes considerable damage not only to fruit stalks causing fruit fall but also bores into palm stems; this is particularly a problem in the Hadramaut area (Fig. 7.8b). They also help the red palm weevil to attack stems. Different pesticides have been tried but failed to control the pest; however, mass trapping using light traps (Fig. 7.9) succeeded in reducing the population of the pest (Ba-Angood and Al-Baity 2006).

Fig. 7.8 (**a**) Adult *Oryctes rhinoceros*, (**b**) damage by *Oryctes* spp.

The Lesser Date Moth This insect, *Batrachedra amydraula* Meyer., is a very important pest causing 20–80 % damage to fruits in Wadi Hadramaut and Tihama (Ba-Angood 1978, 2012) (Fig. 7.10). Larvae bore holes in the fruits at their early stage of development eating the flesh and moving from one fruit to another by larval strands. This leads to fruit fall in early stages, the fruit color changes from green to brown red, and that is why it is called locally *hummairah*. The pest is controlled by several insecticides such as deltamethrin (Al-Ghurabi and Ba-Angood 2011).

Date Palm Dust Mite This mite (DPDM) *Oligonychus afrasiaticus* (McGregor) (Acarina: Tetranychidae) is known to occur in most date-growing areas of Yemen (Fig. 7.11). The pest is very damaging to dates in the early stages of fruit development, attacking fruits and covering them with dust. Studies by Ba-Angood and Bass'haih (2000) have shown that it has an adverse effect on some of the physiochemical characteristics of dates in Wadi Hadramaut. Infested dates of Mijraf, Madini, and Hamra cvs. were smaller in size, malformed, and unripe, compared with healthy ones. They also had a lower content of total soluble solids particularly sugars and lower percentage of water content compared with the healthy fruits. Unless treated by a protective spray before date

Fig. 7.9 Light trap used for trapping fruit stalk and stem borers *Oryctes* spp.

Fig. 7.10 Damage on date palm fruits for the lesser date moth (*hummairah*)

palm flowering and as soon as the symptoms of attack are detected, the damage can be extremely severe. Spraying of dust mites with Vertimec insecticide is a very effective means of control (Ba-Angood 2004).

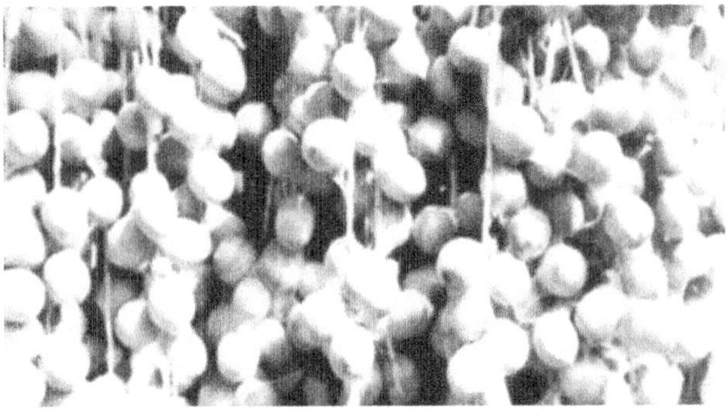

Fig. 7.11 Damage caused *by* date palm dust mite *Oligonychus afrasiaticus*

Fig. 7.12 Infestation by the white scale insect *Parlatoria blanchardii*

Termites *Microcerotermes diversus* and *Microtermes najdensis*, particularly in the Tihama area, are the most serious termite pests which attack date palms, from the roots to close to the tops of the date palms. Chlorpyrifos and reagent insecticides are applied at the bases of offshoots during planting and in basins around date palms for control. But infestation could be very severe in some localities in the Tihama area.

Scale Insects White scale insect (*Parlatoria blanchardii*) is widely distributed throughout most of the date-palm-growing areas of Yemen. Under favorable conditions (high planting density and high humidity), the scale can spread over the surface of the foliage and fruit of date palm, covering them with live and dead insects (Dhouibi 2002). The insect feeds on succulent tissues at the base of the leaf stalk, which is an inaccessible place (for control) on the palm. Heavy infestations cause the pinnae to wither and die (Fig. 7.12). Infestation on fruit reduces its commercial value and may render fruit unfit for human consumption. Sanitation methods by pruning is usually used; sometimes Sumicidin insecticide is sprayed for control (Ba-Angood 2008; Dhouibi 2002). In some areas, the pest is found to be under the biological control of the predator heather ladybird beetle *Chilocorus bipustulatus* (Ba-Angood 2008).

7.2.3.2 Vertebrates

Birds The house sparrow, *Passer domesticus*, and other birds are destructive pests attacking dates on the palm in the sweet khalal and rutab stages of fruit development. Farmers usually cover bunches with jute or cloth bags to reduce bird damage to some extent. Birds can also be driven away by loud-sounding devices or making

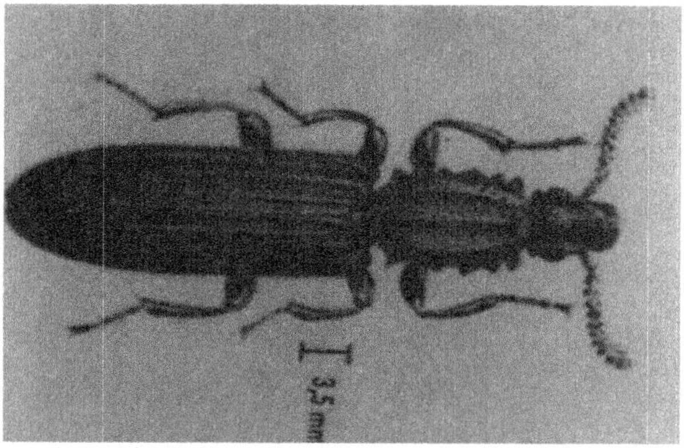

Fig. 7.13 Saw-toothed grain beetle, *Oryzaephilus surinamensis*

some humanlike structures to scare them, but they continue to be a major cause in reducing yield and lowering fruit quality, in some areas.

Rats and Bats Rats were reported recently to attack parts of the date palm in some areas of Tihama and Hadramaut. They also dig underground galleries damaging the palm roots and irrigation systems. Bats are also reported to attack fruits at Taiz, Al Jawf, and Tihama. However, no control method now exists.

Storage Pests Some serious pests are known to attack dates on the palm as well as in storage. The most important one is the saw-toothed grain beetle *Oryzaephilus surinamensis* L (Fig. 7.13). Some moths of *Ephestia* spp. are sometimes recorded. For control of these pests, stored and packed dates are fumigated with Phostoxin tablets to keep them pest-free.

7.2.3.3 Diseases

A survey of date palm diseases has shown that Yemen is still free from the devastating bayoud disease caused by *Fusarium oxysporum* f. sp. *albedinis*. However, other major diseases in the country include the soft rot diseases caused by the fungi *Aspergillus phoenicis*, *Helminthosporium* spp., and a mold, *Alternaria spp.* and followed by Graphiola leaf spot *Graphiola phoenicis* and other leaf spot diseases (Fig. 7.14) caused by the molds *Cladosporium herbarum* and *Alternaria* spp., followed by the black scorch fungus *Thielaviopsis paradoxa* and root rot *Fusarium* spp., *Diplodia phoenicum* fungi, and other physiological diseases (Al-Sakaff 2012). These diseases are not being controlled chemically; infected parts are usually pruned off as a sanitation measure. Sanitation is usually the means of control.

7.3 Genetic Resources and Conservation

Yemeni climatic diversity with its transitions from the extremely dry to humid conditions facilitates a parallel change in date cultivar distribution from dry dates to semidry and soft dates in different governorates.

Fig. 7.14 Leaf spot disease

The biodiversity in Yemeni dates makes them ideal for local selection and cultivar improvement. Morphologically, diversity is exemplified by fruit color, size, shape, and time of maturity, seed, fruit stalks, fronds, leaflets, spines, fiber, crown, and other features.

The names of some date cultivars are based on fruit shape like brides fingers *Asabia el Aroos* or fruit color at khalal (bisr) stage, like Manasif and Ahmer. Some of the cultivars refer to specific areas; Sokotri cv., for example, refers to Socotra Island in the Indian Ocean. The island is a part of Yemen and dates are its primary commercial crop.

Chemical characteristics like reduced sugar concentration in various stages of fruit development are also factors contributing to the biodiversity among date palm cultivars. Kassim and Hussein (2008) found that the percentage of reducing saccharides among soft date cultivars from Hadramaut Governorate varies from 51.5 % (Baglah Safra) to 72.7 % (Gzaz); in dry dates, it ranges from 67.1 % (Sarorah) to 88.2 % (Mijraf). Certain cultivars in Wadi Hadramaut contain more than 90 % reduced sugars, which is similar to the elite cultivars in Saudi Arabia, the UAE, and Libya.

There are no breeding programs for date palms in Yemen, but the effort to maintain outstanding natural selections has continuously been going on. Date palms with shorter stature, stouter stems, fewer spines, higher yield, and better fruit quality are desirable, and some local cultivars have some of these characteristics.

In every date-growing area, there are few palms that are famous for their unique qualities. Fruits of cultivars such as Madini, Barhi, and Mijraf are kept for consumption only by their owners and close friends. Natural selection by adaptation to environment and human selection based on morphological and chemical characteristics has led to the current array of date cultivars in Yemen.

A recent germplasm collection farm of local and tissue culture propagated date palm introductions from Saudi Arabia was established at Seiyun Agricultural Research Centre.

7.4 Plant Tissue Culture

Tissue culture as a quick means of propagation for production of true-to-type, disease-free plants has been introduced recently in Yemen. Two tissue culture laboratories have been established.

The first was established in 2006 at PCAS. The laboratory was propagating mainly potato, strawberry, pineapple, banana, and date palm. Unfortunately, due to technical problems the laboratory is not functioning well. However, there are efforts to provide funds to operate it efficiently and to propagate other tissue culture date palms imported from outside Yemen.

Another tissue culture laboratory was established at Seiyun Agricultural Research Centre to propagate date palms, but unfortunately due to financial and technical reasons, it has not yet produced any tissue culture palms.

The PCAS formerly imported disease-free, true-to-type micropropagated date palm plantlets from Date Palm Developments (D.P.D. Ltd) in the UK. Plants supplied are with minimum 20 cm shoot length and a minimum three juvenile leaves. These plants were kept in the PCAS nurseries for hardening and then distributed to farmers (Fig. 7.15a, b). These cultivars included Khalas, Nabout Safe, Nabout Sultan, Barhi, and others (Ba-Asher and Kasim 2011).

7.5 Cultivar Identification

Date palm cultivars usually are identified using genetic and morphological characteristics. Cultivars vary in crown structure, fruit shape, size of leaflets and thorns, and chemical composition at various stages of fruit development (Ba-Miftah et al. 2010).

Currently in local laboratories, fruits and seeds are weighed using sensitive balances. A vernier scale is used to measure length and width of fruits and seeds. Leaflets, thorns, and inclination angles are also measured to identify some cultivars.

Chemical characteristics are determined by concentrations of total sugars, proteins, and trace elements. To characterize cultivars chemically, fruit samples are examined at successive levels of maturity (khalal, rutab, and tamar stages). Several scientists worked on chemical contents of date fruits from most of the date palm

Fig. 7.15 Date palm plants introduced from the UK hardened in the nurseries of PCAS (**a**), ready for distribution (**b**)

growing locations in Yemen. Recently, Kassim and Hussein (2008) made an assessment of sugar content of Yemeni commercial date cultivars at khalal and rutab stages in the Hadramaut Governorate and found that reducing sugars varied from 51 to 82 %, reaching more than 90 % in a few samples.

Unfortunately, genotyping of biodiversity to characterize variability at the DNA level using molecular marker techniques is not available in the laboratories in Yemen. The genetic resources of Yemeni date cultivars are hard to trace as date cultivation has been carried on since ancient times, and there is so much variation within local cultivars. Some cultivars are believed to have been brought in from Saudi Arabia and the UAE. Indigenous cultivars of seed origin are countless. These have no individual names, but generally are called Baglah, Tos, and Sabyah in the Hadramaut Governorate and account for 5–6 % of the cultivars there. In Tihama, they are known as Mwaleed or Shabat, constituting 0.3 % of the cultivars; in Socotra Island they are called Nakadha, representing only 0.1 % of the total cultivars there. These seedling dates contain unique germplasm and merit further study (Ba-Miftah et al. 2003).

7.6 Cultivar Description

Results of a survey of date palm cultivars in Hadramaut, Tihama, and Socotra Island (Ba-Miftah et al. 2003) have shown that the total number of local date palm cultivars recorded in these areas was 209. Of the total, 65 occur in Wadi Hadramaut (Fig. 7.17b), 59 in Tihama, 48 in Socotra Island, and 37 in the coastal areas of Hadramaut. In addition there are 11 cultivars which have been introduced from neighboring countries. Hamra cultivar is the most widespread in Wadi Hadramaut (35 %), Sokotri (89 %) in the coastal areas of Hadramaut, Serfatah (82 %) in Socotra Island, and Tha'al (Munasif) (45 %) in the Tihama area. Seedling cultivars such as Baglah, Sabyah, and Nos in Hadramaut (5–6 %), were not included. The most productive cultivars are Zargey and Fariedey found in the Tihama area, which yield 100 kg/tree, but unfortunately fruits are classified only as acceptable or poor quality. The least productive ones were Asba'a Al Aroos and Saig in Hadramaut which in some areas yield only 5 kg/tree but the fruits are classified as good.

Ba-Miftah et al. (2004) also carried a similar survey in Shabwah Governorate which showed that there were 20,771 date palm trees, 57 % of them bearing, and males constituted 2.3 %. Shabwah has 32 different cultivars; 8 are considered excellent (Fig. 7.16), 8 are very good, 10 are good, and 5 are acceptable, in addition to one non-fruiting cultivar. Barhi, Khnaizi, and Rthanah cultivars are said to have been imported from Saudi Arabia.

A survey of date palms in Al Jawf and Ma'reb carried out by Ba-Miftah et al. (2009) indicated that the number of date palms had reached 23,797, 48.7 % are fruiting and males constitute 4 %. In Al Jawf Governorate 26 cultivars were identified; 38.5 % of them are classified as excellent, and Abyadh cultivar is widespread. Cultivars Sukkari, Rothata, Khalas, Nabout Saif, Barhey, Nabtat Sultanah, Sag'ee, Zamley, and Maktoum were introduced from Saudi Arabia. In Mareb Governorate, there are 19 cultivars recorded; 32 % are excellent. There are three introduced cultivars (Barhi, Nabout Saif and Nabtat Sultanah); Nabout Saif is the most widespread (29.3 %).

Another survey of date palms in Al Mahra Governorate (Obad et al. 2003) has shown that the number of date palms reached 22,249, 56.5 % are fruiting and 4.9 % males. A total of 97 cultivars were identified; most of them are in the Man'ar area; 24 of them are classified as excellent, 30 are very good, 30 are good, 7 are acceptable, and 6 are of poor quality.

The main commercial date palm cultivars in Yemen are described according to Ba-Miftah et al. (2010) as follows:

Mijraf Classified as excellent, consumed soft and locally as tamar, color of fruit brown, shape oval long. Mean percentage of reducing sugars in date fruits 88.4 %, average yield of 47 kg/tree, and mean offshoot production is 4.

Gzaz Classified as excellent, consumed soft and locally as tamar, color of khalal fruit yellow, tamar brown, shape oval, mean percentage of reducing sugars 90.5 %, average yield of 52 kg/tree, and mean production of offshoots is 7.

Fig. 7.16 Some cultivars of date palm grown in Shabwah Governorate

Hajri Classified as very good, eaten soft and locally as tamar, color of khalal fruit yellow, tamar black, shape inverted oval, mean percentage of reducing sugars 86.8 %, average yield of 37 kg/tree, and mean production of offshoots is 4.

Asbua Alaroos Classified generally as very good, eaten soft and locally as tamar, color of khalal fruit red, tamar black, shape elongate oval, average yield of tree 38 kg/tree, and mean production of offshoots is 4.

Sokotri Classified as good, eaten soft and locally as tamar, color of khalal fruit pale red, tamar black, shape elongated oval, average yield of 49 kg/tree, and mean production of offshoots is 6.

Hamra Classified as good, eaten soft and locally as tamar, color of khalal fruit red, tamar black, shape inverted oval, mean percentage of reducing sugars 69.5 %, average yield of 47 kg/tree, and mean production of offshoots is 6.

7.7 Date Production and Marketing

Most date palm plantings in Yemen are irregular and most gardens are small in size, not exceeding 0.8 ha in most cases. Some more recent gardens have been established, mostly planted with tissue-cultured cultivars, which are relatively large in area. A program to improve date palm production in Wadi Hadramaut in cooperation with FAO and the MoAI was implemented. It involves ten pilot farms (Fig. 7.17a, b) in Wadi Hadramaut (in Saiyun, Tarim, Shibam, Sah, Al-Qatn, and Hraidhah) each 10 ha in area. In these fields, local cultivars, Mijraf, Gzaz, and Jahmi, are grown, and four tissue culture cultivars, Barhi, Khalas, Sultanah, and Sukkari, are grown in regular rows with spacing of 10×10 and 11×11 m. All are irrigated by a drip and bubbler network. The cost of harvest is high in some areas; it reached more than 40 % of the total cost of production in Tihama (Ba-Miftah et al. 2003). Some farmers cover the fruit stalks with perforated sacks to protect against birds, rats, and insects soon after fruit forming. Harvest is manual, climbing the tree and cutting the bunches, sometimes lowering them to the ground by ropes. Handling, storage, and transport usually are done in simple containers such as sacks, bags, and skins.

Dates are sold in a number of ways. Some are auctioned when they are still on the tree. Some are sold in the local market either fresh or more commonly dried

Fig. 7.17 (**a**) A pilot farm in Seiyun, (**b**) date palm trees in Wadi Hadramaut

(tamar stage). Traditional date markets appear to be important to farmers and consumers. Some cultivar dates are sold to packing factories. The farmer usually stores the best dates for his own use at home.

Al-Hebshi (2014) reported that as a result of the action of middlemen, only a small portion of the value-added chain benefits go to the farmers. The prices offered to the farmers in markets or for packing factories often do not cover production costs, meaning little or no profit.

7.8 Processing and Novel Products

Traditional date processing in Yemen started long ago. Dates have been a food security product in periods of hunger that have occurred in some areas of Wadi Hadramaut during periods of war. Dates are a popular edible fruit prepared in various ways as food.

Date paste, with the addition of spices and ghee, is a well-known product that is consumed throughout the country. Some people use it at home in confectionary products, others sell such products as cookies called locally *Ka'k Bu Tamr*.

Date palm parts are used for many purposes such as thatching of houses and for fencing and rope. Making date palm parts into mats, containers, furniture, and other novel items has long been known in Yemen. Mats are woven from leaflets and containers from bunch stalks stitched together with leaflets, beds, furniture, and baskets from the leaf rachis.

There are two pilot packing and processing plants for dates in Yemen. The first was established in 1984 in a fairly old existing facility. It did not operate for almost 15 years but resumed activities in 2004. It still faces major problems since the supply of good quality dates has decreased due to lower prices being offered. Recently, the factory has operated for only 4 months a year. The other factory was established in Al Hudaidah Governorate in 1986, with funding from the French Government. It also faces the same problems mentioned above. It stopped operations for 10 years. Recently, it resumed date processing with 50 % efficiency. Both plants are now packing dates in smaller consumer packs which are sold in local markets (Ba-Asher and Kasim 2011).

Production of alcohol from inferior quality dates is carried out using traditional methods and consumed locally. Although alcohol is prohibited, its use is informally tolerated.

7.9 Conclusions and Recommendations

Date palm has long been recognized as an economic and food security crop in Yemen. Traditionally, date palm production has relied on more than 321 local cultivars; however, only 42 are considered to produce fruit of excellent quality. Recently, a government policy has been adopted to promote diversification by introduction of tissue culture cultivars and local selection from among indigenous seedling dates. The PCAS contracted with the Flood Reconstruction Fund to provide 250,000 tissue culture date palms in 2011–2012, as a relief measure for farmers who lost their farms due to massive flooding in 2008. Cultural operations are mainly traditional but modern techniques are being introduced in field operations, palm and bunch management techniques and fruit handling, and packaging to facilitate marketing and to meet growing consumer demands. Yemen is still free from the devastating bayoud disease, but yield is impaired from the attack of various insect pests, namely, the red palm weevil, Dubas bug, stalk and stem borers, the lesser date moth, white scale insect and the dust mite, vertebrate pests including rats and bats, and fungal diseases including soft rot, black scorch, and leaf spot cause damages.

There are two pilot plants in Yemen for date packing and processing, but they are not operated efficiently. Dates are sold in a number of ways, but marketing services and postharvest operations need to be strengthened.

The government has recognized date palm as one of the strategic crops and there are plans to rehabilitate old orchards, develop new irrigation techniques, as well as to improve storage and packing facilities, and marketing services. Implementation of the following recommendations would contribute to improving date palm production in Yemen:

(a) Rehabilitation of old date palm orchards.
(b) Increasing date production by improving cultivation practices by promoting mechanization for harvesting, modern irrigation techniques, and new systems for crop management and adopting integrated pest and disease management programs for the major insects, mite pests, and diseases.
(c) Establishment of pilot demonstrations for farmers and nurseries to provide improved date palm cultivars.
(d) Strengthening infrastructure through supporting well-equipped tissue culture and research laboratories, as well as modern packing house and packaging facilities.
(e) Strengthening plant protection and quarantine facilities to assure protection of the date palm, especially from bayoud disease.
(f) Upgrading the quality of postharvest handling and packaging techniques of Yemeni dates to meet international standards.
(g) Improving marketing facilities and strategies.
(h) Enhancing local technical staff capacities and involving them in postgraduate programs, training courses, conferences, and seminars.
(i) Promoting agricultural extension services and emphasizing media campaigns, including greater awareness of the importance of date palm among Yemenis.
(j) Building dams to maximize the utilization of water and to protect date palm trees from the risk of floods.
(k) Establishing a National Palm and Date Center as a lead institution to promote the date palm sector in Yemen. National institutions should interact with regional and international institutions and become a member of date palm

regional and international organizations, to exchange information and coordinate collaborative efforts to handle issues encountered in the development of date palms in general.

References

Agricultural Research and Extension Authority (AREA) (2013) Efforts to improve date palm production in Yemen. The Association of Agricultural Research Institutions in the Near East and North Africa (AARINENA), Yemen. Available at http://www.aarinena.org/rais/documents/Newsletter/Vol15No1/10.pdf

Al Bar AM (2003) Effect of nitrogen and phosphorous fertilizers on the yield and quality of Hamra date palm cultivar. Seiyun Agricultural Research Centre, Seiyun, Yemen

Al-Ghurabi AS, Ba-Angood SA (2011) Screening of some chemical insecticides for the control of the lesser date moth *Batrachedra amydraula* Merk on date palm trees at two different sites in the coastal areas of Hadramout, Republic of Yemen. Univ Aden J Nat Appl Sci 15(3):565–572

Al-Habshi K (2014) The red palm weevil *Rhynchophorus ferrugineus* (Olivier) (Coleoptera: Curculionidae). Seiyun Agricultural Research Centre, Seiyun, Yemen

Al-Hebshi MA (2011) Inefficiency in market profit distribution effected date palm production in Yemen. Blessed Tree Magazine, Sept 64–68 available at http://www.iraqi-datepalms.net/Uploaded/file/MagazineDatePalms86.pdf

Al-Sakaff SM (2012) Fungal diseases of date palm in Wadi Hadramout. Yem J Agr Res Stud 26:21–30

Ba-Angood SA (1978) Control of the lesser date moth. PANS 24:29–32

Ba-Angood SA (2004) Recent development in date palm status and suggested ecosystem based (IPM) strategies for their management in Wadi Hadramout. Date Palm Regional Workshop on Ecosystem based (IPM) for Date Palm in the Gulf Countries, Al-Ain, 28–30 Mar 2004

Ba-Angood S (2008) Insect and mite pests of horticultural crops in the Republic of Yemen and their integrated pest management. University of Aden Publishing House, Aden, Yemen

Ba-Angood SA (2012) Technologies and methods that could be adopted for the management of insect pests of date palm in Arab countries, with particular reference to Doubas bug. First regional conference on pest management of date palms, Al-Ain, 23–25 Sept 2012 (in Arabic)

Ba-Angood SA (2014) Emergent assistance to control the red palm weevil outbreak in Yemen -TCP /YEM/3404 – Yemen Progress Report: January– March 2014, FAO, Sana'a, Yemen

Ba-Angood SA, Al-Baity SO (2006) Monitoring date palm stalk borers *Oryctes* spp using light traps in March 2003- February 2004, and its relationship with different environmental conditions at Seiyun area in Wadi Hadramout- Republic of Yemen. The third international date palm conference, UAE University, Abu Dhabi, 19–21 Feb 2006

Ba-Angood SA, Bass'haih GS (2000) A study on the effect of date palm dust mite *Oligonychus afrasiaticus* (McGregor) (Acarin: Tetranychidae) on the physiochemical characters of three different date varieties in Wadi Hadhramout. Yemen Arab J Plant Prot 18(2):82–85

Ba-Angood SA, Alghurabi A, Hubaishan MA (2009) Biology and chemical control of the old world bug Doubas bug *Ommatissus lybicus* on date palm trees in the coastal areas of Hadramout Governorate. Republic of Yemen. Arab J Plant Prot 27:1–9

Ba-Asher AM, Kasim FM (2011) Status and prospects of date palm and development in Yemen. A country paper presented to the first meeting for the establishment for the international council of date palm and date production, Al-Riad, 18–20 Apr 2011

Ba-Miftah MAO, Al Habshi K, Obad AS et al (2003) Results of field survey of date palm varieties in Hadramout, Tihama, Socotra and Al Mahrah. Seiyun Agricultural Research Centre, Yemen

Ba-Miftah MAO, Obad SH, Bin Hadjeh AK, Al Baiti SO (2004) Results of field survey of date palm varieties in Shabwah Governorate. Seiyun Agricultural Research Centre, Yemen

Ba-Miftah MAO, Assagaf MA, Al Gardey HA (2009) Results of field survey of date palm varieties in Al Jof and Ma'reb Governorates. Seiyun Agricultural Research Centre, Yemen

Ba-Miftah MAO, Obad SH, Bin Hadjah AK (2010) Date palm cultivars in Wadi Hadramout. Seiyun Agricultural Research Centre, Yemen

Dhouibi MH (2002). Improvement of date palm production. TCP/Yem/0166(A). Technical Report. FAO, Rome

Kassim YA, Hussein KAA (2008) The quantitative estimation of reduced saccharides, proteins and metallic elements in date samples collected from Hadramout Governorate. Univ Aden J Nat Appl Sci 12(1):13–28

Ministry of Agriculture and Irrigation (MoAI) (2013a) Agricultural Year Book – 2012. Department of Agricultural Statistics and Information. Ministry of Agriculture and Irrigation, Yemen

Ministry of Agriculture and Irrigation (MoAI) (2013b) Promising sector for diversified economy in Yemen: National Agriculture Sector Strategy 2012–2016, Interim Update 31 May 2013, Ministry of Agriculture and Irrigation, Yemen

Obad SH, Hassan AA, Al Habshi, KA, Bin Hadjeh AK (2003) Results of field survey of date palm varieties in Al-Mahrah Governorate. Seiyun Agricultural Research Centre, Seiyun, Yemen

Chapter 8
Date Palm Status and Perspective in Israel

Yuval Cohen and Baruch (Buki) Glasner

Abstract Date palms have been grown in the land of Israel for several thousands of years. Although the date orchards had flourished during the Roman Empire and were famous for their fruit quality, these were lost through hundreds of years without extensive farming in remote dessert oases. During the twentieth century, elite cultivars originating from Egypt, Iraq, Tunisia, and Morocco were imported to form an active industry. Since the 1990s the Israeli date industry has focused on production of the high-quality Medjool fruit. The semidry fruit requires additional management and care before and after its harvest, including laborious fruit thinning and selective and repeated harvests; although the industry is small, in the last 25 years, it has an annual growth rate of 5–10 %. A large portion of Israeli dates are exported, especially into the European markets, where the Israeli Medjool has a very significant market share. Innovation and research have led to physiological and technical solutions improving all aspects of date growing, harvesting, and postharvest treatments. Special efforts have been invested in machines to improve the productivity and safety of the workers on the high trees.

Keywords Automation • Export • Fruit quality • Mechanization • Medjool

8.1 Introduction

The date palm (*Phoenix dactylifera* L.) is a traditional tree crop grown in the Middle East for thousands of years. Although the land of Israel is small, dramatic climate gradients exist along its length (south to north) and width (west to east). While date palms are grown throughout the country, in the Northwest, a Mediterranean climate

Y. Cohen (✉)
Department of Fruit Tree Sciences, Institute of Plant Sciences,
Volcani Research Center, P.O. Box 6, Bet Dagan 50250, Israel
e-mail: vhyuvalc@volcani.agri.gov.il

B. Glasner
Agricultural Manager, Department of Agriculture,
Harutz 6, HADIKLAIM, Tel Aviv, 67060, Israel
e-mail: buki@hadiklaim.co.il

© Springer Science+Business Media Dordrecht 2015
J.M. Al-Khayri et al. (eds.), *Date Palm Genetic Resources and Utilization: Volume 2: Asia and Europe*, DOI 10.1007/978-94-017-9707-8_8

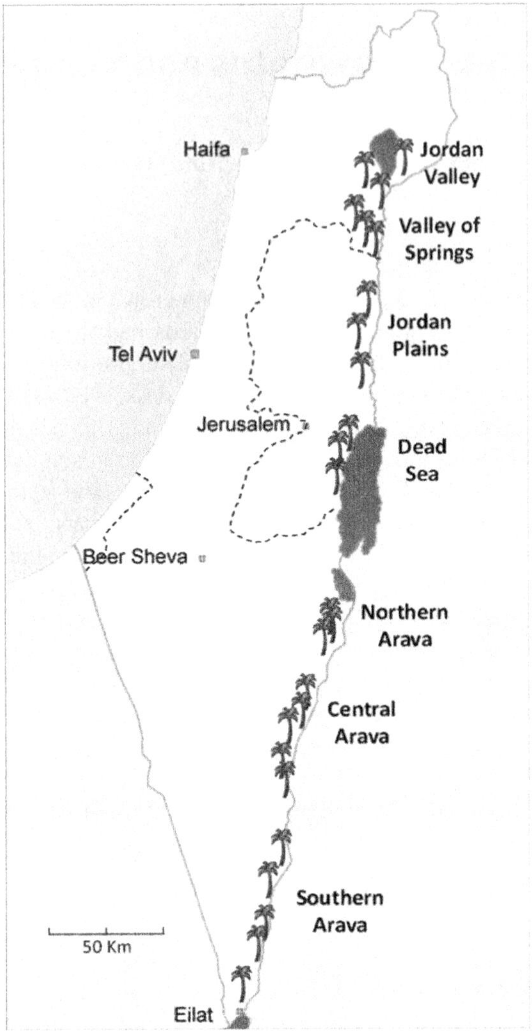

Fig. 8.1 Schematic map of Israel highlighting the commercial date-growing regions

exists that does not favor commercial date production. Commercial plantations are limited to the warmer and arid areas along the Great Rift Valley, where it was and still is a major fruit tree crop. A schematic map with the commercial date-growing regions is presented in Fig. 8.1.

8.1.1 Historical and Current Agricultural Aspects

The history of date palm cultivation in the land of Israel is described in several studies and reviews (Ben-Yehoshua and Ben-Yehoshua 2012; Bernstein 2004; London 2003). Dates have been growing in the Middle East, and in the land of Israel from ancient times. Wild populations of date palms existed in oases and valleys in the desert where high underground water existed. Such populations were previously also near the Dead Sea basin (Zohary et al. 2012; Zohary and Spiegel-Roy 1975). Date palm is one of the first cultivated fruit crops, and its cultivation started as early as 6,000 years ago. Archaeological evidences suggest that in the land of Israel, dates have also been grown for thousands of years. Date stones and kernels as old as 5,000–5,500 years have been found in archaeological sites in or in close proximity to the land of Israel (Zohary and Spiegel-Roy 1975).

During the biblical era, Israel and Judea dates were probably of only minor importance. The date palm is only mentioned 14 times in the Bible, mostly not in the context of its fruit but rather for its high stature and size (Ben-Yehoshua and Ben-Yehoshua 2012; Bernstein 2004). During that era, only few permanent settlements were located in arid regions of the Judean desert and Negev. These desert and subdesert regions were left for nomads, and most of the arid regions were not controlled by the main regimes, and date plantations were not developed.

This situation changed during the Roman era. Stable rulership enabled safer settlements in more remote and arid regions. Personal safety, developments and control of roads, and commercialization had encouraged the development of small agricultural communities along the Dead Sea region. Some of these focused on date cultivation. During that time, the dates from Judea were famous in the Roman Empire for their qualities (Ben-Yehoshua and Ben-Yehoshua 2012; Bernstein 2004; London 2003), especially those grown at the oases of Jericho and Ein Gedi. Several cultivars, with specific high qualities, were reported in ancient scripts; some of these fruits were very large, reaching a length of 11 cm and weight of 50 g (Ben-Yehoshua and Ben-Yehoshua 2012; London 2003). These were not only consumed locally but were also exported throughout the Roman Empire. During that era, the date tree became a symbol of the Israelite. It appears in synagogues as well as on coins from the first centuries BC. The Roman also used the date tree on coins to symbolize their victory over Judea in 71 AD—the Judea Capta coin.

Many date palm seeds were found in archaeological excavations in Israel. Approximately 100,000 ancient date seeds were estimated to be stored in collections (Ben-Yehoshua and Ben-Yehoshua 2012). Many of these seeds' sizes are similar to that of modern cultivars, suggesting they were not collected from wild dates but, rather, from trees in managed orchards. In 2005, a single date seed, from the excavations in Masada (a Herodian fortress from the first century AD, near the Western shores of the Dead Sea), was germinated. Direct carbon dating of the seed shell fragments have dated it, and additional seeds from the same collections, at approximately 2,000 years of age (Sallon et al. 2008). The tree is named Metushelah after the biblical figure who lived a thousand years. This tree was later found to be a male. It is now (2014) growing in Kibbutz Ketura in the Southern Arava.

After the second century AD, the date culture in the land of Israel gradually disappeared. The lack of central control resulted in unsafe conditions in the isolated oases. Cultivars were no longer conserved and were not routinely propagated through offshoots. Palm groves were gradually replaced by seedling date palms. The ancient Judean cultivars were probably extinct by the tenth century. By the

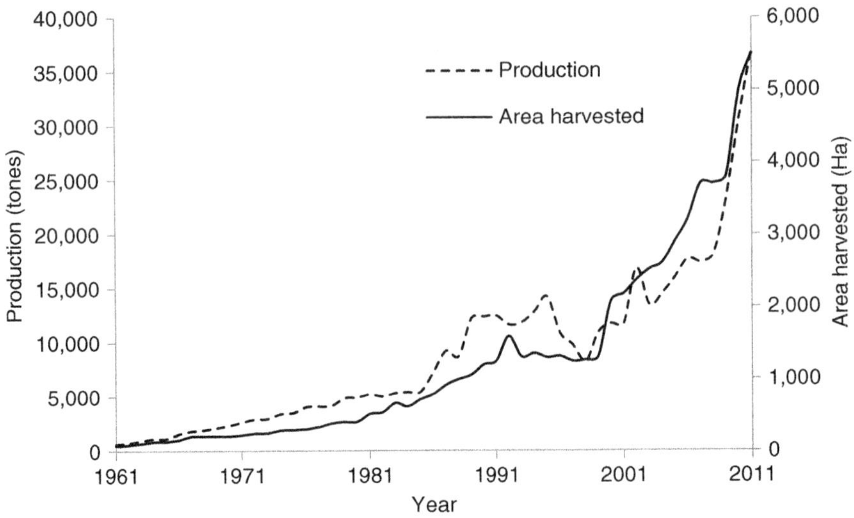

Fig. 8.2 Harvested areas and production of date palm orchards in Israel 1961–2011 (*Source*: FAOSTAT 2013)

nineteenth century, most of the date groves in Israel were based on feral, wild trees or on population of seedlings of lower quality.

The Zionist settlers at the beginning of the twentieth century perceived the date palm romantically. They invested in several expeditions, buying elite date cultivars from Egypt, Iraq, and Iran and growing them near the Kinneret, the Sea of Galilee. Gan Rachel (the garden of Rachel, named after a famous poet), a small botanical collection of date palms, was also established. These represent the beginning of modern date plantations in Israel. Additional expansions occurred during the 1950s with the acquisition of approximately 50,000 offshoots of Iraqi cultivars and in the 1970s with import of additional thousands of Deglet Noor and Medjool trees from California (Bernstein 2004). Expansion of dates into a major crop occurred throughout the Jordan and Arava valleys in Israel (Glasner 2004) (Fig. 8.1).

8.1.2 Importance to Country Agriculture

Fruit crops in Israel are being grown on about 90,000 ha (Israel Cent Bur Stat 2013). Date is one of the important fruit crops cultivated on approximately 5,000 ha (Fig. 8.2; Israeli date palm survey, 1.2014). Commercial date plantations can only grow in warm and arid regions of Israel, along the Great Rift on its eastern border, in the Jordan and Arava valleys. A climatic gradient along this rift makes date growth in the north, near the Sea of Galilee, marginal, However, dates are of very high importance further south in the valleys of the Jordan river and Arava. In these arid regions, the date is the most important and for many settlements the only fruit

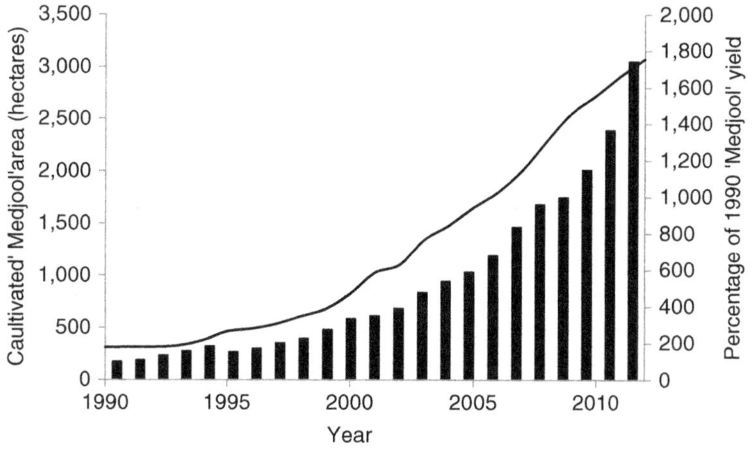

Fig. 8.3 Medjool orchards expansion in Israel during 1990–2012. Harvested area (hectares, line) and increased production (relative to yields at 1990, columns) (*Source*: Israeli date palm survey, 1. 2014)

crop cultivated. The industry's expertise and the high value of the date products make this crop one of the major origins of income for the arid, peripheral regions of the country (Glasner 2004).

More than 30,000 mt of date fruit are annually produced in Israel (Fig. 8.2). The majority of this fruit is of very high quality. About a third of the yield, approximately 13,000 mt of dates, was exported in 2012, mainly to Europe. Export values reach ILS 90,000,000 (Israel Cent Bur Statistics 2013) http://147.237.248.50/reader/?MIval=cw_usr_view_SHTML&ID=430).

8.1.3 Production Statistics and Economics

The Israeli date industry has changed its focus during the last 25 years (Glasner 2004). Once farmers started growing and marketing semidry Medjool fruit, and its success in the European and additional markets, the profitability of date production increased. The entire industry has shifted to Medjool production. Since 1990, the cultivated area of Medjool orchards has multiplied ten times and the total Medjool yield increased by 17-fold (Fig. 8.3). This process resulted in turning the Israeli date industry more and more into a monoculture producing mainly Medjool fruit. Currently, more that 70 % of the trees in the orchards are of Medjool; future planting in the next several years will probably make this process even more pronounced. While new plantations of Medjool are regularly being planted, almost no new plantations of other cultivars are observed, and their share in the industry is declining. This process is accompanied by a dramatic rise in the entire date industry (Fig. 8.2) and its importance to Israeli agriculture.

8.2 Cultivation Practices

8.2.1 Description of Current Cultivation Practices

A schematic diagram of a labor calendar in a typical (Medjool) orchard in Israel is presented in Fig. 8.4. The annual cycle starts with the removal of spines from bases of all leaves developed during the previous year. This task is important since it enables safer work during the pollination, fruit thinning, and harvesting. Removal of spines starts in December and ends by January before the beginning of the flowering season. During flowering, between February and April, farmers collect pollen from male trees and pollinate female flowers. Fruit thinning is performed in May and June, using differing protocols depending on the cultivar. At the end of this period, fruit bunches are supported on leaves, fastened, and for most cultivars covered with bags. During August–October, depending on climatic condition and cultivar, fruit harvesting takes place.

8.2.2 Pollination and Fruit Quality

8.2.2.1 Pollination

Pollination is one of the most important tasks in the orchards. Date palm pollination is performed between late February and the end of April, depending on the growing region and the cultivar. It starts by collecting inflorescences from the male trees. Growers usually tie the spathes of the male inflorescences to prevent

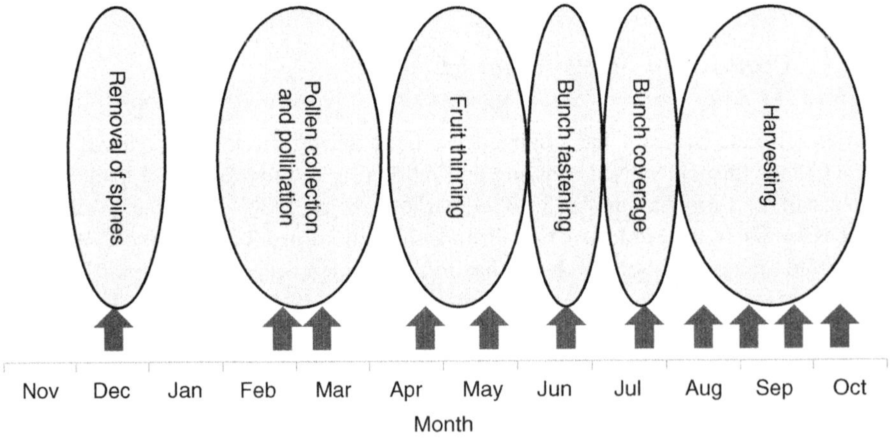

Fig. 8.4 Schematic working calendar for Medjool orchards in southern Israel. *Arrowheads* represent tasks requiring physical climbing to the crown. Some of these tasks can be replaced using advance machinery from the ground

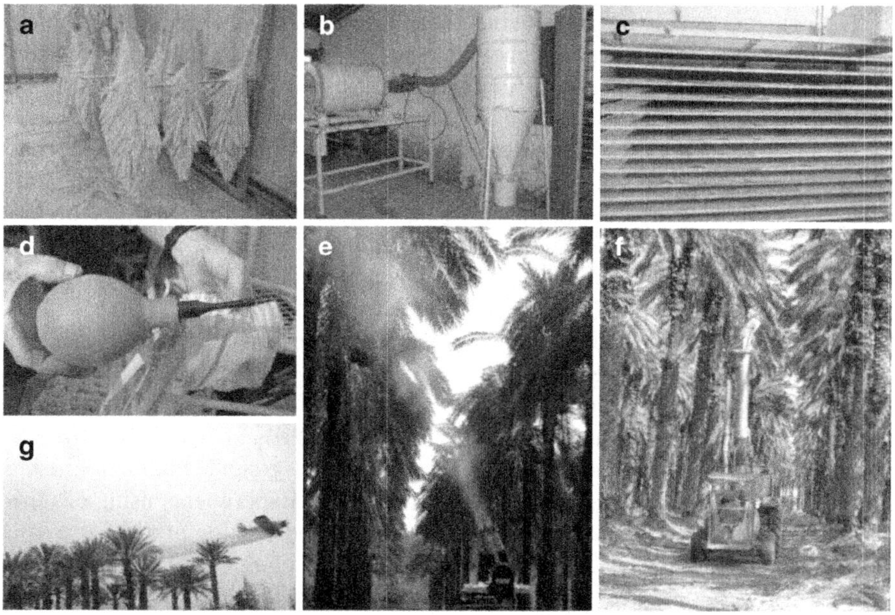

Fig. 8.5 Pollen purification and pollination in Israel. Male inflorescences are left for several days to dry (**a**), and pollen is extracted, purified (**b**), and dried (**c**) prior to storage. Pollination is then performed using hand dusters (**d**) and mechanical blowers from the ground (**e**) and from a higher located blower (**f**) or even applied to the orchard using airplanes (**g**)

dispersion of the pollen upon their opening on the trees. During the flowering season, farmers observe the male trees and collect each male inflorescence once it is mature or split open. Inflorescences are taken into a closed room where they are left for a few days to dry. Pollen is then purified in specifically designed devices that vigorously shake the inflorescences and separate the pollen from the flower debris by centrifugation (Fig. 8.5). Up to 1 kg of purified pollen can be harvested from a single male date tree.

Based on modern agricultural techniques, male trees are usually grown in designated plots. Efficient use of pollen permits the reduction of the number of male trees in the orchard. Pollen harvested from a single tree is sufficient for pollination of 50 female trees. Therefore, an Israeli date plantation has approximately 2 % male trees. Pollen can be stored from season to season at −18 °C, providing pollen for the pollination of the earliest flowering females. Some farmers collect more pollen than they need, and high-quality pollen is a commercial product that is often traded between growers.

Pollen quality is determined routinely by its germination in vitro (Bernstein 2004) that is performed both before storage of pollen and prior to actual pollination. Only pollen with a high germination rate is used. This technique is also used to screen out male trees of poor pollen qualities and to select elite male dates for tissue culture propagation.

The pollen is usually mixed with inert material (mineral talc, or powdered potato starch) to increase the applied pollen volume and reduce the amount of pollen used. Some farmers add charcoal dust to improve the dispersion of the pollen mix through mechanical blowers. The ratio of pollen and added inert materials is dependent on the receptivity of the female cultivar. For cultivars with high fruit setting (like Medjool), a mixture that contains 5 % of pollen is applied, while for sensitive cultivars (as Barhi or Hayany), the mixture contains as much as 30 % pollen. Accordingly, each tree is repeatedly pollinated as often as every second day (in cultivars with short stigma receptivity), or only once a week (for cultivars with longer flower receptivity as Medjool and Deglet Noor) (Bernstein 2004; Reuveni 1970).

Traditional pollination, placing male spikelets within female clusters, is no longer practiced in Israel. Pollen application is performed in several ways: in younger trees where inflorescences are easily accessed, it is applied directly to open inflorescences by hand dusters. Otherwise, it is applied by mechanical dusters from the top of cranes or from larger mechanical blowers from the ground. Pollination using airplanes is also practiced in several regions in Israel. Experiments using electrostatic pollination to conserve pollen and improve the pollination process and fruit setting have been tested (Gan-Mor et al. 2009; Bechar et al. 1999). Although this approach proved to reduce pollen consumption, it did not emerge as a commercial application method for date palms in Israel.

8.2.2.2 Fruit Thinning

Fruit thinning is the main method for controlling fruit size and quality. Fertilization efficiency is dependent on pollen quality, pollination efficiency, and environmental factors. In many seasons, climate conditions, mainly temperature, result in differences in fruit setting of all or specific whorls of bunches. Farmers perform surveys to estimate the levels of fruit set at each of the whorls. Usually, this action starts only after fruit set is clearly visible, by drying and degeneration of two of the three carpels. However, since fruit thinning is time consuming, many farmers with larger orchards start their thinning even before finishing these surveys. The main influence of thinning is to increase the size of the remaining fruits. The thinning has indirect effect on fruit quality, by increasing ventilation surrounding the individual fruits. A heavy load also delays ripening. The thinned bunch can be treated more efficiently against pests in the more open fruit bunch. Fruit thinning may also prevent collapse and breakage of fruitstalks due to extremely heavy bunches.

The level of thinning depends on the potential yield of each cultivar, tree size, and the market preference for quality and size. Fruit yield can be managed by controlling the number of bunches left on a tree, the number of spikelets left per bunch, and the number of fruitlets on each spikelet. Manual thinning to get the best fruit load and density on each spikelet can take up to 3.5 h per tree. More efficient methods, based on shortening of the spikelets and removing the central strands, were developed. Even with this technique, it may take more than an hour to thin a single mature tree.

8.2.2.3 Fruit Thinning in Medjool

Medjool fruit can be especially large. For this cultivar, fruit quality and size are the main factors affecting its price. Heavy bunch loads may also increase the incidence of skin separation, reducing fruit value. Therefore, special care is taken for fruit thinning in this cultivar. It is usually thinned earlier and much more vigorously than other cultivars. Aiming to get large fruits, instead of large yields, most farmers perform harsh thinning to a level of 250–350 fruits per bunch at harvest. This level usually results in an average fruit weight higher than 20 g, with a substantial fraction of fruits weighing up to 30 g and with only very few smaller fruits, lighter than 18 g. Average yields are 80–120 kg per tree.

In principle, manipulating pollen quality and quantity can result in partial levels of fruit set. If the farmer could control fruit set at the pollination stage, most of the labor required for fruit thinning would be saved. Attempts were made in Israel to reduce fruit set by applying only minute levels of pollen. Alternatively, treatments to reduce pollen viability by heat or increased humidity were applied (Bernstein 2004; Dekel et al. 2012). Such treatments were successful in reducing the level of fruit set and dramatically reduced the labor required for fruit thinning. However, these treatments increased the system sensitivity to environmental conditions. Lower temperatures during pollinations resulted in lower natural fruit set, reduced yield, and increased development of parthenocarpic (seedless) fruit.

8.2.2.4 Tying and Covering Fruit Bunches

All bunches are tied to a leaf petiole to support their weight. This task is performed once fruit stalks complete their growth and reach their final length. It usually starts immediately following fruit thinning and is completed in June. Proper tying reduces the breakage of fruit bunches. This practice also reduces the damage of scars or scratches originated from the movement of fruits and their friction among themselves and with neighboring leaves. Another advantage of bunch tying is the organization of bunches in the tree space to enable the most convenient management during bag coverage and harvest.

In parallel to bunch tying, most farmers trim older and dry leaves that touch the bunches. Some 20–30 leaves are removed annually, depending on growth conditions and the trees' growth rate. At the late khalal (basser) stage, when the fruits are still hard, the bunches are usually covered with net sacks. The best material is 25 mesh plastic nets. This provides protection against birds, bats, and fruit moths. It also prevents the fall of individual fruits to the ground prior to their harvesting.

8.2.2.5 Fruit Harvest

Harvest is performed according to the timing of ripening of the different cultivars. Barhi is usually harvested at the khalal stage. Most consumers will not eat the fruit once ripening spots appear. Since fruits do not dry on the bunches, they have a much higher water content and greater weight. Harvesting starts when fruit skin color changes from green to yellow, at the beginning of August in the warmer areas and in September in the cooler ones. Entire bunches are harvested and then sorted; both green fruits and fruits with ripened spots are removed. Barhi fruits are usually packed on their spikelets, although single fruits are also sold. Since fruits have to be kept at hard khalal, they are immediately packed and consumed within few weeks of harvest.

Hayany fruits are harvested as red khalal but consumed at the black rutab stage. They are kept frozen, thawed, and eaten soft. Harvesting is performed into large 400 kg containers. Fruits are removed from bunches using vigorous shaking. Fruit sugar content has to be high, above (30 Brix), but fruit has to remain in the red khalal stage. Fruits ripening on the tree are not marketable and need to be sorted out. Therefore, harvesting time is critical for profitability of Hayany dates. Approximately 150 kg of saleable fruit can be harvested from a single Hayany palm.

Medjool harvesting time starts early in August in the warmer regions and ends in the north by the end of September. Although Medjool is harvested as tamar, the fruit is not completely dry but is kept at a water content of 24–26 %. At this stage the fruit texture is solid yet soft and fleshy. To prevent damage to this delicate and expensive fruit, harvesting is performed very gently. Fruits are collected in specially designed shallow plastic trays, having only a single fruit layer. To harvest most fruit at their proper stage, each bunch is accessed and harvested several times, usually 2–4 harvests per season and sometimes even more, depending on climatic conditions) (Fig. 8.4). Each time, only the fruits that are at the proper stage are collected.

Following harvesting, postharvest disinfection and cleaning are performed. Under strict quality control, fruits having water content above 26 % are dried, and those having water content below 22 % are hydrated. Fruits are then sorted according to size and qualities.

Deglet Noor is harvested and packed with adherence to two different marketing approaches. The first approach is as natural fruit on the spikelets, performed early when the fruit is still soft. Whole bunches are brought to the packinghouses where they are cut and packed preserving their fresh state. The second approach is as single fruits that are harvested drier, many times by mechanical harvesting. Fruits are stored in large containers and treated by hydration prior to shipment. Yields of 100–150 kg/tree are common.

Additional cultivars such as Ameri, Halawi, Dayri, and Zahidi are all grown only in the northern, more temperate regions. They are either harvested into large (12 kg) boxes or into containers or stored frozen until sorted and shipped. Their yields varied from 80 kg/tree (Khadrawy) to 180 kg/tree (Halawi and Zahidi).

8.2.3 Irrigation and Fertilization

All commercial orchards in Israel are based on closed irrigation systems that also apply fertilizer. Most plots are irrigated with drippers, either scattered along the line of the trees or encircling each one. Some farmers prefer mini sprinklers that increase

Table 8.1 Average evapotranspiration, precipitation in different date-growing regions of Israel, and annual irrigation levels (based on Medjool cultivar)

Region	Maximum evapotranspiration per day (mm)	Average annual precipitation (mm)	Annual irrigation (m^3 per hectare)
Southern Kinneret (Sea of Galilee)	10.2	400	9,000
Valley of Springs (Beit Shean)	10.1	300	11,000
Jordan Valley	10.0	100	15,000
Dead Sea region	12.0	50	18,000
Northern and Central Arava	13.7	30	20,000
Southern Arava	13.0	20	22,000

the wet area surrounding each tree. Irrigation is applied by computers, enabling control of water quantities, irrigation frequency, and supply of different fertilizers.

Irrigation levels are calculated by evaporation from a Class A pan. Alternatively, evapotranspiration is calculated from the Penman equation. Farmers use values of 0.5–1 as coefficient for the amount of irrigation supplied, depending on the season and the region of their plot. Total irrigation levels range between 8,000 and 20,000 m^3 per/ha/year (Table 8.1).

Many farmers monitor soil water potential at different depths using tensiometers, to improve water use consumption and productivity. Experiments on water consumption of date palms and on possible reduction of irrigation levels have been performed, including reduction of irrigation during the cold season (Cohen et al. 2010, 2012) and reducing the levels throughout the year. Water consumption of Medjool date trees was measured in trees grown in special large weighing-drainage lysimeters for more than 10 years (Tripler et al. 2007, 2011). Attempts to irrigate according to such parameters enable increased water use efficiency in nearby orchards. Other tools were used in Israel to analyze water use and consumption of date palms. These include stem water conductivity (Nadler et al. 2008), heat dissipation (Sperling et al. 2012), and thermal imaging (Cohen et al. 2012).

As in most date-growing countries, Israel's water resources are limited. Date plantations, especially in the southern arid growing regions, are limited in water supply. Expansion of date orchards in these regions is limited since there are no additional water resources. As a partial solution, as much as 90 % of the Israeli date plantations are irrigated with marginal water of lower qualities.

Israel has invested a lot in water recycling. Almost 78 % of sewage water is recycled for agricultural use, mainly for fruit crops (http://www.mekorot.co.il/Eng/Activities/Pages/WastewaterTreatmentandReclamation.aspx). Currently, as many as 2,000 ha of date palm are irrigated with recycled water, which has undergone secondary and even tertiary purifications. Although orchards have been irrigated in this way for more than 15 years, no health issues or agricultural problems have arisen. Research is being performed to modify fertilization levels according to organic content present in the recycled irrigation water.

Date palm is known for its tolerance to salty soil and water (Chao and Krueger 2007), with a threshold of up to 4.0 dS m^{-1}. Many date orchards in Israel, especially at the Arava region, are irrigated with saline water (EC of 2.8–6 dS m^{-1}) (Tripler et al. 2011). A recent study characterized the effects of increased salinity on growth and productivity of date palms (Tripler et al. 2007, 2011). They found that salinity reduced water consumption and plant growth and recommended the use of water of higher quality.

In Israel due to the nature of the soil, dates are usually fertilized only with nitrogen. In some regions potassium is also added. In most places there is no requirement of fertilization with phosphates. As mentioned above, organic growers use compost for fertilization.

8.2.4 Pest and Disease Control

A recent review that characterizes the arthropod date pests in Israel estimated their damage and reported natural enemies and common protocols for their management (Blumberg 2008). The main pests and diseases affecting date palms in Israel are summarized in Table 8.2.

As an important crop in Israel, efforts are made to characterize date palm pests and diseases and prevent epidemic outbreaks. Two global strategic threats on date cultivation in the world are the bayoud disease caused by the soil-borne fungus *Fusarium oxysporum* f. sp. *albedinis* (FOA) and the red palm weevil (RPW) (*Rhynchophorus ferrugineus*). The bayoud has not been detected in Israel. In order to prevent its import, a reliable detection system was developed in Israel using molecular techniques (Freeman and Maymon 2000). Strict quarantine regulation and prevention of import of date fruit and products (together with disinfection and fumigation of imported spear leaves, lulav) was found successful in preventing the introduction of the disease. Thus far, no other lethal fungal or bacterial disease is known to affect the date orchards in Israel.

The case of the red palm weevil is different. It was first recorded in Israel in 1999 (Kehat 1999; Soroker et al. 2005, 2013). Three phases of infestation of the red palm weevil were identified. The first two phases occurred in limited agricultural regions and were immediately and successfully managed. The third phase started in ornamental palms in urban and suburban regions, far from date-growing areas, and expanded to larger areas and remains uncontrolled. Thousands of infected Canary Islands palms (*Phoenix canariensis*) died, and infected date trees were also detected in commercial orchards (Soroker et al. 2013). RPW represents a threat to both date production and palm landscaping in Israel, and efforts are made to control its spread. Research studies suggest the use of dogs (Nakash et al. 2000) and acoustics (Pinhas et al. 2008; Soroker et al. 2004, 2013) to enable early detection of infected trees. Other methods, including the use of thermal imaging for early detection of infected palms, are also being explored (Soroker et al. 2013).

Table 8.2 Common date pests and diseases in Israel and their risk estimates

Pest/disease	Damage and risk estimates
Fruit pests	
Lesser date palm moth (*Batrachedra amydraula*)	Common in all growing regions. Can cause severe damages
Greater date palm moth (*Arenipses sabella*)	At spring causes damage to the younger leaves and emerging inflorescences. Very rare on fruit but can damage the entire bunches
Sap beetles (*Carpophilus* spp.)	Detected mainly in northern growing regions. Can cause up to 20 % fruit loss. Treated postharvest by heat
Fruit moths: raisin moth (*Cadra figulilella*), carob moth (*Spectrobates ceratoniae*)	Detected only in fruit harvest too late in southern orchards
Date stone beetle (*Coccotrypes dactyliperda*)	Occurs only at the north. Causes fruit drop. Usually there is no need for treatment
Old world date mite (*Oligonychus afrasiaticus* and *Eutetranychus palmatus*)	Occurs mainly in the south. Can cause severe damage at kimri and khalal stages
Crown/trunk/root-damaging pests	
Green scale (*Palmaspis phoenicis*)	Rare in Israel. Usually there is no need for treatment
Parlatoria date scale (*Parlatoria blanchardi*)	Common throughout Israel. Usually does not cause damages
Dubas date bugs (*Ommatissus lybicus*)	Sporadic appearance in the Arava and Dead Sea regions. Usually do not cause major losses
Red date scale (*Phoenicococcus marlatti*)	Common on tree. Can cause damages in young plantings of offshoots.
Rhinoceros beetles (*Oryctes* spp.)	Common, can cause damages to root systems. In cases of large population of beetles, can cause damages to leaves and fruit bunches
Red palm weevil (*Rhynchophorus ferrugineus*)	Causes severe damages in ornamental palms and date palm in orchards
Major diseases	
Black scorch (*Thielaviopsis paradoxa*)	Detected mainly in orchards in the south. Several epidemic events occurred in tissue culture originated date palms
Inflorescence rot (*Mauginiella scaettae Cav*)	Sporadic appearance
Diplodia spp.	Common. Attacks young recently planted trees
Aspergillus niger	Sporadic appearance; reduced following good agriculture practice

8.2.4.1 Pest Control

The increased control of chemical application in Israel and throughout the world limits the use of pesticides. All chemicals need to be approved in both the manufacturing country and its export markets. Strict regulations also limit the time of chemical application, often prohibiting their use on fruits at advance stages of development or during ripening. The limitations on use encourage farmers and researchers to look for different approaches for pest management. These include physical barriers (like net bags to reduce fruit moth damage) and modification of horticultural practices (such as increasing fruit thinning to improve ventilation around individual fruit, frequent harvests to reduce damage from sap beetles (*Carpophilus* spp.) and improve sanitation in the orchard).

Attempts have been made to use integrated pest management (IPM) to control various date pests. Biological control measures were used to control several pests. Predator mites were used to restrict the population of date palm mites (*Oligonychus afrasiaticus* and *Eutetranychus palmatus*) (Palevsky et al. 2004). Other pests are controlled by trapping with pheromones. Recently, isolation of the sex pheromone of the lesser date moth enables (*Batrachedra amydraula*) better monitoring and controlling of this pest by either mass trapping or mating disruption (Levi-Zada et al. 2011, 2013).

8.2.4.2 Weed Control

Weed control is also limited due to chemical approval and registration. In the northern regions, having sufficient precipitation, pre-emergent herbicides are applied before the rainy season. In the southern region, being dryer, postemergent herbicides, mainly contact herbicides, are applied following massive weed growth. Organic farms that cannot use herbicides have developed special trimmers for weed control. Introduction of grazing by donkeys and horses has also been applied in several orchards to reduce and control weed growth.

8.3 Genetic Resources and Conservation

8.3.1 Current Status and Prospect of Genetic Resources

Date palms have been part of the vegetation of Israel for many thousands of years (Zohary et al. 2012; Zohary and Spiegel-Roy 1975). Although the ancient cultivars were lost, date palms were continuously grown in oases, villages, and towns. Wild date populations inhabit many of the desert oases along the great rift. Some of these oases have thousands of trees, while in others only few dates survive. Among these, the oases of Yotvata in the Arava and Zin Wadi south of the Dead Sea are the largest. Date trees and groves are also conserved near springs in the desert and in Beit Shean Valley. It is hard to separate natural date population and the cultivars. Being a wind pollinator, male date trees disperse their pollen for long distances. It is expected that the oases were used for farming. Seed dispersal could have probably occurred also by nomads who visited the date groves, collected fruits, and probably also dispersed the seeds. Although some of the dates in oases appear to be feral, it is expected that both in ancient times, as well as with modern agriculture, contamination of natural populations with date cultivars has occurred. Most of these date trees are currently protected in nature preserves. However, since the expanding date industry is in proximity to them, further mixing of these genotypes is expected.

8.3.2 Germplasm Collections and Genetic Resources

Attempts to collect and conserve date palm germplasm resources in Israel began many years ago. In 1931 at Gan Rachel, a collection of dozens of date cultivars from different countries (mainly Egypt, Iran, and Iraq) were purchased in these countries and transported to Israel and established on the shores of the Sea of Galilee (Bernstein 2004; Stoler 1977). These were screened to select the current nine commercial date cultivars.

Date trees form offshoots only during their early life. Although Gan Rachel still has most of its older trees, they do not produce offshoots and are not being used for cultivar conservation or introduction. In 2005, a new collection was established in Eden Research Station near Beit Shean, with the aim of it becoming a national date palm germplasm collection. This collection has all the date cultivars present in Israel. Additional cultivars from different countries are being imported (as tissue culture plantlets) and grown in the collection. In this collection, offshoots from young trees are routinely propagated in order to maintain the early stage of all cultivars, producing new offshoots.

8.3.3 Quarantine Regulations

Up to the 1970s, many thousands of date palm offshoots were imported. These have become the basis of the Israeli date industry (Bernstein 2004). However, once the industry was established and dates became an important crop, specifically adapted to the arid periphery of Israel, concerns about the safety of the date trees arose. Efforts were made to keep the trees as safe as possible from pests and diseases especially the introduction of lethal diseases such as bayoud that can severely damage date orchards. Thus, quarantine procedures have been developed and enforced on all date products.

The importation of date trees into Israel is no longer allowed. The development and availability of tissue culture plantlets that are pest free and disease free enables the continuous import of date cultivars. However, these are imported only as sterile plantlets in tissue culture tubes and follow a long inspection during their hardening and before planting in the field. Imports of date fruit to Israel are also not allowed, and the local consumption is fulfilled by local farmers. Other date products, as well as crafts and artisans made of date material (e.g., leaflets and fruit bunches), are also not allowed for import.

8.4 Plant Tissue Culture

8.4.1 Role and Importance

As monocot trees, palms cannot be clonally propagated by grafting. Until the last few decades, propagation from offshoots was the only method for clonal propagation of date palms. Tissue culture propagation has the potential to generate a large number of homogenous plants, has no seasonal effect on plant source, and enables easy and safe exchange of plant material.

All date cultivars in the Israeli industry are elite cultivars originally imported from other countries. While currently the industry focus is on Medjool trees, there is interest in the development of new elite products from additional cultivars. Israel lacks large genetic variation and a diverse collection of date cultivars. The limitations of traditional propagation through offshoots restrict the future introduction of new cultivars. Therefore, the use of tissue culture, as well as additional biotechnical techniques for future introduction and cultivar improvement in dates, is extremely important for the Israeli date industry.

8.4.2 Research and Development

The Israeli scientist Oded Reuveni was one of the pioneers in development of date palm tissue culture propagation. He explored the possibility of tissue propagation of date palms in Israel as early as the late 1960s (Reuveni et al. 1972; Reuveni and Lilien-Kipnis 1974; Reuveni 1979). These techniques were based on embryogenesis. The first tissue-culture-originated grove of 100 Medjool dates was planted at the Eden Research Station in 1984 (Bernstein 2004), being one of the first in the world. Zemach Tissue Culture Laboratories together with Zvieli Nurseries started commercial production of date palm plantlets. They also continued the research and attempted to modify the embryogenesis protocols. They reported the use of immature inflorescences from mature trees (Ben-Bassat et al. 1996; Bernstein 2004).

8.4.3 Survey of Research and Commercial Labs

Commercial tissue culture propagation of date palms started at 1989 at Zemach Tissue Culture Laboratories. The first plantlets originating from this company were planted in commercial orchards at 1994. There is stable constant annual supply of several thousands of trees from the beginning of production until today. These were planted in orchards throughout Israel. Another Israeli company, RAHAN (Rosh Hanikra) Tissue Culture laboratories, has also generated thousands of plantlets. Additional plantlets were imported from foreign commercial laboratories. Due to quarantine regulation, they were all imported sterile, in tissue culture tubes and media, and were hardened for 1–2 years in local nurseries prior to their planting in the orchards (Bernstein 2004).

Israeli tissue culture laboratories, as well as commercial import of date plantlets, have focused on the leading commercial cultivars. Most of the production is of Medjool and Barhi cvs. Tissue culture was also used to generate trees from additional cultivars when young offshoots are lacking. These include several thousand trees of Deglet Noor, Dayri, and Argemani (a Hayany-like cultivar, with a purplish color that was identified among 'Hayany' trees in Israel). Several selected male trees, carrying large inflorescences that produced larger quantities of pollen, were also commercially propagated. Tissue culture was also used to safely introduce additional cultivars to Israel. Anbara, Sucary, Nabut Saif, Sultana, and Boufagous were all introduced via tissue culture and are now being tested in Israel.

Tissue culture production is focused on the Israeli market. Approximately 15 % of trees planted in Israel since 1994 have originated from tissue culture. Currently, more than 50,000 trees in commercial orchards are of tissue culture origin (Israeli date palm survey, January, 2014). However, the small volume of the Israeli industry and the abundance of offshoots from young trees restrict the development of the tissue culture companies. Currently, only a single commercial tissue culture laboratory is operating in Israel. A significant fraction of its date plantlets are exported from Israel to several other countries.

The abundance of young date palms that originated from tissue culture enabled studies on their true-to-typeness. Surveys of performance have detected some typical offtypes. These include trees producing parthenocarpic fruits, trees with variegated leaves or with variations in leaf structure, dwarf trees, and trees with variations in overall growth patterns (Cohen et al. 2007). Similar date palm offtypes were detected in other countries (Al Kaabi et al. 2007; McCubbin et al. 2000; Cohen 2011; Zaid and Al Kaabi 2003). The occurrence of offtypes in Israel was dependent on tissue culture protocols and varied among producers and batches (Cohen et al. 2004; Gurevich et al. 2005). While in most cases the fraction of offtypes generated is limited, there have been several reports in which most of the generated trees were found to be abnormal (Cohen et al. 2004, 2007; Gurevich et al. 2005). Several thousands of offtypes were detected in Israel. These include variegated trees, more than 2,000 Barhi and Hallas trees with fruit setting abnormalities, that many times generated supernumerary carpels (Cohen et al. 2004), and approximately 10,000 dwarf trees (most of them of the Medjool cv.) (Cohen et al. 2004). Molecular analysis of the offtype trees detected increased genetic variation, as well as variation in DNA methylation (Cohen et al. 2007, 2004; Gurevich et al. 2005).

8.4.4 Limitations and Recommendations

Although many offtypes were previously detected among tissue-culture-originated date palms, farmers are still enthusiastic about the use of in vitro plantlets. It was found that plantlet survival in the field and their growth patterns are far superior to those usually obtained in plots originated from offshoots. To reduce the occurrence of offtypes, several restrictions and recommendations are being applied by both tissue culture laboratories and farmers.

Material for explants is very strictly chosen, taken from trees of a proven cultivar and quality. Laboratories avoid the use of trees that originated from tissue culture as their source material. Propagation performed from single explants through embryogenesis is limited to only several hundred generated plantlets. To reassure the farmers, companies usually replace abnormal (suspected) offtypes with normal trees free.

The farmers perform detailed surveys of their orchards to detect and report any abnormality. Many farmers do not use offshoots from tissue-culture-originated trees for planting and destroy all offshoots once they appear.

8.5 Cultivars Identification

Only a few studies were performed in Israel on cultivar identification. These were focused mainly on identification of true-to-typeness of trees originated from tissue culture (as described above). Most of the analyses were performed on Barhi and Medjool accessions. Typical AFLP band patterns of Barhi and Medjool were characterized (Cohen et al. 2007; Gurevich et al. 2005). Although some genetic variation was detected between trees propagated from offshoots, cultivar patterns were conserved. Similarly, DNA methylation was found to have a cultivar-specific pattern (Cohen et al. 2007).

8.6 Cultivars Description

Approximately 30 different cultivars from different countries have been tested and grown in Israel during the past 90 years. Currently (2013), the date palm industry is based on nine date cultivars (Table 8.3).

All of these cultivars are of foreign origin: Egypt, Iraq, Morocco, or Tunisia. They are all elite cultivars of high quality. These cultivars are further described in other chapters of this book according to their country of origin. Therefore, only a short description related to their production in Israel is provided.

Ameri Originated in Upper Egypt. Leaves are narrow and delicate. Yield is up to 100 kg/tree. Under Northern Israeli condition, Ameri is especially sensitive to weather during pollination, with flowers of relatively short receptivity to pollen. The fruit is very large and tends to be harvested very dry, requiring postharvest hydration prior to marketing. Ameri is cultivated in Israel only in the Northern growing regions.

Barhi Brought to Israel from Iraq in the 1930s. The tree is relatively slow growing. Leaves are very wide and long. Thorns are moderate. Fruitstalks are long and sensitive to breakage during fastening. The average yield of a mature Barhi tree is higher

Table 8.3 The date palm industry in Israel, number of trees from the different cultivated cultivars in commercial orchards in Israel in 2013

Cultivar	Country of origin	Number of trees
Ameri	Egypt	10,000
Barhi	Iraq	28,000
Dayri	Iraq	30,000
Deglet Noor	Tunisia	30,000
Hayany	Egypt	39,000
Halawi	Iraq	13,000
Khadrawy	Iraq	6,500
Medjool	Morocco	394,000
Zahidi	Iraq	12,000
Total female trees		562,500
Male trees		11,500

Source: Israeli date palm survey, January 2013

than 200 kg and can reach 300 kg/tree. For export, minimal required fruit diameter is 27 mm although many fruits reach a diameter of 30 mm. Barhi is grown mainly in the Northern regions, although some early-ripening orchards are located near the Dead Sea and Arava regions.

Dayri Originated from Iraq. Yields approximately 100 kg/tree. The fruit is small, elongated, and dark at the tamar stage.

Deglet Noor Introduced by import of thousands of offshoots from Indio California, during the 1970s. Deglet Noor is a fast-growing cultivar. Leaves are large but delicate, with large and strong thorns. Fruitstalks are very long and yellow. It is harvested either as soft, almost rutab natural fruit on its spikelets or as a dry fruit that is hydrated prior to marketing. Deglet Noor is cultivated mainly in the dryer Arava and Dead Sea regions. Yields range between 100 and 150 kg/tree. Fruit size is 9–12 g. The fruit's color at basser (khalal) is orange, turning during ripening into a yellowish, almost transparent natural fruit. Although by many it is thought to be the best cultivar according to its taste and texture, Deglet Noor's share in Israeli production is reduced because of the expansion of Medjool orchards.

Hayany Brought from Egypt to the land of Israel in the 1920s, Hayany is cultivated only in the Northern growing regions. Hayany is a fast-growing cultivar. It has delicate leaves and smaller leaves. The fruit is red at khalal stage.

Halawi Originates from Iraq. Grown only in Northern regions. Fruit weighs 7–10 g. It is yellow in khalal and brown in tamar stage.

Khadrawy An Iraqi cultivar, it has an extremely slow growth habit. Yields approximately 100 kg of fruit averaging 9 g.

Medjool Originates from Morocco. It was first introduced to Israel from Indio, California, USA, at the 1950s and mostly during the early 1970s. Consumer preference for a very large and perfect fruit—soft and succulent without any skin damages—resulted in focus of the Israeli date industry on Medjool. Massive plantings of Medjool occurred, making Israel one of its major producers and exporters. Medjool trees are fast growing. They have relatively short leaves with long and strong thorns. The fruitstalk is orange and relatively short. Average yields are 90–130 kg per tree, according to the growing regions and the optimization of fruit quality (size and skin separation). Harvested fruit size averages 20 g, although much larger fruit can be obtained if a high level of fruit thinning is carried out. One of the important physiological disorders of dates is skin separation from the pulp during ripening and drying. While this phenomenon occurs in date fruit of many cultivars (Gophen 2014), it is extremely important in Medjool (Lustig et al. 2014). The price for a fruit with skin separation is only one-half of a similar fruit without it. A gradient in the level of skin separation occurs along the growing regions of Israel, being minimal at the Arava regions and higher near the Dead Sea and affecting the majority of the fruits in the Northern growing regions at the Valley of Springs and Jordan River. This suggests the involvement of climatic conditions in the occurrence of skin separation. Different studies have focused on this phenomenon in Israel, suggesting the induction of anatomical and mechanical changes in the fruit skin (Lustig et al. 2014; Gophen 2014; Adhikari 2010). However, an effective treatment was not discovered. Although Medjool is the most profitable cultivar in Israel, growing it is more laborious and expensive because of both thinning operations and selective harvesting.

Zahidi This is a large and beautiful cultivar from Iraq which produces relatively high yields.

8.7 Dates Production and Marketing

8.7.1 Practical Approaches

Labor in Israel is expensive. Moreover, many farms are limited by the number of workers they can recruit. The Israeli industry and agricultural research teams thus focus on innovations and technical approaches to maximize productivity and reduce labor. Such innovations include all stages of preharvest horticultural practices, harvesting as well as postharvest technologies. Special protocols, devices, and machines were adapted or specially developed for the date industry to maximize labor efficiency.

8.7.2 Working in the Heights

The most important limitation of date plantations is the tall stature of the palm trees. Manual climbing of each tree or the use of ladders is still the most common way to reach the crown of the trees in most date-growing countries. In the past, Israeli workers would climb to the tree crowns using ladders. However, this task limits the accessibility of the crowns to several trees a day and demands a lot of manpower. Recent regulations forbid manual climbing (or with the aid of ladders) to more than 2 m above the ground. As monocots, palm trees cannot be pruned. Bioregulator treatments for the control of excessive trunk growth have been developed (Aloni et al. 2010; Cohen et al. 2013); however, although these reduce the annual growth of the palms, the trees will eventually reach the same height. Special cranes that carry the workers safely to the high crowns were developed. Most of these machines were specifically designed for the Israeli date palm orchards (mainly by Machinery & Engineering—Afron Co., http://www.afron.com/HTMLs/page_1432.aspx?c0=13372&bsp=13112&bss333=13112&bscrp=1). Additional cranes of general use were adapted to work in the date orchard. These cranes have an operator platform that can completely encircle the date tree trunk, enabling safe and efficient work on any tasks on the leaves, inflorescences, and fruit bunches at the high tree's crown. They can carry several workers and few hundred kilograms (to enable efficient handling or harvesting of fruit) (Fig. 8.6). These machines can reach working heights of approximately 4–20 m. An average machine will support the cultivation of only several hundred trees. Farmers having groves of dozens of hectares will use ten and even more of such cranes year round. Since reaching the crown is achieved only using the date cranes, the industry is dependent on these cranes, and they are a major constituent of the growers' expenses.

Being the basic machine for orchard management in Israel, the high cranes form a platform, and additional equipment (Fig. 8.6e) is attached to them to enable complete and safe horticultural management.

8.7.3 Harvest Mechanization

The focus of the Israeli industry on harvesting of semidry Medjool demands selective harvesting, picking each fruit exactly at the right stage, after it has started drying but before it is too dry. This requires accessing each fruit bunch at least 3–4 times within several weeks during the ripening stage. Even with the high cranes, this approach is labor consuming.

A new approach for date harvesting is the use of tree shakers to mechanically harvest the date. Mechanical harvesters are common in many fruit crops, such as walnuts and olives. The first apparatuses for dates were designed for harvesting of dry fruit by completely shaking it from the bunch. However, in the last 15 years, it was adapted specifically for Medjool and is now used in many orchards, mainly in the southern regions where the fruits tend to be dryer and firmer. To prevent damage to the delicate soft Medjool fruit, special sheets or nets were designed to collect the falling fruit (Fig. 8.7). The operator controls the time and strength of the shaking and applies it gradually in order to shake off the ripened fruit but keep the unripe (yellow khalal and rutab fruit) on the bunches. Different variations and apparatuses were designed. In most of them, the shaking unit is attached to the special cranes

Fig. 8.6 High cranes used in date palm orchards in Israel. These machines are designed to work at various maximal heights of 11–20 m but can efficiently work also on shorter trees (**b**). They are either based on a flexible structure (**a–b, d**) or carry the platform on an extendable arm **c**. For the safety of the workers, the operator platform can conveniently approach the date below the crown (**a**) and then close to completely encircle the date tree trunk (**b**, **d**) to form a safe 360° working platform. Additional equipment, such as large bags for harvested fruit, hydraulic pruning tools, or mechanical sprayers (**e**), can be attached to these platforms and safely operated by the crew

and is elevated close to the fruit bunches. The fruits are collected directly, into collection units or conveyers on the cranes (Fig. 8.7), and then semiautomatically or automatically transferred to the one fruit layer picking trays for transport to the packing house. This method requires that the fruit bunches not be covered. Some fruits fall and are lost between harvestings. In order to limit fruit fall, frequent

Fig. 8.7 Mechanical harvesters used in Medjool date palm orchards in Israel. The shaker units (**a**) are usually attached at the end of the high cranes, applying the shaking on the trunk close to its top, near the fruit bunches (**b**–**d**). Fruits can be collected in individual nets covering each bunch (**b**), in special nets below the entire crown surrounding the trunk of each tree (**c**), or in collection units attached to the platform of the high cranes, closing automatically around each tree crown, while shaking the trunk (**d**–**e**). Fruit can be transferred to picking trays on conveyors (**d**–**e**)

harvesting, every couple of days, is performed. By this approach each Medjool tree may be harvested up to 6–8 times in a season. In attempts to prevent losses due to fruit shedding, some units were designed in which the fruits are shaken into large bags which are then collected manually (Fig. 8.7). In these systems, the tree shakers and the actual harvesting and fruit collection are done by different machines. Albeit the great advantages of the mechanical harvesting, it has several limitations: the requirement for the frequent harvest of each tree, the loss of fruits between passages, and the harvesting of some fruits that are not completely ripe. Losses can reach up to 5 % of the yield.

8.7.4 *Mechanization for Pruning of Leaves*

Older leaves can scratch and damage the fruit bunches and limit access to the crown from the crane platform. Removal of the leaves, as well as the older fruitstalks, is part of the orchard management. Usually, this task is performed together with the supporting and tying of the bunches to leaf bases. As much as 60 m^3 dry plant material per hectare is annually removed, requiring approximately 15 days of labor. This material can be mulched and mechanically removed from the orchard (Sagi et al. 2010) or, in orchards with higher precipitation, turned into organic fertilizer; burning the old leaves is no longer allowed.

8.7.5 *Postharvest Operations*

8.7.5.1 Packing Houses

Packing houses serve as intermediaries between growers and consumers. The focus of Israeli industry on export into European markets makes international sanitation and food and health standards extremely important. Almost all Israeli date packing houses comply with the highest standards imposed by the European markets as well as standards demanded by the main consumer chains. These include BRC (British Retail Consortium) and GLABALGAP for the growers. This is especially true for Medjool due to its higher water content and, therefore, higher sensitivity.

The role of the packing houses is to prepare the fruit for the markets according to the specific requirements of each retailer. Procedures include disinfection, washing, drying of fruits of higher water content to improve storability, sorting according to both size and quality, and packing. For long-term storage, the fruits are cooled and stored under very cold temperatures (−18 to −21 ° C).

Fruit are packed in a large variety of boxes, ranging from small packages of 100–200 g, up to bulk ones with 5 kg (11 pounds). Packages are adapted according to the demands of the retailers.

8.7.5.2 Postharvest Treatment for Barhi Fresh Dates

Unlike other date cultivars in Israel, Barhi is harvested and consumed in the khalal stage. Therefore, Barhi is treated in the packing house as a fresh fruit. Barhi harvesting has to be precisely on time. Two contradicting processes determine harvesting time—on the one hand the consumer demand to get fruits with the highest sugar levels and on the other the conception that only hard khalal fruits are salable. Harvest is usually performed on the ideal stage of each individual bunch. Usually each tree is harvested 3–4 times, and only the bunches that are ready will be collected. Since fruits do not ripen uniformly on the bunch, upon harvest, any green fruit that has not reached minimal sugar level as well as any fruit starting the ripening stage, showing brown regions of softening on its skin, are removed.

The Israeli minimal diameter for export quality Barhi fruit is 27 mm. Barhi is usually packed on its strands in ventilated 5 kg carton packs. Export is either by ship or air. Under these conditions, fruit storability is several weeks. It was shown that coating Barhi fruit with wax prolonged its khalal stage, with its slight astringency, yellow color, and fresh appearance for up to 9 weeks at 1 °C (Pesis et al. 2005).

8.7.5.3 Postharvest Treatment for Medjool Semidry Dates

Israel is one of the largest producers and exporters of Medjool fruit. As of 2012, out of approximately 13,000 mt of date fruit exported, 9,000 were Medjool. The higher demands, prices, and quality of the Medjool fruit require special care and postharvest treatments in the packing houses. In the orchards, fruits are harvested directly into special trays, with only a single layer of fruits. This minimizes fruit damage and squeezing, by the weight of additional fruit layers. These special trays allow better ventilation of each fruit. Since Medjool is marketed as a semidry fruit with a water content ranging from 22–26 %, postharvest treatments, including selective drying of fruit with higher moisture content, or hydration of fruit that are too dry, are required. The design of the trays enables convenient application of such treatment in environmental (temperature and humidity) controlled rooms.

The high retail prices of Medjool require sorting to different categories and sizes. Skin separation and size are the two most important criteria for grading qualities and prices. An additional physiological disorder occurring through prolonged storage is sugar leakage and its crystallization on the skin (Shomer et al. 1998). Special efforts are made to control long-term storage and reduce these phenomena. It was found that long-term quality preservation of semidry Medjool dates is related to the level of fruit cell membrane integrity (Borochov-Neori et al. 2009; Shomer et al. 1998). Freezing and storage at extremely low temperatures (between −35 and −50 ° C) reduce the damage. At higher temperature rupture of cellular compartments occurs, whereas in fruits frozen at lower temperature, membranes remain intact even after prolonged storage (Shomer et al. 1998). Further, fruit storage potential is higher in fruits harvested early during the season (Borochov-Neori et al. 2009).

8.7.5.4 Development of Automatic Sorting Systems

Advances in sorting technologies of fruits and vegetables did include development of special systems for sorting date fruits. Some dates are harvested in the khalal stage (Barhi, Hayany); others are harvested and packed still on their spikelets (Deglet Noor, Barhi). Production from most additional cultivars in Israel is small and sold in bulks or small packages. However, the focus of the Israeli industry on Medjool required the development of sorting systems specifically for this cultivar. The small size of dates, relative to other sorted fruit and vegetables, required specific solutions, including improving singularity of the fruits so only single fruits will be loaded on the weighing cups and much higher accuracy of the weighing systems, within grams for each fruit. Both dynamic weighing systems, allowing higher speed of sorting, and static weighing systems with higher accuracy were developed. The soft and sticky nature of the Medjool fruit required special adaptations, treatments to enable easy movement without damage to the fruit and extensive cleaning of the system. High-speed weighing and sorting systems enable sorting of individual fruits according to their weight. These systems increased the capacity of the packing houses, enabling sorting of dozens of tons daily. They also increased uniformity of fruit in each package, permitting development of defined categories and improving farmers' profitability.

Sorting for quality is especially important in Medjools, where external quality affects fruit price. High-quality fruit, without skin separation, will sell for more than twice the price of similar fruit that has skin separation. Several approaches were used to develop algorithms to identify fruit skin areas that are separated from the flesh. Special design of the sorting system allows imaging of each fruit from all sides to define the percentage of the phenomena. The current sorting systems can now sort Medjool fruit by the combination of weigh and quality.

Several additional approaches were developed in Israel for sorting date fruits according to their firmness (Schmilovitch et al. 1995) and maturity indexes. The use of near-infrared (NIR) spectroscopy enabled detecting the moisture content of whole date fruit (Dull et al. 1991; Schmilovitch et al. 1999). A sorting system based on NIR spectrometry was developed to give estimates of the TSS content of khalal Hayany dates in a nondestructive and labor-saving manner. The same machine could also sort Barhi fruit according to their water content (Hoffman et al. 2000). Yet another approach was developed for determining water content of individual semidry Medjool fruit using the dielectric effect using electrodes that gently touch the fruit and form a unique capacitor. An RF signal transmitted through the fruit and its characteristics are correlated to the fruit water content (Schmilovitch 2006).

8.7.6 Survey of Commercial Producers and Major Farms

The Israeli date industry annually produces approximately 35,000 mt. About 15,000 mt are exported, and the rest is directed toward local markets and industrial products. The average annual date consumption in Israel is about 1.5 kg per capita. This is a relatively low level compared to other date-growing countries in the Middle East, where consumption can reach as high as 10 kg per capita. Most marketing efforts are focused on the main cultivar Medjool which comprises two-thirds of the production.

8.7.7 Local Marketing Status

As in most Western countries, the market share of the main chain stores in Israel has increased. This has required adaptation of packages to the retailer size. Most dates are marketed in either 400 g or 1 kg packages, enabling families to purchase fruit according to their consumption. Open markets still have a large share in local date marketing. Dates are sold in stores specializing in a variety of dry fruits. It is important to mention that in the dry fruit category, the date is usually the only fruit that is sold in its natural condition, without any addition of conservatives, sulfur, and other chemicals added in many other dry fruit like raisins and apricots. Promotions are made by the date growers association without mentioning any brand names and usually not focusing on specific cultivars. Commercials focus mainly on introduction of the health additive values of the date fruit and on dates as a natural snack and as energy bar for athletes. Most public relations are performed as general and health-focused articles in Internet sites, by the Israeli date growers association.

8.7.8 Current Export

Israeli date fruits are exported to about 30 countries throughout the five continents. Two-thirds are exported to Western Europe, 11 % to the American continent, and the rest to the Middle East, Eastern Europe, Southeast Asia, and Oceania. Almost all export is performed in large temperature-controlled containers, by ship. The Israeli date growers association invests hundreds of thousands of dollars in promotion to open new markets and expand existing ones for the Israeli date fruit.

Two-thirds of the date fruit export is Medjool. As in local markets, a major part of the fruit is exported to the main chain stores in Europe. Sets of measures and regulations, some general in nature and others assigned by specific chains such as Nurture for Tesco and Field to Fork (F2F) for Marks and Spencer, are strictly enforced. Another demand of the chain stores is withstanding ethical standards informing the consumers that the products were grown under appropriate conditions and that all workers involved in the growth in the orchard, packing, and shipping in the packinghouses were managed accordingly. Most growers and packinghouses are required to be listed in SEDEX site, covering all aspects of labor according to state laws and specific requirements of costumers. Export of date to open markets is also important. It tends to peak approaching the main holidays of the Christians (Christmas and New Year) and Muslims (Eid al Fitr, Eid al Adha, and the month of the Ramadan).

The main Israeli date local and export marketing company is Hadiklaim, a date growers' cooperative that markets fruit of both its partners and allies. Hadiklaim growers cultivate 60 % of the Israeli orchards. The cooperative owns fifteen packinghouses and a central marketing unit. Unlike other companies, Hadiklaim specializes only in marketing date fruit. The second important date marketing company is Mehadrin, managing approximately 25 % of the Israeli fruit. However, this company is marketing many additional fruits and focuses mainly on citrus.

8.7.9 *Nutritional and Health Benefit Aspects of Date Palms*

The beneficial health characters of dates have been promoted in traditional medicine. The fruit composition and nutrition effect is of interest to both the industry and the consumers (Vayalil 2011). Recent studies performed in Israel have characterized the chemical composition, especially the polyphenol composition of the various date cultivars grown in Israel (Borochov-Neori et al. 2013; Rock et al. 2009). These studies also explored the health benefits of date eating. Despite their high sugar content, date consumption did not significantly affect the subjects' body mass index (BMI) or their cholesterol levels and reduced serum oxidative status. It was concluded that date consumption can be beneficial for healthy consumers (Rock et al. 2009). Ethanol and acetone extracts from fruit of different cultivars contained considerable amounts of soluble phenolic compounds and inhibited LDL oxidation, and most extracts also stimulated cholesterol removal from macrophages. Most cultivars also exhibited antiatherogenic potency (Borochov-Neori et al. 2013).

8.8 Processing and Novel Products

8.8.1 *Development of Perfectly Ripe Medjool*

In order to minimize skin separation, attempts were made to generate a new and exclusive Medjool product, based on rutab fruit. The product is called a *fresh, perfectly ripe* Medjool. These fruits are harvested before drying and have a much higher water content (28–32 % and even higher). They have a soft fleshy texture and a fruity-honey taste. They are extremely soft, sticky, and delicate. Since these fruits have lost only small amount of water, they have smooth skin with very few wrinkles (Fig. 8.8). To harvest most fruits in an orchard as perfectly ripe Medjool requires development of different growing protocols, including modifications of fruit thinning and fertilization. Harvesting has to become very frequent, requiring growers to access and harvest from every bunch in all trees every few days. Fruits have to be collected manually (and very carefully), requiring intense labor. Unlike the semidry Medjool the perfectly ripe fruit have a short shelf life and is perishable. Research done in Israel to improve storability and shelf life of the perfectly ripe Medjool has led to a unique premium product that can be stored for several months

8.8.2 *Growing Organic Dates*

In the past 20 years, there are demands in the European markets for dates that are grown organically. More than 200 ha are currently growing organic dates and are supervised by international boards. However, growing organic dates raises special problems and concerns. The use of chemical fertilizers is not

Fig. 8.8 Fresh, perfectly ripe Medjool. The delicate fruit is soft and sensitive. Specific designs of packages for the perfectly ripe Medjool where individual fruit is placed in protective cases (**a**) and in bulk cartons where fruits are separated into single layers and protected from rolling over each other (**b**, **c**)

permitted in the organic orchards. Fertilization is therefore based on compost, requiring transport and dissemination of the compost around each of the trees and modification of the irrigation system. Specific challenges are also related to plant protection and pest control. The limited number of registered and allowed pest control measures in the organic date palm puts the yield under permanent threat. The lesser date moth and date mites are especially hard to control in an organic farm. Currently, there is no effective tool for the treatment of the red palm weevil in organic orchards, and its expansion to the date orchards in Israel threatens future organic date production. The ban on use of herbicides requires allocation of sufficient workers for weed control. It is usually done by using mechanical trimmers. Covering the ground with shredded dry leaves reduces weed germination and growth.

Since organic food customers are driven by idealism, special strict rules are observed during postharvest as well. To prevent any chance of mixing or contamination with residual pesticides, organic date fruits have to be packed and marketed separately and cannot be stored with regularly grown dates. Although the organic market is limited in volume and usually does not pay higher premiums than regularly grown dates, it demands a lot of effort from farmers.

8.8.3 Sylan (Date Syrup) and Other Processed Products

A branch of the date industry has been developed that utilizes the lower quality fruits and the demand for traditional products. In Israel the date industry produces pitted dates, which following the removal of the pits are pressed and vacuum packed, date spread cooked with sugars and additives and traditional natural sylan (rub date syrup) which is cooked and squeezed date syrup. They all serve mainly for baking and cooking.

8.8.4 Date Trees for Ornamentals and Landscaping

Being a Mediterranean country, palm trees have a major impact on landscaping in Israel. Several palm species are very common in gardens, including Canary Islands date palms (*Phoenix canariensis*) and *Washingtonia* fan palms (mainly *W. robusta*). Date palms are also very popular. It is estimated that at least 500,000 date palm trees are planted for landscaping or in private gardens. In the public sectors, date palms are the most popular tree for landscaping on boulevards, junctions, and parking lots. Requirements are usually for uniform trees with trunks of 3–4 m height. About 5,000 trees are annually uprooted and transferred into suburban regions (Fig. 8.9a–b). Some farmers prefer to work with relatively short trees, reducing the cost of machinery to reach the trees' high crowns. They grow the date palms for only 10–12 years before their uprooting and allocation to the public sector. The use of date trees for landscaping also encourages the replacement of cultivars and replanting of plots.

Fig. 8.9 Other uses of date palms in Israel. Uprooting of a date orchard in Northern Israel (**a**) and transport of the trees for landscaping (**b**). Storage of lulavs—young spear leaves—emerging date leaf tips (**c**) and Lulav packing (**d**, **e**) for the ritual of the four species at the Jewish holiday of Sukkot

In an attempt to stop the expansion of red palm weevils, the Israel Plant Protection services have imposed quarantine regulations for several years on date palms transportation from infected areas. Today, when most regions in Israel are infected, this quarantine is no longer in effect. However, the risk of trees falling in urban parks and boulevards threaten the continuation of landscaping with date palms.

8.8.5 Lulav: The Apical Closed Date Leaf as a Special Jewish Date Ceremonial Product

The date tree is important for the three main monotheistic religions. Date leaves serve for ceremonies and rituals in these religions. Lulav, the apical closed leaf of the date (spear leaf), is an important Jewish religious artifact. It is one of the four species used for blessings during a ceremony of the holiday of Sukkot. These include, together with the lulav, etrog (citron fruit), and branches of hadass (myrtle tree) and arava (willow tree). The waving of the four species is a mitzvah prescribed by the Torah and contains symbolic allusions to a Jew's service of God. As a part of a religious ceremony, the four species, including the Lulav, have to be perfect. The lulav should be straight, not bent or broken, and tightly closed, with its entire leaflet staying closely together. An industry of date spear leaves was established (Fig. 8.9c–e). As of 2013 approximately 700,000 lulavs are annually sold within Israel, and 200,000 more are exported (mainly to the USA). Protocols for postharvest lulav conservation have been developed, and currently they can be stored for 6 months, enabling up to 10 lulavs harvested from a single date tree.

Any closed and straight date leaf can serve as a lulav. However, the spear leaves of the Dayri cultivar are long and straight, and its leaflets stay bound to each other. Therefore, it became the perfect lulav and recent planting of Dayri orchards has dramatically expanded; while only 1,600 Dayri trees were planted in Israel in 1975, more than 35,000 trees are currently (2014) being cultivated.

8.9 Conclusions and Recommendations

The date palm is one of the main fruit crops grown in Israel and is very important for the economy of its arid regions. Although the Israeli industry is rather small, it is a profitable and expanding industry. Currently, the industry is focused on a single cultivar, Medjool, and approximately 70 % of the trees in the commercial orchards are of this cultivar. Since many of the trees are still young, Israeli Medjool yields are expected to increase and double within the next 5 years.

The Israeli date industry contributes only a small volume of fruits to international markets. However, it has become a leader in global markets. A significant share of the exclusive Medjool dates in the European market is exported from Israel.

Medjool is not harvested in Israel as a fully dried tamar date but rather as a semidry, succulent fruit. It is marketed as a unique, especially large, delicate, and perfectly looking fruit. This approach makes Medjool a different category of dates, gaining very high prices. This strategy requires modification of growth protocols, strict and massive fruit thinning, and selective, repetitious, and very careful harvesting. Additionally, sorting, storage, and postharvest techniques are implemented to ensure the quality of the fruits in the markets.

Date cultivation in Israel had become a highly technological industry. Research, extension specialists, farmers, and private companies work together to give better, more efficient tools for the date grower and industry. Special equipment enabling efficient and safe work in the heights, mechanization, and automation as well as precision agriculture are being employed. This is evident in any aspect of date growing, harvesting, and postharvesting technologies. These novel tools will lead the Israeli date industry in the near future.

References

Adhikari KB (2010) Study of biotextural behavior in Medjoul date (*Phoenix dactylifera* L.) fruits. Hebrew Univ., Rehovot

Al Kaabi HH, Zaid A, Ainsworth C (2007) Plant-off-types in tissue culture-derived date palm (*Phoenix dactylifera* L) plants. Acta Hort 736:267–281

Aloni DD, Hazon H, Edom U et al (2010) Effects of growth retardants on vegetative growth of date palms. Acta Hort 884:207–213

Bechar A, Shmulevich I, Eisikowitch D et al (1999) Modeling and experiment analysis of electrostatic date pollination. Trans ASAE 42(6):1511–1516

Ben-Bassat D, Bernstein Z, Israeli Y (1996) Plant regeneration from cultured immature inflorescences of date palm obtained from mature trees in situ. Paper presented at the Elche international workshop on date palm cultivation in oasis agriculture of Mediterranean Countries. Zagarosa, 25–27 Apr 1995

Ben-Yehoshua S, Ben-Yehoshua LJ (2012) Ancient dates and their potential use in breeding. In: Janick J (ed) Horticultural reviews, vol 40. Wiley, Hoboken, pp 183–213

Bernstein Z (2004) The date palm. Israeli Fruit Board, Tel Aviv (in Hebrew)

Blumberg D (2008) Review: date palm arthropod pests and their management in Israel. Phytoparasitica 36(5):411–448

Borochov-Neori H, Lutzki B, Judeinstein S, Shomer I (2009) Seasonal variations in cell membrane properties and long-term quality preservation of semi-dry 'Medjool' date fruit. In: VI international postharvest symposium 877, pp 195–200

Borochov-Neori H, Judeinstein S, Greenberg A et al (2013) Date (*Phoenix dactylifera* L.) fruit soluble phenolics composition and anti-atherogenic properties in nine Israeli varieties. J Agric Food Chem 61(18):4278–4286

Chao CT, Krueger RR (2007) The date palm (*Phoenix dactylifera* L.): overview of biology, uses, and cultivation. HortSci 42(5):1077–1082

Cohen Y (2011) Molecular detection of somaclonal variation in date palm. In: Jain SM, Al-Khayri JM, Johnson DV (eds) Date palm biotechnology. Springer, Dordrecht, pp 221–235

Cohen Y, Korchinsky R, Tripler E (2004) Flower abnormalities cause abnormal fruit setting in tissue culture propagated date palm (*Phoenix dactylifera* L.). J Hortic Sci Biotechnol 79(6):1007–1013

Cohen Y, Gurevich V, Korchinsky R et al (2007) Molecular and phenotypic characterization of somaclonal variation in date palm off-types originated from tissue culture. Acta Hort 738: 417–423

Cohen Y, Freeman S, Zveibil A et al (2010) Reevaluation of factors affecting bunch drop in date palm. HortSci 45(6):887–893

Cohen Y, Alchanatis V, Prigojin A et al (2012) Use of aerial thermal imaging to estimate water status of palm trees. Precis. Agric. 13:123–140

Cohen Y, Aloni DD, Edom U et al (2013) Characterization of growth-retardant effects on vegetative growth of date palm seedlings. J Plant Growth Regul 32:533–541

Dekel D, Carmeli D, Strom M, Cohen Y (2012) Use of pollen water suspension and reduced pollen viability for pollination and fruit thinning of date palm trees. Alon Hanotea 66(1):34–38 (In Hebrew)

Dull GG, Leffler RG, Birth GS et al (1991) The near infrared determination of moisture in whole dates. HortSci 26(10):1303–1305

FAOSTAT (2013) Crop Production, Statistics Division. Food and Agriculture Organization of the United Nations, Rome. http://faostat.fao.org/

Freeman S, Maymon M (2000) Reliable detection of the fungal pathogenfusarium oxysporum f. sp. albedinis, causal agent of bayoud disease of date palm, using molecular techniques. Phytoparasitica 28(4):341–348

Gan-Mor S, Ronen B, Vaaknin Y et al (2009) Further studies on electrostatic date pollination-from the laboratory bench to field unit performance test. Appl Eng Agric 25(5):643–646

Glasner B (2004) Growing dates in Israel. Fruit Gardener 36:20–23

Gophen M (2014) Skin separation in date fruits. Int J Plant Res 4:11–16

Gurevich V, Lavi U, Cohen Y (2005) Genetic variation in date palms propagated from offshoots and tissue culture. J Am Soc Horticult Sci 130(1):46–53

Hoffman A, Egozi H, Ben-Zvi R, Schmilovitch Z (2000) Machine for automatic sorting 'Barhi' dates according to maturity by near infrared spectrometry. Acta Hort 553:481–486

Israel Cent Bur Stat (2013) Agriculture Statistics. Tel Aviv, Israel. Available at: http://147.237.248.50/reader/?MIval=cw_usr_view_SHTML&ID=430

Kehat M (1999) Threat to date palms in Israel, Jordan and the Palestinian Authority by the red palm weevil, Rhynchophorus ferrugineus. Phytoparasitica 27(3):241–242

Levi-Zada A, Fefer D, Anshelevitch L et al (2011) Identification of the sex pheromone of the lesser date moth, *Batrachedra amydraula*, using sequential SPME auto-sampling. Tetrahedron Lett 52(35):4550–4553

Levi-Zada A, Sadowsky A, Dobrinin S et al (2013) Reevaluation of the sex pheromone of the lesser date moth, *Batrachedra amydraula*, using autosampling SPME-GC/MS and field bioassays. Chemoecology 2(1):13–20

London A (2003) Date palm plantations in Israel during the Second temple. The Mishnah and the Talmud Bar-Ilan University, Dissertation presented for an M.Sc. degree (in Hebrew)

Lustig I, Bernstein Z, Gophen M (2014) Skin separation in Majhul fruits. Int J Plant Res 4(1):29–35

McCubbin MJ, Van Staden J, Zaid A (2000) A Southern African survey conducted for off-types on date palms produced using somatic embryogenesis. In: The date palm international symposium, Windhoek, pp 68–72

Nadler A, Raveh E, Yermiyahu U, Lado M, Nasser A, Barak M, Green S (2008) Detecting water stress in trees using stem electrical conductivity measurements. Soil Sci Soc Am J 72(4): 1014–1024

Nakash J, Osem Y, Kehat M (2000) A suggestion to use dogs for detecting red palm weevil (Rhynchophorus ferrugineus) infestation in date palms in Israel. Phytoparasitica 28(2): 153–155

Palevsky E, Ucko O, Peles S et al (2004) Evaluation of control measures for *Oligonychus afrasiaticus* infesting date palm cultivars in the Southern Arava Valley of Israel. Crop Prot 23(5):387–392

Pesis E, Feygenberg O, Ben Arie R et al (2005) Extending the storage life of date cv. Barhi. Alon Hanotea 59:18–19 (in Hebrew)

Pinhas J, Soroker V, Hetzroni A et al (2008) Automatic acoustic detection of the red palm weevil. Comput Electron Agric 63(2):131–139

Reuveni O (1970) Pistil receptivity of'Khadrawi','Zahidi'and'Deglet Noor'date flowers. Report 47th annu Date Grs' Inst 47:3–4
Reuveni O (1979) Embryogenesis and plantlets growth of date palm (*Phoenix dactylifera* L.) derived from callus tissues. Plant Physiol 63(Suppl):138
Reuveni O, Lilien-Kipnis H (1974) Studies of the in vitro culture of date palm (*Phoenix dactylifera* L.) tissues and organs. Volcani Institute. Agricultural Research Pamphlet Number 145:1–40
Reuveni O, Adato I, Lilien-Kipnis H (1972) A study of new and rapid methods for the vegetative propagation of date palms. Date Growers' Inst 49:17–24
Rock W, Rosenblat M, Borochov-Neori H et al (2009) Effects of date (*Phoenix dactylifera* L., Medjool or Hallawi Variety) consumption by healthy subjects on serum glucose and lipid levels and on serum oxidative status: a pilot study. J Agr Food Chem 57(17):8010–8017
Sagi I, Kashti Y, Ziv R et al (2010) Development of a method and machinery for collecting pruned palm tree branches. Israel Agriculture. http://www.israelagri.com/?CategoryID=397&ArticleID=651
Sallon S, Solowey E, Cohen Y et al (2008) Germination, genetics, and growth of an ancient date seed. Science 320(5882):1464
Schmilovitch Z (2006) Determination of single-date water content by a novel RF device. Appl Eng Agric 22(3):401–406
Schmilovitch Z, Zaltzman A, Hoffman A, Edan Y (1995) Firmness sensor and system for date sorting. Appl Eng Agric 11:554–560
Schmilovitch Z, Hoffman A, Egozi H et al (1999) Maturity determination of fresh dates by near infrared spectrometry. J Sci Food Agric 79(1):86–90
Shomer I, Borochov-Neori H, Luzki B, Merin U (1998) Morphological, structural and membrane changes in frozen tissues of Madjhoul date (*Phoenix dactylifera* L.) fruits. Postharvest Biol Technol 14(2):207–215
Soroker V, Nakache Y, Landau U et al (2004) Note: Utilization of sounding methodology to detect infestation by Rhynchophorus ferrugineus on palm offshoots. Phytoparasitica 32(1):6–8
Soroker V, Blumberg D, Haberman A, Hamburger-Rishard M et al (2005) Current status of red palm weevil infestation in date palm plantations in Israel. Phytoparasitica 33(1):97–106
Soroker V, Haberman A, Nakache Y et al (2013) History of red palm weevil management in Israel 1999–2012. Paper presented at the AFPP – palm pest mediterranean conference, Nice
Sperling O, Shapira O, Cohen S et al (2012) Estimating sap flux densities in date palm trees using the heat dissipation method and weighing lysimeters. Tree Physiol 32(9):1171–1178
Stoler S (1977) Date palm growing in Israel. Hakibutz Hameuchad, Tel-Aviv (In Hebrew)
Tripler E, Ben-Gal A, Shani U (2007) Consequence of salinity and excess boron on growth, evapotranspiration and ion uptake in date palm (*Phoenix dactylifera* L., cv. Medjool). Plant Soil 297(1):147–155
Tripler E, Shani U, Mualem Y, Ben-Gal A (2011) Long-term growth, water consumption and yield of date palm as a function of salinity. Agric Water Manag 99:128–134
Vayalil PK (2011) Date fruits (*Phoenix dactylifera* Linn): an emerging medicinal food. Crit Rev Food Sci Nutr 52(3):249–271
Zaid A, Al Kaabi H (2003) Plant-off types in tissue culture-derived date palm (*Phoenix dactylifera* L.). Emir J Agric Sci 15(1):17–35
Zohary D, Spiegel-Roy P (1975) Beginnings of fruit growing in the old world. Science 187(4174):319–327
Zohary D, Hopf M, Weiss E (2012) Domestication of plants in the Old World: the origin and spread of domesticated plants in Southwest Asia, Europe, and the Mediterranean Basin, 4th edn. Oxford University Press, Oxford

Chapter 9
Date Palm Status and Perspective in Kuwait

Chellan Sudhersan, Jibi Sudhersan, Jawad Ashkanani, and Lateefa Al-Sabah

Abstract Date palm (*Phoenix dactylifera* L.) has been one of the major agricultural crops of Kuwait for over 90 years; however, large-scale cultivation began only 25 years ago. The Kuwait Institute for Scientific Research (KISR), Public Authority for Agriculture and Fish Resources (PAAFR), and Kuwait University are currently involved in date palm research in Kuwait. Traditional propagation methods by seed and offshoot were practiced initially for cultivation. Tissue culture propagation was developed at KISR in 1995 and the technology used for commercial-scale clonal plant production since 2000. Research activities including date palm cultivar introduction and evaluation, cultivation practices, irrigation and fertilization, pest and disease management, fruit production and postharvest technology, and germplasm maintenance were carried out. Major date palm biotechnological research also is being undertaken. Germplasm maintenance, micropropagation of elite cultivars for farmers, cultivar identification through DNA technology, pest and disease management, crop improvement, irrigation technology improvement, and biodiversity conservation of date palm are in practice. Approximately 601,563 trees are planted in 4,181 registered farms located in the Abdhally, Wafra, and Sulaibia regions of Kuwait. Six major cultivars are used for fruit production with Barhi ranking first in the number of trees. Current annual date production is about 45,000 mt. Monoculture is spreading in Kuwait posing a threat to date palm diversity. A national date palm award was developed by KISR to encourage farmers to diversify planted cultivars.

Keywords Biodiversity • Conservation • Cultivation • Date palm • KISR micropropagation • PAAFR

C. Sudhersan (✉) • J. Sudhersan • J. Ashkanani • L. Al-Sabah
Biotechnology Program, Environment and Life Sciences Research Center Kuwait Institute for Scientific Research, Box 24885, Safat 13109, Kuwait
e-mail: schellan@kisr.edu.kw; sjibi@kisr.edu.kw; neenmod@yahoo.com; lsabah@kisr.edu.kw

© Springer Science+Business Media Dordrecht 2015

J.M. Al-Khayri et al. (eds.), *Date Palm Genetic Resources and Utilization: Volume 2: Asia and Europe*, DOI 10.1007/978-94-017-9707-8_9

9.1 Introduction

In Kuwait, date palm (*Phoenix dactylifera* L.) was introduced from neighboring countries about 90 years ago. The first known introduced cultivar, Braim (Fig. 9.1), was planted in Kuwait in 1920 and is still alive. The ancient peoples of Kuwait had knowledge of date palm cultivation; however, the delay in cultivation was due to the lack of irrigation water. In ancient times, pearl fishing was the major source of income for Kuwaiti families. When pearl fishing declined after 1950, date palm cultivation was initiated for income generation and livelihood. Since 1810, 45 Kuwaiti families have owned date palm farms in Basra, Iraq, with a total area of 46,632 ha planted with 1,865,000 date palm trees (Al-Nisf 2010).

Date palm is given priority in the agriculture system, greenery, and landscape programs in Kuwait due to its high level of tolerance to the harsh arid environment and elevated soil salinity. Date palm is cultivated mainly in the Wafra, Abdhally, and Sulaibia agricultural areas. In addition, it is also grown on farms at Al-Jahra, in residential areas, and along beaches and roadsides all over Kuwait for its fruits and ornamental value (Fig. 9.2). Date palms are the main fruit crop in Kuwait, and about 601,563 date palm trees are planted on 4,181 registered farms (information from PAAFR in 2014). Of the total number of date palms, 386,273 belong to six major cultivars, namely, Barhi, Khalas, Medjool, Nebut Seif, Suckari, and Um Al-Dehn, for fruit production (PAAFR 2003). However, a total of 40 cultivars are reported growing throughout Kuwait (Al-Mudaires 1992).

Due to the recent expansion for commercial production, annual date fruit production in Kuwait has increased from 7,894 mt in 1999 to 15,610 mt in 2003 and 45,000 mt in 2012 (FAO 2013). The date fruit production increase was mainly due to the plantations recently established with micropropagated palms. The unavailability

Fig. 9.1 Date palm cultivar Braim

Fig. 9.2 Kuwait map

of good quality date palm planting material to establish commercial plantations in Kuwait was the major constraint on date fruit production a decade ago, but the problem was solved through developing micropropagation technology. The current annual date production of 45,000 mt. is insufficient to meet domestic demand. Therefore, date fruits are imported, mainly from Saudi Arabia.

Currently, the date cultivation and fruit production in Kuwait is affected mainly by soil salinity, insufficient fresh irrigation water, climate change, pests and diseases, high cost of production, lack of trained manpower for appropriate cultivation practices, and poor postharvest technology.

9.2 Cultivation Practices

Kuwait is a desert country with a harsh and prolonged summer season extending from April to October. The average monthly temperature in summer is 46.2 °C, whereas the winter is mild with an average monthly temperature of 6.9 °C. Rainfall is highly erratic, averaging about 110 mm/yr. Strong winds prevail during the

Fig. 9.3 Date palm tissue culture laboratory at KISR

summer months (MOP 2006). Date palms occupy a unique place in the agriculture systems in Kuwaiti farms and greenery programs due to their high tolerance to adverse climatic conditions. The palm is cultivated extensively in Kuwait for fruit production and ornamental value. Commercial date palm plantations have recently begun to emerge in Kuwait. A 1987 survey in Kuwait indicated that there were about 275,000 date palms, with about 56 % being located in home gardens, 34 % on farms, 1 % at beach houses, and 9 % in public parks. Of the total, only 40 % were at the fruit production stage, and the remaining 60 % were at young growth stages. The total annual production in 1987 was 1,500 mt. A major constraint on commercial plantations was the lack of good quality planting materials on a large scale. A few local nurseries imported tissue culture-derived date palms from the UK and France; the majority of the imports were Barhi cv. Other cvs. such as Khalas, Medjool, Nebut Seif, and Suckari were imported in small quantities. Commercial plantations using tissue culture-derived plants began in 1995 after the Gulf War. In order to overcome the shortage in commercial cultivars for production, KISR gave priority to research and development of micropropagation technology development.

A research project on tissue culture technology development for the micropropagation of date palms was conducted in 1990–1995 at the KISR Biotechnology Department. Followed by the successful completion of this project, a tissue culture facility with laboratories, growth rooms, hardening rooms, greenhouses, and lath houses was established at the KISR main campus at Shuwaikh in 1995 (Fig. 9.3).

Intensive research and development studies on micropropagation of many date cultivars were carried out on a pilot scale in the newly developed tissue culture facility in 1996–2004. A unique culture media and protocol were developed for the

Fig. 9.4 Ten-year-old tissue-cultured date palm plantation

commercial production of clonal plantlets through somatic embryogenesis. A total of 30 cvs. (in alphabetical order), Al-Hammed, Al-Heifi, Anbarah, Ashgar, Awaidi, Barhi, Boyer, Braim, Dayri, Fard, Fersi, Garvis, Ghannamy, Hilali, Jouzi, Khalas, Khasab, Khyarah, Lolwi, Maktoomi, Medjool, Nebut Seif, Quantar, Shekar, Shiek Ali, Shishi, Siwi, Suckari, Sultana, and Um Al-Dehn, were produced via the KISR-owned micropropagation method and all cvs. maintained in two date palm orchards located on the KISR campus at Shuwaikh. All the cultivars produced by KISR were proven to be true to type through DNA fingerprinting and field evaluation (Al-Shayji et al. 1994; Sudhersan and AboEl-Nil 2004). After the pilot-scale plant production study, five major commercial cultivars, Barhi, Khalas, Medjool, Nebut Seif, and Suckari, were multiplied on a large scale and supplied to the farmers for the establishment of commercial plantations. Many plantations established in 2000–2005 with KISR tissue-cultured date palms are in production without exhibiting any genetic malformations (Fig. 9.4). Currently, KISR is capable of commercial-level clonal plant production of any date cultivar for the farmers in Kuwait or neighboring Gulf Cooperation Council (GCC) countries on a contract basis.

Large-scale cultivation of date palms started recently in Kuwait and is identified as a priority crop for commercial production in the Agriculture Master Plan of Kuwait (KISR 1995). The government is encouraging farmers to plant selected commercial cultivars by offering subsidies. Date palm farms are located at Wafra, Abdhally, and Sulaibia regions; there were 2,065 registered farms in Wafra, 1,990 in Abdhally, and 126 in the Sulaibia areas. Currently, about 601,563 date palms are growing in Kuwait (Personnel Communication from PAAFR 2014). Of the total

Table 9.1 The number of date palms in Kuwait

Cultivar	Total no. of trees
Barhi	318,785
Khalas	48,906
Suckari	10,929
Medjool	4,598
Nebut Seif	1,841
Um Al-Dehn	1,214
Other cvs.	194,714
Males	20,576

Table 9.2 Area of date palm cultivation and annual production in Kuwait in 1998–2011

Year	Area (ha)	Production (mt)
2011	5,099	33,562
2010	5,090	32,561
2009	4,665	29,849
2008	3,800	16,000
2007	3,200	16,000
2006	2,200	16,000
2005	2,000	16,000
2004	1,450	15,811
2003	1,350	12,577
2002	1,350	10,376
2001	1,350	10,155
2000	1,350	7,894
1999	1,050	6,484
1998	870	6,662

number of date palms, the majority are Barhi, followed by Suckari, Khalas, Medjool, Nebut Seif, and Um Al-Dehn (Table 9.1).

Cultivation area and annual production have increased gradually in Kuwait (Table 9.2, Figs. 9.5 and 9.6). The area under cultivation increased from 870 ha in 1998 to 5,099 ha in 2011, and annual production increased gradually from 6,662 mt in 1998 to 33,562 the same year (FAO 2013).

One of the major constraints to expanding date production in Kuwait was the lack of good quality planting materials of commercial cultivars. This problem was solved through the development of micropropagation technology for commercial production. Other constraints yet to be resolved are pest and disease management, soil salinity, insufficient fresh irrigation water, lack of modern postharvest technology, and trained manpower for cultivation (Abdul-Salam and Al-Mazrooei 2007).

Under Kuwait's climatic conditions, among 30 different cultivars tested at the KISR date orchard, Barhi, Medjool, Anbarah, Awaidi, Hilali, Khasab, and Siwi cvs. were found to be the highest yielding. Among them, Barhi ranked first and is well suited to Kuwait as compared to other cultivars. This cultivar has good economic

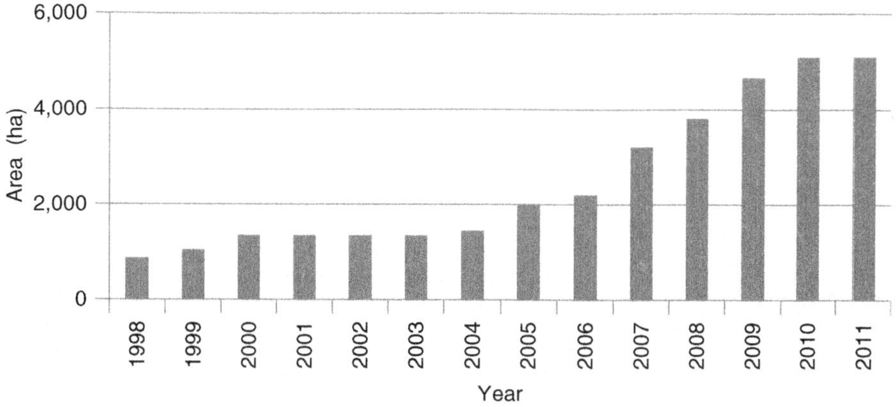

Fig. 9.5 Area of date palm cultivation in Kuwait in 1998–2011

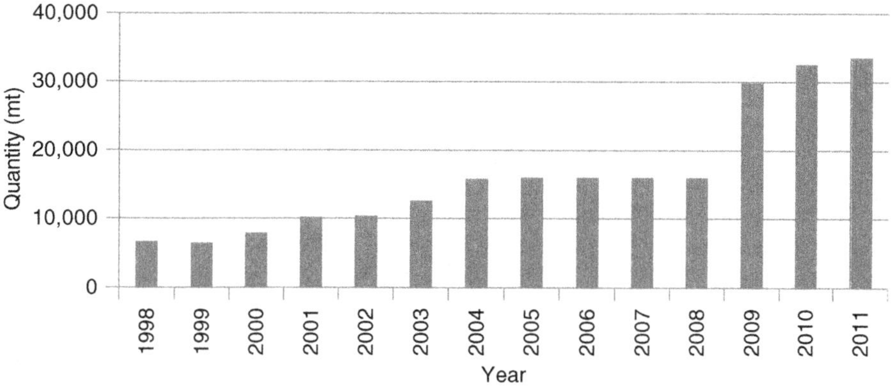

Fig. 9.6 Annual date fruit production in Kuwait in 1998–2011

returns when appropriate good management procedures are adopted and produces an average of 200 kg of fruit annually. However, such good yields cannot be achieved without proper pollination, pruning, irrigation, and fertilization.

Under Kuwait's harsh climatic conditions, potential evapotranspiration (ET_p) far exceeds annual precipitation. Soils are fragile, sandy with organic matter content below 0.1 % and low moisture and nutrient-holding capacities. Therefore, farmers must apply supplementary irrigation for crop production. The irrigation systems used for the date palms are flood (Fig. 9.7), bubbler, and drip irrigation. Although date palm cultivation is a viable agricultural option for Kuwait, current irrigation practices are very inefficient and unsustainable. Therefore, research studies were conducted to calculate the actual water requirements for date palms in Kuwait. The results of the study based on the annual evapotranspiration showed that the annual water requirement of date palms under Kuwait's environmental conditions ranged from 23,392 to 27,251 m^3/ha/y (Abdul-Salam and Al-Mazrooei 2007; Bhat et al. 2011).

Fig. 9.7 Intercropping with fodder crops under flood irrigation

Pruning of date palms is practiced throughout Kuwait and is the most important operation that is carried out regularly. Pruning removes old dead or broken leaves and areal shoots, locally known as *rakoob*. Two types of tools are used: a sickle which is a saw-toothed curved blade used to cut leaf bases and a slanting curved cutting blade with a slight outcurve attached to a wooden handle. The objective of pruning is to clean the tree, allow new leaves to grow and photosynthesize, reduce any rodent or insect infestations, facilitate harvesting, permit the use of leaves as a by-product material and utilize leaf-base fiber, and improve crop quality by reducing shade and avoid bruising of fruits. Thorns are also removed for pollination, bunch thinning, and fruit harvesting operations.

Several operations are performed after pollination and fruit set. These include thinning of heavily laden trees by the removal of unwanted bunches entirely or by bunch thinning, release of bunches from the tree crown, and protection from birds and rodents.

Fruit bunch thinning is commonly practiced in most date-growing regions of Kuwait in order to avoid alternative in fruit bearing and to enhance the fruit quality. Date fruit thinning is carried out by three ways: (a) reducing the number of bunches, (b) reducing the number of strands, and (c) reducing the number of fruits per strands. The common practice of fruit thinning is the removal of whole bunches followed by trimming the central strand of each allowed bunches using a sharp knife to reduce 30 % of the total number of strands. The number of bunches per tree is decided according to the age of the tree: 2–5 during 4–8 years old and 8–10 bunches above 10 years old.

Bunch hanging comes after the date fruits have developed but before the ripening. Three weeks after pollination, the fruit bunches are pulled under the lower leaves and tied to a leaf rachis with a rope or palm leaflets. The main purpose of bunch hanging is to protect the dates from being scratched or bruised by thorns or leaflets during windy conditions, to reduce fruit drop, and to facilitate harvesting.

Date palm yield depends on successful pollination. Male cv. Ghannamy is well known to Kuwaiti farmers, and the majority of farmers use its pollen on all types of cultivars. In addition, unknown seedling males are also used for pollination. Research studies are ongoing at KISR to identify superior male cultivars for specific female date palm cultivars to achieve high-quality date fruit production. Known male cultivars such as Boyer, Dayri, Fard, Ghannamy, Garvis, and a new male KW-1 were selected for the pollination study which involved Barhi, Khalas, Medjool, Nebut Seif, and Suckari cvs. Studies on the use of pollen of other *Phoenix* species (*P. pusilla* and *P. sylvestris*) are also being included in the selection of cultivar-specific superior males.

Although date palms are naturally pollinated by wind, it is not an effective method for maximum yield in cultivated orchards, especially since the groves are composed of predominantly female trees. One male is enough to pollinate 25–30 female palm trees. Therefore, it is recommended to grow one male for 25 female palms in Kuwait.

Date palm pollination is done artificially, either by hand or by mechanical means. As not all female flowers are produced at the same time, pollination of each inflorescence is repeated two or three times for better yield. Pollination starts with collecting male flower bunches. Most male and female trees normally flower during the period of late February to early April. Male inflorescences are collected within a few hours of the sheath splitting open, to prevent pollen loss, and then stored in a cool place. Early-appearing inflorescences are cut and hung upside down in a cool, dry airy area until needed, to prevent mold attack on the pollen. Pollen is collected as it is shed from the hanging inflorescences or extracted mechanically. Male flowers or the extracted pollens are used for pollination. Four types of pollination practices are used: traditional hand pollination by flower spikes, hand pollination by cotton puff, hand pollination with atomizers, and fully mechanized pollination. The first two methods are generally used in Kuwait.

On most farms, fruit quality has not reached international standards. Poor quality fruit is linked to several factors such as improper pollination, poor pollen quality, frequent dust deposition, spider mites and other insect pests, diseases, and finally lack of trained manpower to handle the fruit production procedures properly. Research studies at KISR proved that pollen affects the quality fruit production. Pollen from *Phoenix pusilla* produced seedless fruits in Medjool and Sultana cvs. (Sudhersan et al. 2008, 2010). Further studies to identify superior male palms specific to high-quality date cvs. such as Barhi, Khalas, Medjool, Nebut Seif, and Suckari and pollen from other related species *P. pusilla* and *P. sylvestris* are being carried out at KISR.

Intercropping is a common practice (Fig. 9.7) in Kuwait to make the best use of the date palm farm. Fodder crops such as Rhodes grass, alfalfa, and fruits such as citrus, fig, olive, and pomegranate are intercropped between rows and under date palms. The main objectives of intercropping are to produce other fruits which cannot survive on their own due to extreme temperatures if fully exposed, to generate additional income for farmers throughout the year, to grow fodder crops for livestock, and also to improve the soil nutrient and physical properties (Sudhersan 2013).

9.3 Genetic Resources and Conservation

Date palm is one of the pillars of sustainable agricultural development in Kuwait and plays a role in building and energizing agricultural production capabilities as well as related industries. There are about 40 date palm cultivars reportedly grown in Kuwait (Table 9.3). Although many date palm cultivars are available in Kuwait, date palm diversity is facing a threat due to: cloning and monoculture practices for high yield, new pests and diseases, introduced invasive exotic plant species, modern agricultural practices and techniques, climate change, desertification, and high salinity stress.

Date palm farmers practice de facto conservation and enhancement of date palm genetic diversity by maintaining and multiplying traditional varieties and conscious selection of clones for their various unique plant and fruit traits. However, a complex of interacting factors, mainly consumer preference and market forces, are negatively impacting maintenance of diversity and varietal composition of date palm groves.

Growing a single cultivar on a large scale is termed *monoculture*. Date palm monoculture of elite cultivars certainly enhances crop production. However, this practice will squeeze out the other well-known traditional cultivars and ultimately reduce date palm diversity. Date palm genetic resources are invaluable for present and future generations, providing food security, environmental sustainability, and economic stability. A wide range of genetic variation is needed within the date palm population to help them adapt to changing climate and related biotic factors.

Diversity is the basis of adaptation and is needed to meet the unpredictable environmental changes, including climate change. It is possible to select from diversity but not from uniformity. Therefore, an integrated system based on dynamic ex situ and in situ conservation is necessary for the maintenance and enhancement of date palm diversity to meet future biotic and abiotic challenges. Propagation and cultivation of only a few cultivars on a large scale for economic production is highly dangerous. Date palm cultivar diversification is necessary to guard against potential disease threats and habitat loss.

Kuwaiti date growers need encouragement to develop new cultivars and diversify existing genetic resources. To encourage Kuwaiti date growers toward cultivar diversification and germplasm maintenance, KISR has established a national date palm award for the producer of the best new local date palm variety. The new variety will be multiplied by tissue culture at KISR and 1,000 plants given to the owner as

Table 9.3 Date palm cultivars available in Kuwait

Cultivar	Maturity	Flesh texture	Quality	Origin
Abo Yousef	Middle	Soft	Excellent	Kuwait
Anbarah	Middle	Soft	Excellent	Saudi Arabia
Ashgar	Early	Soft	Excellent	Iraq
Asleyah	Middle	Soft	Good	Saudi Arabia
Awaidi	Early	Soft	Very good	Iraq
Barhi	Middle	Soft	Excellent	Iraq
Blyani	Early	Soft	Acceptable	Iraq
Bobaki	Middle	Semidry	Good	Iraq
Braim	Early	Soft	Excellent	Iraq
Dayri	Middle	Soft	Acceptable	Iraq
Esteraki	Middle	Soft	Excellent	Kuwait
Fersi	Middle	Semidry	Good	Iraq
Hadlah	Middle	Soft	Acceptable	Iraq
Halawi	Early	Semidry	Good	Iraq
Hasawi	Early	Soft	Excellent	Iraq
Hayany	Early	Soft	Acceptable	Egypt
Hilali	Late	Soft	Good	Iraq
Hulwa	Middle	Semidry	Acceptable	Saudi Arabia
Jouzi	Middle	Semidry	Acceptable	Iraq
Khadrawi	Middle	Soft	Good	Iraq
Khasab	Late	Soft	Good	Iraq
Khalas	Middle	Soft	Excellent	Saudi Arabia
Khyarah	Middle	Soft	Excellent	Iraq
Lolwi	Late	Soft	Acceptable	Iraq
Maktoomi	Late	Soft	Excellent	Iraq
Medjool	Middle	Semidry	Excellent	Morocco
Nebut Seif	Middle	Soft	Excellent	Saudi Arabia
Qantar	Early	Soft	Good	Iraq
Ruthana	Middle	Soft	Very good	Saudi Arabia
Sayer	Early	Soft	Acceptable	Saudi Arabia
Sherani	Middle	Semidry	Good	Iraq
Sherawi	Middle	Soft	Good	Iraq
Shishi	Middle	Soft	Good	Saudi Arabia
Shwathe	Late	Soft	Good	Iraq
Silaj	Middle	Soft	Good	Saudi Arabia
Suckari	Middle	Semidry	Excellent	Saudi Arabia
Um Al-Dehn	Late	Soft	Excellent	Iraq
Wannana	Middle	Dry	Acceptable	Saudi Arabia
Zaghloul	Middle	Soft	Good	Egypt
Zahdi	Late	Semidry	Acceptable	Iraq

a gift for further production. Every 2 years, new good quality cultivars will be identified through this award process and added to the KISR date palm germplasm collection bank.

9.3.1 *Germplasm Conservation*

KISR maintains a date palm germplasm collection bank on the main campus at Shuwaikh. Currently the collection holds 34 female date palm cultivars Ajwa, Al-Hammed, Al-Heifi, Al-Shamy, Anbarah, Ashgar, Awaidi, Barhi, Braim, Deglet Noor, Fersi, Hilali, Jouzi, Khalas, Khasab, Khyarah, KF-1, KF-2, KFH-1, Lolwi, Maktoomi, Medjool, Nebut Seif, Qantar, Samaran, Shekar, Shiek Ali, Shishi, Sigai, Siwi, Suckari, Sultana, Um Al-Dehn, and Wannana and 6 male cultivars Boyer, Dayri, Fard, Garvis, Ghannamy, and KM-1. A new date palm gene bank has been proposed for the KISR research station at Kebad in 2015 to accommodate a greater number of cultivars. Among the total of 40 female and male cultivars, Al-Shamy, Al-Heifi, KF-1, KF-2, KFH-1, and KM-1 are new Kuwaiti cultivars; the others were introduced from neighboring countries.

9.3.2 *Germplasm Utilization*

Research studies are being carried out by the KISR Biotechnology Program to develop salt-tolerant date palm cultivars through plant cell and tissue culture methods. A few salt-tolerant cultivars have been selected and are being grown experimentally in a salt-affected area. Trials are also being undertaken at KISR to develop new hybrid date palms for the improvement of fruit quality and also to reduce the tree height, which is a major contributor to the high cost of production in orchards having dates more than 25 years of age. An interspecific hybridization between date palm (*P. dactylifera*) and a dwarf date palm (*P. pusilla*) was carried out to create a new interspecific hybrid in 2009 through in vitro embryo rescue method (Sudhersan and Al-Shayji 2010). The new hybrid is expected to be a dwarf type, but it will require a few more years to confirm the results. Meanwhile, the hybrid date palm has flowered and produced fruits, which are entirely different from the parents (*P. dactylifera* and *P. pusilla*). Fruit size is smaller than the date palm and much larger than the dwarf date palm. Further studies are progressing toward developing the F2 hybrids for the fruit quality improvement.

9.4 Plant Tissue Culture

Date palm cultivation began in Kuwait in 1950 using offshoots and seedlings. However, large-scale commercial plantations were not established due to insufficient numbers of offshoots of the desired cultivars and the high price for offshoots. Most of the date palms and offshoots that were imported came from neighboring countries. In the 1990s, tissue-cultured Barhi cv. plantlets were imported from the UK and France for field trials. During the field growth stage of these palms, many physiological disorders were observed creating confusion among growers and researchers about date palm micropropagation in general. In order to reduce the palm importation, to clear up the confusion over the date palm micropropagation methods, and to make a large number of good quality cultivars available to the farmers on a year-round basis at an affordable price for establishing plantations, KISR has conducted research on date palm tissue culture since 1995. Plant tissue culture laboratory and greenhouse facilities were established in the same period.

Through intensive research carried out on various aspects of date palm clonal micropropagation in 1995–2004 at the plant tissue culture center of the Biotechnology Program of KISR, eight different methods of in vitro plant regeneration using different tissue explants from date palm shoot bud were standardized. The culture media, culture protocol, and acclimatization procedures developed at KISR are unique among laboratories worldwide. A common and simple protocol has been developed for many date palm cultivars (Sudhersan et al. 1993a, b, 2011). Thousands of plantlets of 30 cvs. Al-Hammed, Al-Heifi, Anbarah, Ashgar, Barhi, Boyer, Braim, Dayri, Fard, Fersi, Garvis, Ghannamy, Hilali, Jouzi, Khalas, Khasab, Khyarah, Lolwi, Medjool, Maktoomi, Nebut Seif, Awaidi, Quantar, Shekar, Shiek Ali, Shishi, Siwi, Suckari, Sultana, and Um Al-Dehn were produced through the protocols developed at KISR. Ten plants of each cultivar were planted and are being maintained in two date palm orchards on the KISR campus at Shuwaikh. All the cultivars produced by KISR were proven to be true to type through DNA fingerprinting and field evaluation (Al-Shayji et al. 1994; Sudhersan and AboEl-Nil 2004).

All the 25 female and 5 male micropropagated date palm cultivars began flowering after 3 years of field growth. No flowering abnormalities and genetic variations were observed except a few physiological disorders that are also common in traditionally propagated date palms (Al-Shayji and Sudhersan 2009; Sudhersan and AboEl-Nil 1999; Sudhersan and Al-Shayji 2009; Sudhersan et al. 2001). Among the eight different in vitro regeneration possibilities in date palm developed at KISR (Fig. 9.8), direct somatic embryogenesis from the leaf tissue explant was identified as highly economical and feasible. In 2004, the KISR date palm tissue culture laboratory began continuous production of plantlets of cultivars in demand such as Barhi, Khalas, Medjool, and Nebut Seif at the rate of 3,000 plants per month. During the past 5-year period, more than 100,000 tissue culture-derived date palm plantlets were produced and supplied to Kuwaiti growers.

The tissue culture-derived plants supplied to farmers showed uniform growth, early flowering, and high yield (Fig. 9.9). KISR has supplied tissue culture plantlets to more than 500 customers in Kuwait. The date palm tissue culture facility at KISR can produce and acclimatize 100,000 date palm plants at a time; it is the only laboratory in Kuwait for commercial-scale production and has been functioning since 2000.

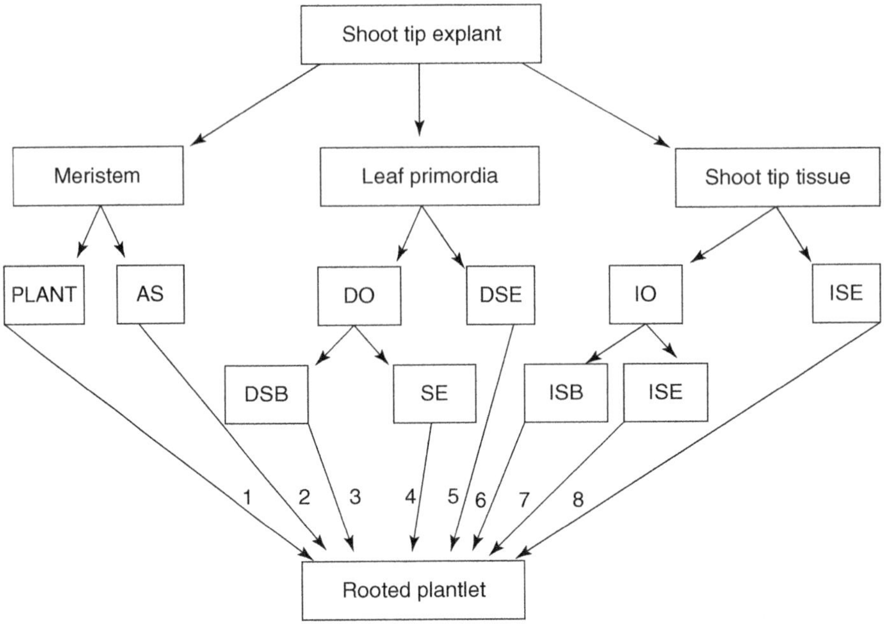

Fig. 9.8 Different regeneration methods of plantlet regeneration in date palm. *AS* axillary shoot, *DO* direct organogenesis, *DSE* direct somatic embryogenesis, *IO* indirect organogenesis, *ISE* indirect somatic embryogenesis, *DSB* direct shoot bud, *SE* somatic embryogenesis, *ISB* indirect shoot bud

Fig. 9.9 Tissue culture-derived date palm plantation of the cv. Nebut Seif

9.5 Cultivar Identification

The date palm holds enormous commercial potential for Kuwait. Date palms are in high demand for fruit production as well as for beautification. Importation alone cannot meet the high demand. Production and propagation of date palms through tissue culture is an alternative that can meet local demand as well as possibly capture a significant portion of the demand in neighboring GCC countries. To gain a market share of this highly profitable business, KISR has initiated a vigorous date palm tissue culture program. An economically successful date culture program is dependent on the ability to identify date cultivars as they are being multiplied. It also depends upon production and marketing elite, true-to-type date palm cultivars. Typed cultivars bring in more revenue than un-typed and allow novel germplasm to be propagated and marketed. The cost to date growers and nurseries can be enormous if errors are made in establishing trueness to type during tissue culture operations.

Traditional typing techniques (based on morphological and biochemical characters) are inefficient, tedious, and unreliable. Therefore, KISR has succeeded in developing true-to-type technology which is a prerequisite to obtain full economic benefits from any commercial date production operation. KISR has developed a rapid and sensitive typing method to correctly identify date palm cultivars.

In recent years, molecular techniques based on DNA assessment have been very successful in typing cultivars of a range of crop plants. At KISR, amplification fragment length polymorphism was revealed by randomly amplified polymorphic DNA (RAPD) and microsatellite or inter-simple sequence repeat (ISSR) assays that were utilized as DNA probes and markers for typing different date palm cultivars. The KISR study revealed that ISSR- and RAPD-based DNA probes were highly sensitive and more effective in revealing polymorphism in date palms. A number of ISSR and RAPD probes and their combinations produced cultivar-specific DNA fingerprints (Fig. 9.10) which allowed the identification and authentication of date palm cultivars. KISR has also developed an in-house procedure for genomic DNA isolation and purification from frozen date leaf material which is an essential prerequisite step for any DNA fingerprinting analysis. The microsatellite ISSR and RAPD fingerprinting technologies require the completion of the following steps: (a) tissue breakage to release the cellular DNA, (b) DNA extraction and purification, (c) DNA characterization and quantification, (d) assemble and optimization of primer amplification reaction, and (e) amplification product analysis.

The DNA fingerprinting technology developed at KISR (Al-Shayji et al. 1994) provides the date palm industry a sensitive and effective genetic marker system to identify, authenticate, and characterize cultivars as well as to ensure the genetic integrity and stability of the micropropagated date plants. DNA fingerprints are available at KISR to identify the date palm cultivars maintained in the KISR date palm orchard. Many Kuwaiti farmers identify cultivars through morphological characters of fruits.

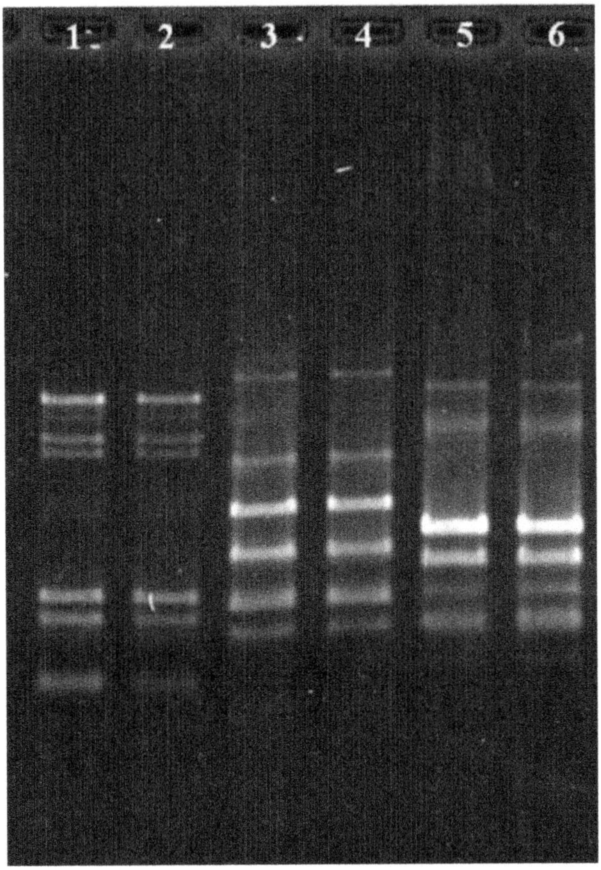

Fig. 9.10 DNA fingerprint of mother and in vitro clone

9.6 Cultivar Description

There are several date palm cultivars with commercial potential in Kuwait. However, Barhi, Khalas, Suckari, and Medjool are considered of superior quality for commercial production. Many attempts have been made to identify and characterize date palm cultivars worldwide (Nixon 1945, 1951; Mason 1915, 1928). Recently, molecular markers have been used for most of the cultivars maintained at the KISR date palm germplasm collection (Al-Shayji et al. 1994). A majority of the dates cultivated in Kuwait were introduced from Iraq, Egypt, Saudi Arabia, and Morocco and are already identified. Only a few cultivars are newly developed in Kuwait through a natural selection process.

Among the well-known cultivars, Barhi is the most important, the fruit yellow in color at khalal stage. Its fruits are mostly consumed at khalal or rutab stages in Kuwait. Fruits of Barhi are very soft and golden yellow in color at the beginning of ripening and later turn dark brown at the fully ripe stage. Barhi cv. originates from Iraq. Khalas is considered the best cultivar in Kuwait for long-term storage. It is

Table 9.4 Date palm cultivars growing in Kuwait and their fruit mean weight

Cultivar	Mean fruit weight (g)
Anbarah	35.1
Ashgar	20.4
Awaidi	21.8
Barhi	23.3
Braim	18.9
Fersi	19.1
Hilali	21.1
Jouzi	23.3
Khalas	24.1
Khasab	19.0
Khyarah	28.0
Lolwi	16.3
Maktoomi	28.3
Medjool	34.3
Nebut Seif	27.4
Shishi	23.4
Shishi	23.4
Siwi	23.6
Suckari	23.5
Sultana	29.2

yellow in color during khalal stage and golden yellow at tamar. This cultivar is of Saudi Arabian origin. Besides these two, other cultivars such as Anbarah, Ashgar, Awaidi, Braim, Fersi, Hilali, Jouzi, Khyarah, Lolwi, Maktoomi, Medjool, Nebut Seif, Suckari, and Sultana are important in Kuwait (Table 9.4).

Recently, three new high-quality female cultivars were identified from seedling populations in Kuwait: Al-Heifi, Al-Shamy, and KUF-1. The first two originated from a Barhi female and unknown males and the third from a Medjool female. The fruits of Al-Heifi and Al-Shamy are superior in quality to Barhi. Currently these two cultivars are being multiplied by tissue culture at KISR. The characteristic features of some of the date palm cultivars growing in Kuwait are shown in Fig. 9.11. Among these familiar cvs. in Kuwait, Awaidi ripens earlier than all others (late June to early August). Khasab and Hilali cvs. ripen late (October to January). If allowed, fruits of both these cvs. can be left on the tree until the next flowering season. This is due to the high humidity and low temperature during September to February. All other cvs. ripen in the middle period (late July to August).

The main features of common cultivars grown in Kuwait are summarized as follows:

Anbarah This is the most important tamar cv. in Madinah, Saudi Arabia. Fruits are dark reddish chestnut in color, very long, tasty, and soft. It matures to tamar stage late in the season.

Ashgar There are several Ashgar cvs. in Iraq and Saudi Arabia. It is early bearing in Kuwait; it ripens in July and is eaten in the khalal, rutab, and tamar stages. The khalal fruit is very light yellow in color, sweet, and juicy. The fruit is shaped like an inverted egg.

Awaidi This originates from Iraq. The fruit is light rose in color, very large, oblong, and eaten at rutab and tamar stages.

Barhi This originates from the Mandali area of Iraq. Fruits are of excellent quality and can be eaten at khalal, rutab, and tamar stages. Barhi is the major cv. in Kuwait.

Braim This cv. originates from southern Iraq. Its khalal fruit is yellow with a tint of orange or apricot color and free of phenolic compounds. Its tamar fruit is light brown with amber flesh. It is early ripening and can be eaten at khalal, rutab, and tamar stages.

Fersi This cv. is one of the desirable medium-quality dates of Shatt Al-Arab, Iraq. The fruit is small, oblong, and light red in color. It matures mid-season and is eaten at rutab and tamar stages.

Fig. 9.11 The characteristic features of fruits of some date palm cultivars grown in Kuwait

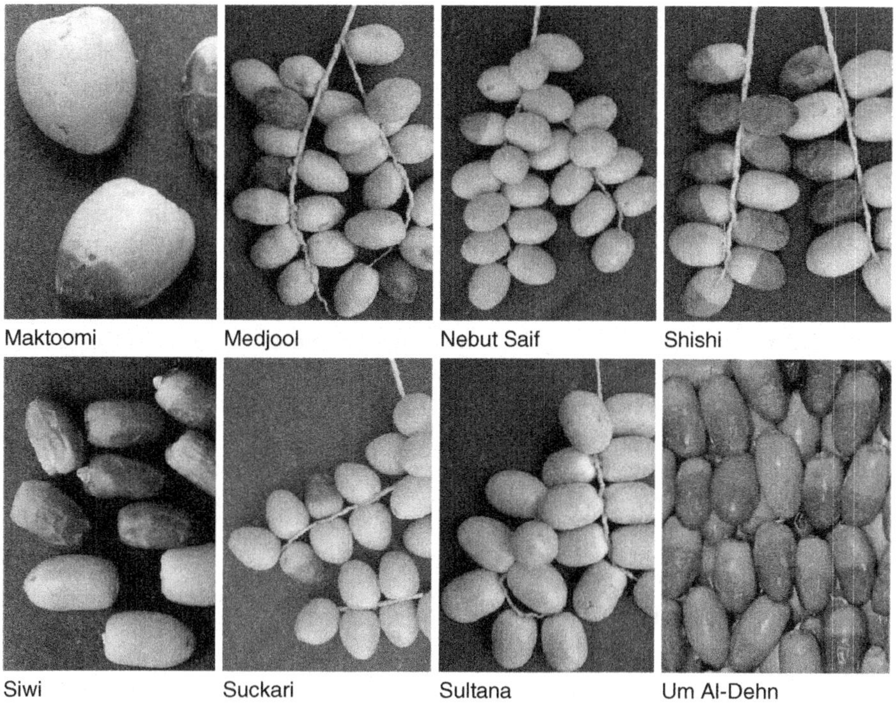

Fig. 9.11 (continued)

Hilali It originates from Saudi Arabia. It is a heavy date producer, ripens late in the season, and is delicious at rutab and tamar stages.

Khalas This originates from Saudi Arabia. It is a mid-season cv. It is edible at rutab and tamar stages; at tamar the fruit is highly valued for its fragrance and light color.

Khasab It is a good cv. originating from Iraq. Khalal stage fruits are red in color and turn black upon reaching tamar stage. In Kuwait, it ripens very late and khalal stage fruits are available up to November.

Jouzi This cv. originates from Shatt Al-Arab, Iraq. At khalal stage the fruit is large and dark red in color, turning dark black at tamar stage. It is eaten at khalal, rutab, and tamar stages.

Khyarah This originates from Basra, Iraq. It is a large-sized fruit, yellow at khalal, light amber color at rutab, and light brownish red at tamar stages.

Lolwi This originates from the UAE. Khalal stage fruit is yellow with a green tint. The fruit is very small and shaped like an inverted egg. Fruit bunch is very large and ripens very late. It is one of the high-yielding cvs. in Kuwait.

Maktoomi Soft large yellow fruit ripens late to an amber color; it originates from Iraq. It is an excellent quality tamar date.

Medjool This famous cv. originates from Morocco. The fruit is semidry, large, and orange yellow, ripening during mid-season to a reddish brown at tamar stage. It is an easy cv. to pollinate.

Nebut Seif This originates from neighboring Saudi Arabia. The fruit is large, light golden color, and tasty. The vegetative growth of this tree is better than all other cvs. growing in Kuwait. It grows to a large size within 15 years. The yield in Kuwait is less compared to Barhi.

Shishi This originates from Saudi Arabia. It is a good cv. for tamar dates. The fruits are soft in texture and eaten only at tamar stage.

Siwi This cv. was introduced from Egypt. It is slightly sweet and yellow in color at khalal stage, but the fruit is eaten commonly at tamar stage. It is a dry type and can be stored at room temperature. It is a heavy producer and a lesser-known cultivar in Kuwait.

Suckari This originates from Saudi Arabia. It is a semidry type and of excellent quality for commercial production. Fruits can be stored at room temperature.

Sultana This originates from Saudi Arabia. The fruits are larger than cv. Nebut Seif. It bears a soft fruit eaten at tamar stage.

Um Al-Dehn This originates from Iraq. It is a good cv. eaten during rutab and tamar stages. Khalal stage fruits are yellow in color, small, and elongated.

9.7 Dates Production and Marketing

Date production in Kuwait is lagging behind the neighboring GCC countries due to various physical and physiological factors (Abdul-Salam and Al-Mazrooei 2007). However, date production has been improving due to government subsidies and interest of the growers. Current annual date production in Kuwait is reported as 45,000 mt according to FAO statistics. Although 45,000 mt of dates are produced annually, a majority of them are damaged due to the lack of proper postharvest technologies. Current annual date production is not self-sufficient. Therefore, most of the fruits are consumed as fresh dates, and chances for marketing are reduced. Most of the date fruits are fresh and processed, and other date products available in the local markets are imported from neighboring countries. Marketing of date fruits and their by-products in Kuwait needs more attention.

Saudi Arabia is the leading country for date fruit and by-product exports to Kuwait. Saudi Arabia, Oman, and the UAE produce surplus quantities of dates, while other countries such as Kuwait have to increase production to achieve self-sufficiency. Market and consumer research has to be undertaken to address consumer demands and develop an appropriate product mix. Quality standards to meet international market requirements are poor and need more attention to strengthen the date agro-industrial chain. Dates and their by-products need to be promoted in international markets. The Government of Kuwait is providing subsidies to growers for commercially viable high-quality cultivars. This will increase the number of good quality productive trees in the near future and will enhance the production of marketable quality dates. KISR is producing good quality date palm planting material through micropropagation as per growers' demand. Postharvest technology is in its infancy, and research is currently being focused on at KISR.

9.8 Processing and Novel Products

Current postharvest technologies including handling, processing, and storage need to be improved to enhance the value of date fruits and their products. In Kuwait, there are no proper facilities, know-how, or trained manpower to handle the fruits during field harvest. Improvement in fruit handling during harvest has to be strengthened. Concerning date processing, the quantity of fruits processed in factories into different products (packed fruits and by-products) represents only a small percent of the total quantity produced. There is also a need for the development of innovative products that cater to dynamic international market needs and demands on an industrial scale. Developments of an organized structure to collect the fruits from the farmers and transport them to processing factories need to be established. Pesticide application is very high, and research is needed to develop alternative eco-friendly natural products to control the pests and pathogens that affect the fruit quality. Production cost is very high due to the labor cost, and necessary steps have to be undertaken to reduce these costs. The development of useful industrial products from the date palm residues, which enhances farmer income, needs more consideration. Overall, date fruit processing and date product improvements are at their infancy in Kuwait at present. More attention is needed to improve postharvest technologies, processing, novel product development, and marketing.

9.9 Conclusion and Recommendations

Date palm is a high priority fruit crop in Kuwait due to its tolerance of the climatic conditions. The Government of Kuwait is providing subsidies for growing selected commercial cultivars. The status of date palm cultivation is improving in Kuwait and moving toward self-sufficiency. However, more support and encouragement are necessary for date palm research. The chief constraints on date production are scarce fresh irrigation water, increasing soil salinity, the high cost of production, insufficient removal of poor quality trees and replacement with high-quality trees, lack of postharvest technologies, a poor marketing system, lack of storage facilities, lack of processing factories, increase in low-quality varieties, lack of research on innovative products, increased losses during postharvest handling, and the absence of quality maintenance for international marketing. In order to improve the date crop in Kuwait, further research and development is required in areas such as cultivar improvement, farm management, environmentally friendly pest and disease management, biotic and abiotic stress tolerance, date fruit and by-product quality improvement, mechanization in crop management, fruit processing, and finally marketing. KISR has established a date palm micropropagation laboratory and developed protocols for commercial-scale plant production to meet the local demand for good quality planting materials. The quantity of imports of planting material is gradually being reduced due to the in vitro micropropagation carried out on a commercial scale. Date palm germplasm conservation and biodiversity improvement are also being practiced at KISR. More attention and support are needed for post-harvest technology and developing eco-friendly pest and disease management research from the government and private sectors.

Acknowledgment The authors hereby acknowledge KISR DG. Dr. Naji Al-Mutairi for his encouragement; KISR management for the support; Mr. Hassan Ali Abdurahman Mahmoud from PAAFR for providing information on a number of registered farms and the total number of palm trees in Kuwait; Dr. U. Saravana Kumar, KISR Research Scientist from Water Research Center for reviewing the manuscript; Mr. Samy Al-Amad, KISR researcher of the Biotechnology program for providing DNA fingerprinting details; and Mr. Sala Al-Melhem from the KISR Biotechnology Program for assisting in data collection.

References

Abdul-Salam M, Al-Mazrooei S (2007) Crop water and irrigation water requirements of date palm (*Phoenix dactylifera*) in the loamy sands of Kuwait. Acta Hort 736:309–315

Al-Nisf YM (2010) Your palm. 10th edn. Safat, Al-Nisf, Kuwait

Al-Mudaires J (1992) Atlas of date palms. Al-Balagh Printing Press, Kuwait

Al-Shayji Y, Saleem M, Al-Amad S et al (1994) Isolation and analysis of total genomic DNA from date palm and related species. Acta Biotech 14:163–168

Al-Shayji Y, Sudhersan C (2009) Pseudodwarf disorder control in tissue cultured date palms. Am Euras J Sci Res 3(2):128–131

Bhat NR, Al-Menaie H, Sulaiman MK et al (2011) Irrigation studies in date palm (*Phoenix dactylifera* L.). Kuwait Institute for Scientific Research, Report No. KISR 10663, Kuwait

FAO (2013) Crop production, statistics division. FAO, Rome

KISR (1995) Master plan for the development of Kuwait's agricultural sector (1995–2015). Kuwait Institute for Scientific Research, Report No. KISR 4615, Kuwait

Mason SC (1915) Botanical characters of the leaves of the date palms used in distinguishing cultivated varieties, USDA Bull 223. U.S. Dept. of Agriculture, Washington, DC

Mason SC (1928) Date varieties at all growth stages shown by vegetative characters. USDA Separate from yearbook of Agr 1067:1–2

MOP (2006) Annual statistical abstract. Ministry of Planning, Statistics and Information Sector, Kuwait

Nixon RW (1945) The need for a monograph of the date varieties of the world. Chron Bot 9(2/3):153–154

Nixon RW (1951) Leaf characters of the Deglet Nour date palm in relation to age and environment. Am Soc Hort Sci Proc 57:179–185

PAAFR (2003) Annual statistical abstracts 2003–2004. Public Authority for Agriculture and Fish Resources, Kuwait

Sudhersan C (2013) Date palm cultivar specific susceptibility to greater date moth infestation. Am Euras J Sust Agr 7:32–36

Sudhersan C, AboEl-Nil M (1999) Occurrence of hermaphroditism in male date palm. Palms 43(18–19):48–50

Sudhersan C, AboEl-Nil M (2004) Axillary shoot production in micropropagated date palm. Curr Sci 86:771–773

Sudhersan C, AboEl-Nil M, Al-Baiz A (1993a) Occurrence of direct somatic embryogenesis on the sword leaf of *Phoenix dactylifera* L. cultivar Barhee. Curr Sci 65(11):877–879

Sudhersan C, AboEl-Nil M, Al-Baiz A (1993b) Direct somatic embryogenesis and plantlet formation from the leaf explants of *Phoenix dactylifera* L. cultivar Barhee. J Swamy Bot Cl 10(172):37–43

Sudhersan C, AboEl-Nil M, Hussain J (2001) Hapaxanthic axillary shoots in date palm plants grown *in vivo* and *in vitro*. Palms 45(2):84–89

Sudhersan C, Al-Shayji Y (2009) Tree crown bending in tissue culture date palms. J Agr Food Envir Sci 3:1–4

Sudhersan C, Al-Shayji Y (2010) Interspecific hybridization and embryo rescue in date palm. In: Jain SM, Khayri JM, Johnson DV (eds) Date palm biotechnology. Springer, Dordrecht, pp 567–584

Sudhersan C, Al-Shayji Y, Jibi Manuel S (2008) Date palm crop improvement via T x D hybridization integrated with *in vitro* technique. Acta Hort 829:219–224

Sudhersan C, Al-Shayji Y, Jibi Manuel S (2011) A simple in vitro protocol for multiple regeneration pathways and control of physiological disorders in date palm. KISR No. 10539. In Vitro Cell Dev Biol 47:S39–40, Abstract P-1014

Sudhersan C, Jibi Manuel S, Al-Sabah L (2010) Xenic and metaxenic effect of *Phoenix pusilla* pollen on certain date palm cultivars. Acta Hort 882:297–302

Chapter 10
Date Palm Status and Perspective in Qatar

Nazir H. Muhammed, Osman E. Ahmed, Talaat A. Ahmed, and Mohammed S. Al-Yafai

Abstract Qatar ranks as the 16 largest date-producing country in the world. There are 581,336 date palm trees growing in an area of 2,469 ha with date production of 21,491 mt as of 2010. It is the major fruit tree grown in the country, and date production is 7.2 % of the total agricultural production. No specialized institution exists for date palm research and development. Average recorded yield for 2010 was 8.7 mt/ha, the second highest in the region and higher than the world average of 6.3 mt/ha, for the same year. Yield has progressively increased since 1980. It is essential that this status should not only be maintained but also increased to reach a higher production rate similar to that being achieved in Egypt and other countries. Good potential exists to expand the area and production of dates, providing existing agricultural lands and water resources are used efficiently. Like other countries, the major constraints of yield are scarcity of good quality water, soil, and water salinity, low-yielding cultivars, poor agronomic practices due to a limited number of knowledgeable growers, insects and diseases, and weak research support. Tissue culture is in initial stages. Many research and development efforts are required to enhance date palm cultivation in the country to increase yield and economic return. For this purpose, several research and development projects have to be planned and implemented. A date palm-based industry needs to be set up to decrease imports of this commodity.

Keywords Constraints • Date palm • Future research activities • Imports • Production • Research • Yields

---

N.H. Muhammed (✉) • O.E. Ahmed • M.S. Al-Yafai
Department of Agricultural Affairs, Ministry of Environment, Doha, Qatar
e-mail: drnazirhussain@gmail.com; osmansoils2013@gmail.com; msabdulla@moe.gov.qa

T.A. Ahmed
Department of Biological and Environmental Sciences, Qatar University, Doha, Qatar
e-mail: t.alfattah@qu.edu.qa

© Springer Science+Business Media Dordrecht 2015

J.M. Al-Khayri et al. (eds.), *Date Palm Genetic Resources and Utilization: Volume 2: Asia and Europe*, DOI 10.1007/978-94-017-9707-8_10

10.1 Introduction

Qatar is not an agricultural country and has to import over 90 % of its food requirements, amounting to QAR 4.68 billion or USD 1.3 billion in 2010. Sovereignty of a state is based on its food security. In 2011, Qatar's self-sufficiency in vegetables, cereals, fruit, and oilseeds was 13.38, 5.78, 13.4, and 0, respectively (Table 10.1). These values are not encouraging. Therefore, the Government of Qatar, because of its critical role in the nation's food security, declared agriculture as a strategic sector. Like other Middle East countries, Qatar faces serious water scarcity and its associated consequences. Hence, a countrywide National Food Security Program was established to increase domestic production, which will lead to scientific and technological development in the following three agricultural areas: (a) water management and desalination technologies, (b) agricultural development, including adoption of technologies such as hydroponics and water-efficient techniques such as drip irrigation, and (c) food processing through the establishment of an agro-industry.

It must be kept in mind that water scarcity and food shortages are likely to increase around the world with climate change, population growth, and rapid economic expansion of different regions. Therefore, risks must be perceived in a way that leads to investing how to cope successfully with future threats to food security of the nation. The Government of Qatar has diagnosed the situation very wisely, and its efforts are very clear and headed in the right direction. The government is now keen to increase the level of self-sufficiency in agricultural produce and livestock and to reduce the dependence on foreign countries for agricultural products. Nearly

Table 10.1 Important characteristics of Qatari agriculture

	Year		
Important parameters	2009	2010	2011
Total arable area, ha	65,000	65,000	65,000
No. of registered farms	1,245	1,275	1,281
Total area of registered farms, ha	43,730	44,422	43,047
Total arable area of registered farms, ha	26,238	26,653	25,828
Number of active farms	795	822	831
Total arable area of active farms, ha	37,010	35,958	34,598
Area of active farms under cultivation, ha	22,206	21,575	20,759
Cropped area in active farms, ha	9,112	10,506	9,021
Crop intensity (ratio net area sown to total cropped area)	41	48.7	43.5
Self-sufficiency of vegetables, %	16.41	15.92	13.38
Self-sufficiency of cereals, %	5.53	5.39	5.78
Self-sufficiency of fruits, %	15.38	14.23	13.4
Self-sufficiency of oil seeds, %	0	0	0
Quantity agricultural commodities imports, million mt	1.342	1.593	1.590
Value agricultural commodities imports, million USD	1.572	1.588	2.031

Source: Annual Bulletin for Areas and Production of Agricultural Crops, Qatar (2011)

50 % of farms are underutilized, and government plans and programs are being implemented to utilize them more fully. Technical and material support to develop farms for poultry and agriculture will be provided. It will not be possible to explore all the natural resources to produce food without addressing the challenge of water. Plans are afoot to reduce exploitation of groundwater and increase the use of sewage water (Qatar Tribune 2014).

Water-saving and water-conserving technologies like mulching, using soil conditioners, application of manure, and improved irrigation systems should be top priorities as well (El-Bably 2002; Erskine et al. 2004; Jaradat 2005; Mulumba and Lal 2008; Ramakrishna et al. 2006; Xie et al. 2005).

10.1.1 Qatar Agriculture

Qatar has a total of 650,000 ha of arable land (Table 10.1). The number of registered farms is 1,281 occupying 43,047 ha, of which 25,828 are arable. The average size of a productive farm in Qatar is 27 ha; on average, about 8 ha are used for crop production. Roughly, equal areas are devoted to fruit trees (mostly date palm), vegetables, and fodder crops. Barley is the most frequently grown cereal crop, but altogether cereals typically occupy less than 1 ha on a typical farm. There are two types of agriculture in the country: open fields and greenhouses. Open field agriculture is currently the core of farming. Farmers do not cultivate the entire available hectarage on their farms. The proportion of cultivated land is on average 30 % of total farmland. The cropping system is dominated by fodder crops occupying 41 % of the cultivated area, followed by vegetables (34 %) and fruit trees (9 %). Cereal crops occupy only 1.2 % of the cultivated land. Farmers use cattle, sheep, and goat manure along with chemical fertilizers for fertilization of crops. Approximately 43 % of water resources are used for fodder crops, about 37 % for fruit trees, and 19 % for vegetables. More than 70 % of farmers use flood irrigation for fodder, while the remaining 30 % use sprinkler or drip irrigation. Date palms, which are one of the best-adapted crops to the desert climate, can contribute to food security. At present, 65 % of farmers use flood irrigation to irrigate their palms. Approximately 12 % of farmers are using protected farming, with greenhouses being the dominant technology, and their numbers ranging between two to 98 units per farm, with 40 % of farms containing only five greenhouses.

10.1.2 Constraints to Qatari Agriculture

Agriculture in Qatar is facing a number of problems. The most obvious is insufficient good quality water due to the absence of surface supplies in the form of rivers, canals, or lakes. The traditional system of flood irrigation loses a lot of water to evaporation and to leaching because of porous sandy soils. The only available water is either saline groundwater or seawater. Alternative water sources like treated sewage water or industrial water currently are not much used. Lack of a trained working force, adaption of new technologies, and above all the lack of interest of the farm owners are the other conspicuous constraints. The consequence is very low yields of crops including date palm.

10.1.3 Date Palm Status in Qatar

The date palm, as in other Arabian countries, is a popular and traditional fruit in Qatar. It occupies the largest area among the fruit trees grown in the country. The production of dates is 7.2 % of the total agricultural production. However, when the area and production are compared to larger countries of the region like Saudi Arabia, UAE, and Oman, it seems to be scanty. Nevertheless, a huge potential exists to increase cultivation area and date production to achieve self-sufficiency in Qatar.

10.2 Cultivation Practices

10.2.1 Present and Recommended Practices

Cultivation practices for date palm in Qatar are an admixture of local traditions and those adopted from the region. As such, there has been little systematic research conducted to devise local techniques based on locally generated data. Present cultivation practices recommended for each calendar month are given in Table 10.2.

10.2.2 Cultivation Practices Research in Qatar

Systematic research is required to formulate recommendations specifically for Qatar conditions. Some recent work has been conducted by the Ministry of Environment and Qatar University in this regard. A brief summary of this research is presented here.

10.2.2.1 Effect of Soil Conditioner

Zeolite is a soil conditioner that can hold moisture in sandy soils for longer periods, lessening loss of irrigation water through evaporation and leaching. A 5-year study began in 2010 at Rawdat Al-Fars Experimental Farm of the Ministry of Environment to investigate the efficiency of Zeolite for safely using saline water to irrigate date palm (cvs. Khalas and Barhi). Zeolite was applied at three rates: 20, 30, and 40 kg/

Table 10.2 Yearly calendar of agricultural operations for date palm in Qatar

Month	Agricultural operations
January	Trimming and pruning and removal of thorns
	Application of compound chemical fertilizer
	Preventive spraying of fungicide and insecticide (to control red palm weevil and pollen disease)
February[a]	Start pollination for early cultivars
	Application of urea fertilizer
	Continue spraying of fungicide and insecticide (greater pollen worm)
March[b]	Continue pollination
	Start fruit thinning
	Start thinning and curving
	Continue spraying of fungicide and insecticide (lesser date moth)
April	Pollination for late cultivars
	Thinning and curving of fruits
	Removal and planting of new cuttings
	Continue preventive spraying of fungicide
May	Preventive spraying for spiders
	Covering and bagging of fruits at turning green color.
June	Continue the same above operations
July	Stop application of pesticides spraying
	Start harvesting fruits of early cultivars
August	Continue harvesting fruits
	Full stop of pesticides spraying
September	Harvesting late cultivars Start establishment of new date palm farms
October	Removal of fruits falling on the ground
	Removal of old fronds and cleaning of the palm
November	It is possible to start application of chemical and organic fertilizer to trees
	Preventive spraying of fungicide and insecticide
December	Application of fermented compost
	Protect newly planted cuttings from cold by sackcloth wrapping

Remarks:
Control of red palm weevil by inoculation method the year-round
Spreading of red palm weevil pheromone traps can be done year-round
Date palm trees should be irrigated year-round by modern irrigation (bubbler) method
Timing for carry out of these operations may vary according to area
[a]Pollination should be repeated if it rains
[b]The period between pollination operation and spray should not be less than 1 week

plant/5 years in addition to control with zero application. Data on plant parameters, soil characteristics, and heavy metal accumulation were recorded. However, only yield data are presented in Table 10.3.

The application of Zeolite soil conditioner increased the fruit yield of date palm significantly in all 4 years of the study. The average of 4 years indicated an increase of 8.8, 7.6, and 5.1 % date yield of Barhi for application of 2, 3, and 4 % Zeolite by

Table 10.3 Date yields (kg tree^{-1}) 2010–2013

	Barhi					Khalas				
Treatments	2010	2011	2012	2013	Mean	2010	2011	2012	2013	Mean
T0 (No Zeolite)	75.6	60.7	86.5	80.6	75.84d	97.4	67.0	88.4	80.4	83.29b
T1 (Zeolite 2 % in 15 cm soil)	85.5	65.3	95.3	84.2	82.57b	100.1	71.0	91.6	87.2	87.47a
T2 (Zeolite 3 % in 15 cm soil)	86.7	65.7	88.3	85.7	81.6b	100.4	71.4	82.8	85.5	85.0 a
T3 (Zeolite 4 % in 15 cm soil)	91.6	69.7	83.5	74.1	79.73c	103.3	71.6	74.2	77.8	81.7 b
Mean for cultivars	84.9 b	65.4 c	88.4 b	81.1 b	–	100.3a	70.3c	84.2b	82.7b	–

Source: Hussain et al. (2014)
Means are of six values in each year
Means for cultivars were compared for all the 4 years together
Means for treatments for 4 years and compared together for both cultivars
(a-d) indicate interactions of cultivars and treatments in each year separately

volume, respectively. The increases were 5.0 and 2.1 by 2 and 3 % application of Zeolite, while 4 % did not produce a positive response. It can also be noted that responses were higher in the initial 3 years revealing effectiveness of the soil conditioner for a 3-year period after which a fresh application may be required. There were no negative effects of Zeolite application on soil and plants because no accumulation of heavy metal was noted. There were slightly more salts in the soil during first year.

10.2.2.2 Appropriate Irrigation Water Requirements

Studies of appropriate water requirements of date palm were conducted under a project entitled Improvement of Date Palm Production and Quality under Qatar Conditions (NPRP3-09-705-4-025) sponsored by the Qatar Foundation and implemented collaboratively by the Ministry of Environment (MoE) and Qatar University (QU). In this experiment, standard irrigation levels (presently recommended in Qatar) were compared with 15, 30, and 45 lesser water quantity per irrigation as well as 10 % more for two date palm cultivars: Naboot Saif and Khalas. Reducing the quantity of applied water per irrigation by 30 % of conventional practice but without disturbing the frequency did not affect the Naboot Saif date yields significantly (Hussain and Ahmad 2014). The yields were reduced by 0 and only 1.7 % for 15 and 30 % water reduction per irrigation (Table 10.4).

However, reduction of irrigation by 45 % decreased the yields by 19 %. This water saving was due to decreased leaching and evaporative losses under Qatar conditions where soils are sandy with excessive drainage. Irrigation water requirements of Khalas

Table 10.4 Date fruit yield (kg/plant) for various water quantities

	Naboot Saif cv.			Khalas cv.			Treatment means		
Treatments	2011	2012	Means for 2 years	2011	2012	Means for 2 years	2011	2012	2011 + 2012
T1	24.94 a	20.97 a	22.96	11.03 c	10.5 c	10.77	17.99 A	15.73 A	16.86
T2	25.80 a	20.27 a	23.04	10.77 c	9.5 c	10.14	18.29 A	14.88 A	16.59
T3	25.63 a	19.47 a	22.55	8.49 d	9.03 c	8.76	17.06 A	14.25 A	15.66
T4	20.49 b	16.43 b	18.46	8.12 d	7.83 d	7.98	14.53 B	12.13 B	13.33
T5	25.94 a	21.93 a	23.94	11.40 c	10.70 c	11.05	18.67 A	16.31 A	17.49
Cultivar means	25.56 A	19.81 B	–	9.96C	9.51C	–	–	–	–
Yearly means	–	–	–	–	–	–	17.31 A	14.66 B	–

Source: Hussain and Ahmad (2014)
Cultivars were compared in each year (2011 and 2012 separately)
Results are means of three replications
Means followed by different letters are significantly different. Capital letters are used to compare significant differences of the overall means

cv. were found to be higher than Naboot Saif. Yield decreases of 5.8, 18, and 26 % were observed due to reduction of applied water per irrigation by 15, 30, and 45 %, respectively. Hence, a reduction of 15 % was safer in the case of Khalas dates. Increasing water application by 15 % did not yield much of an increase, and the difference with conventional practice was nominal. No adverse effect on fruit quality was observed by decreasing irrigation water quantity per irrigation because drought, if any, did not persisted for longer periods. Soil health was not affected at all due to reductions in irrigation water quantity. Rather, there was less addition of salts when the total quantity of water usage was reduced. Decreasing irrigation water did not negatively affect water and nutrient uptake by plants, as indicated by flag leaf analysis.

Thus, irrigation water applied to date palm can be reduced by 30 % per application compared with conventional practice in the case of Naboot Saif cv., while by 15 % in Khalas cv. without significantly decreasing fruit yields. However, the present frequencies of water application must be observed. Water saving will come from partial control over leaching and evaporation losses occurring under Qatar conditions when a large amount of water is applied at one time.

10.2.2.3 Alternative Sources of Irrigation Water

The effects of diluted seawater and treated wastewater on date palm production and quality were compared. Irrigations with treated wastewater affected neither the date palm yields and fruit quality nor caused accumulation of heavy metals in comparison to groundwater (Table 10.5). Hence, it could be an alternative source of irrigation. Diluted seawater of either 10 or 15 dS/m affected the date palm yields significantly (28 and 45 % decrease, respectively) and salinized the soil indicating that a further dilution will be required to bring about safer usage. However, there is the possibility

Table 10.5 Date yield (kg plant^{-1}) due to diluted seawater and treated wastewater

Treatments	2012	2013	Overall treatment means of 2 years
Groundwater alone	41.68 a	38.45 a	40.06 A
Treated wastewater	36.46 ab	34. 68 a	35.57 A
Diluted seawater of EC ≈ 10 dS m^{-1} (Seawater + Groundwater = 1 + 4)	31.62 b	25. 87 b	28.74 B
Diluted seawater of EC ≈ 15 dS m^{-1} (Seawater + Groundwater = 1 + 2)	24.05 c	19. 79 c	21.92 C
Yearly means	33.45 A	29.70 B	–

Source: Hussain and Ahmad (2014)
Results are means of three replications
Means followed by different letters are significantly different. Capital letters are used to compare significant differences of the overall means

of using this water when diluted by proper ratios of seawater and groundwater. There was no heavy metal accumulation observed in leaves and fruits of date palm as well as the soil when seawater, groundwater, and treated wastewater were used for irrigations. Using diluted seawater for date palm irrigation by seawater, groundwater ratios of 1:2 (EC 15 dS/m) and 1:4 (EC 10 dS/m) may cause decrease in yields by 45 and 28 %, respectively. Nevertheless, these were still economical, as these were less than 50 %, the critical value. The losses may increase in later years. Therefore, a ratio of 1:6 or narrower could prove safer (Hussain and Ahmad 2014).

10.2.2.4 Effect on Growth with Seawater Irrigation

Salinity stress is known to retard plant growth through its influence on vital facets of plant metabolism, including disturbing the concentration of endogenous plant hormones. This disturbance can affect the growth and development of plants. At Qatar University, studies were undertaken to improve the endogenous hormone balance through application of exogenous hormones (Aljuburi and Maroff 2007). The experiment consisted of applications of growth regulators: indoleacetic acid (IAA) and naphthalene acetic acid (NAA), mixtures of IAA and NAA, and different concentrations of seawater alone or in combination with IAA and NAA. The application of NAA in combination with salts reduced accumulation of Na and Cl in leaves of date palm (*Phoenix dactylifera* L., cv. Hatamy) seedlings. Salinity in irrigation water reduced leaf and root Mn, Zn, Fe, and leaf K, Ca, Mg concentrations and the ratio of K/Na and Mg/Ca, but increased leaf N and leaf and root P, Cu, Na, and Cl concentrations, compared with untreated seedlings. Compared to seawater alone, growth hormone application reduced the adverse effects of salts (from seawater irrigation) by reducing Na and Cl accumulation in the leaves and roots. There was an increase of Ca, Mg, N, P, and Fe concentrations and the ratio of K/Na in the leaves and roots. Number of roots and shoot dry matter percentage of seedlings with applied hormone treatments were more compared to using seawater alone.

10.3 Genetic Resources and Conservation

There is no specialized institution or body in Qatar for conservation of date palm resources. Cultivars have been imported for growing in different areas. However, a study on genetic diversity was initiated under an ICARDA (International Center for Agricultural Research in the Dry Areas) Project comprised of 12 local cultivars of date palm. DNA markers were established to identify and detect the male date palms. Newly designed SSR (simple sequence repeats) markers developed by ICARDA were verified, and their efficiency in genetic characterization was confirmed.

10.4 Tissue Culture in Qatar

A tissue culture laboratory was established under the Ministry of Environment. A survey of the morphological and physiological diversity among the different date palm cultivars has already been conducted, although propagation methods, explant physiological behavior, and acclimatization of transplants are still under evaluation. A lower response of date palm explants for ontogeny in the media, the differentiation between the different species at the early stages in culture and prefruiting, and a lower percentage of success in rooting and acclimatization of the new transplants were the major constraints faced by the laboratory in early studies.

In vitro studies at the Tissue Culture Laboratory, Ministry of Environment, indicated that micropropagation of date palm cvs. Kapkap and Tharlaj could be successful in medium consisting of full MS and 0.1 mgl^{-1} NNA without charcoal. A mean of 5.5 and 1.0 root per embryo was recorded, respectively, for both cultivars. The other medium did not perform better in terms of root number. The type of medium also affected the root length. Medium containing charcoal at 3 g l^{-1} proved better in increasing root length (Alromaihi and Elmeer 2009).

Date palms are generally propagated by separating offshoots produced by individual trees. Somaclonal variation refers to the phenotypic and genotypic variation observed in plants regenerated through cell culture. On the other hand, some somaclonal variations have desirable agronomic and commercial advantages and on occasion lead to the release of new cultivars. Microsatellites or (SSRs) and Inter-Simple Sequence Repeat (ISSR) markers were used in this study to detect genetic variation based on DNA markers between the regenerated plantlet and its corresponding callus (Ahmed and Al-Qaradawi 2010; Ahmed et al. 2012a). Somaclonal variation between callus and more than one regenerated plant was observed in three cultivars: Kubkub, Merziban, and Dharlag.

SSR and ISSR markers provide a useful technique in the assessment of genetic stability in micropropagated date palm as well as in their genetic characterization, allowing the differentiation of cultivars with a relatively low number of primers, even those corresponding to the same family. SSR and ISR markers demonstrated the genetic stability observed in the date palm cultivars studied in this work and the importance of the genotype in the appearance of somaclonal variation (Fig. 10.1).

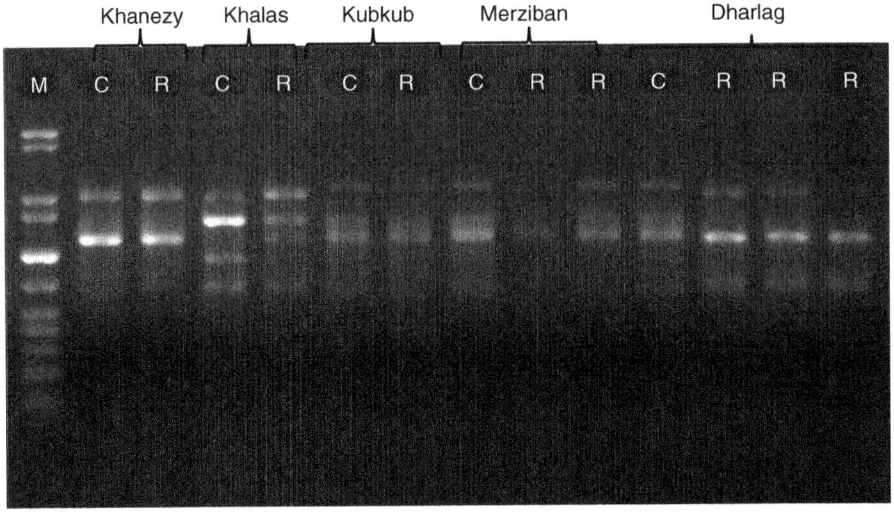

Fig. 10.1 ISSR pattern obtained with the primer ISSRBT23 in five different date palm cultivars during in vitro culture, *M*, *C*, and *R* denote for ladder marker, callus, and regenerated plantlet, respectively

10.5 Cultivars Description

Date palm cultivars have been widely exchanged officially and unofficially in the GCC (Gulf Cooperation Council) region. There has been an active exchange of date palm cultivars (local and exotic) from one country to another. As a result, almost all the cultivars are commonly grown throughout the Gulf. However, the area under each cultivar varies. There are two groups of cultivars in Qatar: cultivars grown in Qatar as well as other countries and cultivars only identified in Qatar. The first group includes cultivars Khalas, Barhi, Bin Saif, Jabri, Hitimi, Khunaizi, Rotanah, Shishi, Naboot Saif, Hilali, Hilaini, etc. The second group includes Azat, Berz, Bashbak, Tarahim, Disky, Zary, Taumai Aswad, Qashmak, Niqal, Ward, and others.

10.6 Cultivars Identification

10.6.1 Molecular Phylogeny Using SSRs

Recently developed techniques based on DNA markers offer new tools for genetic analysis. So far, no detailed research has been conducted to analyze phylogenic relationships among date palm cultivars native to Qatar using DNA markers. The objectives of this study were to analyze the genetic diversity among 15 different cultivars of date palm at the Qatar University Experimental

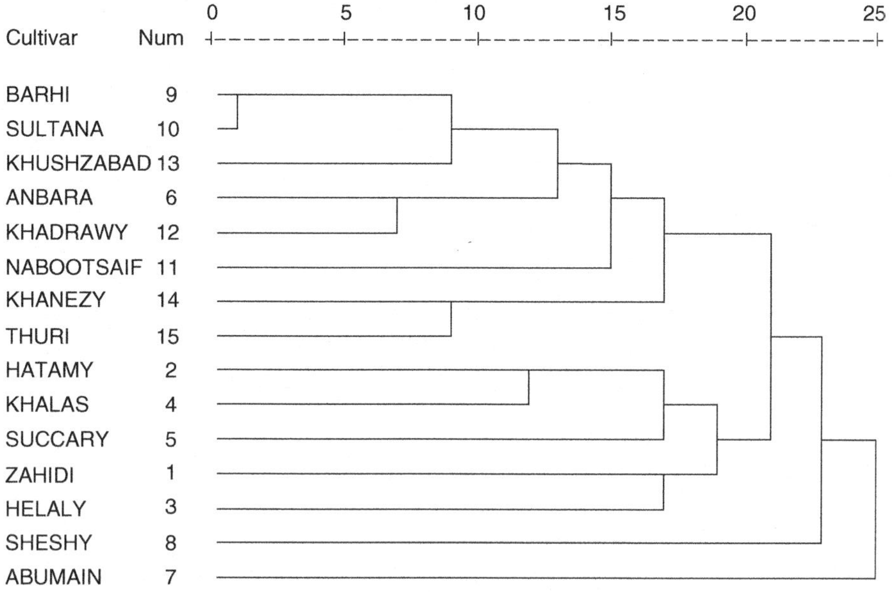

Fig. 10.2 Dendrogram of 15 Qatari date palm cultivars based on Jaccard genetic similarity coefficient using SSR data

Farm using SSR markers and to determine the genetic similarity and/or diversity among the well-known Qatari date palm cultivars. DNA was extracted from fresh young leaves. Among 16 primer pairs tested for their ability to generate expected SSR banding patterns in Qatari date palm genotypes, 10 primers successfully produced clear single bands in most of the studied genotypes. So far, 6 SSR primers did not amplify clear bands in our genetic materials even using different PCR (polymerase chain reaction) conditions. The amplified SSR band sizes ranged from 100 to 300 bp. A total of 40 alleles, with an average of 4 alleles per locus, were scored. A similarity coefficient matrix was computed to cluster the data and to draw precise relationships among the 15 Qatari date palm genotypes studied (Fig. 10.2).

In this study, SSR markers were used to assess the molecular characterization and the phylogenic relationships of Qatari date palm cultivars. Results provide evidence of a genetic diversity among the studied Qatari date cultivars and the ability of SSR markers to detect the genetic diversity in date palm. We may conclude that all date palm ecotypes are interrelated in spite of their agronomic divergence. Genetic similarities and a dendrogram could regroup the Qatari date palm cultivars in a way that one cultivar (Abu Main) was excluded from the group due to its dissimilarity with the others. Two cultivars (Barhi and Sultana) were much closer and could be considered as coming from a single origin. Some cultivars were grouped in different similar pairs (Ahmed and Al-Qaradawi 2009).

10.6.2 Inter- and Intraspecific Genetic Variations Using ISSR

Offshoots separated from individual trees are mainly used for date palm propagation, which maintains the genetic integrity of the cultivars such as fruit morphology and quality; however, some variations are observed. The objective of this study was to determine the genetic similarity or diversity among and within the well-known Qatari date palm cultivars using different 18 primers of ISSR markers. Five commonly cultivated cultivars in Qatar were selected, including Khalas, Sheshy, Rezezy, Barhi, and Khanezy, from three different cultivated locations (Al-Shamal, Al-Khor, and Al-Rayan). All primers have amplified polymorphic bands in the studied cultivars either among the cultivars or within each cultivar in different cultivated areas. These results revealed the existence of genetic variations among the cultivars studied as well as within each cultivar, supporting the observed variation in some morphological and quality characters for different trees that are grown in different environments and derived from the same cultivar. Results reported from this study will help the Qatari date palm community with intra- and inter-fingerprinting of different cultivars leading to identifying new lines or cultivars that are of a high quality and thus may be patented as unique Qatari cultivars (Ahmed et al. 2012b, 2013).

The ISSR marker analysis adopted as a tool to study the inter- and intraspecific genetic variations in date palm is easy and readable. In this current study, we used 18 ISSR primers and the genomic DNA of five date palm cultivars grown in three different locations. All primers have amplified polymorphic bands in the studied cultivars either among the cultivars or within each cultivar grown in different cultivated areas. The results indicate the existing genetic variations not only among well-known cultivars but also within each cultivar, explaining the variation in some morphological and quality characters from different trees within the same cultivar. In addition, data provide evidence of the possibility of using these powerful markers as descriptors in the certification and the control of origin labels of date palm material (Fig. 10.3).

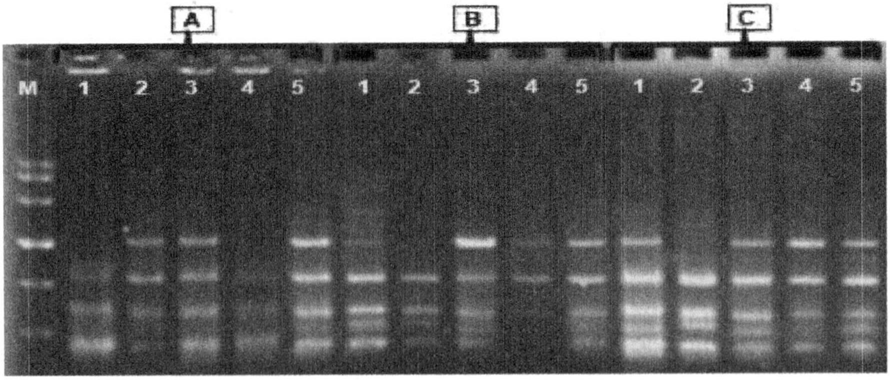

Fig. 10.3 Amplified band patterns of Primer 11 from five studied date palm cultivars in different cultivated areas. *Arrows* show the polymorphic bands within each cultivar

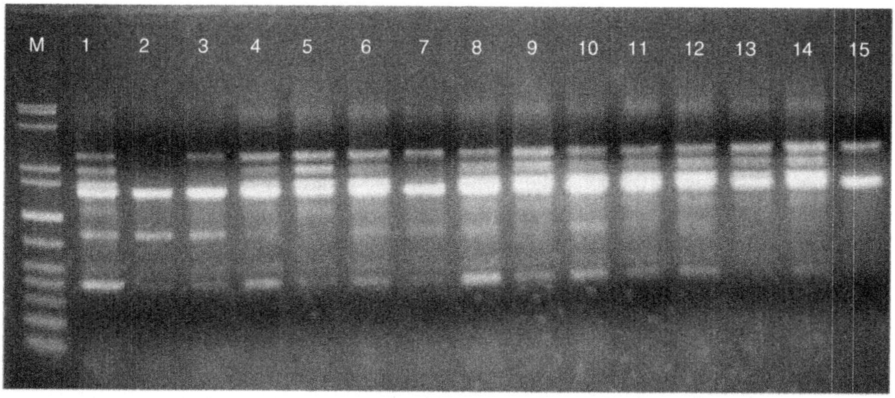

Fig. 10.4 Example of ISSR polymorphism-banding patterns in a subset of 15 Qatari date palm genotypes using primers # 10, M: 50 bp. Standard ladder marker; Lanes (1–15)

10.6.3 Molecular Characterization of Date Palm Using ISSR

To study the genetic diversity among date palm cultivars grown in Qatar, 15 date palm samples were collected from the Qatar University Experimental Farm. DNA was extracted from fresh leaves by using commercial DNeasy Plant System Kit (Qiagen, Inc., Valencia, CA). A total of 18 ISSR single primers were used to amplify DNA fragments using genomic DNA of the 15 samples (Ahmed et al. 2013). First screening was done to test the ability of these primers to amplify clear bands using date palm genomic DNA. All 18 ISSR primers successfully produced clear bands in the first screening. Then, each primer was used separately to genotype the whole set of 15 date palm samples. Total of 4,794 bands were generated using 18 ISSR primers for the 15 date palm samples. On average, each primer generated 400 bands. The number of amplified bands varied from cultivar to cultivar. The highest number of bands was obtained using Primers 2, 5, and 12 for the 15 (470 bands), while the lowest number of bands were obtained by Primers 1, 7, and 8 where they produced only 329 bands. Markers were scored for the presence and absence of the corresponding band among the different cultivars. A similarity matrix was constructed, and the similarity values were used for cluster analysis (Fig. 10.4).

10.6.4 De Novo Genome Sequencing and Comparative Genomics

Collaborative studies between Weill Cornell Medical College in Qatar and the Department of Genetics, University of Georgia, Athens, Georgia, USA, indicated long generation times (5–8 years) and dioecy (separate male and female trees), have complicated date palm cultivation and genetic analysis. A draft genome for Khalas cv. female, the first publicly available resource of its type for a member of the order Arecales, was carried out to resolve this issue. The ~380 Mb sequence, spanning mainly gene-rich regions, includes >25,000 gene models and is predicted to cover ~90 % of genes and ~60 % of the genome. Eight other cultivars, including females of Deglet Noor and Medjool and their backcrossed males were sequenced. More than 3.5 million polymorphic sites, including >10,000 genic copy-number variations, were identified in these studies. A small subset of these polymorphisms can distinguish multiple varieties. Al-Dous et al. (2011) identified a region of the genome linked to gender and found evidence that date palm employs an XY system of gender inheritance.

10.7 Date Production and Marketing

Qatar is a relatively new country where documented date production began in 1980, as compared with other countries of the Gulf region that were producing this commodity even before 1961 (the year after which data are available at FAO). Qatari date palm plantations represent 71 % from the total area planted under fruit trees. Major cultivation is in the north and middle areas of Qatar where environmental conditions are favorable, and the soil has a deep profile with low salinity as compared with other parts of the country. Initially, date palm was grown on 805 ha in 1980 (Table 10.6) with a production of 3,060 mt (Table 10.7). Since 1980, Qatar has made good progress in growing and producing dates. The harvested area achieved 2,469 ha in 2010, which gave a production of 21,491 mt.

In Qatar, the average yield for 2010 was 8.704 mt/ha, being the second highest in the region and higher than the world average of 6.277 mt/ha. Yield has increased progressively since 1980. However, the mean for 2005 (8.823 mt/ha) was slightly higher than in 2010 (8.704 mt/ha) (Table 10.8). Hence, it is vital that this status should not only be maintained but also increased to reach the maximum production being achieved in Egypt and other countries.

The area and production decreased slightly in 2011 (Table 10.9). The number of palms also decreased in 2011 (581,336) as compared to 2010 (617,375). The per

Table 10.6 Harvested area (ha) of date palm in Gulf countries in the last five decades

Years	Qatar	Kuwait	Bahrain	Yemen	Oman	Saudi Arabia	UAE
2010	2,469	5,089	1,603	14,955	31,353	171,975	197,400
2005	1,444	1,400	1,700	13,773	31,352	150,744	185,000
2000	1,931	1,350	823	22,755	35,508	142,450	185,330
1990	998	350	997	15,313	35,000	72,379	22,156
1980	805	0	4,000	11,618	20,100	60,353	5,564
1970	0	0	1,600	10,100	13,000	31,000	640
1961	0	0	1,600	10,456	13,000	22,000	500

Source: FAOSTAT (2010)

Table 10.7 Production (mt) of dates in Gulf countries in the last five decades

Year	Qatar	Kuwait	Bahrain	Yemen	Oman	Saudi Arabia	UAE
2010	23,500	16,700	14,000	57,849	276,400	1,078,300	775,000
2005	19,844	15,800	15,000	29,990	247,331	970,488	750,000
2000	16,116	10,155	16,508	29,834	280,030	734,844	757,601
1990	5,712	1,400	1,000	20,697	120,000	527,881	141,463
1980	3,060	0	38,000	14,240	70,000	342,286	51,157
1970	0	0	15,000	29,000	45,000	240,000	8,000
1961	0	0	15,000	38,100	40,000	160,000	6,000

Source: FAOSTAT (2010)

Table 10.8 Yield (mt/ha) pattern of date palm in Gulf countries during last five decades

Years	Qatar	Kuwait	Oman	Bahrain	Saudi Arabia	UAE	Yemen	World
2010	8.704	6.398	8.816	8.831	6.334	4.181	3.868	6.277
2005	8.823	11.286	7.888	11.286	5.58	4.054	2.177	5.85
2000	8.346	7.522	7.886	20.008	5.158	4.088	1.311	6.2
1990	5.723	4	4.8	5.729	7.293	6.385	1.351	5.496
1980	3.801	0	3.482	9.5	5.671	9.194	1.226	7.444
1970	0	0	3.462	9.375	7.742	12.50	2.871	7.137
1961	0	0	3.007	9.375	7.273	12.00	3.644	7.72

Source: FAOSTAT (2010)

Table 10.9 Some parameters showing performance of date palm in Qatar

	Year		
Parameters	2009	2010	2011
Total area, ha	2,237	2,469	2,366
Production, mt	20,815	21,491	20,696
Value of locally produced dates	374,670	343,861	525,895
Percent of total agricultural production	4.8	5.0	7.2
Number of plants	561,292	617,375	581,336
Average per capita availability, kg	19	18	14

capita availability of dates has been decreasing over time, as the reported values were 19, 18, and 14 kg per person in 2009, 2010, and 2011, respectively.

10.7.1 Comparison of Qatari Production with Other GCC Countries

The date palm-growing region of the GCC, comprising Saudi Arabia, UAE, Oman, Qatar, Bahrain, Kuwait, and Yemen, is the traditional and major home of this fruit tree (Al-Yahyal and Al-Khaniari 2008; Erskine et al. 2004). In 2010, the total

Table 10.10 Comparison of harvested area and production of GCC countries with the World total

	Production (mt)		Area (ha)	
Countries	2010	% of Gulf countries	2010	% of Gulf countries
Saudi Arabia	1,089,350	47.0	171,975	40.5
UAE	825,300	35.6	197,400	46.5
Oman	276,405	11.9	31,353	7.4
Yemen	57,849	2.5	14,955	3.5
Qatar	21,491	1.0	2,469	0.6
Kuwait	32,561	1.4	5,089	1.1
Bahrain	14,156	0.6	1,603	0.4
Total of Gulf countries	2,317,112	–	424,844	–
Total of world	7,683,432	–	1,224,139	–
Gulf % of world	30.16	–	34.71	–

Source: FAOSTAT (2010)

harvested area of date palm in these countries was 34.71 % of the world's total under this crop, while they produced 30.16 % of the total production of the world (Table 10.10). Qatar is not a significant country regarding date production. This crop was grown only in 0.6 % area of the GCC date palm area with a production of approximately 1 % of GCC in 2010 (Table 10.10). However, yield performance was better in comparison with the larger countries of the region: Saudi Arabia, UAE, and Oman. It was in second place (after Bahrain) in this regard in 2010. An interesting characteristic is the progress of yields over time, whereas three major countries were still lower than their respective yields of 1961. Although the total production of dates was also steadily increasing over time in Qatar, nevertheless, the area under this crop and total production could at least be doubled if the major constraint, scarcity of good quality water, is lessened. The government is taking wise decisions for boosting date production, and efforts are under way to develop alternative sources of water like treated wastewater and desalinized/diluted seawater. Research experiments have also been undertaken on this aspect.

10.7.2 Major Constraints of Date Palm Production in Qatar

Date palm is perhaps the oldest cultivated fruit tree. Therefore, it could be the leading plant in the refinement of very advanced cultivation practices coupled with all the modern techniques and be at the peak in yield performance. Unfortunately, this is not so. Many countries have yields that are lower even than they were in 1961. This is surprising and tells the story of ignorance. All plants face one or more constraints in various countries, but some of these may be common to all, though their nature, extent, and impact may vary. Date palm yields progressed in Qatar (maximum 8.823 mt/ha in 2005) but are still far below the actual potential of the crop (34.736 mt/ha in Egypt in 2000, but the 2010 average was 32.255 mt/ha). Thus, not only gap to be bridged is indicated but also the major constraints that should be identified and addressed systematically.

Most of the current agricultural management practices depend on inherited experience and recommendations adopted from different locations outside the country. The most appropriate technologies are always devised by research under local conditions but that has been not the case in Qatar. No studies have been done due to the lack of special departments for irrigation water use (qualities and quantities), fertilizer application, pests and diseases, postharvest management and fruit handling, marketing, packaging, and utilization of by-products. Therefore, major constraints are low-yielding cultivars and many nonproductive male plants in the field, deficiency of good quality irrigation water, and inappropriate irrigation systems (still flood irrigation at many growing sites) and soil and water salinity. Insect and disease attack, poor agronomic techniques, harvest losses, and proper postharvest care are other important factors. Low awareness on the part of date palm growers, the need for training and hampered technology transfer to the end users, lack of advanced packing and processing techniques, and poor coordination/collaboration with high date-yielding countries also contribute to the poor performance.

10.7.2.1 Low-Yielding Cultivars

Cultivars popularly grown in different countries vary broadly not only in their characters but also in their yield potential. When date palm cultivar development work is closely examined, one can clearly infer that there has been no or very little breeding work without any clear-cut direction in the past. The main reason may be propagation through offshoots, also called suckers. A positive development was establishment of a tissue culture laboratory in Qatar a few years ago. Recognized deficiencies of present cultivars are perceived in terms of the following aspects:

(a) Yield potential is generally low, especially among old palms still in the orchards.
(b) Fruit quality keeping is mostly inferior.
(c) Low response to fertilizer therefore gives low yields.
(d) Low drought resistant, comparatively lower disease and pest resistance, and they have less salt tolerance with a threshold at 3–4 dS/m (soil and water salinity is generally higher than this value where these date palms are grown).
(e) Lack of capability to keep the plant height as a dwarf because tall trees have inherent problems like sprays and other agronomic operations.

In addition, there are certain other problems related to date palm cultivars.

(a) Replacement of old cultivars with newer ones is another prominent problem because an orchard is productive for at least 35 years.
(b) Nonproductive plant (mostly male) numbers are very high (reported to be 50 % or so) and pose another crucial problem.
(c) Absence of a gene pool of date palm although this is very important to maintain biodiversity and preservation of plant resources.

(d) Lack of a strong and systematic exchange of breeding material, information, and data with other date palm-growing countries.
(e) Complete genetic fingerprinting, information on ancestors, and enlisting of inherent characters are lacking.

10.7.2.2 Deficiency of Irrigation Water and Appropriate Irrigation

Qatar lies in the MENA (Middle East and North Africa) Region, which is the most water-scarce area in the world. This region is home of 5 % of the world's population, yet it has less than 1 % of the total global renewable freshwater. Qatar, like other GCC countries, depends almost entirely on groundwater and desalinized seawater. The sandy nature of most of the soils coupled with high drainage heightens leaching losses and reduces water-supplying capacity of the soils. High evaporation also requires applying more water for crops including the date palm. Therefore, a sufficient supply of good quality irrigation water has always been a significant problem contributing to lower yields, and this is expected to continue to be significant in the future because of the impact of expected global and regional climatic change. It will become very difficult to provide more water for agriculture given the competition with domestic consumption, which is projected to increase further as a result of increased population and expected development, such as the FIFA World Cup 2022 planned for Qatar.

Water requirements of the date palm vary greatly depending upon soil texture, soil and water salinity, age of the plant, temperature, humidity, evaporation, irrigation technique, and season of the year. On average, date palms require 15,000–35,000 $m^3\ ha^{-1}$ per year which is quite a high requirement and mostly difficult to supply in water-scarce countries. The irrigation system also plays an important role in achieving high irrigation efficiencies. The traditional irrigation system of flood irrigation of basin or furrow is less efficient but cheaper and with lower labor requirements. On the other hand, modern systems like bubbler or drip irrigation are more efficient but installation costs are high, and they require high labor inputs. Keeping in mind the water scarcity, the present irrigation systems need to be improved and converted into subsoil irrigation systems.

10.7.2.3 Soil and Water Salinity

Due to the scarcity of good quality water, the utilization of saline water is significant in date palm irrigation due to the general notion that date palm is a salt-tolerant tree. However, this blanket statement cannot be accepted in full because studies have indicated that growth and yield losses start occurring at water EC of 4–6 dS/m depending upon cultivar potential, soil texture, drainage, and climatic factors. Ayers and Wescot (1985) reported that in general, minimum ECe for 100 % yield of date palm was 4.0 dS/m while there was no yield at ECe of 32 dS/m. The respective values for irrigation water ECiw are 2.7 and 21.0 dS/m.

A 50 % reduction in yield can occur at ECe 18.0 dSm^{-1} and ECiw 12. 0. Date palm was regarded as tolerant by Maas (1990), having a threshold value at 4.0 dSm^{-1}. The quality assessment studies in Qatar indicated that the major portion of groundwater is saline, and there are hidden negative effects on date palm growth and yields. When saline irrigation water is applied to a field, a major portion of it evaporates, leaving a salt load on or just beneath the surface. Thus, the soil salinity, which often is called secondary salinity, slowly and gradually increases and starts to negatively affect the date palms when it exceeds the plant's tolerance potential. The end result is decreased fruit yield.

10.7.2.4 Insects and Diseases

The date palm, like other tree crops, is attacked by a large number of insects, arthropods, fungi, nematode, and phytoplasma bacteria. Some of these cause serious crop damage and most often are difficult to control, thus threatening the very existence of plants in many cases. Considerable expense and effort are expended to control pests and diseases but most often success is limited, and a significant economic loss to the growers occurs.

Date Palm Insects The insects targeting date palm are of many types: insects, borers, scale insects, beetles, moths, and mites belonging to different orders and families. In the past, particular insect attacks were exclusive to individual countries or regions but with recent active exchanges of plant materials under somewhat loose quarantine measures; the situation has changed, and the pests have almost become common throughout the date palm-growing countries. It has been estimated that severe insect attacks cause a 50 % crop loss and in extreme cases, the entire crop may be damaged with ultimate death of the palms, especially during early stages of growth. Insects may infest all the plant parts: roots, stems, leaves, and fruit. For example, unripe fruits are attacked by *Coccotrypes dactyliperda*, which causes premature fruit fall. Ripe fruits are often attacked by nitidulids, *Carpophilus hemipterus*, *C. mutilatus* (*C. dimidiatus*), *Urophorus humeralis*, and *Heptoncus luteolus*, which cause decay. The red palm weevil, *Rhynchophorus ferrugineus*, bores into the leaf bases at the top of the trunk, causing the entire crown to wither and die. A tineid moth and a beetle, *Lasioderma testacea*, damage stored dates. Dates held in storage also are subject to invasion by the fig moth, *Ephestia cautella*, and the Indian meal moth, *Plodia interpunctella*.

Date Palm Diseases Date palms are susceptible to many diseases as well. Like insects, diseases can also cause significant yield losses, and fruit quality may also be damaged. For example, bayoud disease is caused by the fungus *Fusarium oxysporum*. Decay of the inflorescence is caused by *Manginiella scaeltae* in humid seasons. Date palm decline may be a physiological or the result of a species of the fungal genus *Omphalia*. Diplodia disease is a fungus manifestation on leafstalks and offshoots, and it may kill the latter if not controlled. The fungal-caused condition called *black scorch* stunts, distorts, and blackens leaves and adjacent inflorescences. Other fungal diseases include pinhead spot (*Diderma effusum*), gray blight (*Pestalotia palmarum*), and spongy white rot (*Polyporus adustus*). *Ceratocystis paradoxa* (*Thielaviopsis paradoxa*) causes rot of the terminal leaves (buds); if not properly controlled, it usually spreads and kills the palm. At some stage of its prognosis, the leaves of the infected palm are abnormal and grow in different directions creating the phenomenon called *Al-Majnouna*, followed by the death of the palm.

10.7.2.5 Agronomic Practices

The cultural and agronomic practices prevailing in Qatar for the date palm are either traditional or were devised elsewhere but directly adopted without any local research. The major deficiencies at present are as follows:

(a) Site-specific determination of optimum requirements of irrigation water, manure, and fertilizer for all the popular cultivars individually to obtain maximum possible yield using appropriate amount of inputs.
(b) Water conservation techniques for different areas have to be standardized so that scarce and dwindling resources are used appropriately while also being sustained for the future. This is very important to combat future climatic changes when even less water will be available.
(c) Narrow spacing among trees in old plantations makes it difficult for the introduction of mechanization, inefficient hand pollination, leaving many of the date palms unpollinated; minimal fertilization and minimal irrigation that results in lower yield and fruit quality. Little attention is given to fruit thinning and pruning.

10.7.2.6 Harvesting Losses and Postharvest Care

In one way per hectare yields of date palm are low and facing a decline with significant losses occurring either during picking or postharvest handling. Not only the quantity may decrease but quality of fruit may also be diminished. Fruit not harvested at the appropriate time leads to a decline in its quality during storage and for shipping and consumption. Extremely dry weather sometimes causes dates to shrivel on the palm. High humidity or cool temperatures during the maturing period may cause fruit drop or checking, splitting of the skin, darkening, black nose, imperfect maturation, and excessive moisture content or even rotting. Often, harvested fruit is not handled, transported, and stored properly. A tineid moth and a beetle, *Lasioderma testacea*, may damage stored dates. Dates held in storage are subject to infestation by the fig moth, *Ephestia cautella*, and the Indian meal moth, *Plodia interpunctella*. Very little research work is available on this storage problem, and even less has been adopted by growers and other stakeholders. Efforts need to be made to generate advanced harvesting techniques as well as avoiding postharvest losses so that minimum losses occur to the fruit during transportation and storage.

The improvement of keeping quality will ultimately increase the price of the fruit. Serious efforts are required to educate growers about effective handling techniques, adopted in other date palm-growing countries; however, they will not be accepted in Qatar unless or until locally developed recommendations become available.

10.7.2.7 Awareness of Date Growers, Training, and Technology Transfer

The developments of new techniques in date palm production, harvesting, and preservation have been very slow and insufficient to meet the real needs of the growers. In addition, the inadequacy of awareness/training programs is a weak link. Until a technology is adopted in the field, how much useful it may be, any investment of funds and resources goes to waste. An energetic stimulus is direly needed so that the efforts are clearly defined and implemented.

10.7.3 Research and Development Needs

Date palm research and development efforts must be systematic and strengthened to overcome and make up for past deficiencies. Not only site specific, economical, and practicable recommendations have to be formulated but these have to be transferred to the growers ensuring their field adaptation. The role of industry and private sector will also be highly important, especially in the development of new products and by-products, trade, and marketing. The following aspects of all the research and development strategies are recommended. A national level research institute for date palm has to be established. Another body should take responsibility for development work on date palm to create new areas for the crop and alleviate the constraints discussed above. Different research and development projects have to be undertaken keeping in view specified problems, the status of present work and future targets as well as needs. Good planning can provide the basis of the most appropriate and rapid development of date palm in keeping with the traditions and conservation of resources for the future and applying an environmentally friendly approach. These recommendations have been submitted to the government to take the necessary steps.

10.7.3.1 Date Palm Research

Keeping in mind the future requirements of growing date palms more effectively in Qatar, some research is either set to begin in 2014 or is approved and will start as soon as administrative formalities and technical requirements are completed. The ICARDA Project, Development of Sustainable Date Palm Production Systems in the GCC Countries of the Arabian Peninsula, is a research and development project aimed at generating new knowledge and practices for improving the production systems of date palm in the Gulf region. The activities are comprised of improving cultivar productivity, managing natural resources (land and water) for optimal performance, use of different inputs in the cropping process (fertilizers, pollinators, wastewater, etc.), and the study of genetic diversity of date palm. The sharing and transfer of technology and experience among the GCC countries are an integral part of the project. The following activities are in progress or planned in Qatar.

Crop Management Fruit thinning to improve air circulation within the fruit bunch has proved very effective, but on-farm testing is still under progress. Studies on irrigation scheduling also are under way.

Postharvest Handling and Processing Techniques Locally made glass chambers, ventilated and equipped with shelves, have proved effective for drying dates in shorter periods (5 days only vs. 12 days under plastic tunnels or 14–18 days for direct sun drying). The equipment also helps in reducing skin separation and keeps the date fruits much cleaner.

Saline and Treated Wastewater Effects on Date Fruit Quality and Heavy Metal Concentration When Managed with Soil Conditioners This experiment has been approved under Activity 9 (Effect of irrigation with treated water on fruit quality and heavy minerals content) of the ICARDA (2010) project. The objectives are to investigate the effect of saline water and wastewater on date quality under Qatari conditions and to study the concentration of heavy metals, if any, in soil, plant leaves, and date fruits. This is a two-factor experiment of strip-plot design with three replicates, comprising of two soil conditioners: Zeolite at 2 % by volume and Compost at 10 kg/plant/year. Plants will be irrigated with two types of water: groundwater available at the farm and treated wastewater. Fruit mass in each season will be recorded, and soil, leaf, and fruit analyses will be completed.

10.7.3.2 Future Research Projects

A few projects proposed to fill the gaps and deficiencies of locally evolved technology for growing of date palm and increasing production in the country to make it self-sufficient are under consideration. The Government of Qatar should consider this on a prioritized basis to establish a date palm research and development body for preparation, collaboration, and implementation of such projects.

Research Project to Evaluate High Yielding and Stress Resistant Cultivars High-yielding good quality date palm cultivars capable of withstanding different stresses such as biotic (pests and diseases) and abiotic (drought, salinity, high temperature) are the demand of the day. In addition to conventional breeding, modern techniques of biotechnology and genetic engineering should be used to create new cultivars. Breeding material may be brought in from countries possessing cultivars of high potential.

Research Project for Water Conservation and Minimal Irrigation Requirements The circumstances of water availability are not encouraging, and the situation is expected to deteriorate further in view of future climatic changes deemed to occur rapidly. Therefore, water utilization per kilogram of dates has to be minimized for effective and efficient use of the limited precious water resources. For this purpose, water losses through leaching and evaporation have to be minimized, and water utilization has to be justified. Appropriate water requirements are to be determined. This project will encompass research and development activities. Research activities will stimulate design of engineering and agronomic techniques like field leveling, mulching (plastic or harvested and grinded date palm leaves), application of soil conditioners, improvement of soil physical properties (infiltration rate and water-holding capacity) through cultivation practices, and application of organic matter (manures, compost, and raw organic material from crop residues, poultry, and other agricultural allied industries). Research has to be systematic and develop a few but significantly contributing techniques. The development activities will consist of transforming traditional irrigation system (mainly flood irrigation) into bubbler or drip irrigation.

Research Project to Combat Soil and Water Salinity Effects Soil and water salinity are very significant problems hampering date palm growth and yield, and the impact is expected to be exacerbated further with predicted climatic changes. The very limited quantity of good quality irrigation water will consequently require greater utilization of saline groundwater, which will definitely result in salt accumulation in the rhizosphere and secondary salinity buildup. Conducting research on techniques like leaching fractions for various combinations of cultivars, soil textures, climatic conditions, irrigation water of variable salt content, and different soil salinity levels will be another component of the research. The impact of organic matter applications to improve physical properties of soil and for providing relatively better environment conditions to plants even in saline soils and mulching to reduce evaporation and salt accumulation and resultant adverse effects on growth and yield of date palm could be other investigating components.

Research Project for Generation of Improved Agronomic Practices Agronomic practices are key considerations for date palm growth and play a critical role in yield performance. Most of these are traditional, general, and less quantitative in nature. For example, details are not available to recommend specific quantities of fertilizer and manure for a specific cultivar grown in a particular type of soil. Specific stage and time of application and types of fertilizer are also unclear. Similarly, the ideal harvest time for each cultivar is not specified. Therefore, establishing specific recommendations for various agronomic operations is very important for achieving maximum yields. Fertilizer formulations for each important cultivar under a specific set of soil and climatic conditions, investigating effects of growth hormones and conducting research on climatic change, e.g., change in temperature, humidity, evaporation, etc., are important priorities.

Research Project on Integrated Pest Management (IPM) Date palm fruit loss and deterioration of quality caused by insects have become highly significant in the last two decades. Various types of insects such as borers, bugs, beetles, moths, weevils, scale insects, beetles, and mites attack the plant. Dubas bug and red palm weevil have become very important in recent years and are out of control at certain locations. The most common practice against insects is to use chemical pesticides. However, objections are increasingly being raised against such practices. Environmental considerations and concerns about the residual effects of insecticides are emerging rapidly. Therefore, it is logical that an IPM approach be adopted. Collaborative regional efforts to control common insects may be required. Putting strict quarantine measure in force to block entry of new insects and implementing awareness programs among date palm growers will strengthen the efforts.

Research Project on Control Measures Against Date Palm Diseases Diseases are no less harmful to the date palm than insects. Diseases may retard growth, cause withering or mortality of plant or deteriorate fruit quality and thus ultimately decrease yield and the income of growers. Therefore, it is very important to conduct very relevant research to minimize such losses through controlling diseases if they cannot be eradicated totally.

Research Project to Implement Modern Harvesting Techniques and Reduce Postharvest Losses Although proper harvesting of fruit of date palm is very important to minimize losses and deterioration of quality, very little improvement has occurred in traditional techniques. It is highly desirable that fruit harvesting be modernized by adopting new technologies and mechanization such has occurred with other crops. Finding the appropriate harvesting time of each cultivar, single or multiple harvests and date fruit types (soft, semidry or dry) are called for, while keeping in mind prevailing seasonal conditions as well as future climatic changes.

Research Project on Packing, Long-Term Preservation, Value-Added Products and By-products Date production is minimally profitable to growers because of post-harvest losses, few value-added products, and little utilization of residual plant materials such as leaves and empty fruit bunches, which typically are burned. These factors must be considered seriously to add value to date palm growing along with efforts to enhance fruit production. Favorable income can be realized for farmers if new value-added products derived from by-products of date fruits and residual material are introduced. Development of new techniques to store dates for longer periods, the introduction of new products by combining dates with other food items, dry fruit and related foods, along with, using crop residues and stem wood and petioles to make useful and attractive articles for home use, as well as composting and mulching of waste material to enhance crop production are worthy subjects for future research and development.

10.7.3.3 Training Project for Development of Human Resources

Date production by traditional means is centuries old in Arabian countries. However, there is a dilemma because local manpower and human resources are neither well informed nor appropriately or sufficiently trained to carry out modern production, preservation and processing of the crop. Therefore, proper training is of high importance to assure a progressive future, self- reliance, and self-dependence. Qatar has to take a note of this important requirement.

10.7.3.4 Development of New Land Areas for Date Palm Cultivation

Project for Development of Land and Water Resources for Date Growing Land and water resource surveys are a basic requirement to point the agriculture sector of a country in the proper scientific direction. Feasibility studies of land classification and suitability provide critical information to manage these resources effectively while highlighting significant problems to be overcome for growing date palms. Identified problems can best be solved on a prioritized basis.

Development Project to Replace Old Orchards/Cultivars with New Large Modern Orchards by the Government and Private Sectors One of the major reasons for low date fruit yields is the presence of old cultivars having very little potential, as well as low resistance to any kind of stress. Nonproductive male plants (more than needed for pollination) are widespread in the orchards. Therefore, it is urgent that not only old cultivars should be replaced but also the government and private owners/companies should establish some new mega date palm orchards using modern technologies. The following sequential steps are recommended:

(a) Carry out surveys to select orchards at different sites that need replacement.
(b) Continue land classification surveys to select new sites and areas more appropriately suitable for growing date palms.
(c) Prepare a plan for stepwise removal/replacement of old cultivars as well as establishment of new larger orchards in the public sector.
(d) Purchase of land by the government declared suitable for growing of date palm.
(e) Establish mega date palm farms owned by the private sector and maintained through modern techniques.

10.7.3.5 Probabilities of Increasing Date Palm Plantations in Qatar

Presently, date palm is grown on 2,366 ha of land in Qatar, but there is very high potential to increase this area if prevailing problems are resolved. Considering the land evaluation data presented in Table 10.11, it is clear that only Mapping Unit 9 and 12 have problems of slope that are difficult to address. However, all the other mapping units have problems like cleaning (coarse surface fragments), salinity, water availability, and low soil fertility (Anonymous 2004). Thus, about 171,595 ha (Mapping Units 7, 10, 11, 14, 15, and 17) can be brought under date palm by providing irrigation water and application of suitable fertilizers. Hence, there is the theoretical potential to increase date palm area by some 70 times over the present area (see Fig. 10.5).

Table 10.11 Land suitability of Qatari soils for growing date palm

Mapping units	Suitability class	Area (ha)	Area (%)
1	N1c	16,279	1.4
2	N1c	281,860	24.5
3	N1c	242,151	21.1
4	N1c	213,759	18.6
5	N1c	9,989	0.9
6	N1c	21,722	1.9
7	S3m	34,596	3.0
8	N1c	9,849	0.8
9	S3t x c	11,161	0.1
10	N1n	26,355	2.3
11	N1n	61,415	5.3
12	N2t	14,432	1.2
13	N1c	32,429	2.8
14	N1n	3,055	0.3
15	N1m n	24,249	2.1
16	N1c	7,152	0.6
17	N1n	21,924	1.9
18	N1c	22,402	2.0
19	N1c	17,177	1.5

Source: Anonymous (2004)
S3 marginally suitable, *N1* presently not suitable, *C* clearing limitation (surface coarse fragments), *t* due to land grading limitation (topography), *x* due to salinity limitation (leaching water requirement), *m* due to moisture availability limitation (water availability), *n* due to nutrition limitation (NPK application)

10.8 Processing and Novel Products

No processing, packing or novel products production industry has yet been established in Qatar. Most dates are marketed fresh without any formal packaging. However, imported packing materials of all types are available in the market.

10.9 Conclusions and Recommendations

The area under date palm and production (in 2010, 2,469 ha; 21,491 mt of dates) indicates that Qatar is relatively small date producer. Yield has been increasing steadily since 1980. Good potential exists to expand the area and production of dates if existing soil and water resources are used efficiently. Date palm feasibility studies indicate that 171,595 ha potentially can be brought under this important fruit in the country. However, major constraints of low yields, scarcity of good quality water, soil and water salinity, low-yielding cultivars, poor

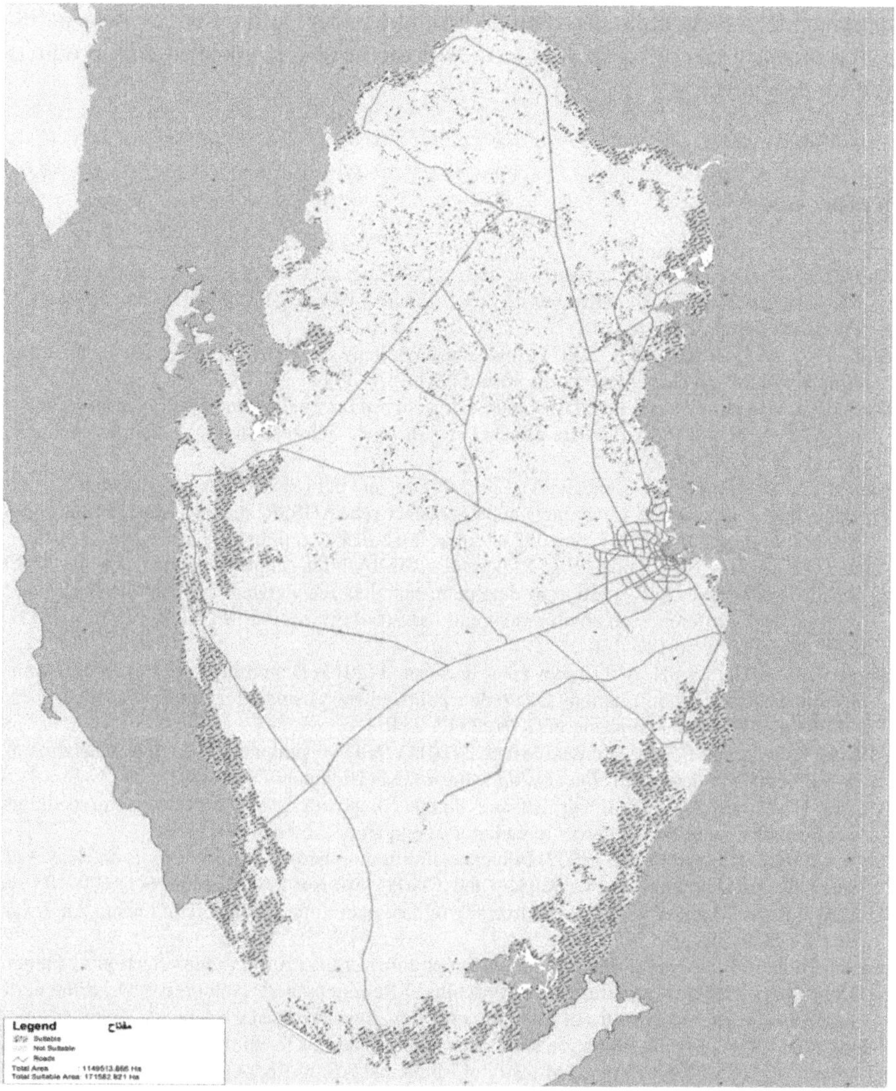

Fig. 10.5 Potential suitable areas for date palm cultivation (shaded in *green*) (*Source*: Department of Agriculture and Water Resources)

agronomic practices due to limited know-how of the growers, insects and diseases, and lack of research support all need to be addressed on a prioritized basis. Tissue culture is to be strengthened to meet needs of the country and stop the importation of plantlets. Research and development facilities for date palm must be established. A considerable amount of research and development work is required to promote date palm cultivation and achieve higher yields. For this purpose, many research and development projects have to be planned and implemented. New areas have to be brought under cultivation. A date palm-based industry has to be created to reduce the need for imported date products and by-products.

References

Ahmed TA (2013) Molecular diversity in date palm using inter simple sequence repeat (ISSR) Markers. Qatar Found Annu Res Forum Proc 2013:EEP 083. doi:10.5339/qfarf.2013.EEP-083. Published online: 22 Nov 2013

Ahmed TA, Al-Qaradawi AYK (2009) Molecular phylogeny of Qatari date palm genotypes using simple sequence repeats markers. Biotechnol 8(1):126–131

Ahmed TA, Al-Qaradawi AYK (2010) Genetic diversity of date palm genotypes in Qatar as determined by SSR and ISSR markers. Proceedings of fourth international Date Palm conference. Acta Hort 882:279–286

Ahmed TA, Al-Hadidi S, Al-Qaradawi AY (2012a) Inter- and intra-specific genetic variations among Qatari date palm cultivars using inter simple sequence repeat (ISSR) markers. Qatar Found Annu Res Forum Proc 2012:EEP87. doi:10.5339/qfarf.2012.EEP87. Published online: 19 Oct 2012

Ahmed TA, Alsamaraee Zaidan SA, Elmeer K (2012b) Inter-simple sequence repeat (ISSR) Analysis of somaclonal variation in date palm plantlets regenerated from callus. In: Second international conference on environment and industrial innovation, IPCBEE, vol 35. IACSIT Press, Singapore, pp 126–130

Ahmed TA, Al-Hadidi SH, Al-Qaradawi AY, Radwan O (2013) Determination of inter- and intra-specific genetic variations among Qatari date palm cultivars using inter simple sequence repeat (ISSR) markers. Afr J Biotechnol 12(19):2540–2546

Al-Dous EK, George B, Al-Mahmoud ME et al (2011) *De novo* genome sequencing and comparative genomics of date palm (*Phoenix dactylifera*). Nat Biotechnol 29(6):521–528

Aljuburi HJ, Maroff A (2007) The growth and mineral composition of Hatamy date palm seedlings as affected by seawater and growth regulators. Acta Hort 736:161–164

Alromaihi KIKB, Elmeer KMS (2009) Influence of different media on in Vitro roots and leaves of date palm somatic embryos cvs. Kapkap and Tharlaj. Am-Eur J Agr Env Sci 6(1):100–103

Al-Yahyal R, Al-Khaniari S (2008) Biodiversity of date palm in the Sultanate of Oman. Afr J Agr Res 3(6):389–395

Annual Bulletin for Areas & Production of Agricultural Crops (2011) Affairs Section of Farms, Doha Qatar. General Department for Agricultural Research and Development, Department of Agricultural Affairs, Ministry of Environment, Government of Qatar

Anonymous (2004) Soil classification and land use specifications for the State of Qatar. Phase one report, vol I, Soil survey report, Ministry of Municipal Affairs and Agriculture, Department of Agriculture and Water Research, Soil Research Section, State of Qatar

Ayers RS, Westcot DW (1985) Water quality for agriculture. Irrigation and Drainage paper 29. FAQ, Rome. http://www.fao.org/DOCReP/003/T0234e/T0234e00.htm

El-Bably AZ (2002) Advanced and integrated approaches for crop tolerance to poor quality irrigation water in Egypt. Soils, Water and Environment Research Institute, Agricultural Research Center, Giza, Egypt. Opt Mediter Série A 50:363–378

Erskine W, Moustafa AT, Osman AE (2004) Regional workshop on date palm development in the Arabian Peninsula. Date palm in the GCC countries of the Arabian Peninsula. Abu Dhabi, 29–31 May 2004. http://www.icarda.org/APRP/Datepalm/Fo&Ac/foreword.htm

FAOSTAT (2010) Dates crop yields and production. http://faostat.fao.org/default.aspx

Hussain NM, Ahmad TA (2014) Final Report QNRF Proposal NPRP 3 09-705-4-025. Improvement of date palm production and quality under Qatar conditions. Date palm crop management component. Qatar University, Doha

Hussain NM, Osman S, Amir K (2014) Effect of Zeolite on date palm production. Unpublished report
ICARDA (2010) Project report 1 of the project Development of sustainable date palm production systems in Gulf Cooperation Council Countries. Damascus, Syria
Jaradat AA (2005) Saline agriculture in the Arabian Peninsula; management of marginal lands and saline water resources. J Food Agr Env 3(2):302–306
Maas EV (1990) Crop salt tolerance. In: Tanji KK (ed) Agriculture salinity assessment and management manual. ASCE, New York, pp 262–304
Mulumba LN, Lal R (2008) Mulching effects on selected soil physical properties. Soil Till Res 98:106–111
Qatar Tribune (2014) Daily Newspaper, Doha, Qatar, 18 Feb 2014, vol 8, no. 2724. www.qatar-tribune.com
Ramakrishna H, Tam M, Wani SP, Long TD (2006) Effect of mulch on soil temperature, moisture, weeds infestation and yield of groundnut in Northern Vietnam. Field Crop Res 95:115–125
Xie ZK, Wang Y, Li F (2005) Effect of plastic mulching on soil water use and spring wheat yield in arid region of North West China. Agr Water Manage 75:71–83

Chapter 11
Date Palm Status and Perspective in Bahrain

Thuraya A. Almansoori, Mohammed A. Al-Khalifa, and Abdulaziz M.A. Mohamed

Abstract Throughout human history, the date palm (*Phoenix dactylifera* L.) has been an integral part of the rich heritage of the Kingdom of Bahrain. Dates have played a fundamental role in the economy and the social life of Bahrain's patrimony. However, the revolutionary changes that have accompanied the discovery of oil have had a massive impact on the cultivation of date palms. A number of factors have reduced interest in making investments in date palm cultivation; among them are socioeconomic changes, climatic conditions, limitation of freshwater, salinization of agricultural lands, and the spread of indigenous and invasive alien pest species. Over the last decades, the number of date palm trees has decreased from 892,000 to 534,600; those remaining are currently distributed over the five governorates of the Kingdom. In spite of the challenges, Bahrain is striving to care for date palms as a state symbol and national treasure. Ex situ germplasm gene banks, in vitro facilities, and quarantine regulations and legislations have been established for the date palm. Nevertheless, Bahrain's annual yield does not exceed 16,000 mt. Consequently, improving date palm industry is a necessity to emphasize that the date palm tree is a valuable asset that unites and integrates the variant parts of the ecosystem and keeps it intact as a stable, well-defined, and sustained entity. Long-term sustainable development demands modernization of the currently performed conventional practices of date palm care, harvesting and postharvesting processes, trading, and marketing, which have not risen to the standards required. In this regard, this chapter describes the current status of the date palm cultivation in Bahrain and the threats confronting the date palm conservation and industry development and aims to highlight potential resolutions and prospects.

T.A. Almansoori (✉)
Department of Biology, College of Science, University of Bahrain,
Box 32038, Sakhir, Kingdom of Bahrain
e-mail: talmansoori@uob.edu.bh

M.A. Al-Khalifa
Department of Agriculture Affairs, Ministry of Municipalities Affairs and Urban Planning,
Box 251, Budaiya, Kingdom of Bahrain
e-mail: shmohamedalkhalifa@gmail.com

A.M.A. Mohamed
Department of Life Sciences, Arabian Gulf University,
PO Box 26671, Manama, Kingdom of Bahrain
e-mail: amamohamed@agu.edu.bh

© Springer Science+Business Media Dordrecht 2015
J.M. Al-Khayri et al. (eds.), *Date Palm Genetic Resources and Utilization: Volume 2: Asia and Europe*, DOI 10.1007/978-94-017-9707-8_11

Keywords Bahrain • Cultivars • Date palm • Genetic resources • Marketing • Practices • Production • Tissue culture

11.1 Introduction

The existence of human settlement in Bahrain Island, which was historically dubbed as Dilmun, Tylos, or Awal, dates back more than 6,000 years as shown by archaeological evidence (Bibby 1972; Larsen 1983). An abundance of natural freshwater from surface wells and natural springs, along with its fertile land, made agriculture an important activity and form of income for the inhabitants of Bahrain at that time.

The agricultural areas of Bahrain are concentrated in the north and northwest coast of the island with isolated areas in the north, central, and along the east coast. These coastal areas boast good water quality and potential irrigation, along with numerous date palm plantations, which constitute the majority of the agricultural land in Bahrain (Al-Basheer and Haroon 1997; Al-Khalifa 2009; Doornkamp et al. 1980).

Date palm, *Phoenix dactylifera* L. (Arecales: Arecaceae), played a significant role within the culture of Bahrain's early Dilmun people. This is supported by many inscriptions and historic texts from ancient Sumaria, which reveal that date palms were utilized in Bahrain since the second millennium B.C. (Nesbitt 1993). The strategic location of Bahrain played a vital role of benefit to traders traveling between ancient Mesopotamia and the Indus Valley Civilizations which made it an ideal trade route stopover between both civilizations to acquire freshwater and dates for expeditions (Nesbitt 1993; Noor Al-Nabi 2012).

Bahrain was known for its vast date palm plantations throughout history. Date palm trees were grown not only for their nutritious fruits but in the pursuit of efficiency, every single part of the palm tree had a worthwhile purpose. Date palms were a fundamental source of life for Bahrainis. They relied on the palm trees in many and various aspects of living. For example, the trunk was used as timber for building structures and as well as for fuel; the leaves were exploited to construct traditional houses, locally known as *barusti*, in addition to fences for farms and chicken coops; and the leaflets were a source of natural material to weave mats, baskets, and hand fans. The leaf midribs (i.e., rachis) were utilized to assemble fish traps and bird cages; they made use of the leaf fibers as fuel and for weaving rope. The fruit spikelets were used to make hand brooms, baskets, and fish cages. Bahrain was and is still very well known for a customary and refreshing drink, which is extracted from the male spathes and pollen. Even the cull date fruits and seeds were used as cattle feed.

From an environmental standpoint, Bahrainis used date palm trees as shelter sanctuaries for birds. In addition to all of the above, currently date palm trees have become an important element of landscaping projects in Bahrain (Ahmed 1978; Al-Khalifa et al. 1994; Barreveld 1993).

Table 11.1 Estimated number of date palms in the governorates of the Kingdom of Bahrain 2010

Governorate	No. date palm trees
Capital	66,763
Muharraq	36,472
Central	113,110
Northern	253,110
Southern	65,145
Total	534,600

Source: Annual Agricultural Statistics Bulletin (2008)

At the beginning of the twentieth century, revolutionary changes of industrialization and rapid modernization accompanied with the discovery of oil inevitably cast their effects on the agriculture sector in general and on date palm cultivation in particular (Al-Rumaihi 1975). Trained workers abandoned date palm cultivations, and replacing them was challenging due to the reluctance of the youth to join that sector and the preference of the industrial sector, which lured them with the promise of a better income (Anon 2011). Additional and harmful effects derive from the modern economic boom, emergence of oil industries, accelerated urbanization, scarcity of freshwater, and increasing soil salinity. All these factors combined have greatly and negatively impacted the agricultural lands and their productivity (Al-Basheer and Harron 1997). As a result, the number of date palms decreased dramatically from 892,000 at the beginning of the 1970s to about 534,600 which are currently distributed over the five provinces of the Bahrain (Table 11.1).

With the current Bahrain government policy toward promoting interest in the agriculture sector, date palms have been receiving increasing and special attention. The government has established ex situ germplasm gene banks and in vitro facilities to conserve the local germplasm, as well as having issued quarantine regulations and legislations to protect date palms, which have and always will be considered the quintessential symbol and national treasure of the Kingdom of Bahrain.

This chapter aims to carefully describe the present status of date palm trees in Bahrain. Date palm cultivation practices, genetic resources, conservation efforts, tissue culture, cultivar identification and description, as well as date production and marketing situations will be discussed. Threats confronting date palm conservation and industry development and potential resolutions and prospects will be indicated.

11.2 Cultivation Practices

The total land area of the Kingdom of Bahrain is about 76,700 ha, of which the cultivable land is estimated to be 6,400 ha (i.e., 8.3 % of the total area). Currently only 3,516 ha, representing (55 %) of the arable land, are under cultivation. The cultivated area is distributed among three major crop plants: date palms (2,330 ha); vegetables, both in open field and protected agriculture (572 ha); and forages (614 ha).

Fig. 11.1 The agroecosystems of date palm in the Kingdom of Bahrain (Credited to PWD, Agric. Affairs). (**a**) Intercropping of the alfalfa between the date palm rows, (**b**) intercropping of leafy vegetables between the date palm rows, (**c**) date palms grown at the edges of the vegetable beds, and (**d**) date palms planted as a border between plastic houses and open vegetable beds in an open field

In old date palm plantations, growers tended to plant the palms close to each other and sometimes in uneven rows, making the use of agricultural machinery difficult. However, the agroecosystem existing today (Fig. 11.1) reveals that the current common practice relies on the intercropping of cash crops like forages, vegetables, and/or other fruit trees between the date palm rows. Agricultural land is divided into beds (2.5×6 m) with date palm offshoots planted in the middle or towards the edges of the beds, and the inner spaces are utilized to plant forages or vegetables. As for monocrop date palm plantations, the land is divided into rows with a space of 6–7 m in between, and the offshoots are planted in these rows at fixed intervals (Ahmed 1978; Al-Khalifa et al. 1994).

Proper date palm cultivation requires following several practices considered important to attain effective growth rate and fruit quality. These practices are performed by skilled farmers with extensive experience. Commonly performed practices (Al-Khalifa et al. 1994; Mohamed 2000) are summarized below.

11.2.1 Dethorning

The process of thorn or spine removal from the base and sides of the new leaves is known as dethorning (i.e., *tarwies*). This process is important to permit easier access to the spathes during pollination and the fruit bunches during harvest. In Bahrain, this practice has been frequently performed at the beginning of the season, which ranges from January to February, prior to pollination.

11.2.2 Pollination

In order to achieve adequate fertilization, date palm growers tend to implement artificial pollination (i.e., *tanbeet*), where they transfer male pollens to the female flowers once the male spathes has opened. Traditionally, as soon as pollen spathes are open, growers collect the male inflorescences and shake a bunch of them over the female inflorescences to release pollen. Then, the male inflorescence cluster is tied in the middle of the female flower bunch using a palm leaflet. The pollination period in Bahrain ranges from February to March.

An investigation was carried out to study the effect of liquid pollination on fruit quantity and quality. However, the statistical analysis revealed no significant variation when compared to the traditional methods in use (ICARDA 2011). Regardless, implementing this practice is highly recommended in Bahrain. In recent years and due to the lack of trained workers and increased wages, some of the growers began the liquid pollination technique, where the female flowers are sprayed with solution containing male pollen (1 g of pollen per liter of water) using either manual or mechanical sprayers. In this way, pollination can be accomplished without having to climb to the top of the tree. This process is performed after the spathes are fully opened, and the whole procedure is repeated three times, with a period of 3–7 days between applications, depending on the cultivar. This process has many advantages compared to the traditional method. In addition to being a quick and efficient procedure, it reduces the risk of climbing the tall trees, helps to overcome the scarcity of the skilled labor, and minimizes the overall cost of the pollination process.

11.2.3 Bending

The growers bend the fruit branches downwards (i.e., *tahdeer*) and tie them to the adjacent leaves. This practice is done to ensure that the fruit branches are secured and will not break in response to the wind and their own weight. This procedure has the added benefit of allowing easier access to the fruits during the harvesting period. Bending is usually performed two months after pollination during April and May when the fruits are small and green at the kimri stage.

11.2.4 Thinning and Pruning

This process is executed to reduce the weight of the fruit bunch and give the fruits more space to grow larger in size. This process also increases aeration and minimizes damage caused by moisture accumulation. This process is achieved either by reducing the number of the fruits bunches per tree (i.e., *alkhaf*) or by reducing the size of the flower spikelets and hence reducing the number of the fruits per spikelet.

Growers tend to remove dry fruit spikelets, dry and damaged leaves, and any remaining petioles (i.e., *taghleeq*). This is usually done in December and January in preparation for the new season.

11.2.5 Harvesting

Date palm offshoots usually start bearing fruits 3 years after planting and reach maximum production at 10 years of age and continue bearing fruits for many years thereafter. In Bahrain, dates are harvested during khalal and rutab stages, which are usually consumed fresh. Due to the multiple cultivars, the harvest season begins in July and ends in November. With young date palms, hand picking is easy; however, as the tree grows older and taller, the harvesting process gets more laborious and time consuming. Since not all of the fruits ripen at the same time, but rather in stages, skilled labors need to climb the palm to reach the crown to harvest the ripe fruits several times per season to avoid yield loss. If this has not been accomplished and the ripe fruits are not harvested in time, they may become infested or ruined by pests such as birds, insects, and rodents, which reduce their market value drastically. For that reason, harvesting is accomplished in several stages that coincide with the maturity of the fruits. This makes the process costly and dangerous, especially if the palms are old and tall.

11.2.6 Pest and Disease Control

In addition to abiotic stresses, date palms in Bahrain are susceptible to biotic stress caused by pests of both indigenous and invasive species that can cause severe damages if left uncontrolled. Pest control measures have not received proper attention from growers. However, the government has taken effective steps to help in combating date palm pests through an Integrated Pest Management (IPM) program implemented by the Plant Wealth Directorate, Agricultural Affairs, and Ministry of Municipalities Affairs and Urban Planning. Other commitments comprise enforcing the implementation of plant quarantine regulations, providing growers with the service of pesticide sprays at low cost, providing technical advice to individual growers for insect identification and control, and conducting surveys of the indigenous and invasive pest species.

A number of date palm pests are recognized locally; however, the burgeoning international trade and the ease of movement of large numbers of people across borders have escalated the problem of unintentional introduction of new invasive species. The introduced species of concern are the red palm weevil *Rhynchophorus ferrugineus*; the rhinoceros beetle, *Oryctes* sp., (*Coleoptera*: *Scarabaeidae*); the inflorescence beetle, *Macrocoma* sp. (*Coleoptera*: *Chrysomelidae*); and the scale insect, *Fiorinia phoenicis* Balachowsky (*Homoptera*: *Diaspididae*), which have all been accidentally introduced into one or more of the Gulf Cooperation Council (GCC) countries (Gassouma 2004) and require considerable efforts to manage and control. The risk from such introductions is extremely high with some have already begun to threaten the existence of the date palms (Mohamed 2010).

Reducing the risk of introducing and spreading of other invasive species is one of the most important steps of quarantine regulations (i.e., legislative control) (Horn 1988). To prevent and slow down the introduction of date palm pests, the government of Bahrain has taken several initiatives to strengthen plant quarantine regulations and services. As a consequence, a number of laws and ministerial resolutions were issued regarding date palm protections, which require continuous political support to enforce their observance to ensure date palm protection and sustainability. The laws and the ministerial resolutions include the Decree by Law No. 21 of 1983 regarding palm protection; Resolution No. 13 of 1984 regarding date palm removal permission; Plant Quarantine regulation (Law) of the Gulf Cooperation Council (GCC) countries issued by Law No. 5 of 2003; Resolution No. 3 of 2006 amending Resolution No. 4 of 2004 prohibiting importing date palms and other palm species from countries affected by palm pests, amended by Resolution No. 9 of 2005; and Resolution No. 6 of 2007 regulating the movement of mature date palms and offshoots within the Kingdom without permission.

These laws and resolutions give regulatory authority to the Plant Quarantine and Protection Department, Plant Wealth Directorate, and Ministry of Municipalities Affairs and Urban Planning, the power to apply quarantine measures on all imported and exported plants and their products, and prohibit the entry of quarantine pests at ports of entry (i.e., Bahrain International Airport, King Fahd Causeway, Khalifa Sea Port, Salman Sea Port, Post Office), coordinating and cooperating with international, regional, and local organization concerned with plant protection, issuing phytosanitary certificates, carrying out control actions against the pests present in the country. These measures played a significant role in reducing and preventing the entry and the spread of date palm pests through importing or through movement of trees within the country.

Implementation of plant quarantine measures provokes major anxiety due to the lack of professional capacity and undeveloped laboratories. Recently, the government of Bahrain has taken various initiatives for strengthening and improving plant quarantine services and facilities to comply with its legal commitments and meet the international trade requirements. The initiatives include the development of quarantine capacities and inspection operational manuals and procedures and laboratory facilities; development and skill upgrading of the technical capacity of the staff; computerization and networking between the plant quarantine and the customs at the ports of entry.

A faunal survey of insects and mites associated with date palm plantations was carried out from 2009 to 2011, in an attempt to classify them according to their importance and to assess the impacts of the invasive species on date palm trees. The survey results revealed the presence of 311 species in date palm plantations, among which 18 species were associated with date palm trees (Mohamed et al. unpublished). The identified species were categorized according to the parts of the tree they attacked (Table 11.2). Three species including the lesser date moth, *Batrachedra amydraula* Meyrick (Lepidoptera: Cosmopterygidae); the Old World date mite (ghobar mite), *Oligonychus afrasiaticus* (McGregor) (Acari: Tetranychidae); and the red palm weevil, *Rhynchophorus ferrugineus* (Olivier) (Coleoptera: Curculionidae) are considered the most detrimental pests of date palm fruits and stems in Bahrain (Mohamed et al. unpublished). Little attention has been directed to the biology, ecology, economics, and the management of the pests with the exception of the red palm weevil. However, further studies are obligatory for better understanding of the life cycle of the pests in question to monitor and verify the threshold sizes of their populations. In addition, further studies on date palm pathogens in Bahrain are needed to determine their economic impact.

The red palm weevil (RPW) is an invasive species posing extremely high risk to the Kingdom's date palms. This weevil is a native of southern Asia and Melanesia and was accidentally introduced into Bahrain in 1995. Immediately upon the discovery of the pest, a task force was set up to minimize its impact on date palms. The surveillance and establishment of a pest control management project to combat the pest was initiated in 2009. The project was based on applying the IPM approaches by utilizing all available technology and management tools so as to minimize the impact on the nontarget organisms, environment, and human health. The project began with the implementation of three programs including surveying, monitoring, and managing and/or controlling *Rhynchophorus ferrugineus* (Mohamed 2012). The first phase of the project focused on the Kingdom's areas most vulnerable to the RPW, which are the western and northwestern portions of the Northern Governorate, where about 47 % of the date palms are growing. Taking the necessary procedures for assessing the date palm conditions and determining the management strategies to control the RPW is essential for the preservation and the prosperity of the date palm trees. The management process of the RPW involves treating and eradicating the infested date palm trees based on the infestation status assessment and the preventive spraying of the date palms to reduce the risk of infestation. Infested palm trees are identified by the existence of tunnels in the trunk, presence of sap oozing

Table 11.2 List of the major insect and mite pests associated with date palm trees in Bahrain

Common name	Scientific name	Family	Order
Major pests of date palm fruits			
Cigarette beetle	*Lasioderma serricorne* (F.)	Anobiidae	Coleoptera
Dry fruit beetle	*Carpophilus dimidiatus* F.	Nitidulidae	Coleoptera
Two dots dry fruit beetle	*Carpophilus hemipterus* (L.)	Nitidulidae	Coleoptera
Saw-toothed beetle	*Oryzaephilus surinamensis* (L.)	Silvanidae	Coleoptera
Greater date moth	*Arenipses sabella* Hmps.	Pyralidae	Lepidoptera
Lesser date moth	*Batrachedra amydraula* Meyer	Batrachedridae	Lepidoptera
Yellow wasp	*Polistes hebroeus* F.	Vespidae	Hymenoptera
Dust mite	*Oligonychus afrasiaticus* (McGregor)	Tetranychidae	Acari
Major insect pests of date palm leaves and stalks			
Parlatoria date scale	*Parlatoria blanchardi* (Targioni-Tozzetti)	Diaspididae	Homoptera
Date palm red (wax) scale	*Phoenicoccus marlatti* (Ckll.)	Diaspididae	Homoptera
Pink hibiscus mealy bug	*Maconellicoccus hirsutus* (Green)	Pseudococcidae	Homoptera
Dubas bug	*Ommatissus binotatus* de Bergevin	Tropiduchidae	Homoptera
Fruit stalk borer	*Oryctes* spp.	Scarabaeidae	Coleoptera
Rhinoceros beetle	*Oryctes agamenon* (L.)	Scarabaeidae	Coleoptera
Leaf borer	*Phonopate frontalis* F.	Bostrychidae	Coleoptera
Sulfurous jewel beetle	*Julodis euphratica* Castelnau and Gory	Buprestidae	Coleoptera
Major insect pests of date palm trunk			
Longhorned date palm stem borer	*Jebusaea hammerschmidtii* Reiche	Cerambycidae	Coleoptera
Red palm weevil	*Rhynchophorus ferrugineus* (Olivier)	Curculionidae	Coleoptera

Source: Mohamed et al. (unpublished)

from the openings, and/or the sawdust near the tunnels or at the base of the trunk (for a review, see Murphy and Briscoe 1999). Under these conditions, the infested tree should be removed, especially if the trunk interior was severely damaged. However, early detection of the infestation mandates simpler treatment. This can be achieved by identifying the tunnels on the trunk, cleaning the inside and outside the tunnels, and inserting three to eight tablets of aluminum phosphide (Phostoxin) per tunnel depending on the severity of the infestation. After that, the tunnels should be sealed to kill the remaining larvae inside. In addition, preventive treatment is carried out by spraying the whole palm trees with a recommended insecticide such as Diazinon 60 % EC, Ethion 48 % EC, Fenpropathrin 10 % EC, and Imidacloprid 20 % EC to reduce the risk of infestation. Over the 3 years of the survey (2009–2011), the

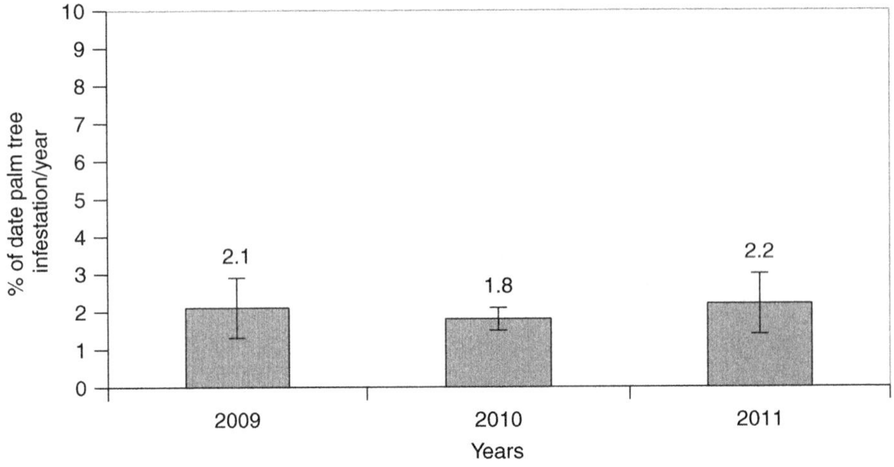

Fig. 11.2 Percentage of date palm infestation by *Rhynchophorus ferrugineus*, Bahrain, 2009–2011 (*Source*: Mohamed 2012)

estimated average of mature date palms and offshoots examined, treated, and eradicated was 264,585, 1085, and 1525, respectively (Mohamed 2012). The average annual infestation rate by RPW was about 2 % over the 3 years of the survey (Fig. 11.2). This infestation rate was close to that reported in Saudi Arabia, which was estimated to be 2.5 % (Vidyasagar et al. 2000a, b), and Oman, with an average of 1.7 % (Azam et al. 2001).

Monitoring adult RPW seasonal activities using baited aggregated pheromone traps was carried out in 2010–2011. The total number of adult weevils collected from the traps (176 pheromone traps distributed among 43 date palm plantations) during the monitoring years was 12,823 and 12,764, respectively. Adult RPW were found to be active throughout the year in Bahrain with a period of increased frequency in January to April (Fig 11.3). After that, the decline of the weevils coincides with the rise of the ambient temperatures (Mohamed 2012). These results support the findings of Kaakeh et al. (2001), Khalifa et al. (2001), and Abbas et al. (2006) in United Arab Emirates. However, in Saudi Arabia, the increase in the number of the adult RPW was found to be more prominent during May, with a second peak during November, and activity was observed to be lowest during August and February (Anon 1998). This variation could be related to the differences in the micro- and the macro-environments of the red palm weevil. Without a doubt, monitoring the weevil via pheromone traps can facilitate locating the spread of the weevil (Gunawardena and Herath 1995; Murphy and Briscoe 1999; Oehlschlager et al. 1995) and assessing the efficiency of the pest control programs in each monitored area within the country. Furthermore, the pheromone traps can help in disrupting the weevils' oviposition by reducing the chances of mating due to the mass traps of the weevils (Abraham et al. 2001; Abuagla and Al-Deeb 2012; Al-Asfoor 2012; Al-Saoud 2010; Faleiro 2000; Kaakeh et al. 2001).

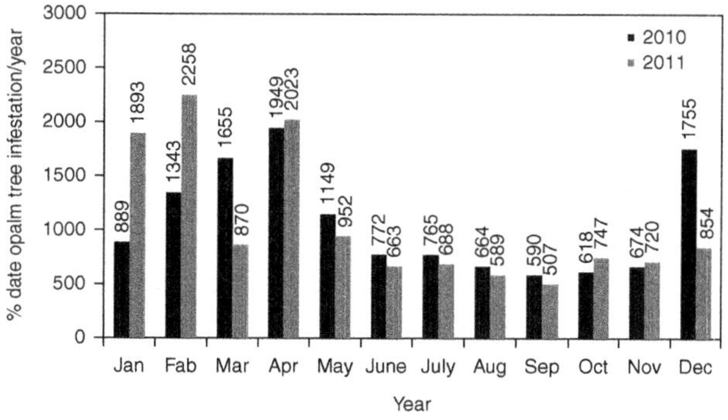

Fig. 11.3 Seasonal abundance of *Rhynchophorus ferrugineus* adults collected by pheromone traps from date palm plantations, Bahrain, 2010–2011 (*Source*: Mohamed 2012)

The survey program was expanded beyond the areas where the red palm weevils were discovered to reach a total of 28 locations in Bahrain by 2011 (Fig. 11.4). The expansion of the infestation in Bahrain was due to two factors: the first related to the ability of the adult red palm weevil to fly and the second factor a result of human activities for illegal translocation of date palm trees to areas free of weevil infestation. In conclusion, the red palm weevil infestation can be controlled through (a) expanded use of the pheromone traps at the rate of two traps per hectare as recommended and (b) intensified awareness campaigns for the growers and nurseries so as to disseminate the practical approach and information that facilitate date palm plantation management, including eradication of the abandoned palm trees, detachment of the offshoots from them and other trees, and avoidance of overwatering. It is also of the utmost importance to publicize that moving date palms and offshoots from infested areas is forbidden without notifying the Plant Wealth Directorate for inspection and treatment prior to translocation.

For best control results, it is recommended to prune during the low insect activity periods, from June to November (Fig. 11.3), spray both the healthy and the infested palm trees during the high activity period, and implement the IPM measures in healthy date palm plantations of neighboring plantations with infestation history. Furthermore, a Geographic Information System (GIS) has been used for the first time as a part of the IPM program in Bahrain to produce digital maps (Mohamed 2012). These maps can be used to assess the weevil distribution and the effectiveness of the control measures on the reduction of the infestation and for improving the red palm weevil control strategies (Al-Asfoor 2012; Massoud et al. 2011; Mohamed 2010, 2012; Pontikakos and Kontodimas 2010; Soroker et al. 2013).

11.3 Genetic Resources and Conservation

Despite its small land area, Bahrain is rich in genetic resources of date palms. There are more than 100 date palm cultivars in Bahrain (Al-Khalifa 2010; Al-Khalifa et al. 1994). It is believed that the structure of the gene pool of date palm has been shaped in response to natural and human selection pressures. Conditions, such as climatic,

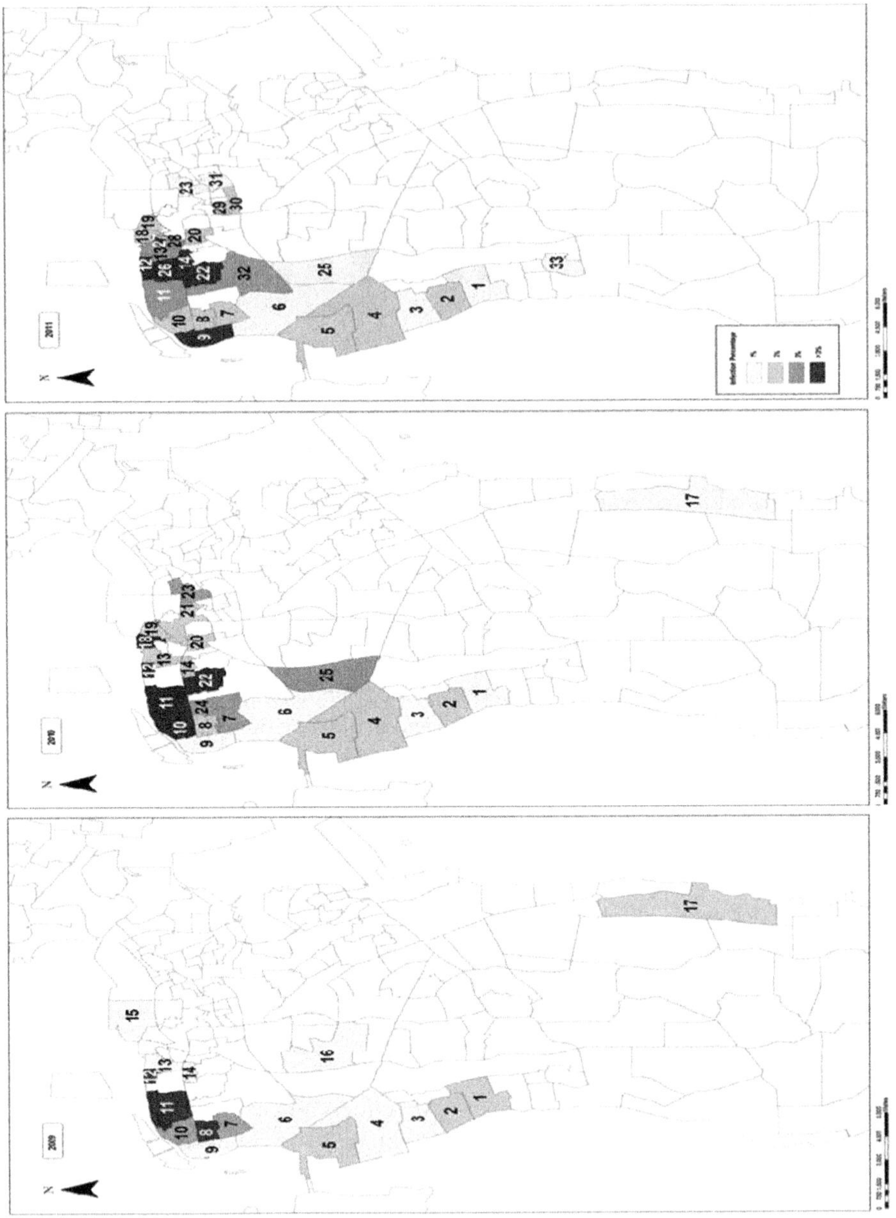

Fig. 11.4 Distribution and infestation levels of *Rhynchophorus ferrugineus* in date palm plantations, Bahrain, 2011 (Modified from Mohamed 2012). *1* Malkiya, *2* Karzakkan, *3* Damistan, *4* Al-Hamalah, *5* Al-Jasrah, *6* Janabiyah, *7* Al-Qurayya, *8* Bani Jamrah, *9* Budaiya, *10* Al-Diraz, *11* Barbar, *12* Jid Al-Haj, *13* Karranah, *14* Abu Saiba, *15* Al-Seef District, *16* A'Ali, *17* Jau, *18* Al-Qal'ah, *19* Karbabad, *20* Al-Qadam, *21* Jidhafs, *22* Muqabah, *23* Alburhama, *24* Al-Markh, *25* Buri, *26* Jannusan, *27* Hillat Abd Alsaleh, *28* Almaqsha, *29* South Sehila, *30* Adhari, *31* Bilad Al-Qadeem, *32* Sar, *33* Dar Kulaib

Table 11.3 Ex situ conservation of date palm genetic diversity in Budaiya Botanical Garden

Cultivar	Ripen maturation	No. of trees
Buchairah	Early	8
Gharrah	Early	8
Hilali	Early	4
Khudari	Early	6
Muwaji	Early	8
Barhi	Middle	1
Khawaja	Middle	4
Khalas	Middle	37
Khunaizi	Middle	34
Rzaiz	Middle	11
Hilali	Late	8
Khasbat-Asfoor	Late	6
Nabtat Saif	Late	1

Source: Directorate of Plant Wealth (unpublished)

edaphic, and socioeconomic, as well as fruit phenotypic traits, yield, and quality, have significantly influenced the choice of cultivars currently grown in the country.

Presently, the native date palm germplasm is threatened by various factors. The scarcity of rainfall, drying of freshwater wells along with the salinization of irrigation water and agricultural soil, combined, has had a dramatic effect on present date palm cultivation. Moreover, the authenticity of the local germplasm is endangered by the invasive cultivars, which have been imported from different countries, neighboring and from afar. These invasive strains are genetically diverse and may dominate and exert pressure on the indigenous genetic resources. Furthermore, some native cultivars, like Salmi, Sitrawi and Tayar, are the most likely to be threatened due to their adverse fruit taste and quality, which does not meet the standards required by the growers and the consumers.

As explained above, conservation of the native cultivars is crucial to maintain the genetic resources of the local date palms. In this regards, to enforce date palm protection and conservation, the Agriculture Affairs, Ministry of Municipalities Affairs and Urban Planning, has established field gene banks in Tubli, Hawarat Aali, and the Budaiya Botanical Garden for ex situ conservation of the genetic resources of date palm cultivars (Table 11.3).

In line with the above ex vitro conservation efforts, Agriculture Affairs has established a tissue culture laboratory and adopted specialized programs for micropropagation and conservation of some of the native cultivars in Bahrain (BFNRCBD 2006).

In order to further an initiative of Her Royal Highness Princess Sabeeka Bint Ibrahim Al-Khalifa, wife of HM the King, and President of the Supreme Council for Women, for agricultural development, the Ministry of Municipalities Affairs and Urban Planning founded a national campaign dubbed *Date palm for every home*. This campaign aims to inform Bahrain's citizens on date palm cultivation and maintenance, dissemination and conservation of the valued Bahraini cultivars, and fostering afforestation and improvement of the environment in residential neighborhoods. The total number of the households which benefited from the campaign reached 3,913 for the years 2011–2012.

In addition, companies interested in environmental management and protection, like the Gulf Petrochemical Industries Co. (GPIC) and the Bahrain Petroleum Company (BAPCO), conserve ex situ date palm genetic diversity either on their company sites, for example, GPIC hosts more than 400 date palms, or in specialized parks, such as the Princess Sabeeka Park in Awali, developed by BAPCO which includes 560 date palms. However, it should be pointed out that not all of the palms cultivated by the two companies are indigenous cultivars; many are imports from neighboring countries.

As presented earlier (Table 11.3), the ex situ conservation facilities comprise a limited number of cultivars compared to the total native genetic resources of the country. In order to protect the overall genetic diversity against degradation, which may be the consequence of biotic, abiotic, and anthropic stresses, and to provide a means to initiate essential breeding programs, it is recommended to implement the modern biotechnology tools available worldwide (Jain 2011; Rao 2004). Researchers should be encouraged to monitor, survey, and identify the genetic diversities among the Bahraini date palm populations using advanced molecular markers. Furthermore, mass production by tissue culture techniques, in vitro cold storage of date palm tissue at various developmental stages, and cryopreservation are additional biotechnological tools which can be adopted to conserve the heritage and the genetic resources of Bahraini date palms. These methods have to be implemented with care and in harmony with relevant molecular aspects to avoid genetic variations that may result from somaclonal variations. This conveys an urgent need to establish a center for date palm genetic resources and conservation in Bahrain. Chief among this center's objectives should be conservation of the genetic diversity of Bahraini date palm trees, providing pertinent field logistics and ultimately conducting, promoting, and emphasizing research focusing on exploring innovative techniques to sustain local germplasm.

11.4 Plant Tissue Culture

Tissue culture refers broadly to the technology that depends on in vitro aseptic culture of plant cells, tissues, and organs under defined chemical and physical conditions; it can be a strategic choice for sustainable date palm development and conservation in Bahrain. This technology has proved its numerous applications both in basic and applied researches, as well as commercial implementation in many countries (Al-Khalifa et al. 2013; Badawy et al. 2005; Zaid and Arias 1999). Mass production of genetically identical, true-to-type, pest and disease-free plantlets of elite cultivars makes tissue culture a dependable technique for rapid development of the date palm production industry (Alkhateeb 2008; Zaid and Arias 1999).

The first Bahraini plant tissue culture laboratory was established in 1988, through a joint venture between the Agriculture Affairs and the Taiwan Technical Mission in Bahrain. In 1993, through the Arab Organization for Agriculture Development (AOAD), date palm tissue culture was initiated in the laboratory. The laboratory was provided with facilities to accommodate 12,000 plantlets, with further expectations to increase the capacity to 50,000 plantlets (AOAD 2004). No significant progress was achieved until 1999, when further steps were taken to improve date palm tissue culture facilities in the Agriculture Affairs (Al-Khalifa 2004). In 1999, a project aiming at mass production of three elite Bahraini date palm cultivars, Khalas, Khunaizi, and Merziban, was initiated. However, due to limited financial resources and the fact that the laboratory was equipped with only very basic facilities that could not support large-scale production, the project resulted in focusing on a single cultivar, Khunaizi (Al-Khalifa 2004). The laboratory results obtained are promising as the plantlets produced were acclimated successfully in the Agriculture Affairs greenhouses and field sites tests. Currently, new cultivars, Khalas, Muwaje, and Hilali, were introduced and micropropagated successfully.

Plant tissue culture research facilities were established at the Arabian Gulf University, Manama, and the University of Bahrain. Date palm micropropagation techniques were investigated in the above laboratories (Al-Issa 1992; Al-Mansoori 2001). The Al-Issa (1992) project aimed for mass production of date palm cv. Khunaizi using tissue culture techniques. The author demonstrated complete protocols for date palm micropropagation via somatic embryogenesis and direct and indirect organogenesis. Al-Mansoori (2001) described an inclusive practical method for mass production of four Bahraini date palm cultivars via somatic embryogenesis. The method executed by Al-Mansoori (2001) is based on Tisserat (1979). This method could be one of the most prominent and central protocols for date palm propagation. However, precautions have to be taken prior to adopting the technique for commercial production as the high level of the 2,4-D added into the induction media can generate undesired genetic variations by increasing the risk of somaclonal variation (Alkhateeb 2008; Saker et al. 2000).

Tissue culture techniques have been implemented to screen date palm genotypes for salinity tolerance and to investigate the physiological mechanisms retaining salt tolerance at the cellular and whole plant levels. The results revealed differential cultivar performance and suggested the validity of implementing tissue culture techniques to screen date palm cultivars for salt tolerance at the cellular and whole plant levels of organization (Al-Mansoori 2001; Al-Mansoori et al. 2007).

However, to the present, plant tissue culture techniques have not been adopted for commercial production of date palms in Bahrain. Limitation of facilities, lack of trained technicians, and the value of the marketable product, in addition to the strong competition from the neighboring countries, have constrained the interest of investors leading them to import acclimated tissue-cultured date palms and trade as distributers rather than being producers (Table 11.4).

Despite the efforts expended on date palm tissue culture in Bahrain, the resources and the technologies available and currently in use are limited and not up to par. In this context, there is an undeniable and urgent need for capacity building and

Table 11.4 Number of tissue-cultured date palms imported into Bahrain and the suppliers, 2008–2012

Year	Saudi Arabia		United Arab Emeritus		United Kingdom		Total	
	No. of shipments	No. of plants	No. of shipments	No. of plants	No. of shipments	No. of plants	No. of shipments	No. of plants
2008	–	–	–	–	1	215	1	215
2009	2	3,060	2	7,000	1	600	5	10,660
2010	–	–	1	4,000	–	–	1	4,000
2011	1	1,000	–	–	–	–	1	1,000
2012	3	155	1	4,000	–	–	4	4,155
Total							12	2,0030

Source: Directorate of Plant Wealth (2012)

establishment of a national center specialized in date palm tissue culture and biotechnology. This center should aspire to introduce the most advanced technology available today in order to save Bahrain's native date palms and furthermore incorporate the date palm industry as a part of the overall economic growth of the country. This calls for the combined efforts and cooperation of all governmental and nongovernmental bodies dealing with date palm industry in Bahrain.

11.5 Cultivars Identification

Although there are more than 100 date palm cultivars in Bahrain, only 33 have been definitively identified. Commonly and traditionally, variations among the cultivars were described based on morphological traits. Some of the most common phenotypic traits used to identify the Bahraini cultivars are summarized in Table 11.5. However, phenotypic traits are not consistent and can be affected by environmental conditions and cultivation practices. In addition, these traits cannot be detected precisely before the maturation (fruiting) stage of the palm (Elshibli and Korpelainen 2009).

In Bahrain, phenotypic variations have been detected within individual cultivars (Al-Khalifa 2010). For example, Khalas cultivar possesses three distinct strains, which vary in mature fruit size, shape, and color. Variation in ripening time was recorded within Hilali cultivar, and this was attributed to the effect of relative humidity, which significantly influences the quality of the yield. In addition, two discrete strains of Khasbat-Asfoor are well known to Bahraini date growers. The Arabi strain bears fruits until the beginning of November, whereas the Hasawi strain extends the fruiting season to the beginning of December. Further variations were recorded in fruit phenotypic traits within Muwaje, Merziban, and Khunaizi cultivars.

Al-Ruqaishi et al. (2008) showed that the phenotypic traits may not be reliable. They reported that due to phenotypic similarities, different cultivars were given the same common name. For example, the Omani Khalas and the Bahraini Khalas have

Table 11.5 Selected morphological traits and descriptors used to identify date palm cultivars in Bahrain

Organ	Phenotypic trait	Descriptor
Fruit	Color	Yellow, red, greenish yellow, pinkish
	Shape	Oval, inverted oval, rectangular-oval, cylindrical, spherical, spherical with flat ends, convex rectangular, elliptical
	Mass	Very low (<7.5 g), low (7.5–10.5 g), medium (10.5–13.0 g), high (13.0–16.5 g), very high (>16.5 g)
	Size	Very small (<7.99 cm^3), small (8.0–10.99 cm^3), medium (11.0–13.99 cm^3), large (14.00–16.99 cm^3), very large (>17.00 cm^3)
	Length	Very short (<3 cm), short (3.0–3.49 cm), medium (3.5–3.99 cm), long (4.0–4.49), very long (>4.49 cm)
	Diameter	Very thin (<2 cm), thin (2.0–2.24 cm), medium (2.25–2.49 cm), thick (2.5–2.74 cm), very thick (>2.74 cm)
	Texture	Soft (>30 % water content), semidry (20–30 %), dry (<30 %)
	Ripening time	Early (June–early July), medium (July–August), late (September–early December)
	Quality	Excellent, very good, good, medium, poor
Fruit bunch	Color	Yellowish green, greenish yellow, yellowish orange, reddish orange
	Length	Short (<90 cm), medium (90–150 cm), long (>150 cm)
Trunk	Diameter	Thick (≥70 cm), medium (50–69 cm), thin (≤49 cm)
Leaf	Length	Long (>427 cm), medium (427–335 cm), short (<335 cm)
Leaflets	Color	Pale green, dark green, shiny green, and waxy green
	Length	Short (<60 cm), medium (60–75 cm), long (>75 cm)
	Width	Thin (<3.8 cm), medium (3.8–4.4 cm), wide (>4.4 cm)
Spines	Number	Few (<20), medium (20–30), many (>30)
	Length	Short (<10 cm), medium (10–15 cm), long (>15 cm)
Petiole base	Girth	Small (<30 cm), medium (30–40 cm), large (>40 cm)
Canopy	Shape	Open, integrated, tilted, and erect

been given the same common name. However, molecular markers revealed genetic variation among the two cultivars, where Omani Khalas was genetically interrelated to Buhabisha, Fard, and Bahlani cvs., whereas the Bahraini Khalas was genetically related to Merziban.

Pathak and Hamzah (2008) found that using morphological and biochemical markers to detect genetic diversity among date palm cultivars was demanding, laborious, and time consuming. Accordingly, they used molecular markers, in particular Random Amplified Polymorphic DNA (RAPD), to study the genetic diversity among various populations of date palm in Manama City, the capital of the Kingdom of Bahrain. The analysis of the molecular variance revealed 52 % genetic similarity between the palms examined. The results indicated moderate genetic variations among the tested palm trees.

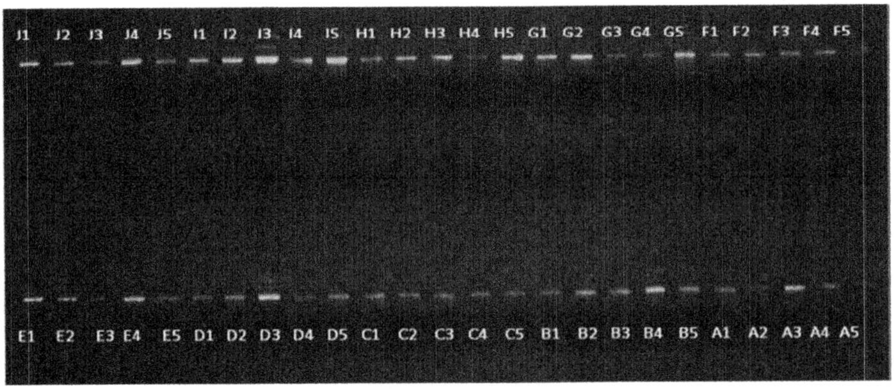

Fig. 11.5 The genomic DNA concentration of five date palm cultivars (*1* Gharrah, *2* Khunaizi, *3* Khalas, *4*Khasbat-Asfoor, *5* Hilali) grown in ten different regions of Bahrain (*A* Al-Budaiya, *B* Karana, *C* Karzakkan, *D* Howarat A'ali, *E* Al-Hamalah, *F* Askar, *G* Sanabas, *H* Al-jasrah, *I* Jid Ali) in Bahrain (*Source*: A. A-Mamari, MOAF, Oman)

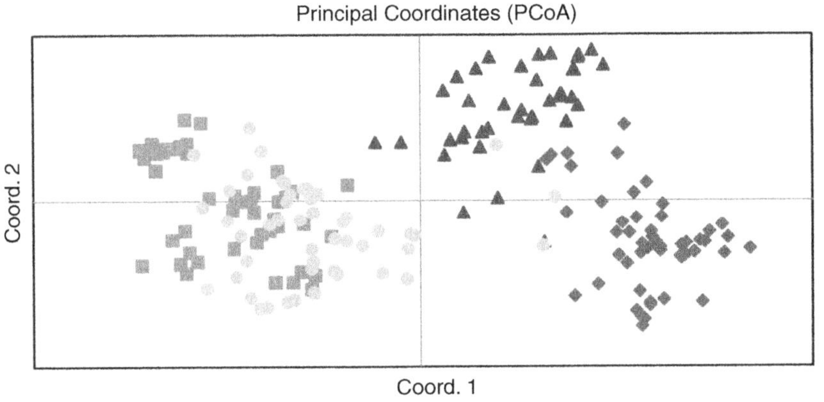

Fig. 11.6 Principle Coordinates Analysis (*PCoA*) of date palm accessions from Oman (*OM*), Saudi Arabia (*SA*), Kuwait (*KU*), and Bahrain (*B*) using 19 microsatellite primers (*Source*: A. A-Mamari, MOAF, Oman)

In Oman, as a part of a Molecular Genetic Diversity of Date Palm sub-project of the Development of Sustainable Date Palm Production Systems in Gulf Cooperation Council Countries Project (ICARDA 2011; MAF 2012), microsatellites, simple sequence repeats (SSRs), have been used to investigate the genetic diversity among five date palm cultivars of four Gulf Cooperation Council (GCC) countries, including Bahrain. The Bahraini cultivars tested were Gharrah, Khunaizi, Khalas, Khasbat-Asfoor, and Hilali (Fig. 11.5). Results indicated that the cultivars of the GCC countries exhibit genetic variation, with Bahraini cultivars being genetically closer to the Saudi Arabia cultivars, followed by Kuwait and Oman cultivars, respectively (Fig. 11.6). The similarities recorded were attributed to the higher rate of date palm germplasm exchanges between the studied countries. In addition, the in vitro derived progeny of the above cultivars proved to be genetically identical when analyzed using SSR markers (ICARDA 2011).

From the above, it can be seen that the desired information pertaining to date palm molecular identification and phylogeny is inadequate in Bahrain. As stated earlier, there is a strong need to establish a center for date palm biotechnology and genetic resources and conservation. The center should be equipped with suitable facilities and empowered with expertise to convey precise analysis and field logistics, and it must endeavor to identify the genetic diversity among the native cultivars using the most recent molecular technologies and to ensure that the derived progeny of the in vitro micropropagation techniques in use is genetically identical and true to type. Cooperation of all related scientific bodies, including research institutions such as the University of Bahrain and the Arabian Gulf University, is crucial to achieve the target goals.

11.6 Cultivar Description

11.6.1 Growth Requirements

11.6.1.1 Climatic Conditions

Bahrain has suitable climatic conditions for date palm growth and yield. The Kingdom is situated in the subtropical arid and hot zone. Its climate is characterized by high ambient temperature and elevated relative humidity and evapotranspiration rates. The average air temperature ranges from 14 to 41 °C. The annual rainfall, which is confined to a rainy season that extends from November to April, is inconsistent and sparse with a range of 39–128 mm (BFNRCBD 2006).

The high relative humidity in summer, which is the fruit ripening season, has shaped date palm germplasm in the Kingdom. It propelled the domination of cultivars that bear soft and tender fruits, which cannot be left on the tree to dry out; instead, they have to be picked as soon as they are ripe to avoid rotting. On the other hand, the pollination and flowering seasons, which extend from February to March in Bahrain, need moderate temperature and no rainfall, as heavy rain impedes the pollination process. Accordingly, date palm growers tend to pollinate the female flowers in the morning but after full sunrise when the flowers are moderately dry and the relative humidity are lower (Al-Khalifa et al. 1994).

11.6.1.2 Soil

Date palms are known for their ability to grow in various types of soil. However, they thrive in nonsaline soils, which are rich in mineral nutrients and have appropriate drainage. Accordingly, date palm plantations are widespread in the northern and the western parts of the Kingdom, which are characterized by having nonsaline sandy soils which are rich in organic matters and where there is a high water table. Although date palms are salt tolerant trees, the continuous exposure to saline soil and water exerts a dramatic effect on their growth and productivity. In Bahrain, salinization of the underground water along with poor drainage increases the salinity of the soil, which in turn affects date palm plantations drastically, especially in the central and the eastern parts of the Kingdom (Ahmed 1978; Al-Basheer and Harron 1997; Al-Khalifa et al. 1994).

11.6.1.3 Fertilization

Like any plant, the date palm requires essential mineral nutrients to grow, thrive, and bear sufficient yield. However, there are no standard schedules for fertilization and compost application of mature date palm trees in Bahrain. Only offshoots are fertilized in their early stages by adding organic manure (cow or domestic animals fertilizers) to the soil. The vast majority of date palms are not fertilized specifically; instead they may acquire their needs from fertilizers added to the intercropped vegetables or forages (Ahmed 1978; Al-Basheer and Harron 1997).

Currently and in collaboration with ICARDA, the effect of chemical and organic fertilizers on date palm growth and yield has been investigated. Results revealed no significant improvement in plant growth and yield. This was attributed to the technique implemented during the fertilization process (ICARDA 2011).

11.6.1.4 Irrigation

Generally, there is not a specific irrigation schedule adopted for mature date palms. Usually, the dates obtain their water needs from the forage or vegetable crop beds when they are flood irrigated or by drawing from water channels used as a part of other plant irrigation systems (Ahmed 1978; Al-Basheer and Harron 1997; Al-Khalifa et al. 1994). High summer temperatures prompt a dramatic effect on the evapotranspiration rate, which affects the amount of water needed by the plant. Accordingly, during the summer, date palms are irrigated once or twice each week, whereas during winter, they are irrigated only once every 2 weeks. It has been recorded that date palms can survive on 10 m^3 of irrigation water per tree per year. However, to reach their optimum potential yield, they need approximately 50 m^3 of water per tree per year (Al-Khalifa et al. 1994). For newly transplanted offshoots, date palm growers tend to irrigate the young offshoots on a daily basis until they harden and establish a well-defined root system (Ahmed 1978; Al-Khalifa et al. 1994).

Currently, modern date palm plantations are characterized by the execution of water saving irrigation practices, such as bubbler irrigation. However, such methods are still poorly operated with no irrigation schedules and imperfect management (Al-Basheer and Harron 1997; Al-Khalifa et al. 1994).

Table 11.6 Estimated numbers and areas of the agricultural holdings in Bahrain

Type of plantation	No. of holdings	Total area, ha	% of total area
Vegetables	7	54	1.3
Forage	4	50	1.2
Date palm plantations	90	437	10.2
Intercropped date plantations	1,114	3,759	87.3
Total	1,215	4,300	100

Source: Al-Basheer and Haroon (1997)

11.6.2 Cultivars Distribution in Bahrain

Although date palms are distributed over all five governorates of Bahrain (Table 11.1), the main plantations are situated on the northern coastline, the northern region of the western coastline towards Zallaq village, and the northern region of the eastern coastline of the main island of Bahrain towards Sitra Island. Al-Khalifa (2010) revealed that the slight variation in the climatic conditions between the coastlines and the dryer inlands affects the distribution of the cultivars to a degree. The author states that early fruits ripening cultivars, like Muwaji, favor the wet coastline areas to avoid fruits curling and crimping, whereas Um-Rahim and Barhi favor relatively dryer areas, such as Al-Hamalah, Aali, Bori, and Riffa.

In Bahrain, there are a total of 1,215 agricultural holdings occupying 4,300 ha (Table 11.6). Among 1,204 of these holdings, which represent 97.5 % of the total area, date palms are intercropped with vegetables and/or forages or as sole fruit trees. The majority of the holdings are located in the Northern Governorates, which is a rural area characterized by water availability and soil fertility (Al-Khalifa 2009; Al-Khalifa et al. 1994; Doornkamp et al. 1980; Nesbitt 1993). It is important to note that none of the orchards are specialized in a particular date cultivar; instead, they comprise various cultivars, which are usually intercropped with other cash crops.

11.6.3 Cultivar Production and Economics

In Bahrain, date harvesting extends from June for early fruit ripening to early December for late fruit ripening cultivars. Nevertheless, the main harvesting period is confined to the months of July and August, when the entire fruits of the medium cultivars ripen simultaneously (Al-Khalifa 2010). During the season, dates are mainly harvested at two distinct stages: khalal (fully colored stage) and rutab (wet and soft ripe stage). In Bahrain, dates cannot be left to dry on the trees and cannot be harvested as tamar dates. Instead, dates at the wet and soft ripen stage are hand-picked and if needed, sun dried to tamar.

The average annual production per tree varies from 40 to 80 kg depending on the cultivar with an actual production that may exceed 150 kg/tree. It was recorded that the most productive cultivars (70–80 kg) are Buchairah, Khunaizi,

Table 11.7 Average annual production of Bahraini date palm cultivars

Maturity season	Cultivar	Percentage of total date palms	Average annual production per tree (kg)
Early	Tayar	0.1	50
	Muwaji	6	60
	Buchairah	1.3	70
	Mubashir	0.3	40
	Gharrah	2.1	55
Medium	Khalas	3.5	60
	Khunaizi	11	80
	Shishi	0.7	65
	Setrawi	0.1	60
	Hatimi	0.2	65
	Sils	0.2	50
	Amari	0.1	60
	Rzaiz	0.2	65
	Banat-Alssyid	0.2	70
	Merziban	13.2	75
	Banat-Alabade	0.1	45
	Shabibi	0.2	65
	Brismi	0.2	55
	Khawaja	1	65
	Hallaw	1.3	75
	HallawTaroot	0.4	65
	Barhi	1.1	70
	Tanjoob	0.1	60
	Humri	0.1	65
	Fardh	0.2	65
Late	Mudallal	0.1	45
	NabtatSaif	0.5	60
	Ashhal	0.1	60
	Um-Rahim	0.5	70
	Jabiri	0.3	70
	Hilali	1.2	70
	Selmi	4.4	65
	Sabo	0.4	65
	Shambari	0.6	65
	Khasbat-Asfoor	2.9	70

Source: Al-Khalifa et al. (1994)

Banat-Alssyid, Merziban, Hallaw, Barhi, Um-Rahim, Jabiri, Hilali, and Khasbat-Asfoor. The least productive cultivars (40–50 kg) include Tayar, Mubashir, Sils, Banat-Alabade, and Mudallal (Table 11.7). Khunaizi and Merziban are the most dominant cultivars. They exemplify 11 and 13.2 % of the total number of date palms in Bahrain, respectively (Table 11.7).

The date fruits of the early and the late cultivars have proved to be the most profitable, as they supply consumers with early and late season dates, whereas the medium cultivars are the least cost-effective, unless they are of high quality. The retail price of dates ranges from 350 Files/kg for the medium cultivar Khunaizi during August to 1,500 Files/kg for the early-bearing cultivar Gharrah (Table 11.8). Similarly the wholesale price ranges from 300 Files/kg for the Merziban during the peak of the season to 1,330 Files/kg for Gharrah during the early part of the season (Table 11.8) (1,000 Files = 1 Bahraini Dinar = USD 2.65).

The lower price of the medium cultivars is related to numerous factors including the nature of the fruits produced by the majority of the local cultivars, which are wet and soft and cannot mature to the tamar stage without further artificial processing. Dependence on the local market, where most of the products are tendered fresh, is another critical element causing a marked reduction in the price of these dates. The large supply within a specified period of time compared to the limited local market demands significantly diminishes their profit. The low profit achieved results from poor postharvest operations and ineffective marketing strategies in Bahrain.

11.6.4 Nutritional Aspects

Physicochemical analyses of date palm fruits have been carried out in many countries (Hasnaoui et al. 2011; Sawaya et al. 1983), but very limited data are available regarding the chemical composition of the various date cultivars in Bahrain. Ahmed (1978) indicated that dates at tamar stage are considered as essential sources of sugars (70.6 %) and fiber (10 %), whereas proteins and lipid constituents are found in considerably lower percentages (2.5 and 1.1 %, respectively).

Allaith (2008) investigated the antioxidant activity, phenolics, and ascorbic acid constituents of 16 Bahraini date palm cultivars at various ripening stages. His findings showed that the antioxidant activity is highest in the khalal stage, followed by the rutab stage, and the least activity was recorded in the dried tamar stage. Variation in antioxidant activity was significant between the various cultivars at the khalal and rutab stages. It was recorded that the average phenolics and ascorbic acid for khalal stage were 196.8 ± 72.1 and 6.6 ± 2.4 mg (mean ± std) per 100 g fresh weight, respectively, whereas the same constituents at rutab stage were 116.7 ± 44.1 and 3.3 ± 1.3 mg per 100 g fresh weight, respectively. The author suggested that phenolic composition plays a significant role in the antioxidant activity of date palm fruits. Al-Laith (2009) reported that the total antioxidant activity, phenolics, and ascorbic acid decline as fruits ripen from khalal to rutab. It was reported that the decline in the above phytochemicals follow a pseudo-first-order reaction.

As in many date-growing countries (Abdillah and Andriani 2012), date seeds have been used as staple food, as well as for an ersatz coffee drink in Bahrain. Ali-Mohamed and Khamis (2004) analyzed the chemical composition of the seeds of six Bahraini cultivars. Results revealed significant variations among the cultivars studied. Merziban cv. seeds exhibited the most abundant mineral nutrients, whereas

Table 11.8 Average retail and wholesale prices (Fils/kg) of some date palm cultivars in 2010

Cultivars	January	February	March	April	May	June	July	August	September	October	November	December	Average
Retail (Fils/kg)													
Muwaji	–	–	–	–	–	1,000	500	–	–	–	–	–	750
Buchairah	–	–	–	–	–	1,350	–	–	–	–	–	–	1,350
Gharrah	–	–	–	–	–	1,500	700	–	–	–	–	–	1,100
Khunaizi	–	–	–	–	–	–	650	380	–	–	–	–	515
Merziban	–	–	–	–	–	–	700	450	–	–	–	–	575
Khalas	–	–	–	–	–	–	1,350	950	–	–	–	–	1,150
A. Asfoor	–	–	–	–	–	–	–	–	500	500	700	–	567
Others	–	–	–	–	–	–	633	555	517	517	750	–	600
Wholesale (Fils/kg)													
Muwaji	–	–	–	–	–	850	450	–	–	–	–	–	650
Buchairah	–	–	–	–	–	1,150	–	–	–	–	–	–	1,150
Gharrah	–	–	–	–	–	1,330	600	–	–	–	–	–	965
Khunaizi	–	–	–	–	–	–	500	350	–	–	–	–	425
Merziban	–	–	–	–	–	–	650	300	–	–	–	–	475
Khalas	–	–	–	–	–	–	1,200	800	–	–	–	–	1,000
A. Asfoor	–	–	–	–	–	–	–	–	450	450	600	–	500
Others	–	–	–	–	–	–	628	471	418	418	500	–	487

Source: Agriculture (2014)

Khawaja exhibited the least concentrations of essential mineral ions content. They revealed that date palm seeds are rich in essential macro- and micronutrients including potassium (K^+), magnesium (Mg^{2+}), sodium (Na^+), calcium (Ca^{2+}), iron (Fe^{2+}), manganese (Mn^{2+}), zinc (Zn^{2+}), copper (Cu^{2+}), nickel (Ni^{2+}), cobalt (Co^{2+}), and chromium (Cr^{3+}). In addition, lead (Pb^{2+}) and cadmium (Cd^{2+}), which are considered pollutants, were recorded in the studied seeds. However, the investigation revealed that their concentration in date palm seeds is lower than that found in coffee and barley seeds. Moreover, the level of the daily intake of minerals from date palm seeds is within an acceptable dietary intake range.

11.6.5 Description of Cultivars Grown in Bahrain

As stated earlier, there are over 100 different date cultivars in Bahrain. Al-Khalifa (2010) defined the differences between cultivars, which include fruit characteristics, such as quality, shape, size, color, length, width, as well as ripening time and tree morphology, including trunk diameter, shape of the canopy, leaf length, color, petioles base, curvature of the leaves and shape of the leaf tip, spine numbers, length, thickness, durability, as well as the thickness of its base. Furthermore, the variations encompass the leaflets position, length and width, and hardness and curvature and color, size, and length of their fruit bunches, which may have different lengths of spikelets as well (Table 11.5). Descriptions of some cultivars are summarized in Table 11.9, whereas Fig. 11.7 illustrates the morphological variations among the fruits of some Bahraini cultivars.

Among the early-bearing cultivars is Gharrah, which is known for its superior fruit quality. This cultivar bears commercially profitable fruits, which are characterized by being large, rectangular, light yellow in color, and sweet in taste both at khalal and rutab stages due to the low tannin levels in their tissue. This cultivar has a thin trunk, medium leaves, leaflets, and spines. Its leaflets are yellowish green in color. Another cultivar that ripens early is Buchairah. It is known for its preferred yellow fruits, which are mainly consumed at rutab stage. It is characterized by having a thick trunk and long leaves, leaflets, and spines. Mubashir has satisfactory fruits, which are mainly consumed at rutab stage as well. This cultivar is characterized by having a medium-sized trunk, leaves, leaflets, and spines (Al-Khalifa 2010; Al-Khalifa et al. 1994).

The medium fruit ripening cultivars include one of the most desired cultivar, the Khalas. This cultivar has fruits which are considered to be the most popular, not just in Bahrain but in several Arabian Gulf Countries as well. The fruits of this cultivar are large, oval, shiny, and yellow in color. They are sweet at the rutab stage and acceptable at the khalal stage. It has an average to large trunk with long leaves and spines. The color of its leaflets is light green. Khunaizi also ripens at around the same time. This cultivar is very widespread, because it tolerates various types of soil and water quality. It is characterized by having oval, dark red, sweet fruits at both khalal and rutab stages. It has a medium to large trunk, dark green leaflets, and

Table 11.9 The main characteristics of the major Bahraini date palm cultivars

Maturity season[a] and cultivar name	Fruits characteristics						Leaf length (cm)	Leaflets length (cm)	Spine length (cm)
	Quality	Shape	Color	Length (cm)	Width (cm)	Weight (g)			
Early maturing									
Buchairah	Medium	Oval-rectangular	Yellow	3.0–3.6	1.4–2.5	9.0	300–350	45–65	12–14
Gharrah	Excellent	Rectangular	Yellow	3.4–3.9	1.9–2.4	7.3–9.2	270–290	65–72	12–15
Mubashir	Medium	Rectangular	Yellow	3.5–4.0	1.8–2.9	5.5–9.9	300–360	50–56	9–22
Muwaji	V. good	Rectangular	Red	4.0–5.0	2.0–2.6	10–13	270–300	60–65	14–18
Tayar	Poor	Rectangular	Yellow	3.2–4.4	1.9–2.8	7.2–9.3	290–310	70–75	5.9–13
Medium maturing									
Amari	Poor	Oval-rectangular	Yellow	3.2–3.5	1.9–2.3	7.5–8.0	310–320	40–60	11–20.0
Banat-Alabade	Poor	Rectangular	Yellow	2.5–3.0	1.8–2.0	6.0–6.5	180–220	48–50	9–12.0
Banat-Alssyid	Medium	Oval-spherical	Yellow	2.8–3.4	1.9–2.5	7.0–8.5	350–360	38–59	7–13.0
Barhi	Excellent	Oval	Yellow	2.9–3.9	2.2–2.5	8.0–8.5	370–400	45–65	7–12.0
Brismi	Medium	Rectangular	Red	2.6–3.2	2.1–2.4	7.0–8.4	310–330	44–60	9–16.0
Fardh	Good	Rectangular-oval	Red	2.8–3.5	1.9–2.3	7.2–8.5	370–375	65–67	8–11.5
Hallaw	Good	Rectangular	Light red	3.5–3.8	1.2–2.3	5.5–9.3	300–310	34–37	9–10.5
Hallaw Taroot	Medium	Rectangular	Red	3.7–3.9	2.0–2.2	5.2–8.7	300–320	43–50	8–14.0
Hatimi	Good	Oval	Yellow	2.8–3.2	2.3–2.5	7.5–8.0	300–320	38–48	11–18.0
Humri	Medium	Oval	Red	2.8–3.1	1.1–2.2	6.8–7.2	330–335	40–60	14–22.0
Khalas	Excellent	Oval-rectangular	Yellow	3.5–3.9	2.2–2.6	9.0–11.5	265–300	56–67	6.5–12.0
Khawaja	Excellent	Rectangular	Yellow	3.1–4.3	2.0–3.0	11–11.5	370–390	47–55	8.5–14.5
Khunaizi	V. good	Rectangular	Red	2.0–3.5	2.0–2.5	6.0–6.5	270–300	49–57	9–18.0
Merziban	Good	Rectangular	Yellow	3.0–4.2	2.0–2.4	10–11	300–312	30–42	4.5–9.0

Rzaiz	Good	Oval-rectangular	Light red	3.2–3.9	1.9–2.2	7.0–8.5	310–320	45–50	8–14.0
Setrawi	Poor	Rectangular	Yellow	2.8–3.5	1.9–2.1	7.5–8.5	310–350	40–64	9–23.0
Shabibi	Good	Rectangular	Yellow	3.0–3.2	2.1–2.4	9.5–10	280–300	40–48	8–16.0
Shishi	Good	Rectangular	Yellow	2.8–3.5	2.1–2.5	10–11	300–310	45–50	10–12.0
Sils	Poor	Rectangular	Red	3.0–3.4	1.8–2.0	6.0–6.5	330–340	34–50	8–17.0
Tanjoob	Medium	Rectangular-oval	Red	3.0–3.8	2.2–2.3	7.9–8.2	308–360	46–59	8–13.0
Late maturing									
Ashhal	V. good	Oval	Light red	2.8–3.1	2.2–2.4	8.3–9.0	360–365	40–60	20–30
Hilali	Excellent	Oval	Yellow	2.5–2.9	2.1–2.6	7.0–9.0	245–265	40–45	9–15
Jabiri	V. good	Oval	Yellow	2.1–2.7	1.7–2.9	6.4–6.8	300–320	30–40	8–12
Khasbat-Asfoor	V. good	Oval-spherical	Red	3.2–3.7	2.2–2.6	7.0–8.5	290–310	47–55	8–19
Mudallal	Excellent	Rectangular	Yellow	3.8–4.1	1.9–2.1	7.9–8.8	270–280	45–55	15–22
Nabtat Saif	Excellent	Oval	Yellow	2.5–3.0	2.2–2.4	7.5–8.5	370–390	60–65	12–25
Sabo	Good	Rectangular	Yellow	3.0–3.3	2.2–2.6	7.5–8.0	290–300	45–50	6–12
Selmi	Poor	Oval	Yellow	2.8–3.0	2.1–2.2	6.5–8.0	300–310	61	9–22
Shambari	Good	Oval-spherical	Yellow	2.2–3.0	1.8–2.3	6.5–7.0	280–300	34–60	9–22
Um-Rahim	Good	Oval	Yellow	2.9–4.4	2.3–2.5	8.0–8.5	330–345	44–62	9–15

[a]Maturity season: early maturing (pollination Jan–Feb, maturation June–July), medium maturing (pollination Feb–Mar, maturation July–Aug), and late maturing (pollination Feb–Mar, maturation Sep–Oct)

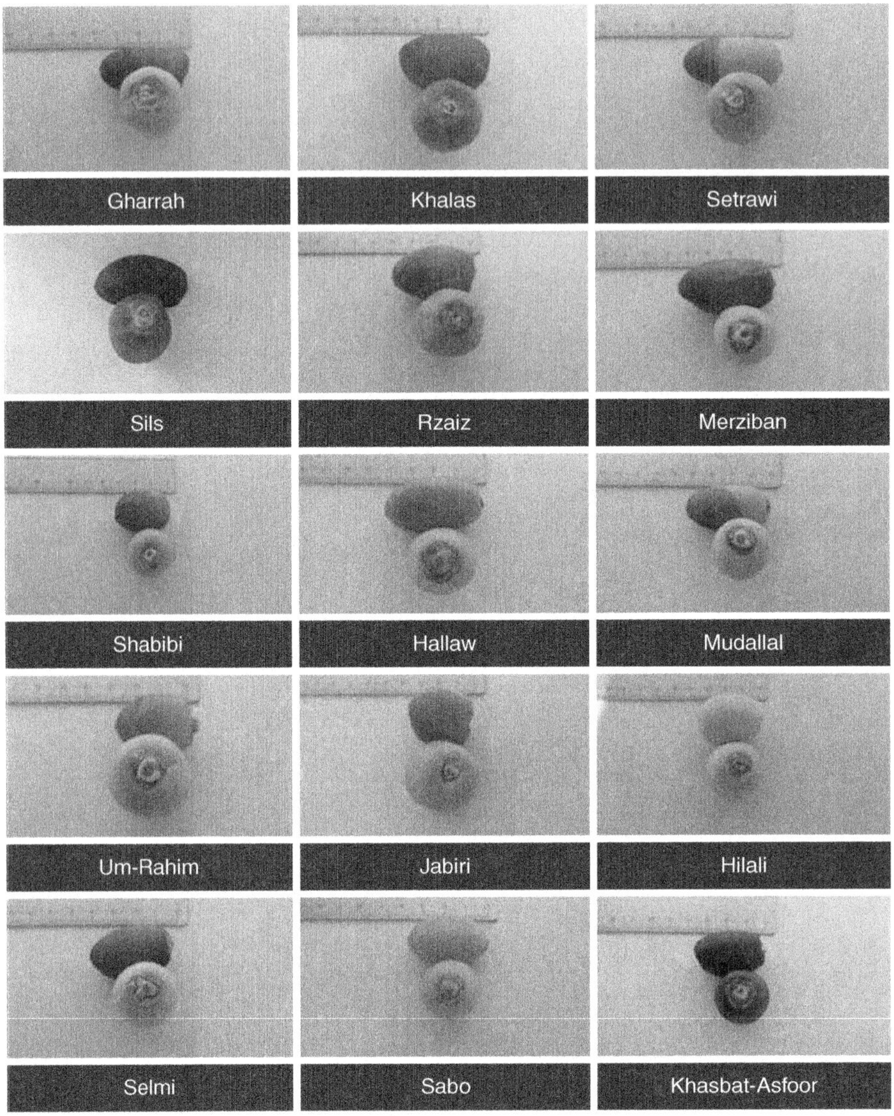

Fig. 11.7 Morphological variations between fruits of some Bahraini cultivars

dense thick spines that are located at the base of the leaves. The other most widespread cultivar is the Merziban. This cultivar is distinguished by its medium to large, yellow fruits, which are consumed at rutab and tamar stages. Trees of this cultivar have an average size trunk along with long leaves, leaflets, and spines. Its extended harvest period makes it a remarkable cultivar.

Among those that ripen late is the most famous cultivar of all, Khasbat-Asfoor, which bears average-sized, oval to round, red fruits. This cultivar is consumed late in the season as khalal and rutab. The tree of this cultivar is very strong with thick trunk and long and strong leaves. It has broad leaflets, which are yellowish green in color and very strong alternate spines that are located mainly at the base of the leaf (Al-Khalifa 2004, 2009, 2010).

11.7 Date Production and Marketing

In Bahrain, the average annual date production reaches 16,000 mt, with a yield range of 50–150 kg per tree, depending on the cultivar, environmental conditions, and the technical practices (Al-Khalifa 2009). Some 70 % of the fruits are consumed fresh, locally at the stage of khalal and rutab (Table 11.10). Due to the high relative humidity during the fruit ripening season, farmers lose a large portion of their yield due to infestation of pests and from diseases. To overcome this problem, some growers tend to dry the ripened rutab of certain cultivars to tamar stage using traditional methods. Of the total annual production, only 10 % is consumed as dry fruits (tamar). In addition to the direct consumption, dates of certain cultivars have been utilized to extract molasses (date syrup = dibs). Also, to the present, this process is being done via traditional methods, and only 1.0 % of the total annual production is used for molasses production (Table 11.10). The remaining 19 % is used for cattle feed due to the quality of the fruits.

Due to the small size of the agricultural holdings, which usually do not exceed 50 ha, harvest mechanization has not yet been adopted. Instead, the normal traditional harvesting approach is followed. For the first 10 years of the palm's life, fruit harvesting can be accomplished manually, by handpicking. However, for older and taller trees, harvesting requires climbing the tree. This process is traditionally achieved via locally made apparatus called an *alkar*, which is made of strong ropes to hold the body of the climber. This procedure is laborious and time consuming and requires high skill and courage. The worker needs to climb the tree several times to harvest the ripe fruits, which do not ripen simultaneously. Failure to accomplish the process on time makes the fruits susceptible to bird pecking, pests, and diseases, making the fruit unsuitable for direct human consumption and greatly reduce the marketing value (Al-Khalifa 2009).

Table 11.10 Date fruit consumption in Bahrain

Consumption	Consumption/mt	% of consumption
Fresh dates	11,200	70
Dry dates	1,600	10
Date syrup	200	1
Animal feed	3,000	19
Total	16,000	100

Source: Al-Khalifa (2009)

Table 11.11 Treated Sewage Effluent (TSE) utilization under various crops

Crops	Area (ha)	TSE utilized in 2013 (m^3/day)
Dates and other fruits	1,720	20,000
Fodder	590	22,000
Vegetables	650	27,343
Total	2,960	69,343

Source: Directorate of Agriculture Engineer and Water Resources (2013)

Among the challenges that face date palm cultivation and achieving optimum yield is the availability of fresh irrigation water. In an attempt to overcome the current status, the government began a program of wastewater recycling for irrigation purposes. It has been reported that the total quantity of Treated Sewage Effluent (TSE) pumped for the agricultural and landscaping sectors increased from 4,192 m^3/day in 1988 to 110,000 m^3/day in 2012 (Directorate of Agriculture Engineer and Water Resources, Agriculture Affairs). Today, there are 540 farms, covering an area of 2,960 ha, all connected to the TSE distribution network. The overall daily TSE utilized for the agriculture purposes represents 69,343 m^3/day, of which 20,000 m^3/day are allocated for date palms and fruits (Table 11.11).

Besides the above and in order to protect and encourage date palm plantation, the government supports farmers interested in date palm cultivation with easy and flexible loans (Al-Basheer and Harron 1997; BFNRCBD 2006). At present, no modern packaging factories are available. In 1981, the government founded a date packaging factory, which was intended to subsidize date palm growers through commercial support of their output. The factory was involved in food industry through frozen and dry dates packaging and molasses extraction. The factory persisted for 10 years prior to privatization and subsequent collapse. The inability of the private sector to sustain a profitable packaging factory could be related to the limitation of the local market place and the strong competition with the neighboring countries, which tend to produce dates with superior quality, along with better packaging and lower prices. Recently, the Ministry of Industry and Commerce has licensed two date production and packaging factories. One factory would sustain a production capacity of 6,000 mt/year, whereas the second would support a production capacity of only 40 mt/year. It is worth mentioning that both factories are under construction.

Due to high internal demand, Bahrain relies on importing dates from various producing countries, with majority from Saudi Arabia, United Arab Emirates, and Oman. The total quantities of dates imported by Bahrain in 2009–2011 were around 678, 618, 976 mt, respectively (Foreign Trade 2014).

In Bahrain, there is no well-structured strategy to ensure that the returns from investment in the date palm industry are in line with the efforts and costs it takes to maintain a date palm plantation and fruit processing and marketing. In order to establish a well-functioning and competitive market that could benefit fully from the ongoing globalization, several actions should be considered. Improving the quality of the fruits by focusing on growing cultivars, which are economically rewarding; capacity building including manpower as well as the infrastructure; and supporting and encouraging scientific research together can boost the return on investment in the Bahrain date palm industry.

11.8 Conclusions and Recommendations

In conclusion, there is the need to find ways to improve the date palm sector in Bahrain. The long-term improvement of the date palm industry requires a radical shift from the traditional cultivation, harvesting, and postharvesting methods toward new practical and well-designed alternatives which can reduce economic, social, and environmental costs.

The following actions are strongly recommended to develop the date palm sector in Bahrain:

(a) Launch a national strategic development plan aimed at reinforcement and supporting the date palm sector.
(b) Establish a center of research and training dealing with various aspects of date palm including cultivation, postharvest management, and marketing.
(c) Establish modern date palm plantations to insure higher field income.
(d) Replace poor cultivars and old trees. In order to boost plant production and yield, it is essential to replace the poor cultivars and old palms with superior more productive cultivars.
(e) Promote capacity building. This includes well-trained manpower and highly innovative modern techniques. Introducing and implementing state-of-the-art technologies and motivating the youth to engage in the date palm sector are key elements to boost date palm industry.
(f) Encourage and support scientific research in various fields concerning date palm production from cultivation to marketing by providing expertise and required funding.
(g) Foster and subsidize entrepreneurs concerned with the date palm industry.
(h) Promote the exchange of experience with other date palm growing countries, in particularly the neighboring GCC countries, in the various aspects of the date palm industry.
(i) Encourage founding new private units for date fruit postharvest packaging, processing, and marketing.
(j) Facilitate the adoption of modern cultivation and postharvest technologies and biotechnologies.

References

Abbas MST, Hanounik SB, Shahdad AS, Al-Bagham SA (2006) Aggregation pheromone traps, a major component of IPM strategy for the red palm weevil, *Rhynchophorus ferrugineus* in date palms (Coleoptera: Curculionidae). J Pest Sci 79:69–73

Abdillah LA, Andriani M (2012) Friendly alternative healthy drinks through the use of date seeds as coffee powder. In: Proceeding of ICEBM-Untar Jakarta. Universitas Tarumanagara, Jakarta, pp 80–87

Abraham VA, Faleiro JR, Al-Shuaibi MA, Al Abdan S (2001) Status of pheromone trap captured female red palm weevils from date gardens in Saudi Arabia. J Trop Agric 39:197–199

Abuagla AM, Al-Deeb MA (2012) Effect of bait quantity and trap color on the trapping efficacy of the pheromone trap for the red palm weevil, *Rhynchophorus ferrugineus*. J Insect Sci 12:1–6

Agriculture (2014) In: Kingdom of Bahrain Central Informatics Organization. Retrieved 4 May 2014, from http://www.cio.gov.bh/CIO_ENG/SubDetailed.aspx?subcatid=604
Ahmed R (1978) The Date palm in Bahrain, 2nd edn. Extension publication no 2. Ministry of Commerce and Agriculture. Directorate of Agriculture, Bahrain
Al-Asfoor AAA (2012) A study of some ecological and biological aspects of the red palm weevil in the Kingdom of Bahrain. M.S. Thesis, Arabian Gulf University
Al-Basheer AA, Harron AI (1997) An analytical study of agricultural systems and assess the economic reflexes of the technical obstacles facing the palm sector in the State of Bahrain. Date Palm Research and Development Network, ACSAD, Damascus
Ali-Mohamed AY, Khamis ASH (2004) Mineral ion content of the seeds of six cultivars of Bahraini date palm (*Phoenix dactylifera*). J Agric Food Chem 52:6522–6525
Al-Issa AM (1992) Vegetative propagation of date Palm (*Phoenix dactylifera* L. Khonaizy) by tissue culture. M.S. Thesis, The Arabian Gulf University, Bahrain
Al-Khalifa MA (2004) The palm tree in Bahrain history. Government Press, Kingdom of Bahrain
Al-Khalifa MA (2009) The status of date palms in the Kingdom of Bahrain: challenges and solutions. In: Proceeding of the date palm "life and heritage" symposium, Kingdom of Bahrain. Isa Culture Center, Manama, pp 311–348 (in Arabic)
Al-Khalifa MA (2010) Date palm the cultivars and the differences between them. Ministry of Municipalities and Agriculture Affairs, Agriculture Affairs, Kingdom of Bahrain
Al-Khalifa MA, Mansour AA, Muslim AA (1994) Date palm cultivation in Bahrain. Extension publication no. 1. Government Press, Ministry of Information, Kingdom of Bahrain
Al-Khalifah NS, Askari E, Shanavaskhan AE (2013) Date palm tissue culture and genetical identification of cultivars grown in Saudi Arabia. National Center for Agriculture Technologies, King Abdulaziz City for Science and Technology, Kingdom of Saudi Arabia
Alkhateeb AA (2008) A review the problems facing the use of tissue culture technique in date palm (*Phoenix dactylifera* L.). Sci J King Faisal Univ Basic Appl Sci 9:85–104
Allaith AAA (2008) Antioxidant activity of Bahraini date palm (*Phoenix dactylifera* L.) fruit of various cultivars Inter. J Food Sci Technol 43:1033–1040
Al-Laith AA (2009) Degradation kinetics of the antioxidant activity in date palm. Arab Gulf J Sci Res 276:16–22
Almansoori TA (2001) Salt tolerance in date palm (*Phoenix dactylifera* L.). Ph.D. Thesis, The University of Reading
Almansoori TA, El-Deen MNA, Caligari DS (2007) Evaluation of *in vitro* screening techniques for salt tolerance in date palm. Acta Hort 736:301–307
Al-Rumaihi MG (1975) Bahrain as study on social and political changes since the First World War. University of Kuwait, Kuwait
Al-Ruqaishi IA, Davey M, Alderson P, Mayes S (2008) Genetic relationships and genotype tracing in date palms (*Phoenix dactylifera* L.) in Oman, based on microsatellite markers. Plant Genet Res Char Util 6:70–72
Al-Saoud A (2010) Effect of red palm weevil, *Rhynchophorus ferruginous* Olivier (Coleoptera: Curculionidae) aggregation pheromone traps, height and colors on the number of captured weevils. Acta Hort 882:419–429
Annual Agricultural Statistics Bulletin (2008) Ministry of Municipalities and Agriculture Affairs, Agriculture Affairs, Kingdom of Bahrain
Anonymous (1998) Final report of the Indian technical team (part A), red palm weevil control project, Ministry of Agriculture and Water, Kingdom of Saudi Arabia
Anonymous (2011) Agricultural development in the states of the Cooperation Council for the Arab States of Gulf, 5th ed. Secretariat General of the Cooperation Council for the Arab States of Gulf Press (in Arabic), Riyadh, Saudi Arabia
AOAD (2004) Technical consultation to the Kingdom of Bahrain in the field of date palm propagation using tissue culture techniques. Arab Organization for Agricultural Development. League of Arab States, Al Khartoum (in Arabic)
Azam KM, Razvi SA, Al-Mahmuli I (2001) Survey of red palm weevil, *Rhynchophorus ferrugineus oliver.* infestation in date palm in Oman. In: Second international conference on date palms, Al-Ain, pp 25–27

Badawy EM, Habib AMA, El-Bana A, Yosry GM (2005) Propagation of date palm (*Phoenix dactylifera*) plants by using tissue culture technique. Arab J Biotechnol 8:343–354

Barreveld WH (1993) Date palm products. FAO agricultural services bulletin 101, Rome

BFNRCBD (2006) Bahrain First National Report to the Convention on Biological Diversity. Public Commission for the Protection of Marine Resources, Environment and Wildlife. General Directorate for Environment and Wildlife Protection, Kingdom of Bahrain

Bibby G (1972) Looking for Dilmun. Penguin, Harmondsworth

Doornkamp JC, Brunsden D, Jones DKC (eds) (1980) Geology, geomorphology and pedology of Bahrain. Geo Abstracts Ltd., Norwich

Elshibli S, Korpelainen H (2009) Biodiversity of date palms (*Phoenix dactylifera* L.) in Sudan: chemical, morphological and DNA polymorphisms of selected cultivars. Plant Genet Res Char Util 7:194–203

Faleiro JR (2000) Investigation of the role of pheromone trapping in the suppression of red palm weevil *Rhynchophorus ferrugineus* Oliv. population in coconut plantations. In: International conference on management national research sustainable agricultural products 21st century, New Delhi, pp 1338–1339

Foreign Trade (2014) In: Kingdom of Bahrain Central Informatics Organization. Retrieved 4 May 2014, from http://www.cio.gov.bh/CIO_ENG/SubDetailed.aspx?subcatid=604

Gassouma MS (2004) Pests of the date palm (*phoenix dactylifera*). In: Regional workshop on date palm development in the G.C.C. Countries of the Arabian Peninsula (ICARDA/MAF/UAEU), Abu-Dhabi

Gunawardena NE, Herath HMWKB (1995) Enhancement of the activity of ferrugineol by N-pentanol in an attractant baited trap for the coconut pest, *Rhynchophorus ferrugineus* F. (Coleoptera: Curculionidae). J Nat Sci Coun Sri Lanka 23:81–86

Hasnaoui A, Elhoumaizi MA, Hakkou A et al (2011) Physico-chemical characterization, classification and quality evaluation of date palm fruits of some Moroccan cultivars. J Sci Res 3:139–149

Horn DJ (1988) Ecological approach to pest management. The Guilford Press, New York

International Center for Agricultural Research in the Dry Areas (ICARDA) (2011) Development of sustainable date palm production systems in Gulf Cooperation Council Countries. Project report-1, Aleppo, Syria

Jain SM (2011) Prospects of in vitro conservation of date palm genetic diversity for sustainable production. Emir J Food Agric 23:110–119

Kaakeh W, El-Ezaby F, Aboul-Nour MM, Khamis AA (2001) Management of the red palm weevil, *Rhynchophorus ferrugineus* Oliv., by a pheremone/food-based trapping system. In: Second international conference on date palms, Al-Ain, pp 325–343

Khalifa O, El Assal AH, Ezaby FAA et al (2001) Database for infestation of date palm by red palm weevil *(Rhynchophorus ferrugineus* Oliver*)* in U.A.E. and Oman. In: Second international conference on date palms, Al-Ain, pp 25–27

Larsen CE (1983) Life and land use on the Bahrain Islands: the geoarchaeology of an ancient society. University of Chicago Press, USA

MAF (2012) Annual report 2012. Directorate General of Agriculture and Livestock Research, Ministry of Agriculture and Fisheries, Muscat, Sultanate of Oman

Massoud AM, Faleiro JR, Abo El-Saad M, Sultan E (2011) Geographic information system used for assessing the activity of the red palm weevil *Rhynchophorus ferrugineus* (Olivier) in the date palm oasis of Al-Hassa, Saudi Arabia. J Plant Prot Res 5:234–239

Mohamed S (2000) Major date palm field operations and their effect on improving fruit quality. In: Hamdan IY, Hegazi NA (eds) Date palm post-harvest processing technology. FAO Regional Office for the Near East, Cairo

Mohamed AMA (2010) Red palm weevil insect. In: Proceeding of the date palm "life and heritage" symposium, Kingdom of Bahrain. Isa Culture Center, Manama, pp 193–207 (in Arabic)

Mohamed AMA (2012) Report on monitoring and control of red palm weevil (2009–2011). Agriculture Affairs, Ministry of Municipalities Affairs and Urban Planning, Kingdom of Bahrain (in Arabic), Kingdom of Bahrain

Mohamed AMA, Al-Asfoor AA, Ghanem IA (unpublished) Faunal survey of insects and mites associated with date palm in Kingdom of Bahrain

Murphy ST, Briscoe BR (1999) The red palm weevil as an alien invasive: biology and the prospects for biological control as a component of IPM. Biocont News Info 20:35–46
Nesbitt M (1993) Archaeobotanical evidence for early Dilmun diet at Saar, Bahrain. Arab Arch Epigr 4:20–47
Noor Al-Nabi M (2012) History of land use and development of Bahrain. Information Affairs Authority, Directorate of Government Printing Press, Kingdom of Bahrain
Oehlschlager AC, McDonald RS, Chinchilla CM, Patschke SN (1995) Influence of a pheromone-based mass-trapping system on the distribution of *Rhynchophorus palmarum* (Coleoptera: Curculionidae) in Oil Palm. Environ Entom 24:1005–1012
Pathak MR, Hamzah R (2008) RAPD analysis of date palm cultivars of Bahrain. Flori Orn Biotech 2:9–11
Pontikakos C, Kontodimas D (2010) A location aware system for integrated management of *Rhynchophorus ferrugineus*. Dies Palmarum, San Remo, pp 18–20
Rao NK (2004) Plant genetic resources: advancing conservation and use through biotechnology. Afr J Biotech 3:136–145
Saker MM, Bekheet SA, Taha HS et al (2000) Detection of somaclonal variation in tissue culture derived date palm plants using isoenzyme analysis and RAPD fingerprints. Bio Plant 43:347–351
Sawaya WN, Miski AM, Khalil JK et al (1983) Physicochemical characterization of the major date varieties grown in Saudi Arabia. Date Palm J 2:1–25
Soroker V, Puma P, La Pergola A et al (2013) Early detection and monitoring of red palm weevil approaches and challenges. In: AFPP-palm pest mediterranean conference, Nice. http://www.sauvonsnospalmiers.fr/IMG/pdf/4_-_soroker_victoria.pdf
Tisserat B (1979) Propagation of date palm (*Phoenix dactylifera* L.) in vitro. J Exp Bot 30:1275–1283
Vidyasagar PSPV, Al-Saihati AA, Al-Mohanna OE et al (2000a) Management of red palm weevil *Rhynchophorus ferrugineus* Olivier, a serious pest of date palm in Al-Qatif, Kingdom of Saudi Arabia. J Plant Crops 28:35–43
Vidyasagar PSPV, Hagi M, Abozuhairah RA et al (2000b) Impact of mass pheromone trapping on red palm weevil adult population and infestation level in date palm gardens of Saudi Arabia. Planter 76:347–355
Zaid A, Arias EJ (1999) Date palm cultivation. FAO plant production and protection paper 156. FAO, Rome

Chapter 12
Date Palm Status and Perspective in Syria

Nadia Haider

Abstract Date palm is one of the holy trees in Syria due to its cultural significance, in addition to its economic and environmental benefits. It is cultivated in the arid regions of the country around Palmyra and in some eastern areas along the Euphrates Basin. The Ministry of Agriculture and Agrarian Reform designed a plan for developing the date palm tree in Syria beginning in 1986 when it defined the optimal belt for this tree according to the environmental requirements of its cultivation and production. This belt constitutes about one-third of the total area of Syria and it includes the majority of the Syrian Badia lands. Since 1986, centers for date palm propagation have been established in provinces within the belt, focusing on elite lines and cultivars to collect offshoots from both elite lines and native and introduced cultivars to serve as mother orchards for selected genetic resources which are propagated to generate offshoots to be cultivated in suitable regions of the country. The Ministry of Agriculture also encouraged farmers to plant date palm within the specified belt to expand its cultivation area. Added to the various cultivars that are propagated within the date palm belt area, there are a set of lines of seedling origin in the Palmyra oasis and orchards that are being characterized for adoption as promising local Syrian cultivars. More research should be carried out for DNA-fingerprinting of date palm cultivars grown in Syria and their propagation by tissue culture.

Keywords Conservation • Cultivation • Date palm • Genetic resources • Syria

N. Haider
Department of Molecular Biology and Biotechnology,
AECS, 6091, Damascus, Syria
e-mail: ascientific@aec.org.sy

© Springer Science+Business Media Dordrecht 2015

J.M. Al-Khayri et al. (eds.), *Date Palm Genetic Resources and Utilization: Volume 2: Asia and Europe*, DOI 10.1007/978-94-017-9707-8_12

12.1 Introduction

12.1.1 Historical and Current Agricultural Aspects

The date palm tree has been prominent in what is today Syria for centuries. Archaeological remains of dates have been found on a number of Neolithic sites, particularly in Syria and Egypt (El-Juhany 2010). Date palms and their culture are depicted in ancient Babylonian and Assyrian tablets including the famous Code of Hammurabi, which contained laws pertaining to date culture and sales (Popenoe 1973). In the first half of the third millennium BC, the date palm was one of the most cultivated trees in the ancient city of Mari in Syria. Babylonians developed planting of the tree along the Euphrates (known in Syria as Al-Furat) River banks about 5,000 BC. Phoenicians also paid great attention to date palm hence it was called Phoenix. This tree was holy to the population of Palmyra for several centuries BC. This is evidenced by the Arabic name of the city, Tad-Mour, which means in the ancient language of Palmyra the place of palm tree (Qatana online reference). In Syria, date palm is cultivated in the arid regions around Palmyra and in some eastern areas along the Euphrates Basin (from Deir Al-Zour to Al-Boukamal), in addition to some dispersed trees in other areas of the country (Kaakeh online reference). It is also cultivated in other Syrian cities as an ornamental tree (Shibli 2008). Date palm cultivation can be expanded in regions east and west of the Euphrates River, west and south of Palmyra, and in the coastal region.

The date palm belt constitutes 30 % of the total area of Syria and it includes the majority of the Syrian Badia lands in which the rainfall level does not exceed 200 mm/year. Statistics for 2009 estimated that 1,900 ha of land are planted with date palm and that the number of trees had reached 500,000, of which 250,000 were fruiting in private orchards. The Ministry of Agriculture and Agrarian Reform has also started to: (a) cultivate date palm instead of olive and pistachio trees in the Syrian Badia because date palm suits the environment of Badia better than the two other trees and has higher economic benefit and (b) establish date palm oases around existing wells in the Badia.

12.1.2 Importance to Syrian Agriculture

Date palm tree is one of the holy trees in Syria due to its cultural significance as well as its economic and environmental benefits. The threat to date palms in some areas has arisen because of their neglect for many years and the lack of effort made to expand cultivation. In 1986, the Ministry of Agriculture became aware of the situation and therefore designed a plan to direct more attention to the date palm and to expand its cultivation (Al-Baba 2009).

The date palm is considered as the life tree in dry regions of Syria because it is the most adapted tree in such regions to drought, high temperatures, dusty wind, and salinity of soil and water shortage to certain levels. It produces the highly nutritive date fruits, leaves, and other residues of the different parts of the tree which have various home and rural industrial uses. Date palm represents a source of income and nutrition to oasis inhabitants and it contributes to providing job opportunities for many rural women (Al-Baba 2009). Cultivation of date palm also creates favorable conditions for improving secondary crop culture such as cereals, alfalfa, and certain vegetables. In 1978, growers started to intercrop date palms (as windbreaks) with citrus in Al-Boukamal to protect the citrus trees from frost and strong wind. It has been also intercropped with other trees such as almonds. The date palm develops roots up to 10 m long which contributes to the maintenance of the soil against erosion (Zabar and Borowy 2012). Unripe dates, seeds, and leaves are used to feed livestock. Another use of the leaves is making compost.

Date palm also helps in creating oases near dry sand dunes in deserts to stop desert encroachment. In such oases, date palms provide shade for smaller trees and annual vegetables and cereals. This helps the farmers make the best use of the soil and water. Oases also provide habitat for animals and even humans if the area is large enough (http://en.wikipedia.org/wiki/Oasis). Since 2009, the Ministry of Agriculture has established several date palm oases in the Badia using offshoots of cultivars grown in the Palmyra and Al-Boukamal propagation centers in order to expand Syria's date palm cultivation. These are the oases of Al-Sawana (1,650 offshoots), Al-Talila (in Palmyra, 1,800 offshoots of major cultivars), Al-Tanf (3,000 offshoots of seedling origin), and Al-Manqoora protectorate (12,600 ha) which includes 1,500 date palm trees and is expanded every year to include 5,000 trees. In these oases and other private oases that are used as productive orchards, date palm trees are allowed enough space (10 × 10 m) and cultivated with other fruit trees such as citrus (Al-Boukamal oases), olives (Palmyra oases), peach, plum, and green plum (of 5–10 years production age) (private oases). The vegetables and cereals that are also grown in the oases are okra, tomato, eggplant, trefoil (clover), alfalfa, barley, cotton, and sugar beet. The private oasis in Deir Al-Zour and Raqqa are irrigated from the Al-Furat River, and those in Hasaka and Palmyra are irrigated from the Khabour River and artesian wells, respectively. During 2008–2011, 1,000 offshoots were planted at the site of well No. 17 in Palmyra Badia, and 20,000 offshoots of seedling origin were also planted around wells and protectorates of Badia in the different provinces within the date palm belt (Al Maalouf 2012). The Palmyra date palm oasis, which lies at the lowest point of the Syrian Desert, collects water from hundreds of kilometers in all directions (Serra et al. 2009); it has an area of 3,000 ha, where in addition to dates, figs and pomegranate are also planted. The Syrian General Commission for Al-Badia Management and Development is planning to establish a 200 ha oasis in Mahin town of Homs countryside to be planted with 5,000 date palm trees, along with 10,000 cactus offshoots as a source of forage for the animals in the Badia (Sana 2012). Since most date palm oases in Syria are recent, trees have not entered the production phase yet.

Table 12.1 Location, cultivation area, and the number of trees cultivated for each date palm propagation center

Center name	Date of establishment	Area (ha)	No. of trees in 2009
Date palm propagation center in Palmyra	1987	200	4,300
Date palm propagation center in Sabkhat Al-Moh	1999	400	6,000
Zanoobya palm oasis in Palmyra	2001	200	1,350
Date palm propagation center in Al-Boukamal	1987	800	19,500
Date palm propagation center in Qahtaniya in Raqqa	2000	400	2,500
Date palm propagation center in Saalo in Deir Al-Zour	2003	640	4,500
Date palm center in Al-Khabour, Hasaka	2005	420	3,400
Date palm propagation center in Al-Balash, Hasaka	2007	400	1,500
Date palm propagation center in Manqoora protectress in Damascus countryside	2007	400	1,000
Total	–	3,860	44,050

Source: Al-Baba (2011)

12.1.3 Production Statistics and Economics

There are several propagation centers in Syria for the production of date palm using traditional methods (offshoots). Table 12.1 shows their locations, land areas, and number of trees cultivated (Al-Baba 2011).

As for the annual date fruit production, it is estimated to be 5,000 mt according to the statistics of the Ministry of Agriculture in 2010. It is worth noting here that most date palm trees in Syria have been planted recently and therefore have not entered the production phase yet and that in 1986, the number of cultivated date palm trees was only 40,000 and the annual production was 500 mt (Al-Baba 2011).

12.1.4 Current Agricultural Problems

Until 2000, there existed a set of barriers to developing date palm cultivation in Syria; one was the unavailability of reliable offshoots that suit the environment in the date palm belt. Another barrier was the low number of qualified and trained technicians with knowledge of date palm cultivation requirements, because the crop had been neglected and had previously been regarded as a secondary tree crop for several reasons, such as (a) most of the old date palm trees were of seedling origin which had a low return and a below-average fruit quality, (b) rarity of cultivars of commercial value, (c) dependence on date imports, and (d) little awareness of the value of the date palm tree. It should be noted that date palm offshoots of seedling origin are used in Syria only for ornamental purposes and as windbreaks around farms and orchards. Seeds are planted in plastic bags which are kept in the nursery until each seedling has 3–5 leaves. The seedlings (50 % will be females) are then transferred to the permanent location. Some elite male and female lines (0.3–0.5 %) can be selected from trees of seedling origin and adopted as local cultivars. Other problems were (a) the absence of involvement of the private sector in establishing date palm orchards for production and trade purposes; (b) the absence of an official decision-making body, with financial and technical independence, to deal with developing date palm in Syria at all levels (Al-Baba 2011; Electronic Economic Newspaper 2007); (c) poor general farm management; (d) pests and diseases and inadequate Integrated Pest Management (IPM) control; (e) underdeveloped harvesting, processing, and marketing practices; and (f) insufficient research and development.

The problem of planting material was resolved in 1995–1997 by importing a collection of high-quality date palm cultivars suitable for the date palm belt in Syria. These cultivars were introduced from Iraq, Iran, Egypt, Algeria, Morocco, UAE, and Saudi Arabia to the different propagation centers and entered the primary phase of fruit and offshoot production, beginning in 2000. The outcomes were superior fruit quality and quantity becoming available. Since then reliable offshoots of elite cultivars are being produced in centers of Palmyra and Al-Boukamal, most of which are sold to farmers at the nominal price of 500 SYP (Syrian pounds) per offshoot, while cost is 3,000–3,500 SPY on the open market. The program was aimed at encouraging farmers to expand date palm cultivation within the designated belt. Added to that, training courses and exhibitions have been carried out to introduce farmers to: (a) the value of the date palm tree, particularly in favorable regions where other fruit trees do not perform as well, and (b) date palm cultivation and care. This has led to an increase in farmers' demand for offshoots from propagation centers and to developing new date orchards, the number of which is continuing to increase each year. One of the other limiting factors for date cultivation in Syria is the shortage of irrigation water and the difficulty in digging wells in suitable regions within the belt designated for the date palm. Another barrier is prohibiting the planting of date palm trees in irrigated plain lands that are defined for growing other economic crops, such as wheat, cotton, and beets. This was resolved after the issuance of decree No. 20 in 2008 by the Prime Minister's Counsel that permitted planting date palm within the date palm belt, of which the eastern provinces lands and Euphrates River Basin constitute the majority, in addition to the Palmyra region and parts of Syrian Badia. The traditional methods used for propagation of date palm (offshoots) are a limitation which could be ameliorated through tissue culture. Added to what is stated above, many Syrians do not appreciate the value of the date palm because their views are influenced by the presence of old date palm trees of seedling origin (*deqel*) and hence of low productivity and producing fruit of below-average quality. This problem has been overcome by providing farmers offshoots at attractive prices.

12.2 Cultivation Practices

12.2.1 Chronological Account of Research and Development

The Ministry of Agriculture designed a plan for developing date palm cultivation in Syria. It defined the cultivation belt for the palm (Fig. 12.1) in 1986, according to the environmental requirements necessary for successful cultivation and fruit production. Five priority areas were specified within the date palm belt: (a) Al-Boukamal and Deir Al-Zour, (b) Raqqa and Marqada in Hasaka, (c) Palmyra, (d) Al-Zulf and Al-Tanf in Badia of the Damascus countryside, and (e) Khanaser (in Aleppo Badia) and East Damascus. In the same year, centers for date palm propagation were established in Palmyra and Al-Boukamal in order to collect offshoots from elite lines and native and introduced cultivars. These centers function as mother

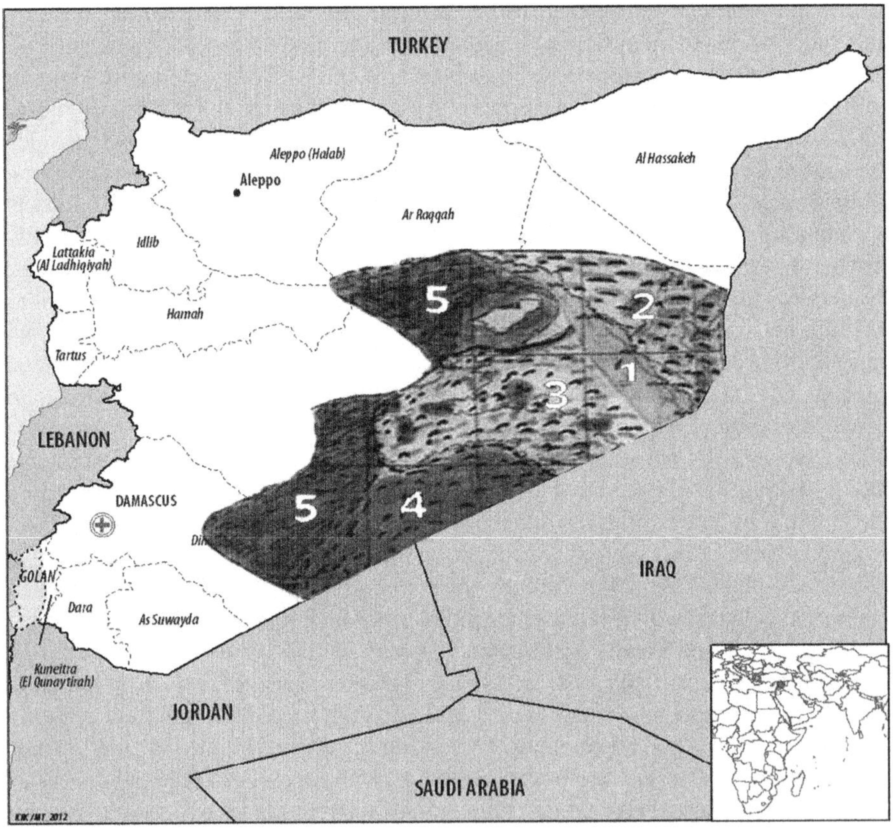

Fig. 12.1 A map of Syria showing the five priority areas in the date palm belt. *1* Al-Boukamal and Deir Al-Zour, *2* Raqqa and Marqada in Hasaka, *3* Palmyra, *4* Al-Zulf and Al-Tanf in Badia of the Damascus countryside, *5* Khanaser (in Aleppo Badia) and East Damascus (*Source*: Modified from Al-Baba (2000))

orchards of selected genetic resources which are propagated to generate offshoots for cultivation in suitable regions of the country. The centers also provide farmers with offshoots and guide them on how to cultivate and manage the date palm. National experts in date palm morphology also characterize and describe the female and male lines that are generated in those centers from seeds of the elite introduced cultivars to obtain local cultivars (Hakkar 2012). At that time, the number of offshoots and the cultivars that could be obtained from farmers' orchards in Palmyra and Deir Al-Zour was very limited because the farmers were neglecting the palms for several reasons, such as (a) imported dates were available in the local market at low prices; (b) the difficulty in dealing with date palm cultivation, pollination, and other services; and (c) the lack of awareness of farmers of the technical agricultural practices the date palm tree needs. In 2005, the number of offshoots generated from the centers of Al-Boukamal, Palmyra, and Sabkhat Al-Moh was 11,000 (Teshreen Newspaper 2005).

12.2.2 Description of Current Cultivation Practices

Date palm propagation in Syria is carried out through offshoots. The offshoots develop as suckers at the base of the mother plant, and consequently the fruits produced will be of the same quality as the mother palm (true to type). The season to separate offshoots either at the propagation centers or by farmers within the date palm belt takes place from April to June which is the best time for date palm cultivation in Syria. The disease- and pest-free offshoots (3–5 years old) are cut and properly separated from the mother palm. The offshoots are either planted in their permanent field location or spend a rooting period of 1 year in a nursery to ensure an optimum survival rate and to avoid uneven development of the plantation (Zaid and de Wet 2002a), as well as to reduce labor costs. In order to facilitate intercultural operations and proper development of the palms, planting is done, generally, at 8 m distance between rows and plants in a square system. A total of 156 palms are accommodated in 1 ha (8×10 m^2), 8 m in the rows and 10 m between rows, 8×8 m, or 10×10 m. Of the palms, 4 % should be males to provide adequate pollen grains. Offshoots are irrigated immediately after planting. Then for 45 days, they are irrigated every day in sandy soils, every 3 days for medium texture soils, and every 5 days in heavy soils, noting that date palm prefers the light soils. During the first year of planting, the irrigation quantity is 100 l per offshoot per irrigation. The irrigation interval, for drip or flood irrigation, is then gradually increased to two times a week and then once a week and then according to the need, growth vigor of offshoots, the environmental conditions, and soil type. In addition to irrigation, date palms need good nutrient management and pest and disease control (Grant online reference). The oasis should be kept clean by turning over the soil and weeding; and care for the date palm extends over almost the entire year according to specific needs, in order to have healthy and highly productive palms (Qatana online reference).

A sufficient number of green date palm leaves is necessary for growth, development, and yield. Insufficient number of leaves results in low-quality fruits and lesser inflorescence production the following spring (Pal online reference). Pruning of leaves is practiced by removal of old dead, infected, or broken leaves using sickle saws. The main purpose of pruning is to clean tree, to allow new leaves to grow and photosynthesize, and to facilitate fruit harvest and other activities. There are two periods for dry leaf pruning, during the pollination process in spring and during fruit harvest in autumn. Each fruit bunch needs nine leaves for normal growth.

Fruit thinning is necessary for several reasons, namely, to ensure adequate flowering in the following year, to improve fruit quality, to prevent delayed ripening, to reduce compacting of the fruit, and to increase ventilation of the bunches. Thinning parameters are the size of the bunch, density of fruits, tree age and vigor, and the cultivation practices (Al-Baba 2000). Only five bunches are left on each adult tree if it is weak due to poor cultivation practices. This number can be increased to 10–12 for a vigorous adult tree that receives proper care. Fruit thinning is done manually in July and August either by removal of some bunches (usually from the top and bottom of the crown) depending on the number of active leaves available for each bunch (9–10 leaves per bunch in full production phase) or thinning by removing a few strands from the middle of the crowded fruit bunches, or shortening the length of strands by cutting almost 1/4 of the female bunch to get rid of the weak terminal flowers (Pal online reference). The three methods of thinning might be carried out during the same thinning operation. From the fifth year onward, three to four bunches are left for each tree. In areas where production is possible, fruits are thinned by one-half for higher-quality fruits (e.g., Medjool). In the heavy production phase, 9–12 bunches are left on the tree.

Nutrient application is important for satisfactory production of quality date fruits. A dose of 50 kg N, 30 kg P, and 20 kg/ha K is recommended every year starting from the third year of offshoot planting. Amounts of P and K depend on the soil type and growth vigor. P and K fertilizers are added at the beginning of winter, while N fertilizer is added in March and April; alternatively, compound fertilizers NPK 20:10:10 are added in spring at the rate of 6 kg per tree in two equal applications. From the third to the eighth year of a plantation, the amount increases gradually to become 2 kg after the age of 15 years. Organic fertilizers (green manures) are also added in winter at the rate of 50 kg per adult tree and at a horizontal distance of about 50 cm from the base of the stem.

As for irrigation, it is done immediately after transplanting the palm for newly planted offshoots, to limit transplant stress. Once the plantation is established, a frequent irrigation schedule is followed to provide a sufficient water supply to the young date palms. In Al-Boukamal date palm propagation center, the irrigation is done by flood irrigation, while the drip system is applied in the remaining propagation centers. Irrigation requirements for palm tree range 15,000–28,000 m^3/year, depending on the age of the palms, soil type, and method used.

12.2.3 Pollination, Fruit Quality, and Metaxenia

Date palm is highly cross-pollinating due to its dioecious nature. In all propagation centers in Syria, manual pollination by climbing the tree is used. For this, four male trees are enough to pollinate 100 female palms. Two methods for pollination are applied. In the first method, about 5–7 strands of selected male flowers are inserted between the strands of female flowers of each spadix within 3 months from start of opening (April to June). The pollinated bunches are tied for a month and then covered with perforated bags mainly to increase the rate of expansion. In the second method, extracted pollen are dried and then mixed with flour at a ratio of pollen/flour 1:9 g. A piece of cotton cloth is immersed in the mixture and then put inside the female bunch which is then tied and covered with a perforated paper bag. If the male spathes open earlier than the female, the pollen grains are dried at a location with low humidity and no wind and then stored for later use. If the pollen grains are to be kept for the next year, they are put in tightly closed glass containers and stored at 0 to −4 °C or −18 to −20 °C if they are to be kept longer. At the biser stage (the beginning of ripening), bunches are enclosed in a net to protect them from birds and insects and to prevent falling of the fruits. Several operations are performed after pollination and fruit set. These include thinning of heavily loaded palms, release of bunches from the tree crown, thinning of the bunches, and their protection from birds and rodents (Nizwa.NET online reference). Swingle (1928) reported on the direct effect of pollen from a male clone on the morphology and other characters of seed and fruit tissues surrounding the embryo and endosperm and defined the phenomenon as metaxenia, which is quite common in date palm; therefore selection of good pollinator is important. The quality of date fruits, particularly fruit size and time of ripening, is influenced by the quality of the pollen. Research is being conducted in the date palm propagation centers on the selection of the best pollinizers based on their effect on fruit quality and ripening time and their compatibility with female cultivars. The same pollinizer is tested on several female cultivars for several seasons. Thus far, no prominent male has been identified in these centers for pollination. Hence, a mixture of pollen grains from several elite date palm males is used for pollination of female trees.

In a study of the influence of elite date palm males on qualitative and quantitative characteristics of some date palm cultivars in Syria, it was revealed that the pollen used for pollination influenced the fruits quality of Khastawi, Birbin, and Zahidi cvs. and on the net rate of productivity. It also had a significant effect on weight, size, length, and width of the fruit (GCSAR 2012).

12.2.4 Pest and Disease Control

The date palm is subjected, during its different life stages, to various pests and diseases which have a pronounced effect on palm growth and date fruit yield and quality. The importance of these pest and diseases and the harm they cause depend on the region where they are cultivated (Shibli 2008). Although each of date palm-growing countries has its own list of pests and diseases, many of them are common to all countries (El-Juhany 2010). The main insects and fungal diseases which affect date palms and dates in Syria are inflorescence rot (al-khamedj), alwijam disease, and black scorch. As for the insects, they are dubas bug (*Ommatissus lybicus* Deberg.), the lesser date moth (*Batrachedra amydraula* Meyr.), the large date moth (*Arenipses sabella* Hamps.), fruit stalk borer (*Oryctes elegans* Prell, *O. agamemnon* Arabicus, and *O. boas* Fab.), frond borer (*Phonapate frontalis* sub), and dust mite (*Oligonychus afrasiaticus* McG.). It is worth noting that there are no viral infections reported on date palms in Syria and that the date palms in the coastal region are grown only as ornamentals.

In 2005, a number of palm trees in the coastal region were infected with the red palm weevil (*Rhynchophorus ferrugineus* Olivier), which entered Syria with offshoots and trees imported illegally. The Ministry of Agriculture took strict quarantine precautions to prevent transfer of this insect to the optimal regions of date cultivation (Palmyra and Al-Boukamal). Therefore, the insect remains under control (Al-Baba 2009). One of the other precautions to control the insect is spraying healthy trees with the insecticide Decis (or Lentrak) once every 21 days during the period from the beginning of spring to the beginning of autumn. If the infection is discovered early, 4–6 Phostoxin disks are inserted within the holes formed in the trunk by the female insect, which is then covered by gypsum. The whole tree, however, is felled, burned, and then buried at a depth of 1 m, in those areas that are severely affected. Another crucial precaution is that wounds should be avoided, and in order to achieve that, pruning wounds and the tree offshoots separation areas are covered with a paste made of insecticide and gypsum. Trees are also sprayed with a selective insecticide after leaf pruning, offshoot separation, inflorescence appearance, and fruit harvest. Infection from the red palm weevil can be prevented by the use of noninfested seedlings which can be achieved through tissue cultivation propagation of seedlings (Mahmoudi et al. 2008). According to Jamal (http://tishreen.news.sy/tishreen/public/read/258450), traditional propagation of date palm through offshoots generates plants more resistant to the red palm weevil as compared to those derived from tissue culture.

These pests and diseases are controlled by using systemic chemical pesticides. Biological control was used only for the dubas bug. In 2009, a specialized committee, which included experts from the Ministry of Agriculture, the Arab Center for the Studies of Arid Zones and Dry Lands (ACSAD), Atomic Energy Commission of Syria (AECS), and the General Commission for Scientific Agricultural Research (GCSAR), was formed to survey local date pests and design efficient mechanisms for prevention and control. The committee plans to release an Atlas and develop an integrated pest control plan (Al-Baba 2011).

Kaakeh (online reference) reported the infection of date palm in Deir Al-Zour with three species of insects: the Parlatoria date scale (*Parlatoria blanchardi* Targ.), the saw-toothed grain beetle (*Oryzaephilus surinamensis* L.), and the lesser date moth (*Batrachedra amydraula* Meyr.). He defined the degree of infestation and studied the effects of various temperatures on some biological features of the insects infesting stored fruits, in order to determine the life cycles of these insects to establish ideal date fruit storage conditions to avoid infestations.

A Syrian quarantine act issued on 6.5.1998 prohibits the import of date palm trees and offshoots; it only permits importation of tissue culture plantlets which are up to 6 months of age after initial hardening (25–35 cm length with 4–5 primary leaves), planted in peat moss and free of all pest and diseases (Afif 2006).

12.3 Genetic Resources and Conservation

In Syria, the date palm is distributed mainly in the northeastern region and some inland areas. The cities of Homs (Palmyra) and Deir Al-Zour (Fig. 12.1) are two traditional areas of seedling date palm gardens. In 2007, there were in Palmyra 60,000 domesticated date palm trees of seedling origin with high genetic diversity. Of these, 20,000 trees were female. Evaluation and selection of date palm genetic resources of seedling origin resulted in the description of a total of 120 new elite lines, some of which will be adopted as Syrian local cultivars after their characterization and evaluation of the stability of their characteristics (Electronic Economic Newspaper 2007). As for commonly grown cultivars (e.g., Zahidi, Deglet Noor), the majority are introduced. Some of these cultivars were recently introduced to Syria and have not begun to bear fruit.

12.3.1 Chronological Account of Research in Genetics, Breeding, and Conservation

Krueger (1998) believes that there are possibly no examples of wild *Phoenix dactylifera* due to the long history of exploitation and selection of date palm, although there may be a few apparently wild groves still growing around oases, springs, or seepage areas, most of which are the end result of an unknown number of acts of selection. In Syria, the Ministry of Agriculture gave special attention to date palm tree starting in 1986. It established three production centers in Palmyra, Al-Boukamal, and Sabkhat Al Moh and a department for date palm. The functions of these centers are to (a) collect, conserve, and breed native or introduced cultivars and new elite lines of date palm that are adapted to the environmental conditions in these regions to act as mother orchards to provide offshoots for the date palm belt and (b) expand date palm cultivation quantitatively within the date palm belt (Al-Baba 2000). Added to that, these centers function as model farms to train date palm technicians (Al-Baba 2009). Experts from these centers and the Ministry of Agriculture cooperated with date palm researchers from ACSAD and did a lot of work to develop date cultivation in Syria (Hakkar 2012). The number of elite adoptive cultivars in these centers is expected to reach 40, including native and introduced. A map was designed for the distribution of these cultivars in priority areas based upon the

Table 12.2 Date palm cultivars grown in Syria and the number of trees of each cultivar as of 2009

Cultivar	No. of trees	Cultivar	No. of trees
Asabe El-Arous	250	Khyara	250
Ashrasi	12,000	Lolo	9,500
Barhi	21,000	Maktoom	8,500
Birbin	28,000	Medjool	20,000
Deglet Noor	15,000	Mumtaza	2,400
Fard	250	Nabtat-seyf	20,700
Gish Rabi	10,000	Shahabi	4,000
Jwahir	2,800	Sheikh Ali	350
Kabkab (red)	7,000	Shishi	2,500
Kabkab (yellow)	7,300	Smitni	2,000
Khadrawy	2,500	Sukkari	2,000
Khalas	22,000	Tafsirt	2,000
Khashram	200	Tagiat	2,200
Khastawi	40,000	Zaghloul	8,000
Khineze	4,300	Zahidi	46,000
Khudri	2,300		

Source: Al-Baba (2011)
Local lines of seedling origin which constitute a large proportion of date palms in Syria. The elite lines of these are characterized and selected to be adopted as local Syrian cultivars

geographical location within the belt and the environmental requirements of each cultivar (Al-Baba 2011). Table 12.2 shows date palm cultivars grown in Syria and the number of trees of each cultivar, as of 2009. Of these, only six cultivars are native: Khastawi, Zahidi, Ashrasi, Birbin, Maktoom, and Asabe El-Arous (Al-Kamour 2006). Areas cultivated with date palm, number of bearing date palms, and cultivated offshoots, by province, as of 2009, are listed in Table 12.3 (Al-Baba 2011). The Ministry of Agriculture is encouraging farmers to expand date palm cultivation area within the specified date belt due to environmental and economic importance as well as observed superiority of date palm over other fruit trees.

Established cultivars were introduced as in vitro plants from Saudi Arabia, UAE, Iran, Libya, and Egypt (Al-Baba 2011). There are also other cultivars that were imported from the date palm tissue culture laboratories of the UAE which are being hardened at the present time for propagating them after they prove suitable for the date palm belt. Haider et al. (2012) assessed, for the first time, the genetic relationships among five prominent male date palm genotypes and 18 native and introduced female date palm cultivars clonally propagated in Syria.

There are three main types of Syrian date cultivars based on fruit moisture content: soft such as Birbin, semidry such as Zahidi, and dry such as Ashrasi. The most important cultivars are Medjool (Mujhoolah), Khastawi, Barhi, Khalas, Nabtat-seyf, Lolo, Gish Rabi, Zaghloul, Kabkab (yellow), and Shahanit.

In Syria, added to the various cultivars that are propagated within the date palm belt area, there are a set of lines of seedling origin in the Deir Al-Zour and Palmyra oases and in orchards. Research is being carried out on elite lines in all the centers

Table 12.3 Areas cultivated with date palm and the number of date palm fruiting trees and cultivated offshoots by province as of 2009

Province	Area (ha)	No. of fruiting trees	No. of offshoots
Aleppo	40	19,350	11,000
Al-Ghaab	Minor area	595	–
Damascus countryside	243	8,500	1,000
Daraa	12	750	–
Deir Al-Zour	3,278	166,000	75,000
Hama	36	1,200	500
Hasaka	162	4,600	4,500
Homs (Palmyra)	3,640	23,000	20,000
Latakia	32	6,000	1,500
Quneitra	Minor area	150	–
Raqqa	283	7,900	6,600
Swaida	Minor area	405	–
Tartous	Minor area	2,350	1,000
Total	7,726 plus minor areas	240,800	121,100

Source: Al-Baba (2011)

(Al-Baba 2000). A set of 32 elite local female lines and 6 males of high quality and quantity of pollen was selected in these centers. The number of individuals of each of these lines is still limited due to the traditional offshoot propagation method used in the centers. Morphological characterization of these lines is in progress for adopting them as Syrian cultivars.

As for breeding of date palm to achieve the highest fruit quality and yield, the most serious drawback is the long time period from seed to flowering of about 6 years (Nixon and Furr 1965) and the 5-year minimum required to produce enough offshoots for trials and 10–15 years to reach full fruit production (Krueger 1998). There are currently some ongoing research projects on breeding of date palm in the propagation centers but there are no results yet.

12.3.2 Current Status and Prospect of Genetic Resources

Date palm germplasm should be better characterized and evaluated for successful utilization of its genetic resources (Krueger 2011). There are specialized committees in the Ministry of Agriculture which deal with characterization and classification of local date palm lines in order to select the elite lines and adopt them as Syrian local cultivars (Al-Baba 2000). Experts in the Ministry also characterize all date palm lines of seedling origin in all Syrian provinces, especially Palmyra, in order to select the elite date palms based on quantitative and qualitative measures, to adopt them as Syrian cultivars and give them distinctive local names. In 2001, the Ministry of Agriculture specified a portion of the offshoots produced from adapted and reliable cultivars in date palm propagation centers for sale to farmers at low prices, and the number of these offshoots is increasing every year. This has led to an increase in the number of date palms in the entire country which has reached 500,000 (as of 2009) and expansion of areas of date palm cultivation within the belt. The number of offshoots produced by these centers reached 21,500 in the 2009 season, in addition to the offshoots produced by the farmers.

12.3.3 Germplasm Conservation

The Ministry of Agriculture is trying to develop the cultivation and production of date palm in Syria by expanding its cultivation via establishing propagation centers in provinces within the date palm belt, focusing on elite lines and cultivars. These centers serve as date palm germplasm banks (the same cultivars in all centers all from the Al-Boukamal center). For expanding date palm cultivation in Syria, currently about 25,000 offshoots are produced per year by the government centers. Some of these offshoots are cultivated in regions within the date palm belt. The Ministry also offers training courses and field days to instruct farmers and technicians in the best practices for planting and caring for date palm.

12.3.4 Threats and Degradation

The genetic diversity that exists in the cultivated date palms is being reduced by continued selection pressures by man and shifts to fewer and more modern cultivars (Krueger 1998). Collections of date palm germplasm are fewer than for most other crops due to the relatively limited geographic area in which cultivation is possible and the relatively narrow base of genetic diversity present. Ex situ collections of date palm (such as the Palmyra oasis) allow for a careful preservation of a specific genotype and documentation of characterization and evaluation data, reducing the chances of disease problems and permitting easier experimentation to be carried out (Krueger 1998). One of the main reasons for the degradation of date palm cultivation in the coastal region of Syria is the red palm weevil. In 2008, as a control measure, 1,228 trees and 4,090 offshoots in Latakia and 86 trees in Tartous were destroyed because of infestation (Syrianews 2009).

12.3.5 Quarantine Regulations

Unfortunately, pests and diseases have been spread with the expansion of trade and travel in the globalizing world system (UN Press Release 2004). Syrian quarantine regulations on the import of date offshoots are intended to prevent further red palm weevils and bayoud disease from entering the country. The fungus that causes bayoud is transferred through the infected offshoots, palm tissues and products, and in infected soil on the roots. All of these materials that may carry the fungus are prohibited entry into Syria from countries where the pest and the disease are present (e.g., Algeria) (El-Melegi 2010). The red palm weevil is regulated through decree No. 389 that prohibits importing trees and offshoots that are propagated traditionally and allows only importing date palm plantlets propagated through tissue culture. Earlier regulations used to completely prohibit importing date palm trees to prevent transfer of this insect to the date palm in Syria (Syrianews 2009).

12.4 Plant Tissue Culture

12.4.1 Role and Importance

The application of tissue culture techniques for date palm propagation, also called in vitro propagation, helps produce the largest number of individuals from the minimum number of explants. Compared to the traditional techniques of date palm propagation, tissue culture has many advantages. It is easier and faster, dormancy problems are eliminated, and juvenile stage is reduced. It is also a suitable tool for propagation of clones that do not produce offshoots and viable seeds or that do not produce seeds at all (Alkhateeb 2008). Other advantages are production of genetically uniform plants which are usually identical to the mother plants and have almost 100 % survival rate, easy and fast exchange of plant material without risk of the spread of diseases and pests (e.g., bayoud and red palm weevil), and generation of disease and pest-free cultivars (UAE University 2005). Alkhateeb (2008) suggested that the development of propagation methods through tissue culture resulted in massive expansion of date palm plantations. Shubat (2012), however, argued that trees generated using this method is less resistant to the red palm weevil as compared to those generated using offshoots.

12.4.2 Research and Development

In 1989, a laboratory for date palm tissue culture propagation was established in the General Organization for Seed Multiplication (GOSM) in Aleppo (Al-Baba 2011). The goal was to develop date palm tissue culture techniques using somatic embryogenesis and direct organogenesis, which are the most popular techniques. In somatic embryogenesis, complete somatic embryos are formed from vegetative cells which behave more like sexually produced zygotic embryos. They are produced from in vitro–produced friable callus cells. As for direct organogenesis, adventitious shoots are developed from the undifferentiated callus masses or directly from the explants (Alkhateeb 2008). Due to the disadvantages of the two techniques, Amin (2001) applied a new method by using the adventive buds for propagation of two cultivars grown in Syria, Khastawi and Zahidi. The method is based on generating renewed buds on the surface of the treated in vitro explants, which do not normally produce buds (Margara 1984). Amin believes that this method may be much safer than somatic embryogenesis in producing in vitro date palm plants identical to the mother plant in their vegetative specifications, because this method does not pass through the callus stage. Amin also mentioned that this technique is being improved now to be applied at a commercial level and for more than one cultivar. Al-Maarri (1995) referred to experiments he carried out on date palm tissue culture and also discussed several issues related to date palm tissue culture. These are (a) date palm tissue culture techniques (embryogenesis, organogenesis, and inflorescence tissue), (b) the role of hormones on date palm tissue culture, (c) some physiological disorders concerning date palm propagation, (d) genetic stability in date palm produced through tissue culture, (e) development of date palm vitroplants in the field, and (f) date palm tissue cryoconservation.

Al-Khayri and Al-Maarri (1997) studied the effect of seasonal variation on the regeneration capacity of date palm. Ibraheem and others also carried out a number of experiments on date palm tissue culture. Examples are: (a) somatic embryogenesis to propagate date palm cultivars of interest for Syria (e.g., Zahdi, Khistawi, Asabe El-Arous, and Barban) by using zygotic embryos (Pinker et al. 2009), (b) a comparative study between solid and liquid cultures relative to callus growth and somatic embryo formation in cv. Zaghloul (Ibraheem et al. 2013), (c) the effect of sodium chloride stress on Zaghloul somatic embryogenesis (Ibraheem et al. 2012a), (d) screening of some date palm cultivars to salt stress in vitro (Ibraheem et al. 2012b), (e) optimizing of date palm in vitro rooting protocol for cvs. Deglet Noor and Asabe El-Arous (Ibraheem et al. 2010a), (f) somatic embryogenesis approach for shoot tips of Zaghloul (Ibraheem et al. 2010b), and (g) propagation of date palm cultivars in vitro by using zygotic embryos and leaflet segments (Ibraheem et al. 2009).

12.4.3 Scaling-Up Production and Tissue Culture Applications

As mentioned above, research carried out by GOSM for optimization of tissue culture on date palm has not succeeded. Therefore, date palm plantlets propagated through tissue culture are imported from Gulf countries, Iran and Egypt. Some other cultivars were imported from the Maghreb countries, such as Medjool and Deglet Noor, which were propagated in other countries using tissue culture. The survival and success rate of these plantlets was 100 %. They are hardened and acclimatized in the propagation centers and then planted in permanent locations within the date palm belt. The cultivars were very successful and identical to the cultivars they originated from in the country of origin. Offshoots of these cultivars are used to expand the plantation of these cultivars in the date belt. So far, about 100,000 offshoots of the tissue culture-derived cultivars have been produced and planted in Syria.

12.4.4 Survey of Research and Commercial Labs

In addition to the study by Amin (2001) on in vitro propagation of date palm using adventive buds, the Department of Tissue Culture in GOSM conducted experiments for optimization of date palm tissue culture techniques (Al-Admeh et al. 2006). Unfortunately, these experiments failed and the laboratory was closed in 2003. Hardening the plantlets produced in tissue culture laboratory in Aleppo used to be one of the main objectives of the branch of GOSM in Latakia. A set of 2,259 hardened offshoots was sent to Al-Jalaa production center in Deir Al-Zour (Al-Wehda 2007). Jamal (http://tishreen.news.sy/tishreen/public/read/258450) mentioned that there are more than 15 protocols for propagation of date palm tissue culture that were developed in laboratories of the College of Agriculture in Damascus University, but they are not yet published. He believes that researchers can use these protocols in Syria.

12.4.5 Recommended Protocols

Various laboratories in the world have made attempts to propagate date palm by tissue culture techniques. According to Zaid and de Wet (2002a), success has been achieved at only a few international laboratories. Protocols of tissue culture of date palm in Syria have to be refined in order to develop a reliable method which can be used for large-scale propagation of date palm in the country.

12.5 Cultivar Identification

12.5.1 Role and Importance

Because the date palm plays an important role as an income and employment source for people living in the growing areas, date palm biodiversity requires a more comprehensive knowledge for its conservation and sustainable use (Jaradat 2011). The phenotypic and genetic description and determination of genetic variability of native and introduced date palm cultivars in Syria are of major importance in date palm breeding and improvement programs, germplasm characterization, and conservation to control genetic erosion (Munshi and Osman 2010). This also helps in collecting and cataloguing the germplasm in established germplasm banks (Bornet and Branchard 2001). Propagation of date palm cultivars using tissue culture provides thousands of plantlets from a single offshoot, which should be true to type to the mother plant. Molecular markers are needed to detect any genetic changes or traits in in vitro–propagated plants. For example, the frequency of somaclonal variation in the resultant plants from two tissue culture techniques, namely, organogenesis and embryogenesis, was estimated and compared for ten date cultivars grown in the UAE using amplified fragment length polymorphism (AFLP) analysis (Al Kaabi et al. 2007). Similarly, randomly amplified polymorphic DNA (RAPD) markers were used to determine genetic changes between callus-derived plantlets of the Iranian cv. Khanizi and its mother plant (Eshraghi et al. 2005). Saker et al. (2006) employed both techniques (RAPD and AFLP) to assess genetic variations in tissue culture-derived date palm offshoots of cvs. Sakkoty, Gandila, and Bertamoda.

12.5.2 Research in Morphological Descriptors

Traditionally, morphological characteristics (leaves, spines, and fruit characters) have been used widely for the description of date palm cultivars. The description of the fruit characters is considered more common than describing the vegetative characters for differentiation between date palm cultivars. Date palm cultivars introduced to Syria from other countries have accumulated changes in some of their features from the original cultivars, after they adapted to the new environmental conditions (Electronic Economic Newspaper 2007). In 2006, a committee of date palm experts from the Ministry of Agriculture and ACSAD was formed for morphological characterization of all local date palm cultivars and elite lines grown in Syria. The committee released an Atlas that can be regarded as the first reference for classification of date palm in Syria (Al-Baba et al. 2013). This Atlas includes an introduction to 40 cultivars and lines, i.e., name, description of vegetative, and fruit characters that differentiate each cultivar and line supported with photos, chemical composition of fruits, fruit-ripening date, stage of consumption, and quality of fruits. The Atlas covers only Palmyra and Al-Boukamal propagation centers because they are the only centers in which date palm trees have been fruiting since 2001. The most informative and reliable physical characters were selected and evaluated according to the states of the character that can be encountered in the cultivars examined in any locality. For each cultivar, the descriptors (see Table 12.5) were applied to 10 aged palm trees in each location. Haider et al. (2009) carried out a morphological characterization of 18 cultivars of those described in the released Atlas. The objective was to obtain an accurate description and knowledge of these genetic resources, based on the floral and vegetative morphometric characters. The results represent the first informative morphological description of Syrian date palm cultivars. The descriptors considered were fruit weight, production level, leaf length, leaf spines, leaf density, leaf color, bunch color, bunch length, bending degree of the bunch, size of inflorescence, flower density, flowering time, fruit ripening time, consumption level, seed color, color of ripe fruit, seed shape, rate of fruit weight and size, color of fruit cap, cap position, seed/fruit percentage, fruit type, fruit skin separation, and the number of offshoots each cultivar generates. In spite of these descriptors, it remains very difficult, however, to identify cultivars especially outside the fruiting period. Moreover, the evaluation of inter-cultivar genetic diversity on the basis of morphological markers, mainly those of the fruit, is difficult because a large set of complex phenotypic data (Elhoumaizi et al. 2002) is required. It is worth mentioning that there is a joint project between the Ministry of Agriculture and ACSAD which is funded and supervised by the latter for selecting and evaluating elite local date palm lines in Syria, Saudi Arabia, and Egypt. The aim of this project is to establish gene banks for selected elite lines including farmers' lines in each country. These lines will be characterized and adopted as new cultivars that have high-quantity production of high-quality dates.

12.5.3 Research in Molecular Descriptors

Despite the outstanding agronomic and socioeconomic significance of date palm, attempts to improve knowledge about the biodiversity of date palm in Syria have been limited to the phenotypic description of those cultivars. Informative morphological characters for describing date palm cultivars can be observed only in mature trees, and they vary due to the environmental effects and stages of growth and development. DNA-based markers provide useful information on the genetic diversity of date palm cultivars, as they remain unaffected by environmental factors and the developmental stage of the plants. Using RAPD and inter-simple sequence repeat (ISSR) markers, Haider et al. (2012) conducted a study for: (a) the identification of five prominent male date palm genotypes and 18 native and introduced female cultivars clonally propagated in Syria and (b) evaluation of the genetic relationships among them. These cultivars are Deglet Noor, Medjool, Zahidi, Birbin, Ashrasi, Maktoom, Khastawi, Barhi, Khalas, Khadrawy, Nabtatseyf, Lolo, Gish Rabi, Khineze, Zaghloul, Shahabi, and Kabkab (yellow and red). The results have produced the first informative DNA-based markers for the genetic characterization of date palm cultivars analyzed. RAPD data allowed the discrimination of five cultivars (Deglet Noor, Kabkab (yellow), Medjool, Khastawi, and Birbin) and one male genotype. The use of ISSR, however, could distinguish only one cultivar (Khadrawy) and the same male genotype distinguished by RAPD. There are no commercial labs for description or characterization of date palm in Syria. Research into accurate identification of date palm cultivars propagated in Syria using molecular markers is in progress. Therefore, there is as yet no recommended protocol.

12.6 Cultivar Descriptions

12.6.1 Growth Requirements

Growing date palm and successful ripening of fruits are achieved in regions that are free of frost and where the climate is hot and dry in summer and the temperature in winter does not go below −9 °C for long periods. There must be a long and hot summer during fruit development from khalal (*balah* or *biser*, partially ripe showing a yellow or red color and crunchy) to rutab (fully ripe, light-brown, and soft) stage of development, then tamar (dark brown and soft, semidry or dry). Date fruits production is dependent on the availability of certain heat requirements according to the cultivar. Most dry cultivars are found in the drier areas, whereas soft and semidry cultivars are confined to the humid and semidry areas (Zabar and Borowy 2012). Date palm trees prefer temperatures above 23.89 °C and can tolerate temperatures as high as 43.33 °C. During winter, date palm tree can tolerate temperatures as low as −9 °C (http://www.convertworld.com/ar/temperature/%D9%85%D8%A6%D9%88%D9%8A%D8%A9.html). It is worth noting here that the biological zero for date palm growth is 9 °C, for flowering 18 °C, for fertilization 20 °C or higher, and for ripening is 32 °C or more. Date palms like the brightest available light, including full sun. It needs sufficient light to grow and produce fruits. In shaded regions, date palms grow very slowly at the early stages and take a very long period to flower and have poor fruit production. Date palms can tolerate drought but does better with regular water availability. Water requirements differ according to factors such as age of the palm, soil type, climate, cultivar, and season. Compared to other plant species, date palm shows no damage under windy conditions (Zaid and de Wet 2002b). As for mineral nutrients, fertilization should be carried out to prevent deficiencies rather than to treat deficiencies in date palm. The date thrives in sand, sandy loam, clay, and other heavy soils. It needs good drainage and aeration, and it is remarkably tolerant of alkali. Excessive salt will stunt growth and lower the quality of the fruit (Al-Bakr 1972) but a moderate degree of salinity is not harmful (Zabar and Borowy 2012).

12.6.2 Cultivar Distribution

The most important cultivars in Syria and their distribution are given in Table 12.4 (see also Fig. 12.1).

12.6.3 Cultivar Production Statistics and Economics

The production of date palm differs according to the cultivar, age of the tree, agricultural practices (irrigation, pollination, etc.), pest resistance, and the region of cultivation. The fruit yield (kg) of the most important cultivars per year is as follows: Barhi (150–300), Zahidi (250–350), Khastawi (250–300), Birbin (300–400), Nabtat-seyf (70–130), Lolo (150–200), Gish Rabi (150–250), Khineze (200–350), Zaghloul (150–250), Shahabi (300), Kabkab (yellow and red, 100–150), Deglet Noor (75–100), Medjool (70–90), Ashrasi (60–90), Maktoom (80–120), Khalas (150–250), and Khadrawy (150).

Table 12.4 The most important cultivars in Syria and their distribution in the five priority areas defined in the date palm belt

Cultivar	Priority area no. 1	Priority area no. 2	Priority area no. 3	Priority area no. 4	Priority area no. 5
Ashrasi	Deir Al-Zour	Palmyra	Raqqa	Hasaka	
Barhi	Deir Al-Zour	Palmyra	Raqqa	Hasaka	Badia (Hama and Damascus countryside)
Birbin	Deir Al-Zour	Palmyra	Raqqa	Hasaka	Badia (Hama, Damascus countryside and Aleppo)
Deglet Noor	Al-Boukamal	Mayadin	Raqqa	Hasaka	
Kabkab, red and yellow	Deir Al-Zour	Palmyra	Raqqa	Hasaka	
Khastawi	Deir Al-Zour	Palmyra	Raqqa	Hasaka	Badia of Hama
Lolo	Deir Al-Zour	Palmyra	Raqqa	Hasaka	
Maktoom	Deir Al-Zour	Palmyra	Raqqa	Hasaka	Badia (Hama and Damascus countryside)
Medjool	Deir Al-Zour	Palmyra	Raqqa	Hasaka	Badia (Hama, Damascus countryside and Aleppo)
Nabtat-seyf	Deir Al-Zour	Hasaka	Raqqa	Palmyra	
Zahidi	Deir Al-Zour	Palmyra	Raqqa	Hasaka	Badia of Hama

Source: Al-Baba (2011), Holy Tree Magazine 3(3) Sept. 2011

12.6.4 Cultivars Description

Table 12.5 shows the morphological characters that were evaluated and used by (Haider et al. 2009) for morphological characterization of 18 important date palm cultivars grown in Syria (see Appendix A). Photos of the adult tree and fruits of some of these cultivars are given in Fig. 12.2.

12.7 Date Production and Marketing

12.7.1 Practical Approaches

Preharvest practices that influence date quality at harvest include covering fruit bunches with paper bags to protect them from dust, pests, and rain and fruit thinning to reduce fruit compacting within bunches and increase fruit size and quality (Kader and Hussein 2009). If diseases and/or pests occur in a date plantation, the overall situation has to be evaluated to identify and address the causes (Mahmoudi et al. 2008). Dates of Zahidi, Medjool, Deglet Noor, Khastawi, Lolo, Zaghloul and Barhi cvs. are marketed on branches (strands) or bunches. Dates are marketed whole, pitted, cut into small pieces, or macerated (ground or chopped). Whole unpitted or pitted dates may be marketed loose or pressed (compressed into layers using mechanical force) (Kader and Hussein 2009). The marketing and consumption of

Table 12.5 Morphological characters that were evaluated and used for morphological characterization of date palm cultivars cultivated in Syria

Character	Class	Cultivar																	
		Deglet Noor	Medjool	Zahidi	Birbin	Ashrasi	Maktoom	Khastawi	Barhi	Khalas	Khadrawy	Nabtat-seyf	Lolo	Gish Rabi	Khineze	Zaghloul	Shahabi	Kabkab (yellow)	Kabkab (red)
Leaf length	Long	√	–	–	√	–	√	–	√	–	√	√	√	–	–	√	√	√	√
	Medium	–	√	√	–	√	–	√	–	√	–	–	–	√	√	–	–	–	–
Leaf thorns	Dense	–	√	√	√	–	–	√	–	–	–	√	–	√	√	–	–	–	–
	Moderate	–	–	–	–	√	√	–	√	–	–	–	√	–	–	√	–	–	–
	Few	√	–	–	–	–	–	–	–	√	√	–	–	–	–	–	√	√	√
Leaf density	High	–	√	–	√	√	–	√	–	–	–	√	–	√	√	–	–	√	√
	Moderate	√	–	√	–	–	√	–	√	√	–	–	√	–	–	√	√	–	–
	Few	–	–	–	–	–	–	–	–	–	√	–	–	–	–	–	–	–	–
Leaf color	Dark green	–	√	√	√	√	–	√	–	–	–	–	–	√	√	–	–	√	√
	Medium	–	–	–	–	–	√	–	–	–	√	√	√	–	–	√	√	–	–
	Faint	√	–	–	–	–	–	–	√	√	–	–	–	–	–	–	–	–	–
Bunch color	Green	–	–	–	√	–	–	–	–	–	–	–	–	–	–	–	–	–	–
	Yellow	√	√	–	–	√	√	–	√	√	–	–	√	√	–	–	√	√	√
	Brown	–	–	√	–	–	–	√	–	–	√	√	–	–	√	√	–	–	–
Length of bunch	Long	√	–	–	–	–	–	–	√	–	–	–	–	–	–	–	–	–	–
	Medium	–	–	√	√	–	√	–	–	√	√	√	√	√	√	–	√	√	√
	Short	–	√	–	–	√	–	√	–	–	–	–	–	–	–	√	–	–	–
Bending level of bunch	High	√	–	–	–	–	–	–	√	–	–	–	–	–	–	–	–	–	–
	Medium	–	–	–	√	–	√	–	–	√	√	–	√	√	√	√	√	√	√
	Low	–	√	√	–	√	–	√	–	–	–	√	–	–	–	–	–	–	–
Size of inflorescence	Big	–	–	√	√	–	–	√	√	–	–	√	√	√	–	√	√	–	–
	Medium	√	√	–	–	–	√	–	–	√	–	–	–	–	√	–	–	√	√
	Small	–	–	–	–	√	–	–	–	–	√	–	–	–	–	–	–	–	–

Flower density	High	–	–	√	√	–	–	√	√	–	–	√	√	√	√	√	√	–	–
	Medium	–	–	–	–	–	√	–	–	√	√	–	–	–	–	–	–	√	√
	Low	√	√	–	–	√	–	–	–	–	–	–	–	–	–	–	–	–	–
Flowering time	Early	–	–	–	√	–	–	–	–	–	–	–	–	–	–	–	–	–	–
	Medium	√	√	√	–	√	√	√	–	–	–	–	√	–	√	√	√	√	√
	Late	–	–	–	–	–	–	–	√	√	√	√	–	√	–	–	–	–	–
Time of fruit ripening	Early	–	–	–	√	–	–	–	–	–	–	–	–	–	√	√	√	√	–
	Medium	–	–	√	–	–	√	√	√	√	√	–	√	√	–	–	–	–	–
	Late	√	√	–	–	√	–	–	–	–	–	√	–	–	–	–	–	–	√
Consumption stage	Biser	–	–	–	–	–	√	–	√	√	–	–	–	–	√	√	–	–	√
	Rutab	–	–	–	√	–	√	√	√	√	√	–	√	√	√	√	√	√	√
	Tamar	√	√	√	–	√	√	√	√	√	√	√	√	√	√	–	–	√	√
Color biser	Red	√	√	–	√	–	–	–	–	–	√	–	–	–	√	√	–	–	√
	Yellow	–	–	√	–	√	√	√	√	√	–	√	√	√	–	–	√	√	–
Color of ripe fruit	Black	–	–	–	√	–	–	√	–	–	√	–	–	–	√	√	–	–	√
	Dark brown	–	√	–	–	√	–	–	–	–	–	–	√	–	–	–	–	–	–
	Brown	–	–	√	–	–	√	–	√	–	–	√	–	–	–	–	–	–	–
	Yellow	√	–	–	–	–	–	–	–	√	–	–	–	√	–	–	√	√	–
Biser shape	Oval	–	–	√	–	√	√	–	√	–	–	–	–	–	–	–	–	√	√
	Cylindrical	√	√	–	–	–	–	–	–	√	√	–	–	√	√	√	√	–	–
	Globular	–	–	–	√	–	–	√	–	–	–	√	√	–	–	–	–	–	–

(continued)

Table 12.5 (continued)

Character	Class	Cultivar																	
		Deglet Noor	Medjool	Zahidi	Birbin	Ashrasi	Maktoom	Khastawi	Barhi	Khalas	Khadrawy	Nabtat-seyf	Lolo	Gish Rabi	Khineze	Zaghloul	Shahabi	Kabkab (yellow)	Kabkab (red)
Average fruit weight and size	Small	–	–	√	–	–	–	√	–	–	–	–	–	–	–	–	–	–	–
	Medium	√	–	–	√	–	√	–	√	√	–	–	√	–	√	–	–	–	–
	Big	–	√	–	–	√	–	–	–	–	√	√	–	√	–	√	√	√	√
Color of fruit cap	Red	√	√	–	√	–	–	–	–	–	√	–	–	–	√	√	–	–	√
	Yellow	–	–	–	–	√	√	–	√	√	–	–	–	–	–	–	√	–	–
	Brown	–	–	√	–	–	–	√	–	–	–	√	√	–	–	–	–	√	–
Status of fruit cap	Gummy	–	–	–	–	√	√	√	–	√	–	–	–	–	–	–	–	–	–
	Superficial	–	√	√	–	–	–	–	√	–	√	√	–	–	–	–	√	√	–
	Embossed	√	–	–	√	–	–	–	–	–	–	–	√	√	√	√	–	–	√
Seed to fruit ratio	Big	–	–	–	√	–	–	–	–	–	–	–	–	√	–	√	–	–	–
	Average	–	–	√	–	–	–	–	–	–	√	√	√	–	√	–	√	√	√
	Small	√	√	–	–	√	√	√	√	√	–	–	–	–	–	–	–	–	–
Date ® classification	Soft	–	–	–	√	–	–	–	–	–	–	–	–	–	–	√	√	–	–
	Semidry	√	√	√	–	–	√	√	√	√	–	√	√	√	√	–	–	√	–
	Dry	–	–	–	–	√	–	–	–	–	√	–	–	–	–	–	–	–	√
Skin separation	Adherent	√	√	–	–	√	–	√	√	√	√	√	√	√	√	–	√	–	√
	Separate	–	–	√	√	–	√	–	–	–	–	–	–	–	–	√	–	√	–
Tree crown	Globular	–	√	√	√	√	–	√	–	√	√	√	√	–	–	√	–	√	√
	Elliptic	√	–	–	–	–	√	–	√	–	–	–	–	√	√	–	√	–	–
Economic importance	Good	√	√	–	–	√	√	–	√	√	√	√	√	–	–	–	–	–	√
	Moderate	–	–	√	–	–	–	√	–	–	–	–	–	√	√	–	√	√	–
	Few	–	–	–	√	–	–	–	–	–	–	–	–	–	–	√	–	–	–
Distribution	Rare	√	–	–	–	–	√	–	–	–	√	–	–	–	√	√	√	√	√
	Moderate	–	√	–	√	√	–	–	√	√	–	√	√	√	–	–	–	–	–
	Good	–	–	√	–	–	–	√	–	–	–	–	–	–	–	–	–	–	–

√ refers to the presence of the character and – to its absence dates in Syria is local because the majority of date palm trees are of recent production, especially the imported cultivars, and the annual production is insufficient to meet local market demands. Syria (along with Jordan, Morocco, Lebanon, and Yemen) has been identified by Al-Shreed et al. (2012) as belonging to a group of date-producing countries which are also date importers. Syria, however, does not export any dates.

12.7.2 Optimization of Yield

Pollination and fruit thinning are critical processes in date palm production. The pollen source in dates affects fruit quality, yield, and annual productivity. Different pollen sources are expected to affect fruit size, flesh/seed ratio, flesh and seed development, and time to fruit maturation. Hence, deciding on appropriate pollenizer cultivars is very crucial. A combination of both organic and inorganic fertilizers is recommended for higher yields and better physical and chemical fruit characteristics. Other

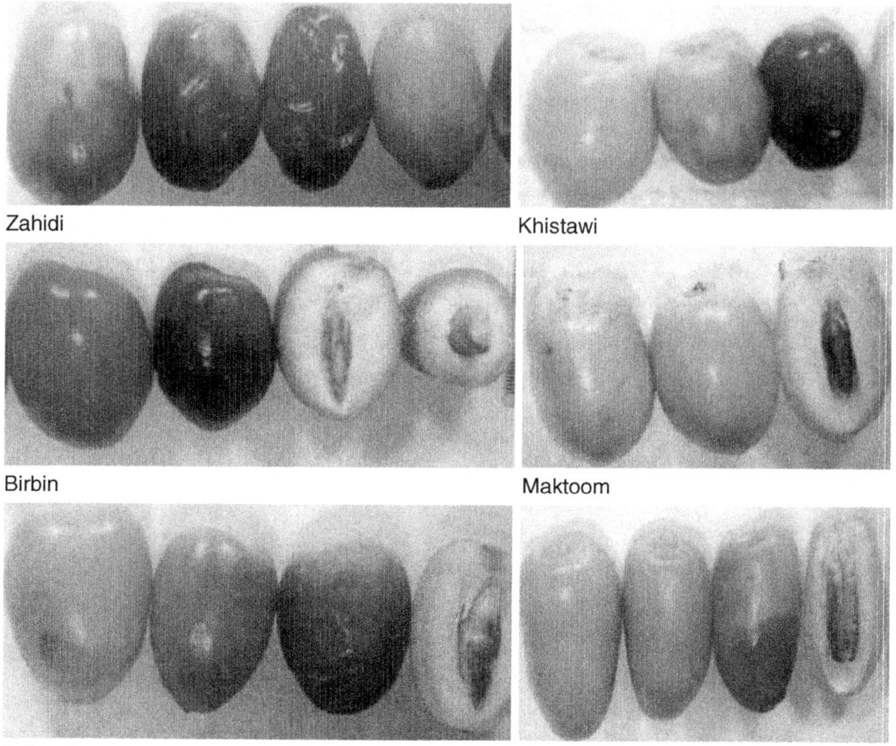

Fig. 12.2 Photos of tree and fruits of the most important date palm cultivars in Syria (*Source*: Al-Baba (2000))

Khalas

Barhi

Khsab

Fig. 12.2 (continued)

Fig. 12.2 (continued)

cultivation services such as fertilizers, irrigation, pest control, and regular pruning by leaving at least nine leaves per bunch are other crucial factors for optimization of date palm yield whose average values ranges between 70 and 150 kg of fruits per cultivar per year. It is worth noting here that the type, quantity, and time of fertilizer application vary according to the type of soil and the development stage of the palm.

12.7.3 Mechanization of Harvesting

This and the following section draw significantly from a study published by ICARDA (International Center for Agricultural Research in the Dry Areas), headquartered in Aleppo, on harvesting and postharvest handling of dates (Kader and

Hussein 2009). The study presents best practices to which Syrian date producers and processors should aspire to adopt.

Defining the time of harvest is important (Shibli 2008) and it depends on the cultivar. Dates are harvested at the end of August at the *biser* stage or in September to November at the rutab and tamar stages. Time of harvest is based on date fruit appearance and texture related to moisture and sugar content (Kader and Hussein 2009). Date bunches are usually enclosed in net covers to collect the fallen ripe fruits and to protect them from insects and birds (Kader and Hussein 2009). Harvest mechanism differs according to the stage at which the dates are harvested. When dates are *biser*, they are harvested manually by cutting the whole bunch such as in cvs. Barhi and Zaghloul. At rutab stage, fruits are harvested by hand, taking into consideration that the fruits of the same bunch do not ripen at the same time. As for tamar dates, they are harvested as whole bunches with a sharp knife by climbing the tree (when the majority of dates are ripe). Fallen dates on the ground are not collected or sold for human consumption (Kader and Hussein 2009). As for cultivars whose fruits differ in ripening time and take a long time to ripen like Deglet Noor cv., they are harvested before the fruits ripen and ripening is completed while the fruits are on the bunches after harvest. Artificial ripening is carried out either by heating fruits through exposure to sunlight for a few days until they ripen, or they are put in special chambers (27–38 °C) with high humidity (85–90 %) for a short time. The fruits that are consumed locally are collected when they reach the rutab or tamar stage; the entire bunches are cut and transferred to the place where they are ripened artificially (Qatana online reference). The fruits are packed in the field after harvest in special boxes or baskets made locally, or the bunches are taken to the shops to be sold directly. These latter dates should be cooled to 0 °C and transported under refrigeration (0–2 °C and 90–95 % relative humidity) to maintain their quality (Kader and Hussein 2009), or the ripe fruits are separated from the bunch and then sorted to remove infected or seedless fruits that are used as animal feed (Qatana online reference). Great care is given to the fruits that are consumed in tamar and rutab stages, regarding the time of collection, and they should be shipped quickly to the shops.

12.7.4 Postharvest Operations

Postharvest operations carried out are: (a) initial sorting to remove defective dates and foreign materials; (b) cleaning to remove dust, dirt, and other foreign materials using air pressure and water followed by air-drying to remove surface moisture; (c) sorting by quality and size into grades; (d) packaging to protect the dates from physical damage and moisture absorption using moisture-proof packaging material; (e) use of insect-proof packaging to prevent reinfestation of the dates during their subsequent storage and handling steps; and (f) cooling the dates to below 10 °C (preferably to 0 °C) before transportation or storage under the same temperatures (0–10 °C) and 65–75 % relative humidity. Forced-air cooling is used for cooling dates (Kader and Hussein 2009). *Biser* dates are stored at 0 °C and 85–95 % relative humidity to reduce water loss, delay ripening to the rutab stage, and maintain their textural and flavor quality. Packaging in well-sealed plastic bags or use of a plastic liner in the box helps in reducing water loss (Kader and Hussein 2009). Optimal temperature for storage of tamar dates is 0 °C for 6–12 months, depending on cultivar. For longer storage durations, temperatures below the highest freezing temperature of −15.7 °C are used. Dates with 20 % moisture or lower can be kept at −18 °C for more than 1 year, or at 0 °C for 1 year, or at 4 °C for 8 months, or at 20 °C for 1 month (relative humidity should be kept between 65 and 75 % in all cases) (Kader and Hussein 2009). The temperature of storage differs according to cultivar and moisture level. The highly moist soft cultivars are stored at 0 °C or less and the semidry cultivars at 0 °C. It may go less than 0 °C when the moisture in the fruits increases or the storage period increases (Shibli 2008).

12.7.5 Survey of Commercial Producers and Major Farms

Date palm propagation centers are the only commercial producers of date palm offshoots in Syria. Because commercial date palm cultivation in the country is relatively recent, there is no organized private sector for the production and commercialization of dates. The main commercial producers of dates are farmers in Deir Al-Zour and Palmyra (see Appendix B).

12.7.6 Marketing Status and Research

Several individual farmers grow, pack, and sell their fruit locally in roadside stands. Under the impact of rising labor costs and increasing height of the palms, cultural practices in particular the timing and method of harvesting need to be mechanized. The dates are left longer on the palm and are harvested in one operation (Mahmoudi et al. 2008) in order to reduce labor requirements. There is no marketing research on dates in Syria.

12.7.7 Current Import and Export

Date production in Syria up to 2010 provided only 10 % of the local market demand in the country. Annual date production is about 5,000 m, and the market demand is 50,000 m. Imported dates cover the shortage. This situation is temporary because most trees in the country are in the preproduction phase and have not yet reached full commercial production. Accordingly, there are no date exports.

12.8 Industry

Because more formal cultivation of date palm in the country is generally recent and the production is, therefore, insufficient even to meet local demand, the present date palm-based industry in Syria is limited to homemade fruit-based desserts and jam and leaf-woven baskets and plates.

12.9 Conclusions and Recommendations

The program of the Ministry of Agriculture in selecting the geographical locations suitable for cultivation and production of date palm was successful. According to the environmental conditions of each location, appropriate cultivars were imported from other countries. These cultivars performed well in date palm propagation centers established by the Ministry, and their yields and fruit quality were similar to those of the same cultivars in the country of origin. When these centers started to produce enough reliable offshoots and sold them to farmers at reduced prices, additional imports of offshoots and tissue culture plants ceased.

After 2001, when the imported cultivars entered the production phase, the farmers started to establish private orchards especially in Deir Al-Zour and Raqqa and planting them with elite and commercial cultivars.

The Ministry of Agriculture also formed a team of specialized and trained technicians on date palm cultivation and care. The duty of this team is to guide farmers on how to establish ideal date palm orchards and provide proper cultivation, protection, and pest control care for the palms. Due to the efforts of this team, a set of date palm male and female local lines of seedling origin were selected and then adopted as elite Syrian cultivars that suit the date palm belt. These cultivars have been propagated in the gene banks of date palm (i.e., date palm propagation centers). The Ministry is also trying to restore the tissue culture laboratories in Syria.

There are several barriers affecting date palm cultivation in Syria. The date-growing belt is located in the Badia and Euphrates River Basin (Fig. 12.1). Establishment of date orchards there is handicapped by the restriction on digging wells in Badia and by the fact that the lands in the Badia specialize in growing strategic crops; besides, there is the difficulty in providing water, the high level of salinity in well water, and the limited area that has freshwater available (Hakkar 2012). The most important requirements for developing date palm cultivation in Syria are permitting the digging of deep wells to reach fresh underground water, developing a mechanism for investing in new lands for date palm cultivation, after conducting a study of the water supply, and to expand the cultivation of date palm to increase yield. In addition, means should be developed to withdraw water from the Euphrates River and channel it to the Badia for irrigation; there is a need for tissue culture propagation laboratories to increase the yield and to provide disease-free offshoots from local elite lines and promote them as elite local cultivars and adopt biological control of diseases and pests (Hakkar 2012).

One of the main limitations to date palm cultivation in Syria is propagation of date palm using offshoots and not producing in vitro–propagated plantlets locally. Tissue culture techniques for propagation of date palm have to be developed in order to increase the yield of this tree (Agriculture World, Egyptian Agriculture Net 2012). Alkhateeb (2008), however, referred to problems that are often encountered during optimization of date palm tissue culture protocols. Some of these are at the laboratory level such as browning of cultured tissues, bacterial and fungal contaminations, early rooting of tissue-cultured buds, deterioration of embryonic callus and its inability to form embryos, and callus formation on bases of rooting plantlets. Other problems are at the field level such as failure to set fruit, dwarfism of the palms, abnormal growth and development of leaves and fruit strands, dryness of apical bud, terminal bud bending, albinism (variegation) of leaves, and changes in fruit quality. One of the common problems that also occur in plants propagated via tissue culture is somaclonal variation (El Hadrami et al. 2011; Jain 2012), defined as genotypic or phenotypic variation in tissue culture-derived plants. These plants may differ in genetic fidelity and are not always identical (Kaeppler et al. 2000). More serious research should be implemented to develop protocols for true-to-type multiplication of date palm by tissue culture and assessment of the behavior of generated plants in the field.

By the year 2000, date palm production in Syria was facing serious problems such as low yields due to the lack of research and the spread of pests and diseases. Expanding Syrian date palm cultivation requires financial support for farmers who are cultivating the tree, providing support for the success of cultivation especially by providing free or low-priced offshoots (Shubat 2012). The current status of date palm cultivation in Syria could be improved. There are extensive marginal lands in the priority areas Nos. 4 and 5 (Fig. 12.1) of date palm that are not exploited for date palm plantations due to a shortage of water. When expanding the cultivation of date palm, the Ministry of Agriculture should concentrate on agricultural assistance through the existing extension departments in the targeted regions, to provide good offshoots, and to develop mechanical methods of pollination (Shubat 2012). The date palm tissue culture system should be well established via somatic embryogenesis and organogenesis. The main processes to date cultivation such as propagation, offshoot planting, irrigation, fertilization, integrated pest control, pruning, pollination, fruit thinning, and harvesting techniques need to be improved (El-Juhany 2010). Suggestions raised by El-Juhany (2010) for rehabilitation of this sector are to mechanize the cultivation processes, organize training programs and workshops for the workers and farmers on all cultivation processes of date palm trees, and establish model date palm orchards and databases for date cultivation information.

There are no limitations for morphological characterization of date palm in Syria due to availability of local experts and equipment. Employment of molecular markers for genetic diversity of date palm cultivars in Syria began with the study by Haider et al. (2012). Although the RAPD and ISSR markers generated by the authors proved efficient to determine the genetic relationships in date palms grown in Syria, they were not sufficient for their full discrimination. For DNA-fingerprinting of date palm cultivars cultivated in the country, research is currently in progress in laboratories of AECS to employ Elmeer et al. (2011) SSR primers that proved useful for assessment of genetic diversity in date palm.

The manual harvest of the fruits increases cost and incurs significant harvest loss. High postharvest losses are due to fermentation, insect infestation, birds, and mechanical damage. The following recommendations (as communicated personally with Mr. Munzer Al-Baba, Head of Date Palm Department, Ministry of Agriculture) will help develop date palm cultivation in Syria:

(a) Resuming research and experiments to define the ideal fertilizer and irrigation water requirements for each cultivar according to the geographical location, growth, and production phase. In addition, to study the metaxenia effect of selected males on the different lines and cultivars in order to identify pollinizers of known identity and characters, as well as to develop cultivar-specific molecular markers for accurate and reliable identification purposes.
(b) Developing environmentally and economically sustainable IPM programs and improvement of postharvest operations such as storage and artificial ripening of fruits for late season-ripening dates.
(c) Mechanization of services provided to date palm cultivation and production.
(d) Development of salinity-tolerant cultivars from selected local lines because the water of the majority of wells and soils in Palmyra and Tanf is highly saline. It is worth noting that there are at the moment in the Palmyra date palm propagation center three lines which were selected and characterized for their tolerance to high salinity (about 12–14 g/l).
(e) Encouraging establishment of small or home-based units for manufacturing date palm products other than the fruits such as leaves (e.g., hats, purses, and trays) and seeds (e.g., beverage, eyeliner, and animal feed).

References

Afif A (2006) The red palm weevil destroys date palm trees in the coastal region. Who is the responsible and how it entered? Teshreen Newspaper (in Arabic). http://91.144.8.68/tishreen/public/read/60449

Agriculture World, Egyptian Agriculture Net (2012) Syria is conducting experiments for date palm propagation in Raqqa to localize date palm cultivation and production of offshoots that are adapted to the region environment (in Arabic). http://www.agricultureegypt.com/NewsDetails.aspx?CatID=cf068128-c380-42dc-b634-ccbcdc6b3cc0&ID=869bd71c-8fb8-4e11-96f6-b0b043390fd6#.UcK8u9iphLQ

Al Kaabi HH, Zaid A, Shephard H, Ainsworth C (2007) AFLP variation in tissue culture-derived date palm (*Phoenix dactylifera* L.) plants. Acta Hort 736:135–159

Al Maalouf F (2012) Expansion of date palm oases and 35 % of the area of Syria is suitable. Al Thawra Newspaper, Syria. http://thawra.alwehda.gov.sy/_print_veiw.asp?FileName=26232180820120421220951

Al-Admeh F, Al-Tawil MW, Al-Bashir M (2006) The structure of scientific research and its primary institutions and their trends: the reality and development requirements. National Conference on Scientific Research and Technical Development, Syria (in Arabic). www.hcsr.gov.sy

Al-Baba M (2000) Date palm tree. http://www.reefnet.gov.sy/reef/index.php?option=com_content&view=article&id=361:2008-06-18-10-27-03&catid=49:farms&Itemid=150

Al-Baba M (2009) A dialogue with the head of date palm department in the Ministry of Agriculture: date palm tree has economic benefits and it lives in harsh environmental condition. Al Orobah Newspaper, Homs, Syria (in Arabic). http://ouruba.alwehda.gov.sy/_print_veiw.asp?FileName=605826625200906l4224510

Al-Baba M (2011) The current status of date palm cultivation and dates production in Syrian Arab Republic (in Arabic). http://abudhabienv.ae/?p=4308

Al-Baba M, Qadmani MA, Ziyada S et al (2013) Atlas of date palm in Syria. The Ministry of Agriculture and Agrarian Reform and ACSAD, Damascus

Al-Bakr A (1972) The date palm. Its past and present status. Alani Press, Baghdad

Al-Kamour B (2006) Date palm propagation center in Al-Jalaa town is the first unit for collecting date palm cultivars in Syria (in Arabic). Al-Furat Newspaper, Deir Al-Zour. http://furat.alwehda.gov.sy/_archive.asp?FileName=16952239420060322150536

Alkhateeb AA (2008) The problems facing the use of tissue culture technique in date palm (*Phoenix dactylifera* L.) – a review. Sci J King Faisal Univ (Basic Appl Sci) 9(2):85–102

Al-Khayri JM, Al-Maarri KW (1997) Effect of seasonal variation on the regeneration capacity of date palm. In Vitro 33(3):22–26

Al-Maarri KW (1995) Date palm propagation through tissue culture technique. Photographic typesetting Foundation (Dibs), Damascus (in Arabic)

Al-Shreed F, Al-Jamal M, Al-Abbad A et al (2012) A study on the export of Saudi Arabian dates in the global markets. J Dev Agric Econ 4(9):268–274

Al-Wehda (2007) The branch of the General Organization for Seed Multiplication in Latakia. Duties, productivity and marketing. What about it. Latakia. http://wehda.alwehda.gov.sy/_archive.asp?FileName=56437005320071028122635

Amin T (2001) In-vitro propagation of date palm by formation adventive buds. In: Proceedings of the second international conference on date palms, Al-Ain, pp 568–587

Bornet B, Branchard M (2001) Nonanchored inter simple sequence repeat (ISSR) markers: reproducible and specific tools for genome fingerprinting. Plant Mol Biol Rep 19:209–215

El Hadrami A, Daayf F, Elshibli S et al (2011) Somaclonal variation in date palm. In: Jain SM, Al-Khayri JM, Johnson DV (eds) Date palm biotechnology. Springer, Dordrecht, pp 183–203

Electronic Economic Newspaper (2007) Invitations to benefit from date palm residues for wood industries (in Arabic). http://www.aleqt.com/2007/08/21/article_105232.html

Elhoumaizi MA, Saaidi M, Oihabi A, Cilas C (2002) Phenotypic diversity of date-palm cultivars (*Phoenix dactylifera* L.) from Morocco. Genet Res Crop Evol 49(5):483–490

El-Juhany L (2010) Degradation of date palm trees and date production in Arab countries: causes and potential rehabilitation. Austr J Basic Appl Sci 4(8):3998–4010

Elmeer K, Sarwath H, Malek J et al (2011) New microsatellite markers for assessment of genetic diversity in date palm (*Phoenix dactylifera* L.). 3 Biotech 1(2):91–97

El-Melegi M (2010) Bayoud disease in date palm. Plant health. http://www.meleigi.com/news.php?action=view&id=147

Eshraghi P, Zarghami R, Ofoghi H (2005) Genetic stability of micropropagated plantlets in date palm. J Sci Islam Rep Iran 16(4):311–315

GCSAR (2012) The annual report of the General Commission for Scientific Agricultural Research for 2011 (in Arabic). http://www.gcsar.gov.sy/gcsarAR/spip.php?rubrique70

Grant BL. Date palm tree care: tips on how to grow date trees. http://www.gardeningknowhow.com/edible/fruits/date/date-palm-tree-care.htm

Haider N, Nabulsi I, MirAli N (2012) Phylogenetic relationships among date palm (*Phoenix dactylifera* L.) cultivars in Syria using RAPD and ISSR markers. J Plant Biol Res 1(2):12–24

Haider N, Nabulsi I, MirAli N, Al-Baba M (2009) Morphological and molecular characterization of Syrian date palm varieties (native and introduced) (in Arabic). AECS-B\RSS 858

Hakkar F (2012) Cultivation and multiplication of date palm in Raqqa. New scientific research and experiments bodes new production. http://thawra.alwehda.gov.sy/_print_veiw.asp?FileName=31411384120120814211451

Ibraheem YM, Pinker I, Böhme M (2009) Propagation of date palm (*Phoenix dactylifera* L.) varieties in vitro by using zygotic embryos and leaflet segments. BHGL – Tagungsband 26, S. 91, http://www.dgg-online.org/tagungsbaende/Tagungsband_45_2009.pdf

Ibraheem YM, Pinker I, Böhme M (2010a) Optimizing of date palm (*Phoenix dactylifera* L.) rooting protocol in vitro for cv. Deglet Nour and cv. Asabe El-Arous. BHGL – Tagungsband 27, S. 85. http://www.dgg-online.org/tagungsbaende/Tagungsband_46_2010.pdf
Ibraheem YM, Pinker I, Böhme M (2010b) Somatic embryogenesis approach for shoot tips of 'Zaghlool'. Acta Hort 882:883–890
Ibraheem YM, Pinker I, Böhme M (2012a) The effect of sodium chloride-stress on 'Zaghlool' somatic embryogenesis. Acta Hort 961:367–374
Ibraheem YM, Pinker I, Böhme M, Al-Hussin Z (2012b) Screening of some *Phoenix dactylifera* cultivars to salt stress in vitro. Acta Hort 961:359–366
Ibraheem Y, Pinker I, Böhme M (2013) A comparative study between solid and liquid cultures relative to callus growth and somatic embryo formation in *Phoenix dactylifera* L. cv. Zaghlool. Emir J Food Agric 25(11). http://ejfa.info/index.php/ejfa/article/view/16661
Jain SM (2012) *In vitro* mutagenesis for improving date palm (*Phoenix dactylifera* L.). Emir J Food Agric 24(5):400–407
Jaradat AA (2011) Biodiversity of date palm. Land use, land cover and soil sciences. Encyclopedia of Life Support Systems (EOLSS), Developed under the Auspices of the UNESCO, Eolss Publishers, Oxford. Available: http://www.eolss.net [1 Apr 2013]. http://afrsweb.usda.gov/SP2UserFiles/Place/36450000/Products-Reprints/2011/1579.pdf
Kaakeh N. Preliminary biological studies on some date-palm insects in Syria. www.pubhort.org/datepalm/datepalm1/datepalm1_22.pdf
Kader AA, Hussein AM (2009) Harvesting and postharvest handling of dates. ICARDA, Aleppo
Kaeppler SM, Kaeppler HF, Rhee Y (2000) Epigenetic aspects of somaclonal variation in plants. Plant Mol Biol 43:179–188
Krueger RR (1998) Date palm germplasm: overview and utilization in the USA. In: Proceedings of the first international conference on date palms, Al-Ain, pp 1–37
Krueger RR (2011) Date palm germplasm. In: Jain SM, Al-Khayri JM, Johnson DV (eds) Date palm biotechnology. Springer, Dordrecht, pp 313–336
Mahmoudi H, Hosseininia G, Azadi H et al (2008) Enhancing date palm processing, marketing and pest control through organic culture. J Org Syst 3(2):29–39
Margara J (1984) Bases de la multiplication vegetative. Les meristemes et lorganogense. I.N.R.A, Paris
Munshi A, Osman G (2010) Investigation on molecular phylogeny of some date palm (*Phoenix dactylifra* L.) cultivars by protein, RAPD and ISSR markers in Saudi Arabia. Austr J Crop Sci 4(1):23–28
Nixon RW, Furr JR (1965) Problems and progress in date breeding. Date Growers Inst Rep 42:2–5
Nizwa.NET Dates. Oman agriculture. http://www.nizwa.net/agr/dates/dates/palm/datepalm3.html
Pal RN. Date palm (*Phoenix dactylifera*). http://www.fruitipedia.com/Date%20Palm.htm
Pinker I, Ibraheem YM, Böhme MH (2009) Propagation of some date palm cultivars by using tissue culture methods. Acta Hort 839:71–77
Popenoe P (1973) The date palm. Field Research Projects, Miami
UN Press Release (2004) Global date palm production at risk due to pests, diseases. SAG/276, FAO, Rome. http://www.un.org/News/Press/docs/2004/sag276.doc.htm
Qatana H. Date palm. Arab Encyclopedia. p 535 (in Arabic). http://www.arab-ency.com/index.php?module=pnEncyclopedia&func=display_term&id=1653
Saker MM, Adawy SS, Mohamed AA, El-Itriby HA (2006) Monitoring of cultivar identity in tissue culture-derived date palms using RAPD and AFLP analysis. Biol Plant 50(2):198–204
Sana (2012) Homs province, date palm oasis in Badia of Homs within the plan of development of the Syrian Badia during the current year. Sana Newspaper (in Arabic). http://sana.sy/ara/346/2012/01/09/393275.htm
Serra G, Nahaz MM, Idan M, et al. (2009) Assessment and characterization of the ibis protected area in the Palmyra desert: a proposed 5-year management and development framework. http://www.gianlucaserra.com/IPA_Assessment_2009.pdf

Shibli S (2008) A complete course on date palm (in Arabic). http://www.alkherat.com/vb/showthread.php?6579-%CF%E6%D1%C9-%E3%CA%DF%C7%E3%E1%C9-%CD%E6%E1-%C7%E1%E4%CE%ED%E1-%E3%DD%ED%CF%C9-%CC%CF%C7%F0

Shubat A (2012) The environmental belt can accommodate five million trees but the cultivated are only 500 thousands! Teshreen Newspaper (in Arabic). http://tishreen.news.sy/tishreen/public/read/258450

Swingle WT (1928) Metaxenia in the date palm, possibly hormone action by the embryo or endosperm. J Hered 19:257–268

Syrianews (2009) Steps the Ministry of Agriculture is taking to control the red palm weevil (in Arabic). http://www.syriandays.com/index.php?page=show_det&id=12817

Teshreen Newspaper (2005) Preparation of 11 thousands date palm offshoots to plant them and a plan for developing date palm cultivation (in Arabic). http://www.tishreen.news.sy/tishreen/public/read/36050

UAE University (2005) Date Palm Tissue Culture Laboratory (DPTCL). http://datepalm.uaeu.ac.ae/subpages/Laboratory.html

Zabar AF, Borowy A (2012) Cultivation of date palm in Iraq. Ann Univ Mariae Curie Sklodowska 22(1):39–54

Zaid A, de Wet PF (2002a) Chapter V: date palm propagation. In: Zaid A (ed) Date palm cultivation. FAO plant production and protection paper, vol 156. Rev. 1, Rome. http://www.fao.org/docrep/006/y4360e/y4360e09.htm

Zaid A, de Wet PF (2002b) Chapter IV: climatic requirements of date palm. In: Zaid A (ed) Date palm cultivation. FAO plant production and protection paper, vol 156. Rev. 1, Rome. http://www.fao.org/docrep/006/y4360e/y4360e09.htm

Chapter 13
Date Palm Status and Perspective in Palestine

Hassan Abu-Qaoud

Abstract Cultivated date palms (*Phoenix dactylifera* L.) have existed in Palestine for 5,000 years. The Mediterranean climate conditions dominant in the area provide optimal conditions for growth and development of date palm. Date palm cultivation in the Palestinian territories exists in the regions of Jericho and the Jordan Valley in the West Bank and in the Gaza Strip. This chapter reviews studies and research achievements on date palm in Palestine and addresses the importance of this sector: the status, production, and developmental constraints. After 1967, considerable areas of the Jordan Valley were cultivated by Israeli farmers with new high-quality date palm cultivars including Medjool. The adaptability of the new cultivars and the use of advanced techniques for propagation and cultivation of dates produced crop of excellent quality and yield to supply local and international markets. The total harvested area of dates in both the West Bank and Gaza was 725 ha in 2012. In the West Bank, there were 85,000 date palms spread over 600 ha, with a production capacity of 2,300 mt in 2012. However, the total date fruit production in Gaza was about 1,300 mt, with Hayany as a major cultivar in Gaza and Medjool in the West Bank. The average annual per capita consumption of dates in Palestine is 0.6 kg. Israel still remains a major supplier of dates to the Palestinian market. Several constraints face date cultivation and development, including high investment costs, poor marketing, and inequitable competition with Israeli products, limited water, pests, and diseases. However, there is a trend by formal governmental agencies as well as by national NGOs to support the cultivation of date palms in Palestine. The Palestinian Ministry of Agriculture has initiated several programs to support date palm cultivation in the Jordan Valley since 2000 with local NGO participation. Expansion of the knowledge about date palm will result in improving the productivity of this crop in Palestine.

Keywords Cultivar • Date palm • Gaza Strip • Jericho • Medjool • Palestine • West Bank

---

H. Abu-Qaoud
Department of Plant production and Protection, Faculty of Agriculture,
An-Najah National University, 707, Nablus, Palestine
e-mail: hassan@najah.edu

© Springer Science+Business Media Dordrecht 2015
J.M. Al-Khayri et al. (eds.), *Date Palm Genetic Resources and Utilization: Volume 2: Asia and Europe*, DOI 10.1007/978-94-017-9707-8_13

13.1 Introduction

Date palm (*Phoenix dactylifera* L.) cultivation is a worldwide agricultural industry producing about 7.3 million mt of fruit in 2011 (FAOSTAT 2012). Date fruit is produced largely in hot arid regions. The climatic conditions that prevail in certain Palestinian areas during spring and summer provide optimal conditions for growth and development of date palm and fruit maturation. Date palm cultivation has been known in Palestine for thousands of years. Recently, date cultivation has rapidly increased in Palestine, the area cultivated in the West Bank has doubled several times from 2000 to 2012, and the production has also increased to a projected quantity of 5,000 mt in 2015. The objective of this chapter is to provide current information on the status of date palm cultivation in Palestine, which will help to improve the productivity of this important crop.

13.1.1 Historical and Current Agriculture

The Palestinian territories (West Bank and Gaza Strip) cover 602,351 ha, distributed between the West Bank (566,082 ha, 94 % of the total area) and Gaza Strip (36,269 ha, 6 % of the total area) (PCBS 2011). In 2005, the Palestinian population was 3.8 million, with 63 % in the West Bank and 37 % in the Gaza Strip (ARIJ 2007; PCBS 2006). The total area of cultivated agricultural land currently used by Palestinians covers 30.5 % (97,517 ha) of the Palestinian land area and 54.4 % of the total land suitable for cultivation, of which 92.1 % is in the West Bank and 7.9 % is in Gaza (PCBS 2010). Rainfed agriculture is practiced in 87.0 % of the total cultivated area, while only 13.0 % is irrigated agriculture. The total cultivated area in fruit trees, vegetables, and field crops is 91,155 ha (82,630 ha in the West Bank and 8,525 ha in Gaza).

The Mediterranean climate of long, hot, and dry summers, and a rainy winter, enables the Palestinian territories to grow many crops at different time periods throughout the year. Agriculture is distributed among eight ecological zones. Five are in the West Bank: Southern (Bethlehem, Hebron Governorate, and East Jerusalem), Jordan Valley (Jericho and eastern parts of Tubas), Central (Ramallah, Nablus, and Salfit), Northeast (Jenin and Tubas), and Northwest (Tulkarm and Qalqiliya). There are three zones in the Gaza Strip: Northern, Middle, and Southern (Fig. 13.1).

13.1.2 Palestinian Agriculture

The value of the major agricultural commodities in Palestine is shown in Fig. 13.2; vegetables are the highest, followed by olives and grapes. Even though date fruit production is relatively low, its value is higher than that of oranges and almonds, indicating a rapid shift to date cultivation in recent years, mainly in the West Bank.

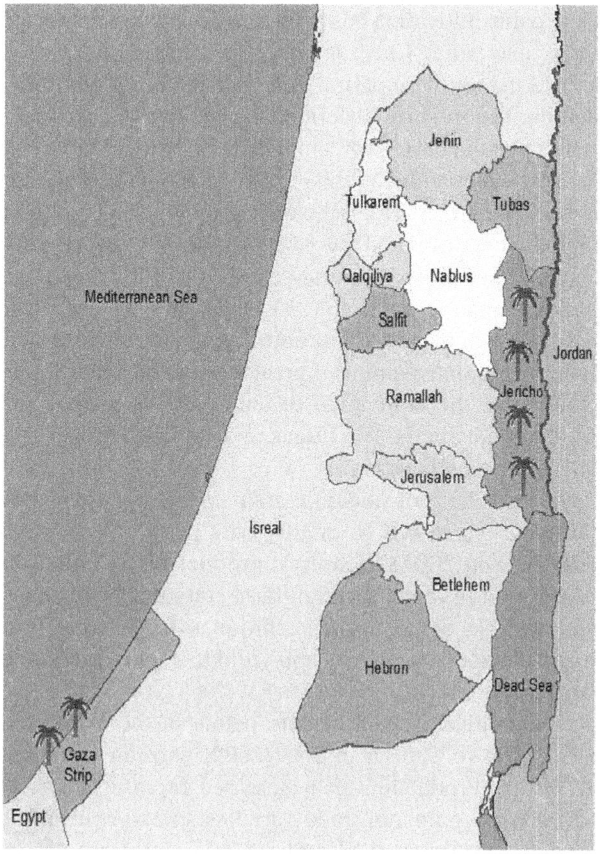

Fig. 13.1 West Bank and Gaza, Palestine

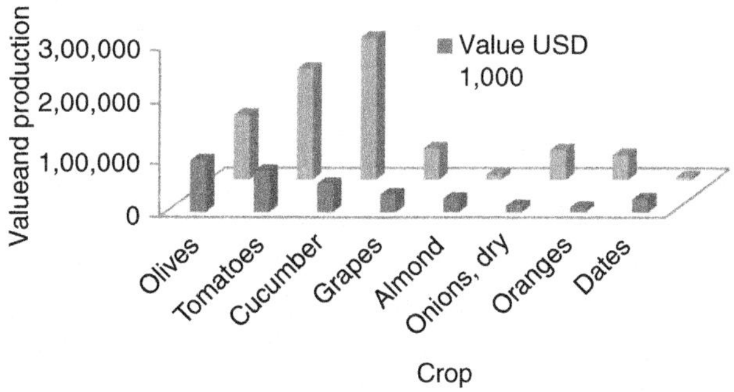

Fig. 13.2 Value and production of the main plant commodities in Palestine (PCBS 2011)

13.1.3 Production Statistics and Economics

Date palm cultivation has been known in Palestine for thousands of years. The Jordan Valley has been cultivated by Palestinian farmers for 5,000 years, especially surrounding the city of Jericho, which is considered the oldest city in the world.

Date palm cultivation has been practiced through the ages in the Jordan Valley, Gaza Strip, and Sinai since ancient times (Zaid and Wet 2002). Greeks and Romans named the northern part of Palestine the *Land of the Date* (Goor 1967). It is known that the historic city of Jericho was dubbed *Palm City*; Arab travelers from Jericho stated that the basic problem which faced the Arab cavalry was traversing date palm forests, which covered the region. Most of the date cultivation in the West Bank is now concentrated in Jericho and along the banks of the Jordan River. The extremely high temperatures and low relative humidity that prevail in those areas during spring and summer provide optimal conditions for growth and development of date palm and fruit maturation (Table 13.1). Being a tree of great economic, nutritional, and religious value and with its ability to grow in various climatic conditions, including saline soils, date palm is of great interest for Palestinian farmers (Kalbouneh 2011). Nowadays, the cultivation of date palm is located in regions of Jericho and the Jordan Valley in the West Bank and the Gaza Strip, especially near the cities of Deir al-Balah and Khan Younis.

After 1967, considerable areas of the Jordan Valley were cultivated by Israeli farmers, and new high-quality date palm cultivars, mainly Medjool, were introduced (Daiq 2007). Date fruit production by Israeli farmers in the Jordan Valley is reported in Israel's agricultural statistics. The adaptability of the new cultivars and the use of advanced techniques for propagation and cultivation produced fruits of excellent quality and yield to supply both local and international markets (EQA 2006a).

The cultivated area of date palms in the West Bank increased from 76 ha in 1993/1994 to 130 ha in 2001/2002, with an increase in production from 880 to 1,700 mt. Production also increased rapidly between 2006 and 2012. There are 85,000 date palms of good varieties spread over 600 ha. By 2012 the production capacity had increased rapidly to 2,300 mt, from only 60 mt in 2000. It is expected that this number will double by 2015 with a projected production level of 5,000 mt.

The total harvested area of dates in both the West Bank and Gaza was 725 ha in 2011, representing only 0.06 % of to the total harvested area in the world (FAOSTAT 2012). In comparison with Middle Eastern countries, this is a very small contribution (Table 13.2). However, there has been a significant increase in the harvested area of dates in the West Bank and Gaza in the last 10 years. Compared with olives, the oldest fruit grown in Palestine (Fig. 13.3), the cultivated area of date palm has increased rapidly from 2000 to 2011 (Fig. 13.4), and consequently the total production has notably increased (Fig. 13.4).

In the Gaza Strip, date palm production is relatively stable. Production reached about 2,000–3,000 mt during the 1970–1980 decade; this amount increased to 5,000–6,000 mt in the period 1995–2002 and declined to about 1,300 mt in the 2003–2008 period. The climatic conditions in Gaza (Table 13.3) are more suitable for producing soft dates. Therefore, 90 % of the date palms are cultivated for rutab stage fruits (Abu-Qaoud 1996). The production percentage of dates to overall fruit tree production in Gaza declined rapidly in recent years. In the period 1969–1984, the contribution was 25 %, decreased to 12.2 % between 1985 and 1993, fell to 8.6 % during the 1994–2002 period, and finally dropped to 4.4 % during in the

Table 13.1 Climatic data for Jericho district

Element	Month											
	Jan.	Feb.	Mar.	Apr.	May.	Jun.	Jul.	Aug.	Sep.	Oct.	Nov.	Dec.
Mean max temp, °C	19.1	20.9	24.3	29.3	33.7	36.7	37.8	37.6	36.1	32.2	26.4	20.4
Mean min temp, °C	7.4	8.3	10.5	14.2	17.6	20.4	22.1	22.4	21.2	17.9	12.9	9
Absolute max temp, °C	25	27.6	33.8	41.4	46.4	45	44	45.6	43.4	40.6	34.8	28.8
Absolute min temp, °C	0.2	0.4	2.8	2.4	10.4	15.4	18	19	13.2	11.4	4.3	2.1
Mean temp, °C	13.2	14.6	17.4	21.7	25.6	28.5	29.9	30	28.6	25.1	19.6	14.7
Mean sunshine, h/day	5.5	5.9	7.7	9.3	9.4	11.8	11.7	11.6	10.5	8.7	6.5	5.6
Mean RH, %	70	65	57	45	38	38	40	44	47	51	60	70
Total rainfall, mm	36	31	25	10	2	0	0	0	0	7	22	33
Evaporation, mm	78	76	128	189	261	289	298	276	227	135	94	59

Source: Palestinian Meteorological Department (2013)

Table 13.2 Date cultivation in selected Middle East countries

Country	Harvested area (ha)	Percentage of world total	Production (mt)	Production Values (USD1,000)
Saudi Arabia	172,297	14.4	1,122,820	573,429
Iran	154,274	12.9	1,016,610	519,187
Iraq	123,230	10.3	619,182	261,573
Egypt	41,652	3.5	1,373,570	701,488
Israel	5,500	0.45	37,008	18,900
Palestine	873	0.07	4,688	2,394
World	1,200,006		7,504,984	3,832,825

Source: FAOSTAT (2012)

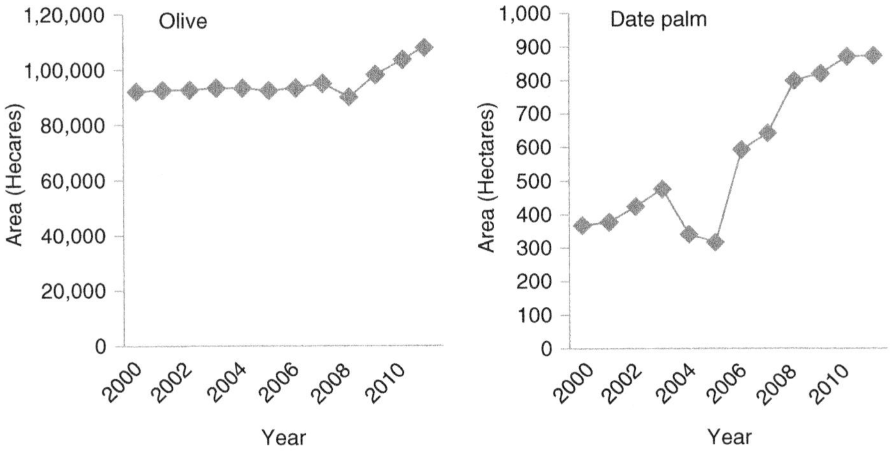

Fig. 13.3 Comparison between the harvested area (ha) of olives and date palms in Palestine

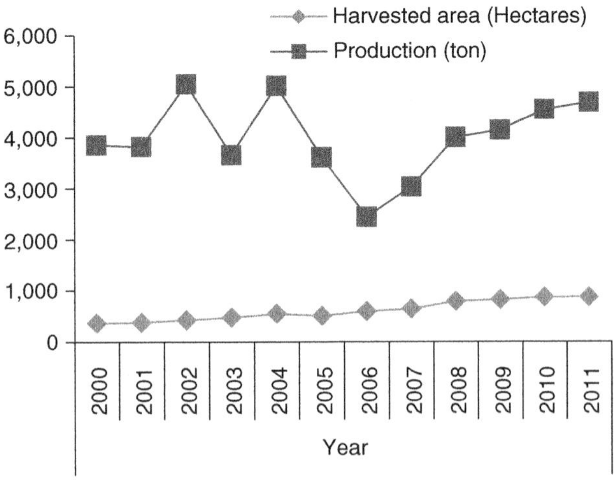

Fig. 13.4 Total harvested area (ha) and production (mt) of dates in Palestine (West Bank and Gaza)

Table 13.3 Climatic data of the Gaza Strip

	Month											
Element	Jan.	Feb.	Mar.	Apr.	May.	Jun.	Jul.	Aug.	Sep.	Oct.	Nov.	Dec.
Mean max temp, °C	17.5	17.5	19.5	23	24.5	27	29	29.5	27.5	26.5	23	19
Mean min temp, °C	9.4	10	11.7	14.5	16.9	19.7	21.8	22.2	21.2	19.4	14.5	11.3
Absolute max temp, °C	31.2	34.4	34.8	41.2	43.5	40	36	32.8	38.8	37.4	35.4	31.6
Absolute min temp, °C	2	2.6	3.6	7.4	11.4	14.8	18.5	19.2	16.2	12.2	7.5	3.4
Mean temp, °C	13.4	13.7	15.6	18.7	20.7	23.3	25.4	25.8	24.3	22.9	18.7	15.1
Mean sunshine, h/day	5.2	5.9	7.3	8.2	8.9	9.7	10.5	10.4	9.3	8.9	6.5	5.1
Mean RH, %	67	67	70	70	73	75	76	75	73	69	67	68
Total rainfall, mm	105	88	37	9	1	0	0	0	0	36	71	99
Evaporation, mm	68	76	115	142	162	190	193	183	165	132	87	69

Source: Palestinian Meteorological Department (2013)

period 2002–2009. The reduction in local demand, weak postharvest infrastructure, marketing difficulties, and a shift to other cash crops were the main reasons for the change (PCBS 1993–2009).

13.2 Cultivar Descriptions

13.2.1 Nutritional Aspects

Date fruits are among the most nourishing natural foods available to humankind and are known to have numerous health benefits. Rich in several vitamins and minerals, the date fruit is a good source of high nutritional value food. In general, there are three types of date fruits: soft, semidry, or dry. The semidry dates are the most popular; they are less sweet but more aromatic and distinctive than the others. Soft dates have higher moisture content, a mild flavor, and relatively low sugar. Dry dates are indeed dry, extremely hard, and intensely sweet.

13.2.2 Important Cultivars

Medjool This is one of the finest dates produced in Palestine, regionally and globally. Because it is grown in areas below sea level, there is an increase in the proportion of oxygen available to the palm which aids respiration and that in turn adds flavor and a distinctive color; the fruit is consumed at the tamar stage (Fig. 13.5a). The Palestinian Ghor region (the Palestinian area of the Jordan Valley) possesses a comparative advantage for growing Medjools. The climate in the Ghor region, where temperatures range from 12 °C in March up to 50 °C between July and October, is considered ideal for Medjool date production, which requires hot and dry weather. In fact, Medjool dates are produced in only certain other select areas of the world: the Maghreb region of North Africa, Southern California and Arizona in the USA, Namibia, and South Africa. However, the entry of Palestinian Medjools into European markets depends not only on the growing capacity but the ability to compete with other countries, especially Israel. Medjool market prices are significantly higher than for any other date cultivar (El-Jafari and Lafi 2004) (Table 13.4).

The Medjool trunk is medium in size, leaves are short with average curvature and of medium width, and the number of spines on each leaf ranges between 30 and 38. Fruit size varies from small to large; shape is mostly oval, ranging between orange and yellow, topped with fine reddish-to-brown stripes that develop during the khalal stage. The fruit becomes reddish brown in color when fully mature and has a waxy coat and a light skin; the shrunken mesocarp is wrinkled and rough and has a flesh thickness of 0.5–0.7 cm, a soft texture, little fiber, and a delicious taste; and the palm is early bearing. The palm produces 70–90 kg per tree per year.

Fig. 13.5 Fruits of the three major date cultivars in Palestine. (**a**) Medjool consumed as tamar, (**b**) Barhi consumed mainly as khalal, **c** Hayani consumed mainly as rutab

Table 13.4 Farm gate prices for export quality dates in Israel in 1996

Cultivar	Export price at farm (USD/kg)
Medjool	3.5
Deglet Noor	2.5
Barhi	1.5
Hayany	0.6
Iraqi varieties	0.7

Source: Botes and Zaid (2002)

Barhi One of the most famous date cultivars, Barhi is characterized by its sweet taste and high per tree productivity, making its cultivation highly attractive. It is mainly consumed as khalal when still crisp in texture (Fig. 13.5b). The palm has a spreading crown, green leaves, and an aesthetically pleasing shape. The fruit has lower calories than other cultivars, making it ideal food for diabetics. The tree produces a low number of offshoots which is attributed to the cultivar itself.

Hayani This cultivar is mainly cultivated for rutab consumption in the Gaza Strip. Hayani, the most commonly grown cultivar, performs best over the three regions with an average fruit set of 87.6 %. It was concluded that the common climatic conditions in the

Gaza Strip are suitable for planting new cultivars along with Hayani, the traditional variety. Also, semidry cultivars could be introduced. The possibility of success for the dry cultivars under the conditions of the middle and south districts would depend on specific treatment (Al-Bana 2007). Hayani is a large palm with large leaves of average curvature, long thorns, medium-sized fruit, a length of 4–5 cm, and a diameter of 2.5–3 cm. The color is dark red at the completion of growth. Shape is cylindrical with a conical end. The fruit turns black at rutab (Fig. 13.5c). In this phase the exocarp easily separates from the mesocarp; annual production of this cultivar is about 90–120 kg per tree.

13.2.3 Economics of Date Palm Cultivation in Palestine

The successful cultivation of the date palm has a positive impact on Palestinian agriculture, especially in the Jordan Valley and the Gaza Strip. It has great importance in creating job opportunities for farmers and needy families, finding an alternative for the products of Israeli settlements, and providing food security to poor families, especially during political crises, border closures, and curfews related to the Israeli occupation.

Several studies have been conducted on the feasibility of date palm cultivation in Palestine (Al-Assi et al. 2003; Daiq 2003, 2007). These studies aimed at shedding light on the economics of date palm in Palestine and the creation of appropriate conditions for its cultivation. Dates are a basic food and are capable of being stored and therefore an important strategic commodity in the life of Palestinians, due to constant disruptions of their food supplies.

The date palm requires 4–5 years to become established and begin fruit production. It continues to produce fruits for 20 years over a productive life span of the tree of approximately 25 years.

The estimated cost and revenue of a Medjool palm tree are shown in Table 13.5. The total cost of establishment of an offshoot is about USD100 with production beginning after 4 years, with a revenue of USD50 from both fruit and offshoots. The amount increases afterwards until the tenth year when revenue will be USD140 per tree. The date palm is planted at a spacing of 9 × 8 m, which is equivalent to 14 seedlings in a given *dunum* (=1 decare); thus, the revenue of planting one dunum of Medjool date after 10 years will be 140 × 14 = USD1,960. This is considered as one of the highest revenue rates as compared to other agricultural crops.

13.3 Date Production and Marketing

13.3.1 Production and Demand

The local demand for dates in Palestine is relatively stable. During the late twentieth century, the average per capita consumption was 0.6 kg, which is close to the average domestic consumption globally of 0.9 kg per capita per year but much lower

Table 13.5 Estimated cost and revenue of one tree of cv. Medjool date palm in Palestine

	Tree expenses		Tree production				
Year	Item	Value (USD)	Dates (kg)	Dates value (USD)	Number offshoots	Offshoots value (USD)	Net per tree (USD)
First	Seedling price	55					
	Planting	15					
Second–Third	Water, labor, fertilizers, pesticides	30					
	Total 1st–3rd year	100	0	0	0	0	−100
Fourth	Running cost	30	20	40	1	40	50
Fifth	Running cost	50	30	60	1	40	50
Sixth	Running cost	60	40	80	1	40	60
Seventh	Running cost	60	50	100	1	40	80
Eight	Running cost	60	60	120	1	40	100
Nine	Running cost	60	70	140	1	40	120
Tenth	Running cost	60	80	160	1	40	140

Source: Daiq (2003, 2007)

than that of Saudi Arabia (38 kg) (El-Jafari and Lafi 2004). The total amount consumed in 1994 was 1,466 mt; this amount increased in 2000 by 28.4 %. Moreover, in 2010, this amount increased to 3,019 mt. The projected amount consumed in 2020 is 4,258 mt (El-Jafari and Lafi 2004; PCBS 2011). About 85 % of the Palestinian date palm production goes to the domestic market (approximately 4,000 mt), and only 15 % of production is exported (PCBS 2011). The date fruit consumption of a Palestinian family is no more than 0.17 % of the total food consumption. This low percentage is mainly due to the consumption habit, the high price of dates, and the presence of several alternative fruits. The development of the local date market requires a change in consumption habit, developing processes to promote this product, emphasizing health and nutritional benefits, and providing adequate marketing infrastructure (Daiq 2007).

13.3.2 Current Imports and Exports

Israel still remains the major supplier of dates to Palestinian markets, including in some cases low-quality fruit. The majority of the locally consumed dates are during the month of Ramadan. From 2002 to 2008, Ramadan coincided with the local production period; therefore, the amount of dates consumed was partially covered by local production. However, as the lunar calendar shifts backwards 11 days a year and Ramadan approaches in the months of May to July in the years 2011–2020, the required amount will be covered from Israeli dates from the previous year. Israeli storage management policy depends largely on the Palestinian market to absorb

Table 13.6 Production and consumption cycle of dates in the West Bank

	Month		
Condition	May–July	August	September–October
Status and maturation	Israel's stored dates from last years	Maturation of Israeli dates	Maturation of Palestinian dates
Consumption	Peak of dates consumption (Ramadan 2011–2020)	–	Low local consumption
Export	–	–	Exporting market

Fig. 13.6 Washing, sorting, and packing in a new modern date factory in Jericho (MADICO)

low-quality dates. Generally speaking, Israel depletes its inventories of dates 2 months ahead of the harvest season. In fact, the optimal length of time to store dates does not exceed 10 months. Israeli dates usually mature in August (1 month before the maturation of the Palestinian dates), thus the Israeli producers market stored dates before the maturation of the new crop. Since there is a lack of storage facilities in Palestine, the only other market for Palestinian dates is through exportation. The European market, which imports 50 % of total world exports of dates, has great potential for Palestinian Medjool dates. They are marketed as a high-quality product and supplied at high prices similar to those exported from Israel and the USA to Europe (Table 13.6).

13.3.3 Postharvest Operations

Exporting Palestinian dates to Europe requires taking into account several marketing functions to compete. These include washing, cleaning, drying, sorting, packaging, labeling, and storing in addition to production functions such as fertilizing, decreasing cluster weights, and packaging clusters (El-Jafari and Lafi 2004). In 2010, a modern processing facility, with the most up-to-date technologies for washing, drying, grading, sorting, packaging, and storing dates, was established in Jericho by the private company MADICO (Fig. 13.6). The production capacity of the factory is 1 mt per hour with an electronic grading system monitored by cameras. The production line is able to sort up to ten different sizes. The facility meets the requirements of the BRC Global Standards, Global Gap, ISO 22000, and the Palestinian Certificate for Excellence. The Palestinian Dates Center practices equal employment opportunities and diversity, with a specific focus on providing employment opportunities for women from rural areas in the Jordan Valley. The center offers services to all Palestinian date farmers in the Jordan Valley and provides them with the opportunity to process, package, and store dates. This provides local farmers with an opportunity to utilize resources otherwise unavailable. It also strengthens the local farmers' market position and allows higher quality production and greater time flexibility (MADICO 2011).

13.4 Constraints Facing Date Palm Development in Palestine

Several constraints face the cultivation and production of date palm in Arab countries, among them: poor farm management, pest and disease control, harvesting, processing, and marketing, shortage of qualified and trained national staff and laborers, and insufficient research and development (Erskine et al. 2011). However, for Palestine, the following are the major obstacles facing date palm cultivation:

(a) Lack of water resources is the most significant obstacle to expanding date palm cultivation. Although the date palm is able to survive under arid conditions, they require sufficient water of acceptable quality to reach their potential yield; water requirements in such climates are higher when the water quality is low. Saline soils are a problem in this area because of insufficient annual rainfall to flush accumulated salts from the crop root zone. The high evapotranspiration rates of date palm calculated in the Jordan Valley emphasize the high water requirements (Table 13.7). The figures are based on the use of drip irrigation with an efficiency of 90 % (Kalbouneh 2011). Therefore, date farming in the Jordan Valley was found to be the best option for farmers in that region. Date palm production requires only one-third of the water needed to produce bananas or oranges, so when Medjools are substituted for banana plants and orange trees, two-thirds of irrigation water formerly required can be applied to other crops. In addition, dates can be produced in association with other cash crops. The hot and dry weather in the Jordan Valley is the optimum climate for date farming (El-Jafari and Lafi 2004); as a result there is high expansion in date cultivation.

(b) High production costs: date production is labor intensive and requires workers with sufficient experience in this area; therefore, operating costs are relatively high. In addition to the high cost prices of lifts used in cultivation and harvest, which range between USD15,000–20,000, limitations of availability of tissue-cultured seedlings of good quality and the high cost of local offshoots add to the production cost. The cost of an offshoot may reach USD50–60.

Table 13.7 Calculated evapotranspiration and irrigation demand for date palm in the Jordan Valley

Month	ET^0 (mm)	Crop coeff. (kc)	Evapotranspiration (mm)	Irrigation req. (mm)
January	52.4	0.77	40.4	0.0
February	65.5	0.87	57.1	17.5
March	109.4	0.81	88.2	53.4
April	128.4	0.98	126.0	108.7
May	171.4	0.94	161.2	160.9
June	189.0	1.01	191.1	191.1
July	234.1	0.85	198.2	198.2
August	213.6	0.86	183.1	183.1
September	165.0	0.92	151.5	151.5
October	102.9	1.08	111.4	110.9
November	64.8	1.13	73.0	52.8
December	42.5	1.21	51.5	6.9
Total	1,539.0		1,432.6	1,235.0

Source: Kalbouneh (2011)

(c) High shipping and transportation costs: generally the cost of shipping and transportation for exports and imports in Palestine is more than 30 % above the norms; for instance, this is equivalent to four times that of Jordan and more than 50 % above the transportation cost of Israeli goods.

13.4.1 Pest and Disease Challenges

Over the last decade, productivity of date palms has declined in many areas. As much as 30 % of production can potentially be lost as a result of pests and diseases. Different pests and diseases attack date palm, and many of the pests are common in producing countries, mainly in the Mediterranean region. The red palm weevil (*Rhynchophorus ferrugineus*) has recently become one of the major date palm pests (El-Juhany 2010). Human activity is the culprit of the high rate of the pest spread, mainly by transporting infested date trees and offshoots from contaminated to uninfected areas. Usually the damage caused by the red palm weevil larvae is detectable only long after infection occurs. No safe and effective technique for early detection of the pest yet has been devised (Ferry and Gomez 2002). This insect was first detected in the Jordan Valley and especially in the Jericho area and the Gaza Strip in 1999 (Al-Jaghoub et al. 2000). Different measures have been reported to be effective to control this dangerous insect, including chemical, biological, and other methods. Ayedh and Rasool (2010) used gamma radiation to sterilize red date palm weevil males; they reported that egg hatchability was significantly reduced using gamma radiation; the results indicated that 15 Gy is the optimum dose for sterilizing the red palm weevil.

Another important pest is the dubas date bug (*Ommatissus binotatus* Fieber). This insect has been recorded in the Jordan Valley and Jericho; other insects reported in Palestine include the white palm cortical insect, red date scale, Parlatoria date scale insect, mealy bug, frond borer, lesser date moth, greater date moth, and date seed beetle.

A survey of insect pests attacking date palm trees at Al-Arish region, northern Sinai, Egypt, which borders Gaza (Al-Arish city), identified eleven insect pests belonging to nine families from the orders Isoptera, Homoptera, Lepidoptera, and Coleoptera as causing damage and occurring frequently in the area. The most dominant and economically important pests were two scale insects (*Parlatoria blanchardii* and *Phoenicococcus marlatti*), a mealy bug (*Dysmicoccus brevipes*), the lesser date moth (*Batrachedra amydraula*), and the termite, *Psammotermes hypostoma*. Attacks of desert locust (*Schistocerca gregaria*) were of rare occurrence especially on offshoots (El-Sherif et al. 2000). Other pests reported by Blumberg (2008), who wrote a review on the date palm arthropod pests and their management in Israel, included the sap beetle, date stone beetle, red scale, Parlatoria scale, and green scale.

Regarding diseases, several diseases were reported including Graphiola leaf spot, Diplodia leaf base rot, inflorescence rot, black scorch or terminal bud rot, leaf spot, fruit rot, and bending head (Lahaam 2005).

13.5 Conclusions and Recommendations

In spite of the feasibility of date palm production, cultivation in the Palestine is still below expectations. Several constraints impede progress, including high investment costs and underdeveloped marketing structures. Palestinian agricultural products, in particular those from the date palm, suffer from weak and inadequate marketing and processing services. This is mainly attributable to the absence of specialized firms. Marketing mainly depends on individual assessments and not on a thorough analysis of the local and international markets. Regarding date production, the following factors are major obstacles facing marketing and production: (a) weak competition in the domestic market and the absence of any control of the flow of date imports to the Palestinian market; (b) despite the high quality of Palestinian dates, Medjools in particular, the process of marketing to Europe still encounters many difficulties; (c) weakness and lack of modern marketing procedures and the adoption of standardized methods of washing, sorting, packaging, and drying; (d) farmers' lack of knowledge about exports and the requirements of global markets and consumer patterns; (e) lack of commitment by farmers and exporters to specifications and quality standards in terms of production and postharvest operations; (f) poor agricultural extension services at all stages of production; and (g) limited policies to support date palm cultivation.

The absence of official support for this sector is one of the important factors which have led to the weakening of the competitiveness of Palestinian agricultural products, including dates. The lack of support is reflected in production cost and subsequently the ability to enter foreign markets, as well as the continuity of overall agricultural practices.

El-Jaffary and Lafi (2004) proposed several points to overcome marketing problems of the Medjool date, some of which have been implemented. Their recommendations were:

(a) Stimulate and encourage investment to establish marketing infrastructure and facilities. Since the expansion in production takes a certain time period (up to 8 years) between planting the palm trees and the first harvest, investment in marketing facilities could be implemented gradually to absorb the annual increases in date production.
(b) Invest in cold storage facilities to maintain the postharvest quality of dates and to control the supply of date products to both local and export markets.
(c) Medjool palm growers should comply with international standards and requirements to be able to market their product through export channels. To accomplish that, training and rejuvenation programs should be provided to Medjool date producers and exporters to integrate them efficiently with local and export markets. Marketing surveys and research should be carried out on retail consumers through the export market chains.
(d) Concerning the efficiency of marketing Medjool dates in local markets, the following two steps should be applied: Firstly, protect Palestinian markets through appropriate laws and regulations; this will enable producers, wholesalers, and retailers to market local production efficiently in the domestic market. Secondly, wholesalers should design appropriate marketing packages to promote local consumption of date products as healthy food, not only during the Ramadan season but throughout the year.

Regardless of these difficulties, a trend by formal governmental agencies, as well as by national NGOs, to support the cultivation of date palms in the Jordan Valley has been initiated (EQA 2006a; LRC 2004). The Palestine Ministry of Agriculture has initiated three programs to support date palm cultivation in the Jordan Valley since 2000 with local participation of NGOs (EQA 2006b). Most of the seedlings distributed to farmers were Medjool and Deglet Noor cultivars. Regardless of constraints, the cultivated area of dates is increasing, and the projected production will exceed consumption; additionally, the production in the coming years will not coincide with the local high-demand period of Ramadan, and therefore dates should be exported to foreign markets. Several measures need to be taken for exporting dates, including washing, cleaning, drying, sorting, packaging, labeling, and storing and packaging clusters. Therefore, investment in postharvest infrastructure should be supported; other difficulties facing the production process include pollination, fertilization, pest management, and postharvest problems; research should be initiated and supported in these fields.

References

Abu-Qaoud H (1996) Status of date palm in Palestine. In: Ferry M, Greiner D (eds). Le palmier dattier dans l'agriculture d'oasis des pays méditerranéens, Options Méditerranéennes, Série A. Séminaires Méditerranéens, CIHEAM. vol 28. pp 81–84

Al-Assi S, Abu Ghazaal S, Ahmidat M (2003) Pre-feasibility study for the investment in a dates factory in Jericho. Palestine economic policy research institute (MAS)

Al-Ayedh HY, Rasool KG (2010) Determination of the optimum sterilizing radiation dose for control of the red date palm weevil Rhynchophorus ferrugineus Oliv. (Coleoptera: Curculionidae). Crop Prot 29(12):1377–1380

Al-Bana MF (2007) The effect of climatic conditions and geographical distribution on the success of new date palm varieties (Phoenix dactylifera L.) in the Gaza Strip. Acta Hort 736:59–69

Al-Jaghoub N, Hisham A, Fayez H (2000) Red palm weevil Rhynchophorus ferrugineus (Oliver, 1790) (Coleoptera; Curculionidae). In: Proceedings of the date palm international symposium. Windhoek, pp 211–214

ARIJ (2007) Applied research institute in Jerusalem. A review of the Palestinian agricultural sector. Jerusalem, Agencia Española de Cooperación Internacional

Blumberg D (2008) Review: date palm arthropod pests and their management in Israel. Phytopar 36(5):411–448

Botes A, Zaid A (2002) The economic importance of date production and international trade. In: Zaid A (ed) Date palm cultivation, vol 156, FAO. FAO, Rome, pp 45–56

Daiq I (2003) Palm cultivation development project in Palestine. Study report. Agricultural relief committee. (PARC) Ramallah, Palestine

Daiq I (2007) Date palm economics and cultivation circumstances in Palestine. Acta Hort 736:97–104

El-Jafari M, Lafi D (2004) The competitiveness of the Palestinian dates (Medjool) in the local and export markets. Palestine economic policy research institute (MAS), http://www.MAS.org.ps

El-Juhany LI (2010) Degradation of date palm trees and date production in Arab countries: causes and potential rehabilitation. Austral J Basic Appl Sci 4(8):3998–4010

El-Sherif SI, Elwan EA, Abd-El-Razik MIE (2000) Insect pests of date palm trees in the northern Sinai. The first international conference on date palm, Egypt

EQA (2006a) Environment Quality Authority Fara'a watershed baseline study report. Ramallah

EQA (2006b) Environment Quality Authority Fara'a watershed selection of interventions report. Ramallah

Erskine W, Moustafa AT. Osman AE et al (2011) Date palm in the GCC countries of the Arabian Peninsula. http://www.icarda.org/APRP/DatePalm/introduction/intro-body.htm

Ferry M, Gómez S (2002) The red palm weevil in the Mediterranean area. Palms 46(4):172–178

FAOSTAT (2012) Food and Agriculture Organization of the United Nation. Rome, Italy

Goor A (1967) The history of the date through the ages in the Holy Land. Econ Bot 21(4):320–340

Kalbouneh S (2011) Cropping patterns as a tool for water resource management in Palestine: date palm cultivation in Jiftlik, Jordan Valley. Int J Env Studies 68(4):447–460

Lahaam S (2005) Date palm in Jericho. Report, Ministry of Agriculture, Palestine

Land Research Center (LRC) (2004) Land use technical report for Fara'a watershed. technical report submitted to EQA. Jerusalem

MADICO (2011) Manasrah development and investment company. http://www.madico.ps/products.html

Palestinian Meteorological Department (2013) Climatic data for Jericho and Gaza. http://www.pmd.ps/ar/english.htm

PCBS (1993–2011) Palestinian central bureau of statistics agricultural statistical data. Ramallah. http://www.pcbs.gov.ps/

Zaid A, Wet PF (2002) Origin, geographical distribution and nutritional values of date palm. In: Zaid A (ed) Date palm cultivation, vol 156, FAO. FAO, Rome, pp 29–44

Zaid A (ed) (2002) Date palm cultivation, FAO Plant Production and Protection, Paper No. 156. FAO Plant Production and Protection, Rome

Chapter 14
Date Palm Status and Perspective in India

Sunil Pareek

Abstract Cultivation of the date palm in India goes back to the fourth century BC when Alexander the Great brought this fruit to India. The date palm has great socioeconomic importance and nutritional value in the Thar Desert, particularly in the Kutch region of Gujarat and western Rajasthan. Before 1947, India was a major date producer in the world. However, with independence and the partition that year, most of the date palm-growing areas became part of Pakistan. Subsequently, India became the largest world date importer. Over the last five decades, significant research has been done on date palm cultivation led by the Indian Council of Agricultural Research (ICAR) through the All India Coordinated Research Project (AICRP). Almost all the date cultivars grown in India were introduced from the Arabian Gulf and Middle Eastern countries. There is high potential for increasing the production area of date palm in India to fulfill local demand and to produce dates for export. Presently, the Rajasthan and Gujarat State governments and the private sector are convinced of the potential of date production and are striving to establish commercial date plantations and to promote date production. A public-private partnership has been established to boost production through tissue culture of date palms. Despite its important role in arid lands of the country and the efforts made, date palm cultivation, production, processing, and marketing are still beleaguered with several problems. The low quality of cultivars, non-availability of tissue-cultured plants, poor farm management, pests, diseases, lack of organized marketing, and insufficient applied research are major constraints of the date palm in India. This chapter discusses the current status, constraints, and approaches that can be used to develop the date palm industry in India.

Keywords Agricultural constraints • Date palm • Development • India • Prospects

---

S. Pareek
Department of Horticulture, Rajasthan College of Agriculture,
Maharana Pratap University of Agriculture and Technology,
Udaipur, Rajasthan 313001, India
e-mail: sunil_ciah@yahoo.co.in

© Springer Science+Business Media Dordrecht 2015

J.M. Al-Khayri et al. (eds.), *Date Palm Genetic Resources and Utilization: Volume 2: Asia and Europe*, DOI 10.1007/978-94-017-9707-8_14

14.1 Introduction

Date palm (*Phoenix dactylifera* L.) is a crop of desert oases and it is one of the oldest domesticated fruit crops. It is believed to be indigenous to countries around the Arabian Gulf. Date cultivation has historical records going back to 6,000 BC along the Tigris and the Euphrates rivers in present-day Iraq and to 2,000 BC during the Mohenjo-daro civilization, along the Indus River. Before the partition of India, the major date palm-growing areas were Multan, Muzaffarpur, Jhang, Dera Gazi Khan, Bahawalpur, and Khairpur, which now are part of Pakistan. In India, it is believed that date palm was introduced by soldiers of Alexander the Great in the fourth century BC in the Indus Valley. Babar, the first Mughal Ruler of India, has mentioned in his memoires that the date was grown in India in the early sixteenth century (Randhawa 1980). Dates were also grown in India during the regime of Akbar (AD 1555–1605) (Randhawa 1980). Groves of seedling date palms are found today on the coastal belt from Anjar to Mandavi in the Kutch district of Gujarat. At present, over 1.2 million date palm trees are in the Kutch area. In Rajasthan, date cultivation was first introduced by the then ruler Ganga Singh of the erstwhile Bikaner State through the planting of a few offshoots of Halawy, Khadrawy, and Zahidi at Sri Ganganagar, a part of the Thar Desert, after the development of the Gang Canal in that state. In northwestern Rajasthan, scattered plantations of a few thousand date palms are confined mainly to the districts of Sri Ganganagar, Bikaner, Jodhpur, and Jaisalmer. The date palm was of interest to the British during the colonial period. Bonavia (1885), a British military surgeon, wrote a book on dates in India, describing its introductions from the Persian Gulf, the locations in India where dates were grown, a scheme to establish plantations, and descriptions of cultivars. This book represents perhaps the earliest scientific study of dates in South Asia.

Date palm is one of the important horticultural crops of the Kutch region which enjoys a monopoly of commercial cultivation of this crop in India (Fig. 14.1). The total area under cultivation increased to 16,000 ha in 2010–2011, from 8,973 ha in 2000–2001, with production of 120,000 mt (Shandilya 2012). The Kutch region is the largest producer of dates in India because it has the most suitable climate. The date groves of the Kutch region are probably about 500 years old and are believed to have been developed from seeds and offshoots planted by settlers. The settlers often visited Middle Eastern countries for the Hajj or for trade from where they brought a lot of plant material. It is also probable that the Arab gardeners working in the palaces of the former rulers of Kutch might also have contributed to the importation of date seeds and offshoots from Arab countries. There was some trade between the Kutch and Arab countries resulting in substantial imports of date fruits and date products. In the course of time, the initial plantings multiplied and developed into groves (Pareek and Sodagar 1986).

Dates have potential for successful cultivation in Sirsa, Mahendragarh, Hisar, and Bhiwani districts of Haryana State; Abohar and Firozpur areas of Punjab State; Ganganagar, Bikaner, Churu, Jodhpur, Jaisalmer, and Barmer districts of Rajasthan State; and the Kutch region of Gujarat State. Dates can be grown in locations where

Fig. 14.1 *Phoenix* palms cultivated in India. (**a**) Date palms along the boundaries of cotton fields (Nagar Pol near Anjar, Kutch), (**b**) modern date palm grove (Devpur, Kutch), (**c**) wild date palm *P. sylvestris* on the edges on cultivated fields at Udaipur, (**d**) *P. sylvestris* on Mount Abu, Rajasthan (*Source*: Newton et al. (2013))

an assured source of irrigation is available, besides other climatic requirements. Pareek (1984) considered Jaisalmer and western parts of Barmer, Bikaner, and Jodhpur districts, extremely dry areas, as very suitable for date cultivation. In addition, dry areas with storm-type rains comprising Jodhpur, Bikaner, eastern parts of Barmer, and western parts of Nagaur, Churu, and Ganganagar districts. Other dry areas having rains followed by cloudy weather comprising of Abohar, Sirsa, and Ganganagar, the eastern part of Churu, and western part of Sikar district. These locations are potential date palm growing areas in North India.

Dates are very delicious and a favorite fruit in India and are consumed in various ways. Fruits of cvs. Halawy and Barhi are eaten fresh (table dates) during the hard ripe stage (*doka* or khalal stage) having low tannin content which is responsible for the astringent taste. Dates are cured as dry dates (*chhuhara*) and soft dates (*pind khajoor*). Both dry dates and soft dates are in great demand and find a ready market in India. Date fruits have religious significance too. These are eaten during Ramadan month by Muslims. Other parts of date palm tree are also used. Date palm leaves are used for making baskets and brooms and also provide raw materials for furniture making and thus promote cottage industries, providing many employment opportunities to the unemployed masses. Also, date sap provides sugar of excellent quality and wine. Date palms are very valuable as a canopy for ground crops in hot and dry regions. Under the canopy, vegetables, field crops, fodder crops, and small fruit trees such as pomegranate can be grown successfully as they get protection from the hot winds and sun. The date trunk is used as timber.

Date fruits have unique nutritive value, are rich in sugars, and are a good source of energy. Easily digestible sugars in date fruits amount to 75–80 % on a dry weight basis. Jawanda et al. (1972) estimated the total sugar in date fruits of cv. Khadrawy as 63.63, 84.61, and 77.50 % at early *doka*, late *doka*, and early *dang* (rutab) stages, respectively. Besides, dates are a good source of minerals like iron, potassium, calcium, phosphorus, magnesium, sulfur, and copper. Date fruit is also a good source of vitamins such as thiamine, riboflavin, biotin, and folic and ascorbic acid. Pulp protein content is 1.75–2.75 % and fat content is 0.31–1.9 % (Katyal and Dutta 1976).

Although India ranks first in date imports, information on the cultivation of this crop in western India is insufficient, and not much is known to the rest of the world. However, there are about 1.9 million date palms with an annual production of 85,000 mt of fresh fruit in western India.

The cultivation of this crop is unique in the region considering the local adaptation to climatic, edaphic, and socioeconomic conditions. The majority of the groves have developed or were multiplied by seeds, and the farmers identify date palms based on fruit color, size, shape, and taste. Dates are generally grown on the farm boundaries associated with different intercrops. The fruits are harvested at khalal stage and consumed as fresh fruits due to limiting climatic conditions. On the basis of socioeconomic considerations, an elite date palm should have sweet flesh at khalal stage; be brightly colored, preferably red; be early maturing; have tolerance to rain; be high yielding; and produce a high number of offshoots.

14.2 Cultivation Practices

14.2.1 Irrigation

The date palm tree is drought resistant to a great extent and also salinity tolerant compared to other fruit trees, but it does require a reliable source of water from subsurface sources or irrigation for growth and fruit production. Frequency of irrigation depends on the soil texture and weather conditions. In light soils, date palms are usually irrigated at 7–10-day intervals in midsummer and 15–30 days in winter. An irrigation prior to spathe initiation is critical for normal fruiting. Irrigation after fruit set should be given regularly. Insufficient irrigation during fruiting may result in premature fruit drop, shriveling, and reduction in fruit size, weight, and quality. Studies conducted at Bikaner on conservation of soil moisture showed that the use of a black polythene sheet provides better soil moisture retention, suppresses weed growth, and improves fruit quality. Drip irrigation is useful as an efficient utilization of irrigation water. If drip irrigation is used from an early age, the root distribution of date palm will be within the area of moist soil surrounding the tree.

Coarse-textured soil requires a total of 7–9.5 ha-m of water per year and 0.75–1.1 ha-m per month during summer. However, on heavy soil, the water requirement is reduced by 50 %. The penetration of irrigation water should keep the soil moist to a depth of 2–2.5 m. Mulching with cuttings of locally available weeds such as *bui*

(*Aerva javanica* syn. *persica*) or *shaniya* (*Crotalaria burhia*) helps retain maximum soil moisture even when the irrigation interval is prolonged for up to 1 month during April–May. The available soil moisture (ASM) was 1.52 % with a thick layer of *bui*, followed by 0.99 % with *shaniya* mulch and 0.002 % in control. In addition, *bui* mulch produced fruits of better quality of cv. Halawy with 5.2 mm pulp thickness, 47.5 % total soluble solids (TSS), and 0.13 % total acidity at *doka* stage (Chandra et al. 1994).

14.2.2 Fertilization

Date palm orchards in general should be provided with adequate farmyard manure (FYM) which not only results in better growth of the date palms but also improves physical conditions of the soil. Considering the poor soil fertility status of potential date palm-growing areas in western Rajasthan, where the soils are coarse textured, the application of FYM may help significantly to improve moisture retention characteristics. Application of nitrogen, phosphorus, and potassium also may be helpful in providing better vegetative growth and fruit yield of superior quality. Bajwa and Bakhshi (1961) recommended application of an annual dose of 50–60 kg FYM or 1–2 kg ammonium sulfate per bearing tree, applied in two equal parts, one during January–February and another in August–September. According to Randhawa (1980), FYM in date palm orchards may be applied at 10 mt ha^{-1} and its manurial requirements work out to 33 kg of N, 21.75 kg of P_2O_5, and 50 kg of K_2O per hectare.

In the sandy soils of western Rajasthan, FYM must be applied yearly at 20–40 kg $tree^{-1}$ during September–December for the proper bearing of date palms. Besides this, each bearing tree should be given 0.5–1 kg N, 0.5–1 kg P_2O_5, and 0.5 kg K_2O per year. Studies in Bikaner State revealed that the addition of 1,500 g N per palm per year in two split doses, viz., 60 % before flowering and the remaining amount after fruit set, in combination with 500 g each of phosphorus and potash as a basal dose provided better vegetative growth, yield, and quality fruits. In integrated nutrient management, application of 50 % NPK along with 50 g sulfur, biofertilizer (125 g *Azotobacter* + 50 g phosphorus solubilizing bacteria (PSB)), and 25 kg FYM per palm in 10-year-old plantations gave higher yield and better fruit quality. Foliar application of micronutrients such as $FeSO_4$ (0.5 %) and thiourea (0.1 %) in 10-year-old plants started after fruit set gave higher fruit yield, weight, and TSS (Pundir et al. 2004).

14.2.3 Flowering, Pollination, Fruit Set, and Fruit Development

Flowering in date palm begins with spathe emergence, which occurs in North India at the start of the spring season. The timing of spathe initiation is governed by various factors such as the growth site, climatic conditions mainly temperature fluctuation, cultivars, and sex of plant. Bakhshi and Singh (1972) reported flowering in

Abohar by the middle of March. Flowering in male palm trees was observed 15 days earlier than females; fresh pollen was available in abundance for pollination. Jawanda (1964) observed spathe emergence under Abohar conditions from the third week of February and continuing to the middle of April. Most of the spathes, however, emerged during March. Spathe opening starts by the fourth week of February and continues to the third week of April (Jawanda 1964; Jawanda and Kalra 1972). Anthesis of flowers is completed between the fourth week of March and the second week of April. Mertia and Vashishtha (1985) observed variations in spathe emergence in different years in cultivar Halawy. It was noted from February 3 to the end of the month, February 3–23, and March 7–9 during different years at the Chandan Research Station, Jaisalmer, Rajasthan. Vashishtha (1987) reported variations in initiation of flowering during 1985 and 1986 in different date palm cultivars at Jodhpur. Initiation of flowering was earliest in cvs. Halawy and Barhi (January 8), whereas it was on February 10 in 1986 in cvs. Migraf, Pakistan, Muscat-2, and Abdul Rahman. Flowering ended on February 28 in Zahidi in 1985 and on March 11 in Medjool in 1986. Under Bikaner conditions, Chandra et al. (1989) reported that out of 24 cultivars at the Date Palm Research Centre, in 1987, spathe initiation commenced early in cvs. Medjool, Muscat, Khadrawy, and Bint Aisha, while it was delayed in Gizaz and Barhi. Maximum duration of flowering was in cv. Khadrawy (35 days) followed by Medjool and Khalas (32 days). A date palm usually carries 6–12 clusters and 1 cluster may carry 50–100 strands with 16–31 fruits on each (Jawanda and Kalra 1972). At Bikaner, spathe emergence takes place much earlier in male palms than in females. Spathe initiation began early in cvs. Zahidi, Nagal, Khadrawy, and Khunezi (first week of March), while it was delayed in cvs. Khasab, Barhi, Umshok, and Sakloti (last week of March) at Bikaner.

Pollination of female trees is done by hand; strands of freshly opened male inflorescences are cut from the tree and 3–4 strands placed between the strands of female flower clusters in an inverted position, as soon as flower anthesis in female palms takes place. It is necessary to tie the pollinated inflorescence with a slip knot to hold the male strand in place for pollination. The second method of pollinating female clusters consists of dusting them within 2 days of anthesis. The pollen may be collected in abundance as a creamy powder from the freshly opened male inflorescences. Collection should be done in the morning hours to obtain better fruit set. If pollen is in short supply, it may be mixed with talcum powder. Pollen storage can help in pollination of female palm trees where spathe opening takes place much after flowering in male trees. The pollen can be stored at room temperature or in a refrigerator in a dry airtight container and used later in the pollinating period (Kalra and Jawanda 1975). Pollen stored in a refrigerator (9 °C) gave higher fruit set (42.9 %) as compared to that stored at room temperature (36.8 %) (Gupta and Thatai 1980).

Hand pollination is commonly practiced in western Rajasthan. It is done by dusting pollen using cotton balls on freshly opened female spathes in the early morning hours, for at least 2–3 days, followed by placing pollen-impregnated cotton balls in the female flowers or by hanging strands of male inflorescence in an inverted position on the female spathes. To collect pollen from male spathes, they are swept on paper. The collected pollen (after 6 h in sunlight followed by 18 h in shade) can be

Table 14.1 Fruit developmental stages

Stage (India)	Corresponding Arabic term	Characteristics
Gandora	Kimri	Fruit is still hard and green
Doka	Khalal	Fruit is fully grown but remains hard; color becomes yellow or red
Dang	Rutab	Softening of fruit begins from the tip and finally the whole fruit is soft
Pind	Tamar	Fruit fully ripe, weight decreases from dehydration

stored in airtight glass vials in a cool place at ambient temperature for 8 weeks or refrigerated at 9 °C for a year.

Upon loss of two unfertilized carpels (hababouk stage) after fertilization, fruit set takes place and growth commences. There are four developmental stages (Table 14.1): *gandora* or kimri (fruit attains pea size and remains hard green), *doka* or khalal (fruit attains full growth, it remains hard, and their color changes to yellow or red), *dang* or rutab (softening of fruit begins from the tip and finally the whole fruit is softened), and *pind* or tamar (fruit is fully ripened and its weight decreases as a result of dehydration) (Fig. 14.2). Fruit becomes edible from *doka* stage except for those cultivars which are highly astringent. At Bikaner, preharvest application of Ethrel (2-chloroethylphosphonic acid) at 100 ppm on fruit bunches at the color break stage (green to yellow or red) hastens ripening and also increases fruit size and weight.

At Abohar, Ganganagar, Churu, Sirsa, and Hisar, fruits are usually harvested at *doka* stage due to early commencement of monsoon rains. Fruit at the late *doka* and *dang* stages can be obtained at Bikaner, Nagaur, and Jodhpur of western Rajasthan, and *dang* to early *pind*/full *pind* stages can be obtained in extremely dry areas like Jaisalmer and western parts of Barmer, Bikaner, and Jodhpur districts (Pareek 1984). The foregoing indicates the suitability of date palm cultivation in the Thar Desert of western Rajasthan where good quality fruits may be harvested. Vashishtha (1987) reported the earliest *doka* fruits in cv. Halawy on June 9 closely followed by Medjool. At Bikaner, Halawy attained *doka* stage earliest in the first week of June followed by Medjool and Khadrawy in the second week of June (Chandra et al. 1989). Halawy fruits were sweet and edible at *doka* stage indicating its suitability as a table date. *Pind* stage was obtained earliest in Sedami (third week of July) whereas most other cvs. produced soft dates during the fourth week of July. Under Abohar conditions, *doka* stage is normally attained, but afterward, most of the cultivars are seriously damaged due to monsoon rains and humidity in July–August (Bakhshi and Singh 1972; Kalra 1976; Katyal and Dutta 1976).

14.2.4 Pest Control

Date palm groves perpetuate several insects. Scale insects and termites are two major insects which pose a major challenge for successful date palm cultivation in India and make questionable its extension into new areas.

Fig. 14.2 Indian terms used for developmental stages of dates in India

14.2.4.1 Scale Insects

Among the insects, *Parlatoria blanchardi*, commonly known as white scale, seems to be one of the most injurious pests in the Thar Desert. The nymph and adult cause damage to the date palms by feeding on white succulent tissue at the base of the leaf stalk. Infestation affects normal tree growth and productivity declines. Spraying acetamiprid (0.3 g L^{-1}) or dimethoate (2 ml L^{-1}) has been found to control scale insects. Heavily infested leaves should be removed and destroyed. Studies at the Date Palm Research Centre, Bikaner, revealed that scale insect populations might be reduced with release of the predatory beetle *Chilocorus nigritus*.

14.2.4.2 Termites

Termites (*Odontotermes obesus*) occur on all parts of the date palm the year around and considerably affect it. Termites attack the root zone, feed on the shoots, and bore into the main trunk. The young establishing trees under severe attack wither and die. Termites infest offshoots as well as date palms with maximum infestation reported in cv. Halawy. The root zone should be treated with chlorpyrifos (5 ml L^{-1} water) periodically. Termite galleries on the palm trunk should be scraped off and the area treated with the abovementioned insecticide.

14.2.4.3 Lesser Date Moth

A new insect pest identified as the lesser date moth *Batrachedra* sp. was observed recently for the first time in Bikaner. Larvae of this moth attack flowers and fruits. Infestation occurs during the reproductive period, i.e., between March and August. To control this pest, Decamethrin (0.3 ml L^{-1} water) or malathion (1 ml L^{-1} water) should be applied 1 week after fruit set and repeated again 15 days after the first spraying, if required.

14.2.4.4 Bird Damage

The period of fruit growth and development from May to July coincides with the period when there are few field crops available for birds to feed on in the areas where dates can be grown in India. Thus, birds cause severe damage to fruits. The date fruit bunches at the time of ripening are very much damaged by crows, parrots, sparrows, etc. The maximum damage to the fruit is caused at full *doka* and *dang* stages. Protection from bird damage can be done by covering fruit bunches with wire net gauge (3 × 3 mm mesh) at the start of *doka* stage until harvesting (Fig. 14.3).

14.2.5 Disease Control

14.2.5.1 Graphiola Leaf Spot

Among the diseases, Graphiola leaf spot, *Graphiola phoenicis* (Moug.), is the one which causes economic losses by attacking the leaves to reduce their number and affect fruit production. It has been observed to severely affect the leaves, and it is the

Fig. 14.3 Protection from bird damage with wire net gauge (*Source*: Courtesy of Dr. R. Porwal, KVK, Ajmer)

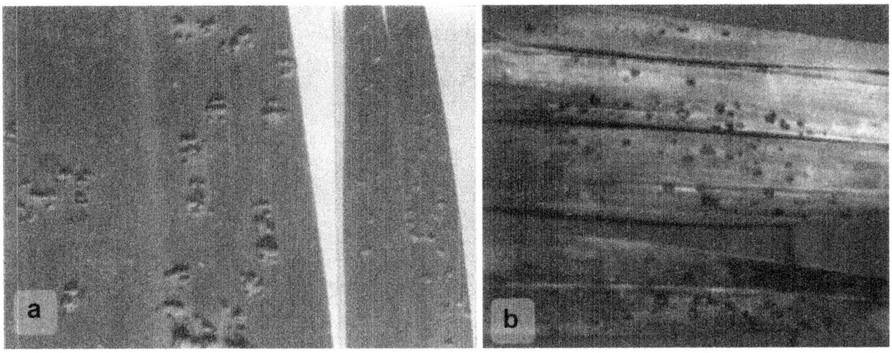

Fig. 14.4 Graphiola leaf spot (**a**) and Alternaria leaf spot (**b**) (Photo courtesy: Dr. JPS Pundir, Date Palm Research Station, Bikaner)

most widespread disease of date palm and probably occurs wherever the palm is cultivated under humid conditions which prevail in many subtropical to marginal date-growing areas. Leaves of date palm are marked with yellowish light brown spots with small black scales or warts having a gray brown to dark brown horny outer surface and long sterile flexuous hyphae protruding from them and an inner membrane containing yellow powder (Fig. 14.4a). In case of heavy infestation, leaves dry and consequently die. It causes great damage to the plant growth and reduces fruit yield. Studies on the epidemiology, cultivar screening, and disease management were carried out at SK Rajasthan Agricultural University (SKRAU), Bikaner, Rajasthan, and at the Fruit Research Station (FRS), Mundra, Gujarat. Graphiola leaf spot occurs widely in the hyperarid partially irrigated western plains of Rajasthan at Bikaner, Jaisalmer, Barmer, and Jodhpur. The disease progressively develops on older leaves as compared to new ones. The highest infection rate was recorded in the month of December due to low temperature and higher atmospheric humidity. Incidence of the disease has been recorded at 2.50–76.25 % in the field. A total of 34 cultivars were screened under natural as well as artificial conditions at Bikaner; among them, cvs. Sayer, Halawy, and Khalas were found to be highly susceptible and only Hatemi to be resistant. Gupta and Mehta (1985) reported the incidence of this disease in cvs. Medjool, Khadrawy, Shamran, Halawy, Barhi, and Zaghloul, but the incidence of the disease was very low in Khadrawy in which it was only up to 4 % under Haryana conditions. Two sprayings of copper oxychloride (0.4 %) were effective to minimize the Graphiola leaf spot followed by Carbendazim (0.1 %) (Manohar and Chandra 1995). Foliar application of Bavistin and Benlate at 0.2 % has been suggested as a control measure (Gupta and Mehta 1985). However, Raj Bhansali (1987) reported Dithane M-45 (0.2 %) and Fytolan (0.2 %) as the most promising chemicals for controlling the disease up to the extent of 71.20 and 65.20 %, respectively, when 4 sprayings at 15-day intervals under Jodhpur conditions.

Studies of changes in green pigments (chlorophyll and carotenoids) at different levels of severity of the disease indicated that as the disease severity increases, the green pigment decreases proportionately in all the cultivars. The percent reduction in chlorophyll and carotenoids in diseased plant was obviously much more in highly susceptible cultivars as compared to resistant ones.

14.2.5.2 Alternaria Leaf Spot

Alternaria leaf spot disease of date palm caused by *Alternaria alternata* (Fr.) Keissler is also a destructive disease of date palm in India. It has been observed in severe form in Rajasthan over the past several years. Symptoms were observed on the leaves in the form of small isolated pale or dark gray to black circular spots measuring 2–8 mm in diameter (Fig. 14.4b). Later spots increase in size and become irregular, straw colored, and coalesced. With a severe attack, the diseased leaves turn yellow and defoliate. Two sprayings of Carbendazim + Mancozeb (0.2 %) at an interval of 15 days were found effective to control this disease up to 52.69 %.

A total of 34 cvs. were screened against this disease (Pal et al. 2006, 2007). Of the 34 cvs., 31 were infected by Alternaria leaf spot and disease ranged between 4.44 and 62 %. Cvs. Medjool, Nagal Hillali, Sri Ganganagar, and Medini were observed (44 %) to be free from infection and fall into the resistant category; 21 cvs. were recorded as moderately susceptible to the disease with 5.1–20 % infection. Cvs. Khadrawy, Barhi, Khalas, Tayar, Barshi, and Bikaneri were observed to be susceptible, while cvs. Sayer, Halawy, and Zahidi were recorded as highly susceptible (Pal et al. 2006).

14.2.5.3 Diplodia Sucker Rot

This disease is caused by *Diplodia phoenicum* (Sacc.) H.S. Fawc. & Klotz, a fungus that sometimes affects leaf stalks and offshoots. This disease is very serious and is a limiting factor for further propagation of date palm. This disease is characterized by the death of offshoots either while they are still attached to the mother palm or have been detached and transplanted. The disease also causes premature death of leaves in older palms. In offshoots, the disease may be expressed in distinct types;

(1) the fungus may infect and kill the outside leaves, leaving the younger shoots and bud alive for a time before finally causing their death as well, or (2) the central leaf cluster and terminal bud may die. The fungus usually enters the palm through wounds made during pruning or through cuts made during removal of the offshoots from the mother palm. Faulty irrigation causing some roots to die back at the base of the palm may also contribute to the infection. For control of this disease, infected leaves and dead tissue should be removed. Pruning tools should be disinfected by dipping them in formalin or in bleach solution after each palm is treated. Offshoots at the time of transplanting must be treated with Carbendazim 0.1 + chlorpyrifos 0.1 + indole butyric acid 1,000 ppm coupled with alternate day irrigation for up to 30 days for better establishment and survival (Anonymous 2011).

14.2.5.4 Fruit Rots

Preharvest fruit rots are another problem since their incidence is governed by the occurrence of rain and high humidity during khalal and later stages of ripening. Considerable loss occurs from fruit rots caused by *Aspergillus niger*, *A. flavus*, *Rhizopus* sp., and *Botryodiplodia* sp. Of 31 date cultivars screened against fruit rots, before and after rains, none of them was found immune, while cvs. Khunezi, Halawy, Medjool, Khalas, and Barhi were susceptible to the disease. Only cv. Zaghloul was found resistant (<5 % incidence). Two sprayings of *Trichoderma viride* 1 % + Azadirachtin (0.3 %) at an interval of 15 days coupled with covering of fruit bunches were found effective to control fruit rots up to 62.70 % (Anonymous 2011).

14.3 Genetic Resources and Conservation

No hybrids are reported in date palm so far. Furthermore, no breeding work has been taken up on date palm in India except evaluation of cultivars/genotypes against rain damage and selection of some promising female seedlings from the Kutch region of Gujarat. Most of the cultivars of date palm were introduced from other countries from time to time, e.g., Halawy, Barhi, Medjool, Khalas, Sayer, and Zahidi (USA); Khadrawy, (Iraq); Barshi, Khunezi, Nagal, and Khasab (Oman); Hatemi, Tayar, and Ruziz (Saudi Arabia); Amari, Sakloti, Agolani, Chip Chap, and Braim (Iraq) in 1998; and Sewi and Amhat (Egypt) in 2009. All the commercial date cultivars developed through selection of chance seedlings have been based on local needs. From the rich genetic diversity of nearly 1.66 million seedling date palms in the coastal belt of the Kutch region of Gujarat, 20 promising palms have been selected, most of which yield nonastringent fruits at *doka* stage (Muralidharan et al. 2008). One of them bears coconut-shaped fruit. These selections flower twice a year. An early ripening date seedling has been identified at Abohar.

14.3.1 *Current Status and Prospect of Genetic Resources*

In India, the wild date palm (*Phoenix sylvestris*) and the true date palm (*P. dactylifera*) are commonly found. There is an established relationship in the evolution of *P. dactylifera* with the other species in the Indian subcontinent. *Phoenix sylvestris* is most commonly found in Rajasthan and Gujarat states, on low ground in the sub-Himalaya tract, along river banks of the Deccan Plateau (south-central India), in forests up to 1,350 m in Himachal Pradesh, and especially in lower hill slopes in Haryana (northwestern India). It survives in disturbed areas, such as wastelands or seasonally inundated areas (Parmar and Kaushal 1982). Apart from its natural distribution in the wild, *P. sylvestris* is also cultivated in parts of South Asia, mostly in its eastern and southern parts such as West Bengal, the Coromandel Coast of Andhra Pradesh, and Chittagong (eastern Bangladesh) (Chowdhury et al. 2008; Parmar and Kaushal 1982; Pattnayak and Misra 1982).

Phoenix dactylifera populations in India are almost entirely composed of seedlings, and the palms are not propagated by offshoots but by seed; eliminating male individuals after the first flowering limits the male population. The quality of dates is therefore highly variable. Morphologically, *P. sylvestris* is close to the date palm and has long been considered as the wild progenitor of the cultivated date palm. However, a recent genetic study challenges this hypothesis (Pintaud et al. 2010), and with the discovery of truly wild date palm populations, it is now completely rejected. Nevertheless, the two species are interfertile, as are all species of *Phoenix*, and their relationship remains to be investigated. Indeed, the two species are believed to occur in sympatry in Pakistan and northwestern India, and natural crossbreeding is possible.

Field surveys conducted in 20 representative date-growing villages of the Kutch region in 2010 and fruit samples from 50 elite date palms/village were collected. Qualitative and quantitative traits varied greatly indicating a broad diversity in color, shape, size, and taste of the fruits in groves. This variability can be better exploited for selection of superior date palm cultivars for resistance to biotic and abiotic stresses and also for genomic studies to preserve genetic material (Muralidharan and Baidiyavadara 2013). Most of the germplasm evaluation reports are based on the study of introduced date palm cultivars from the Arabian Gulf region and other parts of India. There is an urgent need to carry out an extensive survey of Indian seedling date palm groves to identify local germplasm of superior quality so that it can be collected, evaluated, characterized, and released as cultivars.

14.3.2 *Threats and Degradation*

Sometimes strong attributes for cultivation of a crop conceal weaknesses. This is true for date palms in India. Construction of the Indira Gandhi Canal in the Thar Desert boosts agricultural production and also opens new avenues for date cultivation. However, it causes salinity problems in the region and the level of salinity is beyond control which is detrimental to good production of dates. Apart from this irrigation facility providing possibilities for other agricultural crops, it represents a threat to decrease in area the date palms in some regions. The other threat for date palm expansion is Graphiola leaf spot disease, which is very severe in this region. Sometimes it causes intolerable loss to growers.

All the germplasm blocks of the date palm such as the centers at Bikaner, Abohar in Punjab, Jaitsar, Mundra, and others are situated on the country's border area. It is feared that in the case of an outbreak of hostilities between the two countries, the entire introduced and indigenously collected germplasm could be destroyed. Therefore, backup germplasm banks should also be established at other locations away from the border area. Further, urbanization, construction of buildings and bridges, and floods in some areas also cause germplasm drift. The Kutch region of Gujarat is an earthquake-prone region, and it experienced a severe earthquake in the last decade and this is an added threat factor. Furthermore, many existing date palms are aging and have ceased producing offshoots for propagation. Therefore, a number of neglected and underutilized cultivars of date palm are expected to be lost forever and consequently this will reduce date palm diversity in the country. This situation makes conservation of genetic resources of date palms an imperative; there is a great need to put practices to protect important Indian date palm cultivars.

14.3.3 Conservation Efforts

A wide range of genetic diversity is present in date palms which are mainly grown in arid and semiarid parts of India. However, this vast genetic pool has so far largely remained unstudied and underutilized. Conservation of plant genetic resources is essential for future crop improvement programs (More and Singh 2008). This genetic variability can effectively be utilized for the development of desired cultivars. The common objectives in improvement until now have been to improve the quality attributes and appearance of fruits in addition to yield. Most cvs. grown, presently, are seedlings or introductions. There are two principal methods of germplasm conservation, in situ and ex situ. Experience shows that diversity is only secure when diverse conservation strategies are employed. Ex situ and in situ approaches are not mutually exclusive; no single method of conservation is optimal for all situations, and no single method alone can succeed in the longer term. Different complementary conservation systems provide insurance against the shortcomings of any one method (Singh et al. 2012). In India over the last three decades, much more attention has been given to the evaluation of introduced date palm cvs. under local conditions, including foreign germplasm used commercially after their involvement in crop improvement programs. Date palm germplasm has been collected and conserved in field gene banks at the AICRP on Arid Zone Fruit (AZF) centers located in different parts of the country (Dhandar and Singh 2004). The Central Institute for Arid Horticulture (CIAH), Bikaner, is the nodal agency and serves as the national gene

Table 14.2 Date palm germplasm established and maintained at various research stations in India

Central Institute for Arid Horticulture		SK Rajasthan Agricultural University	
Abdul Rahman	Medini	Abdul Rahman	Migraf
Amhat	Medjool	Agolani	Mundra Sel-3
Bikaner local	Muscat	Amri	Muscat
Braim	Nagal	Barshi	Nagal
Chip Chap	Nubsully	Barhi	Nagal Hillali
Dayari	Punjab Red	Bikaneri	Ruziz
Dhamas	Sabiah	Bint Aisha	Sakloti
Fard	Sayer	Gizaz	Sayer
Halawy	Sedami	Halawy	Sedami
Hamra	Sewi	Hamra	Sewi
Hayani	Shamran	Hatemi	Shamran
Hillali	Siwi	Hayani	Sri Ganganagar
Khadrawy	Suria	Khadrawy	Suria
Khalas	Tayar	Khalas	Tayar
Khasab	Umshok	Khasab	Umshok
Khastavi	Yaqubi	Khunezi	Zaghloul
Khunezi	Zahidi	Medini	Zahidi
		Medjool	

bank for date palm. Another important center is the FRS, Mundra, of Sardarkrushinagar Dantiwada Agricultural University (SDAU), Gujarat. Various cvs. have been introduced from the Middle East and conserved in these centers. To conserve the variability from Rajasthan and Gujarat states, some explorations have been done under the AICRP on AZF and Central Arid Zone Research Institute, Jodhpur. The National Bureau of Plant Genetic Resources (NBPGR), New Delhi, is the apex institute in India for introduction, quarantine, in vitro propagation, and cryopreservation of fruit crops. Recently, the NBPGR also began conservation work on date palms in India.

14.3.4 Germplasm Banks

Rich genetic diversity of the date palm is available in the coastal belt of the Kutch region, Gujarat, which should be exploited (Singh et al. 2009). In other parts of the country, seedling dates are grown in meager population. At the CIAH, Bikaner, Rajasthan, 60 genotypes/cultivars have been conserved in a national field gene bank collected from different available sources as well as introduced from abroad (Table 14.2). Further, SDAU, Dryland Research Station, Mundra, Gujarat; Central Arid Zone Research Institute (CAZRI), Jodhpur Rajasthan; SKRAU, Bikaner, Rajasthan; Regional Fruit Research Station (RFRS), Abohar of Punjab Agricultural University, Punjab; and Central State Farm, Jaitsar, Sri Ganganagar, Rajasthan, have also maintained and evaluated date palm germplasm.

14.4 Plant Tissue Culture

14.4.1 Role and Importance

Date palm is traditionally propagated by offshoots which are produced at or below the ground level at the stem base. Small offshoots that appear above the ground level on the trunk are usually destroyed due to difficulty in rooting. Offshoots are produced in a limited number for a certain period in the lifetime of a young date tree. Offshoot formation is dependent on the genetic makeup of the cultivar and environmental factors. The number of offshoots produced by an individual date palm is highly variable and differs from one cultivar to another (Munier 1973). The traditional method of vegetative propagation through offshoots is slow, laborious, time consuming, and expensive (Aboel-Nil 1986; Sudhersan et al. 1993; Tisserat 1981). Transmission of disease-causing pathogens and insects is another disadvantage of conventional offshoot propagation (Aboel-Nil 1986). This factor has led to emphasis on a micropropagation technique during the past 30 years as rapid clonal reproduction of selected cultivars is cheaper and especially important in cultivars which produce a few offshoots only once (Sudhersan and Abo El-Nil 2004). Raj Bhansali (2010) described the advantages of micropropagation technology over conventional propagation. These are the following: (1) only a small laboratory space is required; (2) a large number of plants can be produced in short duration; (3) cloning of selected material is possible; (4) there is no seasonal effect on plants because they can be multiplied under controlled conditions in the laboratory throughout the year; (5) genetically uniform plants are produced; (6) easy and fast exchange of plant material between different regions of a country or between countries is ensured without any risk of the spread of diseases and pests; (7) large-scale production is economically reliable; and (8) plantlets are easy to handle and transport and do not require complicated phytosanitary regulation.

14.4.2 Research and Development

During the last two decades, intensive research and development efforts have been initiated to produce plantlets by tissue culture in the major date-growing countries. In the UAE, Jordan, Egypt, Iraq, Morocco, Spain, France, the UK, and Israel, success has been achieved in exploitation of this technique on a commercial scale, and millions of plants have been produced for planting on a commercial scale. There are, however, reports of serious morphological abnormalities such as dwarfism, deformed leaves and inflorescence infertility, and poor fruit quality and yield in some orchards established using tissue-cultured plants. This is due to somaclonal variation arising in the in vitro–propagated clones of the identified cultivar. A few studies were made to produce plantlets of date palm through the axillary bud culture method, which is reported to be less risky with respect to regeneration of somaclonal variants. Efforts have also been made to exploit this technique to commercialize date palm production.

In India, attempts have been in progress since the 1980s to develop tissue culture protocols for some cultivars; initial successes were achieved at some Indian laboratories but their field application was either unsuccessful or economically unviable. The following laboratories were involved: Chaudhary Charan Singh Haryana Agricultural University (CCSHAU), Hisar; SKRAU, Bikaner; Anand Agricultural University (AAU), Anand; and CAZRI, Jodhpur. Work at these institutions has either been discontinued or the developed protocol still requires improvement. Thus at present, there is in India a technology gap in commercializing micropropagation of elite date palm cultivars, and therefore the farmers and Rajasthan State are dependent on the import of tissue-cultured plants at exorbitant cost through agencies/ companies. A reproducible efficient tissue culture protocol capable of large-scale multiplication for distribution of planting material of suitable cultivars and elite palms is yet to be developed in the country.

14.4.3 Scaling-Up Production and Other Applications

CIAH, Bikaner, has initiated an intensive research program on date palm tissue culture, considering this as one of the priority research area of the institute. The program has the following objectives: (a) standardize and refine micropropagation techniques of date palm; (b) screen useful somaclonal variants; (c) standardize technique of hardening of tissue culture plantlets for better rooting, growth, and development under extreme arid environment; (d) provide planting materials for commercial farmers; and (e) develop a regeneration protocol for gene transfer and crop improvement.

Research is in progress on different aspects of date palm tissue culture through somatic embryogenesis and axillary shoot proliferation using different media, plant growth regulator concentrations, and explants. The standardized regeneration protocol is intended to be further extended to transfer genes responsible for quality and resistance to abiotic stresses. Attempts are under way to resolve problems of explant contamination, browning, and poor response of cultured explant to asexual embryogenesis. The slow-growing/recalcitrant nature of in vitro cultures is another problem. No method seems to be known to accelerate growth responses.

Several publications from Indian researchers have dealt with standardizing and refining the technique with respect to various date palm cvs. by adopting different methods. There are two main methods of in vitro propagation adopted for date palm tissue culture: direct organogenesis and somatic embryogenesis (Raj Bhansali 2010). Explants like zygotic embryos, root segments (Sharma et al. 1980), young leaves (Sharma et al. 1984), shoot apices (Raj Bhansali et al. 1988), and inflorescences (Bhaskaran and Smith 1992) have been used for this purpose.

Protocols for in vitro somatic embryogenesis in date palm have been developed by several researchers (Raj Bhansali et al. 1988; Sharma et al. 1986;

Sudhersan et al. 1993). Explants are incubated in complete darkness for 3–6 months, in culture rooms for the production of embryogenic callus. The phenolic interference problem is overcome by the use of activated charcoal, polyvinylpyrrolidone, cysteine-HCl, ascorbic acid, and citric acid (Dass et al. 1989; Raj Bhansali and Kaul 1991; Raj Bhansali et al. 1988). Frequent subculturing is also adopted to overcome a browning problem. Regeneration of somatic embryos occurs when the callus is subcultured using suitable media, often hormone free. Suspension culture of friable date palm callus for rapid somatic embryogenesis has been established (Bhaskaran and Smith 1992; Sharma et al. 1986). Refinement of the protocol for direct organogenesis in date palm has been accomplished (Sudhersan et al. 1993). However, the rate of multiplication is less as compared to somatic organogenesis (Raj Bhansali 2010).

14.4.3.1 Somatic Embryogenesis

Asexual embryogenesis has been achieved from somatic tissues (Bhaskaran and Smith 1992; Dass et al. 1989; Raj Bhansali and Kaul 1991; Raj Bhansali et al. 1988; Sharma et al. 1984, 1986; Sudhersan et al. 1993). Callus tissues are induced in cvs. Muscat, Medjool, Sayer, Samran, Jaglool, and Khadrawy when explants are incubated in complete darkness for 3–6 months at 25 °C (Bhargava et al. 2003; Raj Bhansali et al. 1988; Sharma et al. 1984, 1986; Yadav et al. 1998).

During growth of embryogenic callus, date palm explants release excessive browning substances, which cause serious problems with this technique (Raj Bhansali 2010). The inhibitory effects may result from the bonding of phenols with proteins and their subsequent oxidation into quinines. These are highly reactive and toxic to the tissues. Presoaking of explants in ascorbic acid and citric acid solutions and adding these compounds to the culture medium helps to curtail the oxidation of phenols. Incorporation of polyvinylpyrrolidone (PVP), cysteine-HCl, and ascorbic acid also minimized browning problems in several date palm cultivars (Dass et al. 1989). Raj Bhansali and Kaul (1991) used antioxidant solution (150 mg L^{-1} citric acid and 100 mg L^{-1} ascorbic acid) for 30–60 min in cold storage (0–4 °C). Furthermore, use of nutritionally balanced media containing activated charcoal (3 g L^{-1}) has significantly checked the browning problems in date palm explants. Raj Bhansali et al. (1988) found that shoot tips and lateral bud cultures grew successfully if transferred frequently (after periods of incubation of 7–15 days) to fresh medium.

14.4.3.2 Suspension Culture

Several researchers have attempted suspension culture of friable date palm callus for rapid embryogenesis (Bhaskaran and Smith 1992; Sharma et al. 1986). Embryogenic callus tissues are cut into as small pieces as possible with a sterile

Fig. 14.5 Date palm micropropagation via somatic embryogenesis. (**a**) Embryogenic callus. (**b**) Development of somatic embryo. (**c**) Germination of somatic embryos. (**d**) Regenerated plantlets. (**e**) Hardened plants in nursery (*Source*: Kumar et al. (2010))

scalpel and then transferred to 50 ml liquid medium in 250 ml Erlenmeyer flasks. The flask contents are filtered through a sieve (500 μm diameter); the filtrate obtained is incubated on a rotary shaker (100 rpm) at 25 °C under the same light conditions. The liquid MS culture medium is diluted to half strength and supplemented with 2,4-D (0.1 mg L^{-1}) and sucrose (3 %). The proembryo masses develop into embryos after passing through several sequential and distinct embryo developmental phases. Hundreds of embryos can be developed from suspension culture within 3 weeks. These embryos are subcultured for 1 month to promote further growth. Approximately 1,000 embryos can be obtained from 200 mg of embryogenic friable callus cultured per vessel (Bhaskaran and Smith 1992). Up to 40 % of these embryos germinated into normal plantlets on solid medium (Fig. 14.5). The somatic embryogenesis method in date palm will help in studies on developmental embryogenesis and could be used for encapsulation of embryos for long-term storage and shipment for export.

14.4.3.3 Genetic Variation in Tissue-Cultured Plants

Random amplified polymorphic DNA (RAPD) and inter-simple sequence repeat (ISSR) markers require only a small sample of DNA without involving radioactive labels and are simpler as well as faster. RAPD has proven to be quite efficient in detecting genetic variations, even in close proximity, such as two nearby isogenic lines. ISSR technique is also very simple, fast, cost effective, highly discriminative, and reliable (Reddy et al. 2002). RAPD and ISSR markers have been successfully applied to detect the genetic similarities or dissimilarities in micropropagated material in various plants (Joshi and Dhavan 2007; Santos et al. 2008; Venkatachalam et al. 2007). Kumar et al. (2010) monitored the genetic stability of long-term (168 weeks) micropropagated plants of date palm using the RAPD and ISSR markers. RAPD and ISSR marker assays were employed to validate the genetic stability of date palm plants multiplied through somatic embryogenesis with up to 42 in vitro subcultures. Out of the 160 RAPD and 21 ISSR primers screened, 13 RAPD and 12 ISSR primers produced a total of 347 (246 RAPDs + 101 ISSR) clear, distinct, and reproducible amplicons, which were monomorphic across all micropropagated plants (27) studied. Thus, a total of 8,592 bands were generated which exhibited homogeneous banding patterns with both RAPD and ISSR markers (Fig. 14.6). The micropropagation protocol developed for in vitro multiplication is appropriate for clonal propagation of date palm and corroborated the fact that somatic embryogenesis is one of the safest modes for production of true-to-type plants (Kumar et al. 2010).

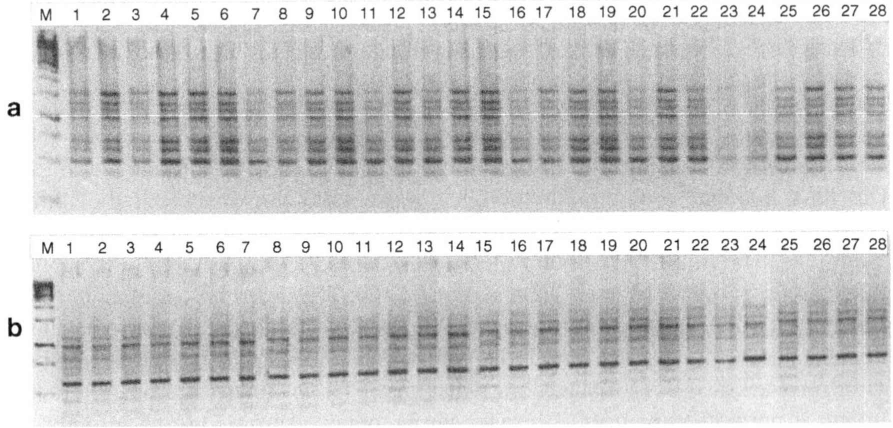

Fig. 14.6 PCR amplification products obtained with (**a**) RAPD primers OPE-15 and (**b**) ISSR primer UBC-835. *Lane M* represents one kb ladder, *lane 1* represents mother plant, *lanes 2–28* represent tissue cultured raised plants at the 42nd cycle of transfer (*Source*: Kumar et al. (2010))

14.4.3.4 Field Establishment of Tissue-Cultured Plants

A method of clonal propagation for date palm through repetitive somatic embryogenesis (RSE) was developed at the CAZRI, Jodhpur. The principle involved multiplication of date palm plantlets through somatic embryogenesis. Free living plants were established in pots and in the field at CAZRI. Tissue-cultured plants raised from explants taken from female mother plants produce female flowers, whereas zygotic seedlings produced male/female flowers with different plant characteristics and flowering behavior. Flowers are pollinated by conventional methods, as fruit setting and development of the fruits are normal. This indicated the true-to-type behavior of plantlets developed from the RSE process, which has now been developed sufficiently for certain cultivars to be highly efficient in raising date palm from tissue culture (Raj Bhansali and Singh 2000, 2003; Raj Bhansali et al. 1988). Comparison of the quality of fruits of Hillali cv. grown from tissue culture and from offshoots revealed no significant differences in either the chemical or physical characteristics of the fruits, clearly indicating that tissue culture techniques are a viable method of date palm propagation (Raj Bhansali 2010).

14.4.4 Research and Commercial Laboratories

India has only a few research laboratories working on date palm tissue culture. These are (1) CIAH, Bikaner; (2) Biotechnology Centre, SKRAU, Bikaner; (3) CCSHAU, Hisar, Haryana; (4) CAZRI, Jodhpur; and (5) AAU, Gujarat. However, none of these laboratories are involved in mass multiplication of planting material and are not functioning as commercial laboratories.

A joint venture company, Atul Rajasthan Date Palms Ltd (ARDP), between the Atul Ltd of Lalbhai Group and the Rajasthan Horticulture Development Society, Government of Rajasthan, has been formed to set up a state-of-the-art tissue culture date palm laboratory at Jodhpur, Rajasthan, based on overseas technology. This laboratory was inaugurated on April 1, 2012. A memorandum of understanding was signed between the UAE University, Al Ain, and Atul to transfer technology of tissue culture of date palm to Atul, since tissue-cultured plants are not available in India. Atul has adopted an all-inclusive approach with major tissue culture date palm suppliers in the Middle East and a link to the Date Palm Research and Development Unit of the UAE University, Al Ain; Al Rajhi Tissue Culture Laboratories, Saudi Arabia and UAE; Al Wathba Marionnet LLC, Al Ain; and Green Coast Nurseries, Fujairah, for sourcing of planting material (Rajmohan 2011). Atul and the Government of Rajasthan have outlined a plan for tissue culture date palm cultivations of 2,000 ha within 5 years. Also, there is a long-term plan to cultivate 100,000 ha in the subsequent 10 years. This is expected to generate more than 15 million tissue-cultured date palm plants in India over the next 10–15 years (Rajmohan et al. 2010).

Fig. 14.7 Date palm demonstration plantation at Jaisalmer, Rajasthan, in an area of 100 ha (*Source*: Rajmohan (2011))

Fig. 14.8 Date palm demonstration plantation at Bikaner, Rajasthan, in an area of 40 ha (*Source*: Rajmohan (2011))

So far, about 100,000 tissue-cultured plants of promising cultivars have been distributed to small and marginal farmers of Rajasthan. Atul has established model demonstration plantations in key locations to demonstrate to and convince farmers of the superior performance of tissue culture raised date palms. A plantation consisting of seven superior cvs. (Barhi, Khadrawy, Khalas, Khunezi, Medjool, Saggai, and Zamli with pollenizers of Madsari and Ghannami) has been established in an area of 100 ha at Jaisalmer, Rajasthan (Fig. 14.7). Another plantation of 40 ha has been established using five cvs. at Bikaner, Rajasthan (Fig. 14.8). The second phase aims to mobilize quality planting materials of superior cultivars, through tissue culture laboratories for the establishment of plantations.

Atul has linked up with internationally reputed firms in the Gulf Region to source tissue-cultured plants. Primary hardened tissue-cultured date palm plants of promising cultivars have been imported from Arab nations and subjected to secondary hardening at Jodhpur, Rajasthan (Fig. 14.9). The third phase targets capacity building for the generation of tissue-cultured date palm plants in India, adopting the best available protocol. The objectives of the fourth phase include large-scale scientific cultivation of date palms in the arid regions of western India. The fifth phase encompasses setting up cooperatives which in turn will do buyback arrangements with farmers to purchase and market date fruit. To support date production, infrastructures for collection of fruit, grading, processing, packing, storage, logistics, branding, distribution, and marketing will be established.

Fig. 14.9 Secondary hardened tissue-cultured plants in the net house of Atul at Jodhpur (*Source*: Rajmohan (2011))

14.5 Cultivar Identification

14.5.1 Molecular Descriptors

RAPD DNA and ISSR markers were used to characterize 8 date palm genotypes grown in the Kutch region of Gujarat. Amplification of genomic DNA of 8 genotypes using 13 RAPD analyses yielded 88 fragments, of which 35 were polymorphic, with an average of 2.69 polymorphic fragments per primer. Two ISSR primers produced 13 bands of which 3 were polymorphic. RAPD markers were more efficient than ISSR assay with regard to polymorphic detection; RAPD markers detected 39.77 % as compared to 23.07 % of ISSR markers. Cluster analysis by unweighted pair group method with arithmetic mean showed that the dendrograms obtained by RAPD and RAPD + ISSR were similar. Cluster A consisted of Early maturing, Ghanshyam, and Late maturing female genotypes with 0.81–0.88 Jaccard's similarity range. Cluster B consisted of Seasonal female, Male-1, Male-2, Male-3, and Male-4 genotypes with 0.82–0.91 similarity range. Genotypes Male-1 and Male-2 were most closely related with the highest value in similarity for Jaccard's coefficient (0.91). Principal coordinate analysis differentiated one group of genotype Male-1 and Male-4 while other genotypes were randomly distributed (Srivastav et al. 2013).

14.5.2 Nutritional Aspects and Utility

Katyal and Dutta (1976) reported Halawy, Khadrawy, Shamran, Medjool, Barhi, and Hayani as the most promising cvs. at Abohar. Halawy was reported as an early, high-yielding cv. and suitable for fresh fruits, dry dates, and soft dates, whereas Khadrawy is suitable for both dry and soft dates. Other suitable cvs. for use in the fresh form were Barhi, and Hayani. Medjool was reported to be appropriate for preparing dry dates of attractive bold size and quality. Similarly, Kalra and Jawanda (1975) from Abohar reported cvs. Halawy, Barhi, and Hayani as raw eating dates. Halawy, Khadrawy, and Shamran cvs. are suitable for soft dates and Medjool for good quality *chhuhara* (dry dates).

The physicochemical characteristics of date fruits were recorded at Abohar by Kalra and Jawanda (1975). They reported the maximum fruit weight in cv. Medjool with an average of 17.62 g and size of 4.20×2.62 cm. Maximum pulp content (92 %) was recorded in cv. Barhi followed by 91.4 % in Medjool and Halawy which were observed to be the best raw eating dates due to low astringency and sufficient sugars.

The performance of date palm cultivars has been evaluated at Bikaner and Jodhpur. Cvs. including Halawy, Khadrawy, Medjool, Barhi, Zahidi, and Shamran were found promising under Bikaner conditions especially Halawy and Sewi (Chandra et al. 1989). At Abohar, cv. Zahidi has been found to be resistant to rain damage and Barhi is more tolerant than Shamran. It was also found that cv. Medjool escapes rain damage because it is late maturing, ripening after the rains. A large number of cvs. and some promising selections (Selection-9, Selection-13, Yaqubi, Kotho, Trofo, Gulachati, Bhugoso, Madhepura, Khedoi-7, Sopari, and Saidy) have been made in India from natural populations in the Kutch region (Muralidharan et al. 2008). Furthermore, these yellow and red fruit color types are suitable for making different processed products (Singh et al. 2012). An elite type of green color and sweet fruit at *doka* stage has been identified from a seedling population in the Kutch region of Gujarat.

The entire Kutch region exhibits great diversity in many phenotypic traits – color, size, shape, and taste of the fruits – and each date palm is a distinct and separate ecotype owing to its seed origin. A study by Muralidharan and Baidiyavadara (2013) revealed the presence of 10 basic colored fruits with 47 color variants in which red and yellow-orange color and its variant shades were prominent. Out of 5 shape groups observed, the majority were of oblong type. Most of the fruits were of medium size (60.5 %) to large (26 %). The TSS value of the fruits was 17–47 %, with the majority of the fruits in the very sweet category (TSS 21–30° B).

14.6 Cultivar Description

Recognizing the potential of date palm cultivations in India, the first main date palm center was established at Abohar by the ICAR in 1955. It was necessitated by the partition of Colonial India and Indian independence in 1947, because the dry districts of southwestern Punjab, namely, Multan, Muzaffargarh, Jhang, and Dera Ghazi Khan, where commercial date orchards existed, became part of Pakistan (Katyal and Dutta 1976). At Abohar, in the period 1956–1962, 49 date cvs. were planted. Of the 49 cvs., 21 were imported from the USA, 17 from Muscat, 5 from Egypt, 3 from Pakistan, and 3 were collected in India. There exists 32 date cvs. which have been successfully established; 20 genotypes were evaluated for growth and fruit parameters (Table 14.3) at Bikaner (Singh et al. 2006)

14.6.1 Growth Requirements

There are three prominent *Phoenix* species grown in India, *P. dactylifera* L., *P. sylvestris* Roxb., and *P. canariensis* Chabaud, the latter only as an ornamental. *Phoenix sylvestris*, called wild date palm, date sugar palm, or *khajoor*, is 10–15 m tall, is indigenous to India, and is cultivated to a large extent for the sap from which a number of goods of monetary value are prepared. *Phoenix* sp. palms have been depicted in carvings from Bharhut and Sanchi and are also seen in the ritual arts pertaining to different nonscriptural vows (Chandra et al. 1992).

It is only the Jaisalmer district in India with a long dry period of 180–185 days and a cumulative heat summation above 3,000 °C which can produce tree-ripened soft dates every year (Chandra et al. 1992; Chundawat 1990; Mertia and Vashishtha 1985; Mertia et al. 1995a, b; Pareek and Sodagar 1986; Singh et al. 2009).

The arid zone of Rajasthan or the Great Indian Thar Desert, popularly known as the Thar, is a vast tract of dry land of about 2.34 million km^2 (Fig. 14.10). The entire tract is distinguished by a low and erratic rainfall, low humidity, high solar radiation, strong dust-raising winds, scanty vegetation, and dry landscape dominated by sand dunes. The climatic features found in the Thar Desert are compatible with the requirements of successful date palm plant production. Climatic data and date palm growth parameters go hand in hand, from planting to production.

In the Thar Desert, a rainless summer occurs only rarely. The June–September monsoon coincides with the fruit-ripening season. This shortens the period of date ripening from 180 to 200 days to 100–170 days (Chandra et al. 1992). Climatic features of potential date palm-growing areas of the Indian arid regions are shown in Table 14.4, and the four zones classified based on the main climatic parameters are shown in Table 14.5.

Cvs. Medjool and Dayan are suitable for dry dates but good quality soft dates are also formed. The early appearance of spathes in the last week of January to the first week of February and dry periods of 180–185 days lead to formation of soft dates on the tree. Attaining the full stage of soft dates on the tree results in maximum loss of weight of 30–35 %, which may be avoided by harvesting fruits a little early at the beginning of the tamar stage and allowing them to ripen in a few weeks without much loss of weight, and they can then be marketed (Mertia et al. 2010).

Chandra and Chaudhary (1990) observed that the requirement of heat summation units (HSU) differs from one cultivar to the other in reaching the color turning and *pind* stage (Table 14.6). To reach the color turning stage from spathe initiation, 1,951 HSU are required in Halawy, as compared to 2,411 HSU in cv. Shamran and

Table 14.3 Growth and fruit parameters of date palm genotypes

Genotype	Tree height (m)	Stem girth (m)	No. of offshoots/ palm	Fruit weight (g)	Fruit length (cm)	Fruit width (cm)	Fruit yield (kg/ plant)	Seed weight (g)
Halawy	3.5	1.4	12	7.85	3.33	2.40	25.2	0.9
Khadrawy	2.9	1.0	12	7.42	3.23	2.15	21.0	0.76
Shamran	2.6	1.0	10	8.30	3.33	2.10	9.0	0.74
Zahidi	3.0	1.0	9	7.25	3.00	2.16	3.0	0.86
Braim	3.0	1.0	11	7.74	3.34	2.10	6.0	1.09
Chip Chap	3.0	1.0	8	8.32	3.56	2.26	1.0	1.25
Sewi	2.6	1.2	6	7.50	2.82	2.00	10.0	0.89
Khuneji	2.8	0.9	6	7.40	2.80	1.70	6.0	0.61
Medjool	2.0	0.8	10	10.5	4.90	3.30	3.0	0.78
Sabiah	2.9	1.0	10	8.60	3.03	1.90	4.7	0.60
Dayari	3.2	1.0	14	9.25	4.10	2.30	15.5	1.14
Muscat	3.3	1.2	16	7.30	3.23	1.57	6.0	0.75
Tayar	2.3	1.0	12	7.00	2.93	1.80	1.6	0.80
Umshok	2.5	0.9	8	6.10	3.10	2.10	6.0	0.76
Hayani	2.3	0.8	7	7.50	3.00	1.67	1.7	0.56
Hamra	2.2	0.9	14	7.42	3.20	1.80	1.6	0.79
Sedami	3.0	1.4	5	7.00	3.00	1.60	4.5	1.10
Medini	2.5	1.1	12	8.50	3.76	2.00	4.0	0.96
Sayer	2.2	0.7	8	5.60	3.00	1.75	13.8	0.70
Bikaner Local	3.0	1.2	15	6.40	2.83	2.10	10.0	0.86

Source: Modified from Singh et al. (2006)

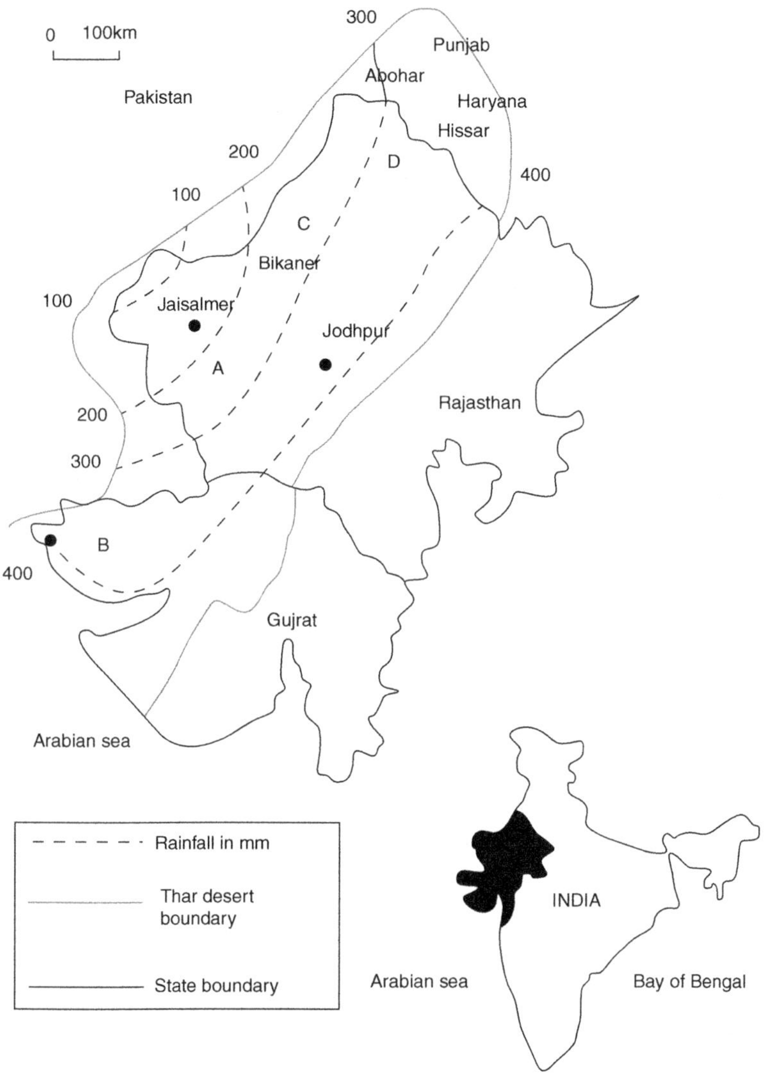

Fig. 14.10 Potential date palm-growing regions in Indian arid zone (*Source*: Raj Bhansali (2010))

2,648 HSU in Medjool. To attain *pind* stage, cv. Ganganagar requires the most HSU (3,843) with the least (3,101 HSU) in Halawy. Due to variations in climate from place to place, the time of maturity as well as the quality of fruit produced in diverse date palm-growing regions differs considerably.

HSU during the period March–August works out at 2,000–2,400 above 18 °C and 3,500–4,000 above 10 °C. On the basis of the base temperature of 10 °C, the HSU requirement for date palm varies in the range of 1,950–3,650 depending upon the cultivar. It is obvious, therefore, that the Thar Desert meets this requirement (Chandra 1990; Chandra et al. 1992).

Table 14.4 Climatic features of potential date palm-growing areas of the Indian arid zone

Region	Potential evapotranspiration (mm)	Heat summation units (>base 18 °C)	Fruiting period (days)	Rainy days	Humidity (%)	Rainfall (mm)	Temperature range (mean) °C
Kutch	1,897	2,900–3,056	165	17	52	350	20–33 (26)
Jaisalmer	2,069	2,900–3,240	145	13	64	215	19–34 (26)
Bikaner	1,772	2,900–3,200	145	19	58	305	18–33 (26)
Jodhpur	1,843	2,900–3,200	145	20	65	350	20–33 (26)
Ganganagar	1,662	2,400–2,900	122	16	56	296	17–33 (25)
Hisar	1,615	2,400–2,513	122	25	64	446	17–33 (25)
Firozpur	1,362	2,400–2,600	122	19	65	300	16–31 (24)

Source: Raj Bhansali (2010)

Table 14.5 On the basis of main climatic parameters, the potential of date palm cultivation in arid zones can be categorized into four zones

Zone	Areas	Edaphoclimate	Harvesting stage
I	Jaisalmer, western parts of Barmer, Bikaner, and Jodhpur district	Extremely dry areas with lowest rainfall and highest heat summation units (HSU)	*Dang* or *pind*
II	Arid coastal areas of Kutch and part of Saurashtra. Divided into 3 subzones		*Doka* or *dang*
	Mundra, Dhrab, and Zarpara	Deep coastal soil with very high water table (1–3 m)	
	Anjar and Khedoi	Deep sandy soil with low water table (10–30 m)	
	Kera, Tera, Vada	High water table (1–2 m)	
III	Jodhpur, Bikaner, eastern part of Barmer, western part of Nagaur, Churu, and Ganganagar	Dry areas with storm-type rains	*Doka* or *dang*
IV	Abohar, Sirsa, Ganganagar, eastern Churu, and western Sikar districts	Dry areas with rains followed by cloudy weather	*Doka*

Table 14.6 Heat summation units required for different date palm cultivars

Cultivars	Heat summation units, base 10 °C	
	Color turning stage	*Pind* stage
Medjool	2,648	3,650
Halawy	1,951	3,101
Khadrawy	2,213	3,342
Shamran	2,411	3,567
Zahidi	2,322	3,479
Khalas	2,357	3,281
Hayani	2,460	3,168
Bint Aisha	2,368	3,650
Sedami	2,264	3,115
Nagal Hillali	2,114	2,985
Zaghloul	2,227	3,631
Sewi	2,244	3,281
Medini	2,500	3,589
Migraf	2,500	3,589
Abdul Rahman	2,116	3,446
Gizaz	2,397	3,412
Suria	2,431	3,446
Hamra	2,444	3,459
Muscat	2,368	3,342
Umshok	2,370	3,151
Barhi	2,323	Not attained

Source: Chandra et al. (1992)

Date cultivars have developed over thousands of years by selection of seedlings, propagating via offshoots only those possessing desirable characteristics. Several date specialists have attempted to list and to describe the cultivars botanically (Kalra and Jawanda 1975; Pundir and Porwal 1998; Srivastva and Dhavan 1981). Prominent date palm cultivars suitable for the arid climatic conditions of India are Halawy, Khadrawy, Shamran, Medjool, Barhi, Zaghloul, Hayani, Zahidi, Khalas, and Sewi. The wild date palm, *Phoenix sylvestris*, which is found growing in almost all states of India, produces inferior quality fruit with little flesh. However, this palm is used for production of jaggery (sugar) and a sweet drink known as *neera*, obtained by tapping the tree for its sap. The salient features of commercial cultivars of date palm in India (Fig. 14.11) are described below:

14.6.1.1 Barhi

This Iraqi cultivar is also suitable for cultivation in Rajasthan State. The trees of this cultivar are semivigorous and medium in height. Fruits are golden yellow at *doka* stage and oval in shape. Average fruit weight is 7.5 g and size 2.9 × 2.3 cm. It has a pulp thickness of 0.7 cm and a TSS of 32 % at *doka* stage. Fruits are nonastringent at *gandora* stage also. This cultivar is most suitable for raw eating dates but not good for soft dates. It is late ripening as fruits reach full *doka* stage by mid-August, and hence, it is moderately damaged by rains and high humidity. Average annual yield potential is about 120 kg *doka* fruits per palm.

Halawy This Iraqi cultivar is one of the most suitable for Rajasthan State. The trees of this cultivar are tall and vigorous in growth. The fruit color is light orange with a yellow shade. The fruit is oblong elliptical with an obtuse apex and an almost horizontal base. Average fruit weight is about 9.5 g and size 3.5 × 2 cm. It has a pulp thickness of 0.6 cm having 31 % TSS at *doka*, and the fruits are sweet at full *doka* stage. The dates may be eaten raw. Fruits have a tendency to shrivel during the ripening stage particularly in case of insufficient irrigation. It is early ripening as the fruit of this cultivar reaches full *doka* stage by the second fortnight of July. Halawy has an average annual yield potential of 100 kg *doka* fruits per palm.

Khadrawy The trees of this Iraqi cultivar are dwarf in growth. Fruits at *doka* stage are greenish yellow in color and oblong in shape with a broad apex and a slightly slanting base. Average fruit weight is 7.7 g and size 3.1 × 2.1 cm. It has a pulp thickness of 0.66 cm having 30 % TSS at *doka* stage. The fruits are highly astringent at full *doka* stage and are badly spoiled due to rains at time of ripening. This cultivar is suitable for both soft dates and preparation of dry dates. It is medium ripening as the fruits reach full *doka* stage by the end of July. Average annual yield potential is 60 kg *doka* fruits per palm.

Khalas The trees are Saudi Arabian cultivar are semivigorous in growth. Fruit is yellow at *doka* and golden brown at *pind* stage. Fruits are oblong in shape. Average fruit weight is 11 g and size 3.2 × 2.4 cm. It has a pulp thickness of 0.83 cm having 28 % TSS. Fruits are sweet at full *doka* stage and suitable for raw eating as well as

Fig. 14.11 Popular commercial date palm cultivars grown in India (*Source*: Courtesy of Dr. R. Porwal, KVK, Ajmer)

for soft dates. It ripens early, reaching full *doka* stage in the second fortnight of July at Bikaner. Average annual yield potential is 75 kg *doka* fruits per palm.

Khunezi This Omani cultivar is emerging as a potential date in India. The palms are semivigorous having a spreading growth habit. Fruit is dark red at *doka* stage with crispy pulp and dark brown at *pind* stage. Fruits are oval shaped with a pointed apex. Average fruit weight is about 10.2 g and size 3.5×2.2 cm. It has a pulp thickness of 0.75 cm having 43 % TSS at full *doka* stage. The fruits are sweet at full *doka* and thus suitable for raw eating but unsuitable for soft dates. The cultivar is also early ripening as fruits reach full *doka* stage by the end of the first fortnight of July. Average annual yield potential is 50 kg *doka* fruits per palm.

Medjool This is a popular Moroccan cultivar with potential for commercial cultivation. The palms are vigorous and tall with an erect growth habit. Fruits are large but variable in size having attractive golden yellow color and are firmer than Barhi or Khadrawy. The irregularity in shape is common and associated with ridges on the seed. Average fruit weight is 15 g and size 3.9×2.8 cm. It has a pulp thickness of 0.8 cm and has 29 % TSS at *doka* stage. Fruits at *doka* stage are highly astringent. It is suitable for dry dates (*chhuhara*) because of its large-sized fruits and good pulp thickness. It is tolerant of rainfall and high-humidity conditions. It is late ripening as fruits of this cultivar reach full *doka* stage in mid-August in Rajasthan. Average annual yield potential is 90 kg *doka* fruits per palm.

Sewi This Egyptian cultivar is vigorous in growth. Trees are tall and erect with moderate canopy. Fruits are yellowish green in color at *doka* stage and oblong in shape. Average fruit weight is 6 g and size 2.9×1.9 cm. It has a pulp thickness of 0.6 cm having 32 % TSS at full *doka* stage. Fruits are sweet at full *doka* stage and suitable for both raw eating and soft dates. It is late ripening as fruits reach full *doka* stage by the first fortnight of August. A fully grown tree of Sewi has an average annual yield potential of 50 kg *doka* fruits per palm.

14.6.1.2 Shamran

This is also known as Sayer or Sayir. Its place of origin is not clear. The date palms of this cultivar are semivigorous. Fruits are yellow in color with a slightly pink-shaded base. Fruits are medium in size and oblong in shape with an obtuse apex. Average fruit weight is 6.5 g and size 3.5×2.18 cm. It has a pulp thickness of 0.56 cm and 30 % TSS at full *doka* stage. Fruits are astringent at *doka* stage. It is suitable for dry and soft dates. Its soft dates have an aroma. It is medium ripening as fruits reach full *doka* by the end of July. It is a prolific bearer and has an average annual yield potential of 100 kg *doka* fruits per palm.

Zaghloul The trees of this Egyptian cultivar are vigorous in growth with an erect growth habit and are promising and suitable for the Thar Desert. Fruits are red in color and large in size. They are oblong, oblique, and asymmetrical in shape. The calyx is small in size and slightly depressed at the base. Fruit bunches are large in size. Average fruit weight is 9.5 g and size 3.9×2.2 cm. It has a pulp thickness of 0.64 cm having 28 % TSS at *doka* stage. Fruits are astringent at *doka*. The *dang* and *pind* fruits are relatively less sweet but have a pleasant taste. It is suitable for soft dates and for preparation of fruit products such as date juice and preserved products. It is medium ripening as fruits of this variety reach full *doka* stage by the beginning of the second fortnight of July. Average yield potential is 120 kg *doka* fruits per palm.

Zahidi The trees of this Iraqi cultivar are semivigorous. Fruits are yellow at *doka* stage with a smooth and hard surface. Fruits are ovate in shape with the broader end toward the apex. Average fruit weight is 10 g and size 3×2.2 cm. It has a pulp thickness of 0.68 cm having 30 % TSS at *doka* stage. Fruits are astringent at *doka* stage. The partial *dang* fruits are eaten as raw dates. It is late ripening as fruits of this cultivar reach full *doka* stage in mid-August. It is a high-yielding cultivar and has an average annual yield potential of 125 kg *doka* fruits per palm.

14.7 Date Production and Marketing

14.7.1 Practical Approaches

Fruit thinning is the most critical cultural practice affecting fruit development, quality, and yield. Excess fruit load may cause shriveling of fruit, breaking of bunch stalks, more damage due to rain and humidity, and delayed ripening. It also reduces the size and quality of fruit. It is therefore necessary to keep the optimum quantity of fruit and thin out the rest (Raj Bhansali 2010). This is usually accomplished by reducing the number of fruit on each bunch and/or by removing some bunches. The number of fruit that a palm can safely carry depends on the cultivar, age, size, and vigor of the palm and the number of green leaves it has. Under normal conditions, 1–2 bunches in the fourth year and 3–4 bunches in the fifth year may be left. Normally 8–10 bunches per palm are retained in India.

In short-stranded cultivars like Khadrawy, the strands are usually cut back to even up the bunch from the top. Most fruit thinning is done by removing 1/2–2/3 of the strands from the center. In long-stranded cultivars like Deglet Noor, 1/3–1/2 of strands are cut in a similar way as in Khadrawy. In addition, strands are also cut back to remove about 1/3 of the flowers. The optimal number of fruits is 1,300–1,600 per palm depending on the cultivar. The thinning percentage is generally 40–50 in Khadrawy, 50–55 in Halawy, and 50–60 in Zahidi and Barhi.

14.7.2 Optimization of Yield

Various recommendations based on Indian research have been made to date palm growers for optimization of yield. The important recommendations are as follows:

(a) Cvs. Halawy, Barhi, Khunezi, and Zahidi are recommended for cultivation for fresh fruits, whereas Medjool is recommended for dry date preparation.
(b) Thinning of fruit bunches after fruit set (1/3 portion from center) helps in early ripening and also improves physical-chemical fruit characteristics.
(c) Preharvest application of Ethrel at 100 ppm at color break stage hastens ripening of fruits and also increases size and weight of date fruits.

(d) Mulching of the basin from the trunk to the canopy area by black polythene sheet or 10 cm-thick layer of the cut local weed *bui* (*Aerva javanica* syn. *persica*) increases soil moisture retention, suppresses weed growth, and improves fruit quality.
(e) Application of 1,500 g nitrogen per palm with 500 g each of phosphorus and potash as basal doses gives maximum yield and growth of fruit. Some 60 % of the N should be applied 1 month before flowering and 40 % at fruit set.
(f) Fruit bunch and leaf ratio should be 1:6 for maximum fruit quality and yield.
(g) Application of 50 % NPK + biofertilizer + FYM per palm per year gives maximum fruit yield, fruit weight, and TSS content.
(h) Foliar application of micronutrient $FeSO_4$ (0.5 %) + thiourea (0.1 %) gives higher fruit yield, fruit weight, and TSS.
(i) Offshoots at transplanting must be treated with Carbendazim 0.1 % + chlorpyrifos 0.1 % + IBA 100 ppm with alternate dry irrigation up to 30 days. This treatment gives better establishment and survival of date offshoots.
(j) Offshoot transplant survival is better when done in spring (February–March) as compared to post-rainy season (September–October).
(k) For integrated management of fruit rots, two sprayings of Carbendazim 0.1 % at 15-day intervals coupled with covering bunches were effective to control disease up to 65 % and minimize yield loss (33 %) as compared to control.
(l) Two sprayings of Carbendazim + Mancozeb (0.2 %) at an interval of 15 days are effective to control the Alternaria leaf spot disease up to 56.56 %.
(m) Spraying of acetamiprid (0.3 g L^{-1}) or imidacloprid (0.3 ml L^{-1}) or dimethoate (2 ml L^{-1}) is effective for the control of scale insect.
(n) For the management of lesser date moth (*Batrachedra* sp.), two sprayings of Decamethrin at 3 ml per 10 l water starting from fruit set and repeated 15 days after first spraying are effective.
(o) For control of scale insect (*Parlatoria* sp.), release of predatory beetle *Chilocorus nigritus* at 10 beetles per palm is effective.
(p) Postharvest storage studies of fresh *doka* fruits revealed that at room temperature, fruits cannot be stored for more than 4 days, whereas they can be stored for 30 days under refrigeration and to 50 days at freezing temperatures.
(q) For preparation of dry dates, full *doka* fruits should be washed and dipped in boiling water for 5–10 min and dried either in an air circulation oven at 48–52 °C for 70–90 h or sun dried for 80–120 h if weather is dry.
(r) Protection from bird damage can be done by covering fruit bunches with wire net gauge (3 × 3 mm mesh) at the start of *doka* stage until harvest.

14.7.3 Harvest and Postharvest Operations

Pareek (1984) recommended that dates should be harvested after 110–170 days compared with 180–220 days or more required for complete ripening. At Hisar, Abohar, Ganganagar, and Churu, fruits must be harvested at *doka* stage and in

Jodhpur-Bikaner area, at late *doka* or *dang* stage. In Jaisalmer-Barmer only, fruits are to be harvested at *dang* or early *pind* stage. Harvesting of date fruits at *doka* stage is done by cutting the entire bunch and removing fruits from the bunch. This should be done at a stage when more than 80 % of fruits in a bunch attain desirable quality. When fruits are harvested at *dang* stage for soft dates, only those fruits which have attained 1/3–1/2 *dang* are picked and thus the harvesting is done of selected fruits. For dry date making, fruits are to be harvested at full *doka* stage (Chandra et al. 1992).

Postharvest losses in date palm fruits in India are as high as 32–40 % due to heavy rains during fruit maturity. Postharvest losses can be minimized to a great extent by harvesting at the optimal stage of maturity. Determining the maturity standards of different cultivars is therefore important for proper management, handling, harvesting, drying, packaging, and storage.

Storage studies of fresh *doka* fruits revealed that at room temperature, fruits could not be stored for more than 4 days, whereas these could be stored for up to 3 weeks under refrigeration and up to 50 days at freezing temperatures. All fruits of cv. Halawy turned to *dang* stage at freezing, whereas only 75 % *doka* fruits turned *dang* under refrigeration (Manohar and Chandra 1995).

Processing involves curing of dates for preparation of *chhuhara* or soft dates (*pind khajoor*). For preparation of *pind*, fully ripened date fruits of *doka* stage are taken and dipped in boiling water for a fixed period which varies with cultivar and then dried in an air circulation oven. Katyal and Dutta (1976) reported the dipping of fruits in boiling water for about 5 min and then drying them either in an air circulation oven at 48 °C or in the sun. Cvs. Halawy, Khadrawy, Medjool, and Thoory are suitable for making dry dates. Good quality *chhuhara* is made by dipping *doka* fruits of Medjool in boiling water for 10 min and drying them at 50 °C for about 120 h in a temperature-controlled air circulation oven (Kalra 1976). However, for cvs. Halawy, Khadrawy, and Shamran, which have smaller fruits, the required boiling time is 6 min and subsequent drying for about 80 h; 5 min boiling for Medjool and 10 min for Khadrawy and Zahidi proved to be the best at Abohar (Gupta and Thatai 1983). The best quality *chhuhara* is prepared from *doka* fruits of Khadrawy by dehydration in an oven at 50 °C, after 5 min boiling water treatment, which gave a recovery of 33.3 % and organoleptic rating of 33 out of 36 followed by Shamran which gives a recovery of 35.3 % and organoleptic rating of 28.8 (Godara and Pareek 1985). Kulkarni et al. (2008) evaluated the processing and dehydration conditions for the preparation of dehydrated dates from immature date fruits. Processing by blanching in water at 96 ± 1 °C and subsequent dehydration at 60 ± 2 °C for 18–20 h resulted in good quality dehydrated dates, as compared to dates dried without heat treatment. Dehydrated dates were found to be acceptable in color, flavor, taste, and overall quality. Dehydrated dates contained total sugars of 520 g/kg, reducing sugars of 415.1 g/kg; tannins, 13.5 g/kg; and ascorbic acid, 33.7 mg/kg. Equilibrium relative humidity of the dehydrated dates was found to be 75.9 % with an initial moisture content of 159 g/kg. Dehydrated dates packed in 75 μ low-density polyethylene packages were shelf stable for 6 months at room temperature (Kulkarni et al. 2008).

14.7.4 Current Import and Export

India is the world's largest importer of dates. In 2011, total date imports were 244,367 mt with a value of USD 52,786,000. Import quantity has been more or less constant since 2001 and it is around 250,000 mt (Table 14.7). India imports from more than 22 countries but 10 (Pakistan, Iraq, Iran, the UAE, Oman, Saudi Arabia, "Afghanistan," Tunisia, Algeria, and "Indonesia") contributed more than 90 %. Pakistan was the most important (Table 14.8). In comparison to European countries, Indian imports were of inferior quality.

Table 14.7 Import of dates in India since 2001

Year	Quantity (mt)	Value (thousand USD)
2001	256,295	141,713
2002	193,467	95,042
2003	298,423	99,059
2004	230,26	81,318
2005	253,341	72,359
2006	286,317	74,760
2007	240,399	55,753
2008	247,875	46,407
2009	193,755	33,011
2010	171,523	27,798
2011	244,367	52,786

Source: FAOSTAT Database http://faostat.fao.org/

Table 14.8 Import of fresh and dried dates in India in 2012–2013 from the top ten countries

Rank	Country	Quantity (thousand kg)	Value (million USD)
1	Pakistan	115,758.31	85.67
2	Iraq	43,137.34	54.78
3	Iran	27,257.71	12.74
4	UAE	12,857.59	8.23
5	Oman	3,052.60	3.12
6	Saudi Arabia	2,331.15	1.95
7	Afghanistan[a]	2,063.07	1.47
8	Tunisia	1,447.52	1.39
9	Algeria	589.00	0.46
10	Indonesia[b]	412.89	0.14

Source: Ministry of Commerce, India

[a]Presumed to originate from neighboring Pakistan since Afghanistan has no commercial date production

[b]Presumed to originate from a third country since Indonesia has no commercial date production

14.8 Processing and Novel Products

14.8.1 Industrial Processing Activities

Dates are highly delicious and a favorite fruit of the Indian people and are consumed in various ways. Fruits of cvs. Halawy and Barhi are eaten as fresh table dates at the hard ripe stage (*doka* stage), when they have low tannin content; tannin is responsible for the astringent taste in unripe dates. Dates are cured as dry dates (*chhuhara*) and soft dates (*pind khajoor*). Both dry and soft dates are in great demand and find a ready market in India. Therefore, in the event of higher production of dates, unlike other fruits, no marketing problems are anticipated. Date fruits have religious significance as well, eaten during Ramadan by Muslims. Other parts of date palm trees are also used. Date palm leaves are used to make baskets and brooms and also provide material for furniture and thus promote cottage industries providing considerable employment opportunities. Besides, it provides sugar of excellent quality and wine. Date palm wine in India is known as *neera*. Date palms are very valuable as a canopy of ground crops in hot and dry regions; vegetables, field crops, fodder crops, and small fruit trees such as pomegranate can be grown successfully as they are protected from hot winds and sun. Its trunk is used as timber.

Date fruits have unique nutritive value. These are rich in sugars and provide a source of energy. Easily digestible sugars amount to 75–80 % on a dry weight basis. Jawanda et al. (1972) estimated the total sugar in fruits of cv. Khadrawy as 63.63, 84.61, and 77.50 %, at early *doka*, late *doka*, and early *dang* stages, respectively. In addition, dates contain minerals like iron, potassium, calcium, and phosphorus. They are also a good source of vitamins such as A, B_1, B_2, and B_6. Protein pulp content is 1.75–2.75 % and fat content 0.31–1.9 % (Katyal and Dutta 1976).

Dates are ideal fruits to substitute for added sugar in foods, and they play an important role in the diet of many people in arid regions (Jain 2012). Manickavasagan et al. (2013) tried to develop acceptable *idli* (traditional Indian breakfast) with chopped dates, date paste, and date syrup. Total phenol and vitamin C contents of *idli* dates were significantly higher than control *idli* with added sugar. The sensory properties of four *idli* products (with date paste, with date syrup, with chopped dates, and a control served with white sugar) were evaluated. The sweetness and aroma of the *idli* with chopped dates got a significantly higher score than the other three *idli* products with no difference among them (Manickavasagan et al. 2013).

14.8.2 Local Uses of Date Palm

Date palm sap is drunk fresh, boiled down into sugar, or allowed to naturally ferment into a mild alcoholic beverage, toddy. Leaves are used for mats and botanical ties, and young leaves are cut for basket weaving. Dried leaves with their stiff, woody rachis are used as fencing. Old leaves are used as thatch. Leaves with a minimum of 6–8 years of age are cut and sun dried for a week, bundled, and stored for future use. The leaf is cut into different shapes as required, colored, and woven to give the final structure. Various artisanal craft items are made from the date palm plant (Fig. 14.12a–d). The mats after use by humans are used for cowsheds. The whole leaf or the leaf rachis is one of the preferred fencing materials in the areas where date palms abound. The leaves can also consumed by cattle after being ground (Fig. 14.12e). The wood is used in the construction of indigenous huts or for water pipes. The trunks are strong and resistant to termites and provide valuable construction timber (Panda et al. 2014).

14.8.3 Uses in Ayurvedic Medicine

Since antiquity, where date palms were grown, both fruits and seeds have been used in traditional and folk systems of medicine (Khare 2007). It is now considered a valuable source of unique natural products for the development of medicines and used against various diseases. Fruits contain high amounts of tannins and are used in the treatment of different maladies such as hemorrhage, dyspnea (shortness of breath), cough, burning sensation, syncope (fainting), injury, and tuberculosis. At present date fruit is used in Ayurvedic formulations for the treatment of various diseases. It is believed to possess unique pharmacological actions such as antibacterial, anti-inflammatory, antidiabetic, antiasthmatic, nephroprotective (kidney health), hepatoprotective (liver health), and aphrodisiac activities (Ateeq et al. 2013).

Regular consumption of date palm pollen and the male flowers is believed to act as an aphrodisiac and to enhance fertility (Khare 2007). Fruit pulp is regarded to be antitussive (cough medicine), expectorant (mucus clearing), demulcent (soothing mucus membranes), laxative, diuretic, and restorative (Khare 2007). Consumption of dates is believed to be strengthening for the body, to prevent premature graying of hair, to retard wrinkle formation, and to give the skin a lustrous healthy look. The date pulp is boiled in milk until soft and used as a tonic, especially for pregnant and lactating mothers (Puri et al. 2000). Dried fruits are pounded and mixed with almonds, quince seed, pistachio nuts, spices, and sugar. The mixture is highly nutritious and given to pregnant women and new mothers (Chandra et al. 1992; Puri et al. 2000). Date seeds are reported to have antiaging properties and to reduce wrinkling of the skin in women (Bauza 2002). Regular consumption of date fruits in the human diet is said to be beneficial in ameliorating cough, rheumatism, burning sensation, nephropathy (kidney disease), gastropathy (stomach problems), bronchitis, and sexual debility (Selvam 2008). It is purportedly good for the heart and is prescribed for respiratory diseases, asthma, chest pains, fevers, high blood pressure, and fatigue (Khare 2007; Selvam 2008).

Immune system activation is an effective and protective approach against infectious diseases. Immunostimulants enhance overall immunity of the host and present

Fig. 14.12 Finished products from date palm: (**a**) brooms, (**b**) brooms, (**c**) mats used for cowsheds, (**d**) mats for domestic purpose, and (**e**) leaf grinding for cattle feed (*Source*: Panda et al. (2014) **a**–**d**, Newton et al. (2013) **e**)

a nonspecific immune response against microbial pathogens. They also heighten humoral and cellular-mediated immune responses, by enhancing cytokine secretion or by directly stimulating B or T lymphocytes. Ingestion of phytochemicals to support the immune system or to combat infections has been a traditional practice. Feeding of ethanol extract of dry dates to parturated mice enhanced both cell-mediated and humoral immunity (Puri et al. 2000).

14.8.4 Bioenergy

An expanding world population, increasing energy demand, depleting reserves of fossil fuels, and increasing effects of pollution from burning these fuels demand more eco-friendly alternatives which can substitute for fossil fuels. Ethanol derived from biomass has the potential to be a substitute for fossil fuel which is renewable, nontoxic, biodegradable, and more eco-friendly. The three major classes of feedstocks for ethanol production are sugars, starches, and lignocelluloses (Gupta and Kushwaha 2011). The principal constituent of the date fruit is sugar and total sugar content at harvest is 70–80 %. High sugar content also is present in date palm sap, as in all palm saps, and could be used as a good source of fermentative microorganisms. In almost all tropical locations in Asia, palm sap is obtained from the decapitated inflorescence of various palm species. There is an art in binding the flower spathes, pounding them to cause the sap to flow properly by cutting the spathe tip and collecting the sap into the earthen pitchers which contain yeast and bacteria from the leftover toddy of the previous lots. Fermentation begins as soon as the sap flows into the pitcher.

Palm wine is either consumed fresh as it is brought down from the tree or fermented up to 24 h; sap is generally a dirty brown sweet liquid having 10–18 % w/w sugar, a pH of 7–7.4, and traces of ethanol, which after fermentation results in formation of a product containing as much as 9 % (by volume, v/v) ethanol and a pH of 4–5.5 (Joshi et al. 1999).

Palm wine is produced by alcoholic-lactic-acetic acid fermentation, involving mainly yeasts and lactic acid bacteria. In the fermenting sap, *Saccharomyces cerevisiae* is invariably present but also lactic acid bacteria such as *Lactobacillus plantarum, L. mesenteroides*, or other bacteria like *Zymomonas mobilis* and *Acetobacter* sp. Other yeasts include *Schizosaccharomyces pombe*, *Saccharomyces cerevisiae*, *S. chevalieri*, *S. exiguus*, *Candida* sp., *Saccharomycodes ludwigii*, *S. pombe*, *Kodamaea ohmeri*, and *Hanseniaspora occidentalis*, which are characterized as maximum ethanol producers in toddy (Joshi et al. 1999).

Palm wine is a good source of B vitamins. Various indigenous strains of *Saccharomyces* sp. were isolated from date palm sap and were evaluated for alcohol dehydrogenase (ADH) enzyme activity, ethanol production, and alcohol tolerance limits. Alcohol content in juice samples fermented with different yeast strains varied considerably (8.9–12.5 %, v/v). Yeast culture showed varied in vitro ethanol tolerance (3–12 %). Isolate SCP-1 was found superior showing 12.5 % ethanol production, high ADH enzyme activity (4.38 units/ml), and higher alcohol tolerance, maintaining cell viability at 12 % ethanol in yeast extract peptone dextrose (YPD) medium up to 48 h (Gupta et al. 2009).

14.9 Conclusions and Recommendations

Given the foregoing retrospective and future prospects of date palm cultivation in India, particularly with respect to growing areas, climate, cultivars, and tissue-cultured plant availability, the following conclusions and recommendations can be drawn:

(a) The Thar Desert is the most suitable region for date palm cultivation. New irrigation works make it possible to grow dates in western Rajasthan and to expand cultivation in the Kutch region of Gujarat. Public-private partnerships in Rajasthan and Gujarat states have been successful in establishing tissue-cultured plants in the field, and 405 ha have been planted, mainly of cv. Barhi. The Gujarat State, Kutch Crop Services Ltd., and Excel Crop Care Ltd. in Bhuj are playing a vital role and such efforts should be continued and expanded.
(b) The potential of Indian states not growing date palms, particularly Tamil Nadu and Andhra Pradesh, should be explored and exploited.
(c) Cultivar Barhi has a huge market potential in Gujarat and on the world market.
(d) In Rajasthan, Halawy cultivar has the best suitability and should be promoted.
(e) Local germplasm field banks should be established for conservation and use.
(f) Despite efforts by Indian and international agencies in recent decades to introduce new cultivars and cultivation practices, India still lacks sufficient planting materials of desirable cultivars. Traditional propagation by offshoots has the potential to be improved and expanded as a method for farmers to expand date production if tissue culture plantlets cannot be afforded or are not available. Famers also need to be encouraged to develop their own seedling date cultivars which have a potential for commercial fruit production.
(g) Protocols for date palm tissue culture were developed at four centers in India, namely, CCSHAU, Hisar; AAU, Anand; RAU, Bikaner; and CIAH, Bikaner. These were, however, for different cultivars, and these technologies have not been tested for additional cultivars and large-scale multiplication.
(h) Adoption of the tissue culture protocols from other countries may not be feasible under prevailing international property rights; therefore, Indian protocols focused on the primary cultivars are needed to produce tissue culture plantlets and avoid their purchase at high prices from international sources. This would reduce the drain of foreign exchange and eliminate subsidies such as those provided by the Government of Rajasthan to farmers to purchase imported tissue-cultured plants. The goal should be to make tissue-cultured plants available to a larger group of farmers.

(i) Certain semi-wild seedling date palms in the Kutch region produce unique quality fruits which if multiplied would enable India to have its own unique cultivars. This can be done only with specific protocols to involve farmers.
(j) There is an urgent need for a concerted effort and commitment to provide the required inputs of infrastructure and manpower.
(k) A collaborative approach is recommended among the four Indian laboratories and should be exploited to identify, test, and refine the tissue culture protocols for commercial application at affordable cost in a time frame of 3–4 years.
(l) Training of technicians and farmers is essential, particularly in Rajasthan, in nursery care of offshoots and tissue culture plantlets to reduce the risk of losing the costly plant material in the absence of proper care.
(m) Indian research is deficient in postharvest management, processing, novel product development, and fruit marketing. There is an urgent need for a research center for date palm postharvest and supply chain management.

References

Aboel-Nil M (1986) Proceedings of second symposium on date palm. King Faisal University, Al Hassa, pp 29–40

Anonymous (2011) Production technology of date palm in Rajasthan. Date Palm Research Centre, Swami Keshwanand Rajasthan Agricultural University, Bikaner

Ateeq A, Sunil SD, Varun SK, Santosh SK (2013) *Phoenix dactylifera* Linn (*Pind Kharjura*): a review. Int J Res Ayurv Phar 4:447–451

Bajwa BS, Bakhshi JC (1961) The date palm cultivation in India. ICAR Farm Bull 63

Bakhshi JC, Singh KK (1972) Experiences with date palm cultivation in arid irrigated regions of Punjab. Punjab Hort J 21:142–146

Bauza E (2002) Date palm kernel extract exhibits antiaging properties and significantly reduces skin wrinkles. Int J Tiss Reac 24:31–136

Bhargava SC, Saxena SN, Sharma R (2003) In vitro multiplication of *Phoenix dactylifera* (L). J Plant Biochem Biotech 12:43–47

Bhaskaran S, Smith RH (1992) Somatic embryogenesis from shoot tip and immature inflorescence of *Phoenix dactylifera* cv. Barhee. Plant Cell Rep 12:22–25

Bonavia E (1885) The future of the date palm in India (*Phoenix dactylifera*). Thacker, Spink and Co., Calcutta

Chandra A, Chaudhary NL (1990) Performance of date palm cultivars in Thar Desert. Part II. Curr Agric 14:1–4

Chandra A, Swaminathan R, Chaudhary NL et al (1989) Extended abstracts. Part II. Int Sym Manag Sandy Soils, CAZRI, Jodhpur, pp 708–709

Chandra A, Chandra A, Gupta IC (1992) Date palm research in Thar Desert. Scientific Publishers, Jodhpur

Chandra A, Swaminathan R, Chaudhary NL et al (1994) A note on the performance of date palm cultivars in the Thar Desert. Indian J Hort 41:28–33

Chowdhury MSH, Halim MA, Muhammed N et al (2008) Traditional utilization of wild date palm (*Phoenix sylvestris*) in rural bangladesh: an approach to sustainable biodiversity management. J For Res 19:245–251

Chundawat BS (1990) Arid fruit culture. Oxford & IBH, New Delhi

Dass HC, Kaul RK, Joshi SP, Raj Bhansali R (1989) In vitro propagation of *Phoenix dactylifera* L. Curr Sci 58:22–24

Dhandar DG, Singh RS (2004) Germplasm collections of arid zone fruits in India. All India Coordinated Research Project on Arid Zone Fruits. Central Institute for Arid Horticulture, Bikaner

Godara RK, Pareek OP (1985) Effects of dehydration methods and boiling water dip treatments on the recovery and quality of chuhhara from doka fruits of Halawy, Khadrawy and Shamran date palm cultivars. Indian J Hort 42:199–205

Gupta N, Kushwaha H (2011) Date palm as a source of bioethanol producing microorganisms. In: Jain SM, Al-Khayri JM, Johnson DV (eds) Date palm biotechnology. Springer, Dordrecht, pp 711–727

Gupta PC, Mehta N (1985) Recommendations regarding various diseases of arid zone fruit crops from Hisar centre. No. 17, AICRP Arid Zone Fruits, HAU, Hisar

Gupta MR, Thatai SK (1980) Storage of date pollen. Punjab Hort J 20:211–214

Gupta MR, Thatai SK (1983) Curing of dates. Haryana J Hort Sci 12:189–193

Gupta N, Dubey A, Tewari L (2009) High efficiency alcohol tolerant *Saccharomyces* isolates of *Phoenix dactylifera* for bioconversion of sugarcane juice into bioethanol. J Sci Ind Res 68:401–405

Jain SM (2012) Date palm biotechnology: current status and perspective – an overview. Emir J Food Agric 24:386–399

Jawanda JS (1964) Date palm research in Punjab. Punjab Hort J 4:1–9

Jawanda JS, Kalra SK (1972) Fruit characters and quality assessment of some promising date varieties at Abohar. Punjab Hort J 12:33–38

Jawanda JS, Munshi SK, Pal RN (1972) Nutritive value of dates. Punjab Hort J 12:54–57

Joshi P, Dhavan V (2007) Assessment of genetic fidelity of micropropagated *Swertia chirayita* plantlets by ISSR marker assay. Biol Plant 51:22–26

Joshi VK, Sandhu DK, Thakur NS (1999) Fruit based alcoholic beverages. In: Joshi VK, Pandey A (eds) Biotechnology: food fermentation. Educational Publishers, Ernakulam, pp 647–744

Kalra SK (1976) Date palm research in India. Ind Farm 26:11–12

Kalra SK, Jawanda JS (1975) Identification and description of date varieties from Egypt. Punjab Hort J 26:11–12

Katyal SL, Dutta CP (1976) Scope and problems of date industry in India. Ind Farm 26:3–5

Khare CP (2007) Indian medicinal plants: an illustrates dictionary. Springer, Heidelberg

Kulkarni SG, Vijayanad P, Aksha M et al (2008) Effect of dehydration on the quality and storage stability of immature dates (*Phoenix dactylifera*). LWT Food Sci Tech 41:278–283

Kumar N, Modi AR, Singh AS et al (2010) Assessment of genetic fidelity of micropropagated date palm (*Phoenix dactylifera* L.) plants by RAPD and ISSR markers assay. Phys Mol Biol Plants 16:207–213

Manickavasagan A, Mathew TA, Al-Attabi ZH, Al-Zakwani IM (2013) Dates as a substitute for added sugar in traditional foods – a case study with idli. Emir J Food Agric 25:899–906

Manohar MS, Chandra N (1995) Date palm culture in Rajasthan. Directorate of Research Rajasthan Agriculture University, Bikaner

Mertia RS, Vashishtha BB (1985) A note on performance of date palm cultivar Halawy at Chandan (Jaisalmer). Ann Arid Zone 24:263–264

Mertia RS, Singh HP, Dass HC, Panwar HS (1995a) For the field planted date palm suckers how to conserve moisture in desert soils. Indian Hort 39:16–17

Mertia RS, Singh HP, Panwar HS, Dass HC (1995b) Moisture efficient techniques for the establishment of date palm. Agric Env Int 47:30–32

Mertia RS, Birbal, Kumawat RN (2010) Performance and field management of six prominent cultivars of date palm (*Phoenix dactylifera*) in extreme arid regions of Thar desert, India. Acta Hort 882:65–68

More TA, Singh RS (2008) Conserving biodiversity in different areas. The Hindu Survey of Indian Agriculture, Hyderabad, pp 50–54

Munier P (1973) Le palmier dattier. Maisonneuve & Larose, Paris

Muralidharan CM, Baidiyavadara DA (2013) Variability and diversity of elite date palm *Phoenix dactylifera* L. in date groves of Kachchh (Gujarat), India. Acta Hort 994:263–270

Muralidharan CM, Tikka SBS, Verma P (2008) Date palm cultivation in Kachchh. Technical bulletin no. 2, SDAU Date Palm Centre, Mundra

Newton C, Gros-Balthazaed M, Ivorra S, Paradia L, Pintaud J, Terral J (2013) *Phoenix dactylifera* and *P. sylvestris* in northwestern India: a glimpse on their complex relationships. Palms 57:37–50

Pal V, Rathore GS, Godara SL (2006) Screening of genotypes of date palm against *Alternaria* leaf spot. Ind J Arid Hort 1:62–63

Pal V, Rathore GS, Godara SL (2007) Physiological studies on mycelia growth and sporulation causing *Alternaria* leaf spot of date palm. Ind J Arid Hort 2:71–72

Panda T, Mishra N, Mohanty RR (2014) Indigenous knowledge and utilization of Khajuri plant (*Phoenix sylvestris*) in Bhadrak district, Odisha, India. Res Plant Biol 4:12–19

Pareek OP (1984) Date palm growing potential of Indian arid zone. Ind Hort 29(2–5):8

Pareek OP, Sodagar NN (1986) Date palm grooves in Kachchh. Ind Hort 31:21–27

Parmar C, Kaushal MK (1982) Phoenix sylvestris. In: Parmar C, Kaushal MK (eds) Wild fruits of the Sub-Himalayan region. Kalyani Publishers, New Delhi

Pattnayak PK, Misra MK (1982) Energetic and economics of traditional gur preparation: a case study in Ganjam district of Orissa. Ind Biom Bioen 26:79–88

Pintaud JC, Zehdi S, Couvreur TLP et al (2010) Species delimitation in the genus *Phoenix* (Arecaceae) based on SSR markers with emphasis on the identity of the date palm (*Phoenix dactylifera* L.). In: Sebreg O, Petersen G, Barfod AS, Davis JI (eds) Diversity, phylogeny and evolution in the monocotyledons. Aarhus University Press, Denmark, pp 267–286

Pundir JPS, Porwal R (1998) Performance of different date palm cultivars under hyper arid supplementary irrigated western plains of Rajasthan (India). In: Proceeding of first international conference on date palms, Al-Ain, 8–10 Mar 1998, pp 329–336

Pundir JPS, Porwal R, Verma IM (2004) Effect of ferrous sulphate and thiourea spray on date fruits cv. Medjool. Ann Agri-Bio Res 9:47–52

Puri A, Sahai R, Singh KL et al (2000) Immunostimulant activity of dry fruits and plant materials used in Indian traditional medical system for mothers after childbirth and invalids. J Ethnopharm 71:89–92

Raj Bhansali R (1987) Control of *Graphiola* leaf spot of date palm and screening of cultivars against diseases. Technical document 20, AICRP on Arid Zone Fruits, HAU, Hisar

Raj Bhansali R (2010) Date palm cultivation in the changing scenario of Indian arid zones: challenges and prospects. In: Ramawat KG (ed) Desert plants. Spinger, Berlin, pp 423–458

Raj Bhansali R, Kaul RK (1991) Into future-date through tissue culture. Ind Hort 36:6–10

Raj Bhansali R, Singh M (2000) Somatic embryogenesis in fruit and forest trees of arid zone. In: Jain SM, Gupta PK, Newton RJ (eds) Somatic embryogenesis in woody plants, vol 6. Kluwer, Dordrecht, pp 141–168

Raj Bhansali R, Singh M (2003) Micropropagation of arid zone fruit trees of India. In: Jain SM, Katusaki I (eds) Micro propagation of woody trees and fruits. Kluwer, Dordrecht, pp 381–432

Raj Bhansali R, Kaul RK, Dass HC (1988) Mass cloning of date palm plantlets through repetitive somatic embryogenesis. J Plant Ant Morph 5:73–79

Rajmohan K (2011) Date palm tissue culture: a pathway to rural development. In: Jain SM, Al-Khayri JM, Johnson DV (eds) Date palm biotechnology. Spinger, Berlin, pp 29–45

Rajmohan K, Mohanan BN, Batra AM (2010) Date palm development mission of Atul Ltd. in India. Acta Hort 882:289–292

Randhawa GS (1980) Date palm. In: Handbook of agriculture. Indian Council of Agricultural Research, New Delhi, pp 1075–1076

Reddy MP, Sarla N, Siddiq EA (2002) Inter simple sequence repeat (ISSR) polymorphism and its application in plant breeding. Euphy 128:9–17

Santos MDM, Buso GCS, Torres AC (2008) Evaluation of genetic variability in micropropagated propagules of ornamental pineapple [*Ananas comosus* var. *bracteatus* (Lindley) Coppens and Leal] using RAPD markers. Gen Mol Res 7:1097–1105

Selvam ABD (2008) Inventory of vegetable crude drug samples housed in botanical survey of India, Howrah. Pharmac Rev 2:61–94

Shandilya A (2012) Kutch's date with prosperity. The Gujarat 2:21
Sharma DR, Kumari R, Chawdhuri JB (1980) In vitro culture of female date palm (*Phoenix dactylifera L.*) tissue. Euphy 29:169–174
Sharma DR, Dawara S, Chowdhuy JB (1984) Somatic embryogenesis and plant regeneration date palm (*Phoenix dactylifera* L.) cv. Khadrawy through tissue culture. Ind J Exp Biol 22:763–766
Sharma DR, Deepak S, Chowdhuri JB (1986) Regeneration of plantlets from tissues of date palm (*Phoenix dactylifera* L). Ind J Exp Biol 24:763–766
Singh RS, Bhargava R, Dhandar DG (2006) Evaluation of date palm germplasm under hot arid ecosystem. Ind J Plant Gen Res 19:91–95
Singh RS, Bhargava R, Vashishtha BB, Ramdevputra MV (2009) Genetic diversity in date palm (*Phoenix dactylifera*). Curr Agric 33:67–72
Singh RS, Krishna H, Bhargava R (2012) Conservation and management of plant genetic resources of arid fruits: a review. Ind J Arid Hort 7:1–11
Srivastav VS, Kapadia CV, Mahatma MK et al (2013) Genetic diversity analysis of date palm (*Phoenix dactylifera* L.) in the Kutch region of India using RAPD and ISSR markers. Emir J Food Agric 25:907–915
Srivastva HC, Dhavan S (1981) Performance of some date varieties in Haryana. Agric Sci Digest 1:76–78
Sudhersan C, Abo El-Nil M (2004) Axillary shoot production in micropropagated date palm (*Phoenix dactylifera*). Curr Sci 86:771–773
Sudhersan C, Abo El-Nil M, Al-Baiz A (1993) Occurrence of direct somatic embryogenesis on the swords leaf of in vitro plantlets of *Phoenix dactylifera* L cultivar Barhee. Curr Sci 65:887–888
Tisserat B (1981) Production of free living date palms through tissue culture. Date Palm J 1:43–54
Vashishtha BB (1987) Evaluation of date palm germplasm. In: Technical document no. 20, AICRP on Arid Zone Fruits, HAU, Hisar
Venkatachalam L, Sreedha RV, Bhagyalakshmi N (2007) Micropropagation in banana using high levels of cytokinins does not involve any genetic changes as revealed by RAPD and ISSR markers. Plant Grow Reg 51:193–205
Yadav NR, Singh J, Yadav RC et al (1998) Somatic embryogenesis and plant regeneration from cell suspension cultures of *Phoenix dactylifera* L. cv. Khadrawy. Phys Plant 36:7–20

Part II
Europe

Chapter 15
Date Palm Status and Perspective in Spain

Diego Rivera, Concepción Obón, Francisco Alcaraz, Encarna Carreño, Emilio Laguna, Asunción Amorós, Dennis V. Johnson, Gisela Díaz, and Asunción Morte

Abstract The date palm (*Phoenix dactylifera* L.) is an archaeophyte in Spain and is the iconic species of an introduced mixed irrigated agricultural system of food crops, animal feed, grasses, and livestock. Traditionally, date palms are propagated from seed and grown for their fruits, as well as for products derived from the leaves and stem wood. The seedling date palm populations in Spain represent a diverse pool of genetic resources with potential to improve the crop in general. Depending on the local microclimate, 20–50 % of date fruits in Spain do not fully ripen on the tree because of cooler temperatures late in the growing season. Artificial ripening techniques have been devised to deal with this situation. Local genotypes are under study for the best fruit types for expanded marketing, as a gourmet fresh fruit product in Europe, to enhance farmer income and to sustain the traditional production system. Tissue culture of selected genotypes is contributing to sustainability of the palm groves. New commercial date products are under study to broaden economic possibilities. The exotic red palm weevil pest represents the main threat to the palm grove and control measures are being pursued. Historic palm groves persist in Spain, especially in Elche and Orihuela, as well as presumably naturalized feral palms exhibiting distinctive morphological characteristics described as

D. Rivera (✉) • F. Alcaraz • E. Carreño • A. Morte
Department of Plant Biology, Faculty of Biology, University of Murcia, Murcia 30100, Spain
e-mail: drivera@um.es; falcaraz@um.es; ecarreno@um.es; amorte@um.es

C. Obón • A. Amorós
Department of Applied Biology, Escuela Politécnica Superior de Orihuela, Miguel Hernández University of Elche, Orihuela 03312, Spain
e-mail: cobon@umh.es; aamoros@umh.es

E. Laguna
Centro para la Investigación y Experimentación Forestal – Servicio de Vida Silvestre, Generalitat Valenciana, Quart de Poblet, Valencia 46930, Spain
e-mail: laguna_emi@gva.es

D.V. Johnson
Consultant, 3726 Middlebrook Ave, Cincinnati, OH 45208, USA
e-mail: djohn37@aol.com

G. Díaz
Department of Applied Biology, Torreblanca, Miguel Hernández University of Elche, 03202 Elche, Spain
e-mail: gdiaz@umh.es

© Springer Science+Business Media Dordrecht 2015
J.M. Al-Khayri et al. (eds.), *Date Palm Genetic Resources and Utilization: Volume 2: Asia and Europe*, DOI 10.1007/978-94-017-9707-8_15

P. iberica. *Phoenix* palms are protected and continue to provide commercial fruits and palm leaves for religious ceremonies and to enhance the local landscape. Elche, the largest palm grove, is a World Heritage Site and a major tourist attraction.

Keywords Artificial ripening • Elche • Genotype • Germplasm • Red palm weevil • Seedling dates • Tissue culture

15.1 Introduction

15.1.1 Historical and Current Agricultural Aspects

The date palm, *Phoenix dactylifera* L., is generally considered an archaeophyte in Spain; it is unknown when and where it was first introduced, but certainly in ancient times. It is almost impossible to trace the origins of date palm cultivation in Spain. Although archaeological remains in the form of carbonized or mummified date seeds exist, these were never radiocarbon dated and thus their adscription to a chronology is hypothetical. Furthermore, it cannot be ruled out that some of these findings may correspond to more recent imports (Rivera et al. 1988). Transformed or simplified images of date palm trees are frequent in late Iberian pottery (third to first century BC) (Fig. 15.1) and in silver and bronze

Fig. 15.1 Representation of a date palm tree on an Iberian pottery fragment from Zama, Hellín, Albacete (Photo: Museo de Albacete, by Marian Vencesla (with permission))

Fig. 15.2 The *Palmeral de Elche*, Alicante, a palm grove of one of the numerous rural parishes into which the municipality of Elche is divided. Only a few among these were included in the historical area declared by UNESCO as a World Heritage Site (Photo: D. Rivera)

coins from *Quart Hadast* (presently Cartagena), *Baria, Sisapo*, and *Tagilit* (third century BC). As Rivera et al. (2013) have shown for the introduction of date palm to the Americas after the fifteenth century, it is likely that introduction of different seedling cultivars into Spain occurred repeatedly since the prehistoric period and through intentional or unintentional germination of seeds from dates imported for food.

The historic *Palmeral de Elche* contains about 180,000 adult date palms (Fig. 15.2). Total date fruit production for Spain was reported at 4,000 mt in 2012 (FAO 2013); however, since 1995 only about 100 mt are marketed from Elche for human consumption (Ferry et al. 2002). The *Palmeral* covered 550 ha in 2004 which represented about 3 % of the agricultural land of Elche. Other date production areas in Spain are Abanilla and Huerta de Murcia in Murcia and Albatera, Alicante, Callosa, Crevillente, and Orihuela, Comunidad Valenciana, which contribute to the overall date production of Spain. The traditional Elche palm grove was inscribed as a UNESCO World Heritage Site in 2000 and intended to be maintained as a unique European example of an introduced oasis system, with the date palm as its key agricultural species (Johnson et al. 2013).

15.1.2 Importance to Country Agriculture

In terms of cultivated area, the date palm is almost marginal in Spain. Of the 30.6 million hectares cultivated in Spain in 2009, only 830 were reported as palm groves producing dates (FAO 2013; INE 2013). Since the 1970s, thousands of adult date palms have been transplanted from traditional palm orchards (e.g., Abanilla, Cuevas de Almanzora) to new urban developments and resorts for ornamental use along the littoral of continental Spain and the Balearic Islands; this strongly reduced the expansion and relevance of palm cultivation for date-fruit production. During the last four decades, the great demand for adult date palms as ornamentals led to decreased local availability and higher prices for date palms, resulting in the importation of mature Egyptian date palms.

15.1.3 Production Statistics and Economics

During the last 50 years, date production in Spain has steadily decreased, with slight fluctuations, from 12,000 mt in 1961 to 3,741 mt in 2011 (FAO 2013). Almost 100 % of Spanish dates are sold fresh in local markets (Fig. 15.3). Depending on quality and fruit size, retail prices reach EUR 4–15/kg.

At present, it is impossible in Spain to earn a stable and sufficient income for a family or company based solely on the production and marketing of date fruits. Therefore, this activity is complemented by the exploitation of other parts of the palm. The leaves are *encaperuzadas*, whereby young shoots are covered with black plastic bags (*caperuzas*) and protected against sunlight for months, yielding pale leaves without chlorophyll, which afterwards can be whitened, processed, and fashioned into traditional handcrafts. Most of this production is specifically made for religious and other ceremonial purposes. The shoots are referred to as *palma blanca* in Spanish and *white palm leaves* in English.

Most of the date palm growers who produce white palm leaves in Elche and surrounding areas do so because they do not make a profit from the fruits and still have the expense of pruning old leaves. With white palm production no additional tree maintenance is necessary and in 1983 each palm generated ESP 1,000 annual income, which in 2003 equaled about EUR 42 (in Elche, Alicante) but in 2010 came down to EUR 14 (in Abanilla, Murcia). In 1983, about 25,000 Elche date palms were *encaperuzadas*, producing approximately 250,000 white palm leaves of different size and quality. Some of the production was used locally for Elche's elaborate Easter pageant (Fig. 15.4), some were shipped to

Fig. 15.3 Dates sold at Fira de Sant Blai (or *porrat'* de Sant Blai), in Albal, Valencia (Photo: E. Laguna)

Fig. 15.4 Selling palms for a religious festival, Elche, Alicante (Photo: C. Obón)

Barcelona, and about 80,000 leaves were exported, particularly to the UK. In 2003, the number of palms *encaperuzadas* in Elche reached 70,000 (Brotons 1989; Gracia 2006).

15.1.4 Current Agricultural Problems

There are three very serious issues affecting date palm agriculture in Spain leading to the abandonment of palm groves: the lack of markets interested in stocking locally produced dates, the significant impact of recently introduced red palm weevil pest (discussed below), and the appreciation of the agricultural land occupied by the palm groves as area for urban growth and/or conversion to more profitable crops.

15.2 Cultivation Practices

15.2.1 Chronological Account of Research and Development

Cavanilles (1793) described, under the name *Phoenix excelsior* Cav., the main date type grown in Elche with the vernacular name Candits and its cultivation practices. Date palm groves in Elche and Orihuela were traditionally irrigated with water unsuitable for use on other crops because of its high salt content particularly sulfates. The *palmerales*, large areas of palm tree cultivation, are often associated in Spain with and take advantage of runoff thermal springs of medicinal waters.

For decades, there was no interest in promoting genetic improvement of the date palm trees. Lack of a breeding program can be explained by taking into account the marginal climatic conditions in the area of cultivation and traditional orchard management. In the 1970s, with the creation of ICONA (Nature Conservation Institute) and IRYDA (Institute of Agrarian Reform and Development), there arose an interest to improve date palm cultivation, although following the classical route of introducing well-known exotic cultivars. By 1974, selected seedlings and offshoots were introduced from the USA, presumably from Indio, California, and planted in the nurseries of the ICONA forest service at Santa Faz, Alicante (García Lidón A, personal communication, 2012). By 1984 dozens of individual palms had survived, notably several which produced sweet green fresh dates, although it appears that this introduction had little impact in the large *palmerales*. Introductions by the Phoenix Station in Elche included Sphinx (an American cv.) as well as cvs. Barhi, Medjool, and Thoory.

15.2.2 Description of Current Cultivation Practices

The structure of a typical Elche palm grove (*tahúlla*) consists of an area of about 1,000 m^2; on average, each usually has 35 adult date palms irrigated with salty water. The few new palm groves in the Elche vicinity have imitated the structure of the old orchards (Brotons 2001).

Date palm is usually propagated by seed, leading to high phenotypic diversity. Date palm cultivation can be problematic in areas such as Elche, due to low temperatures for fruit ripening as well as high humidity at the time of ripening which is detrimental to the fruit quality; some years the dates may not ripen fully. Thus, date palms are usually combined with other crops in a two- or three-layered arrangement, allowing for a higher output than a monoculture would provide. The date palm is considered a multipurpose tree and its value resides not only in fruit for human consumption but also for products such as animal food, *palma blanca*, and building materials (Ferry et al. 2002); consequently, date palms which do not bear palatable dates may still be profitable.

Trends appear to have changed in the late twentieth century, thanks to the combined efforts of the local Phoenix Station, INRA (National Institute for Agricultural Research-France), the University of Alicante, Miguel Hernández University of Elche, and local administrations. Tissue-culture protocols were developed as well as the selection of genotypes for replacement plantings in the Elche palm groves. Genotypes for tissue culture involved Confitera and the introduced cv. Medjool (Ferry et al. 2002), as well as genotypes Lucerga and Léon. This led to clonal populations on small monoculture plantations of date palms which is a novelty within the traditional agricultural system of the local *palmeral*.

15.2.3 Pollination, Fruit Quality, and Metaxenia

The traditional *palmereros* (farmers or farm workers who care for the date palms) of Abanilla and Elche, in selecting male inflorescences for pollination, usually look for the frequency of bees around the flowers. Apparently, large numbers of bee visitors are an indication of fully developed male flowers with no defective pollen. Traditional management of palm groves includes the practice of *macheo*, where certain male individuals are much sought after because of the high quality of their pollen and its positive influence (metaxenia) on the quality and size of date fruits; those male palms are therefore selected for artificial pollination (Rivera et al. 2008).

15.2.4 Pest and Disease Control

Date palm can be damaged by numerous pests and diseases and comprehensive world reviews have been published (Abdullah et al. 2010; Blumberg 2008; El-Shafie 2012; Howard et al. 2001; Zaid et al. 2002). Historically in Spain, no major incidences of pathogens or pests are known, probably due to the relative isolation of the date palm groves, local climatic conditions, genetic heterogeneity of seed-derived palms, and the distinctive agricultural practices used. However, this situation has dramatically changed in recent decades.

New exotic pest threats have appeared as a result of date palm importations from other countries (Gómez 1999). Most serious are the red palm weevil (*Rhynchophorus ferrugineus* Oliv.), palm borer (*Paysandisia archon* Burm.), and a few others. The two named pests reached Spain when infested adult date palm trees were imported for landscaping and ornamental use, especially in new urban developments. Among other pests, red date scale, *Phoenicococcus marlatti* Cock., was first detected in Spain at Elche in the winter of 1993/1994, an explosive development being reported a few years later (Gómez et al. 1996). Because *P. marlatti* attacks the leaf bases and may cause tissue necrosis, the production of the traditional white leaves can be affected (Ferry et al. 2002). It may also cause dryness and mortality of the infested palms. Although application of thiamethoxam by trunk injection has been successfully assayed (Muñoz-Irles et al. 2008b), chemical control of *P. marlatii* is difficult and sometimes ineffective because of the concealed sites in which the scales are located. For that and environmental reasons, biological control of red scale has been attempted in the area. Local predators *Rhyzobius lophanthae* Blais. and *Chilocorus bipustulatus* L. have been identified and a protocol for mass rearing under controlled conditions using *Aspidiotus nerii* Bouche as a host has been developed for *R. lophantae*. The first release assays in the palm grove at Elche showed promising results (Gómez 2002; Muñoz-Irles et al. 2008a).

The most harmful pest of the date palm is the red palm weevil which was first detected in Spain on *Phoenix canariensis* H. Wildpret in 1995; between 2003 and 2009 it explosively spread on the Spanish mainland and to the Balearic and Canary

Islands. Evidences suggest that its expansion in Spain was due to date palm importation from Egypt and movement of infested palms and offshoots (Ferry and Gómez 2002). During 2003–2006, around 15, 000 mt of palms were annually introduced into Spain. Red palm weevil has caused severe damage and destroyed thousands of trees. Damage to date palms is mainly produced by larvae but visible symptoms are not apparent until infection is extensive. The main symptoms are holes in the crown or trunk, usually covered with fibers, oozing brown viscous liquid, a rotting odor, extrusion of chewed up fibers, destruction of vascular system, and, eventually, collapse and death (Zaid et al. 2002). Despite intensive efforts, management of this pest is complicated, mainly due to difficult early detection, lack of quarantine treatment, or ineffective destruction of affected palms. Preventive and curative chemical control has been made by foliage spraying with fenitrothion, chlorpyrifos, diazinon, or methidathion and trunk injection with carbaryl and imidacloprid, all with variable results. Mass monitoring and trapping of adults with pheromones has also been successfully assayed (Esteban-Duran et al. 1998; Martínez-Tenedor et al. 2008; Sansano-Javaloyes et al. 2008), but risks may arise from using it in uninfected areas. Biological control with *Steinernema carpocapsae* Weis (Gómez et al. 2008; Llácer et al. 2009) and *Beauveria bassiana* (Bals.-Criv.) Vuill (Dembilio et al. 2010; Güerri-Agulló et al. 2010, 2011) has been recently developed with promising results. Prophylactic measures such as quarantine regulations, early detection by visual checking and regular monitoring by trained experts or bioacoustic sensors (Gutiérrez et al. 2010), pruning (in winter, only dry leaves), and removal and destruction of damaged palms (cutting, transportation to specific areas, and shredding) are strongly recommended. Legal restrictions against palm movement such as the current EU plant passport are also important. In summary, although intensive efforts are being made from local, central, and European governments and from research institutions, a completely efficient red palm weevil control is still unavailable. It is necessary to adopt an integrated pest management strategy including measures to avoid new palm devastation and consequent economic and cultural losses.

Paysandisia archon (Burm.) is a moth belonging to the family Castniidae, which occurs naturally in Argentina, Brazil, Uruguay, and Paraguay, where its larvae feed on palms, mainly of the genus *Trithrinax*. In Europe, it has become a serious threat for numerous species of ornamental palms. Adult individuals of this species have been observed flying in and around Valencia in mid-summer. The Generalitat Valencian in June 2003 issued an official declaration of the pest and prescribed specific phytosanitary treatments. However, there is no evidence yet concerning the impact on individuals of *Phoenix* species in the zone (Montagud 2004). On the contrary, the impact on palms of the native genus *Chamaerops* seems to be significant and tentative treatments include biological control with the nematode *Steinernema carpocapsae* (Soto and Duart 2013).

Other minor damaging pests detected in Spain (Gómez 1999) are the date stone beetle *Coccotrypes dactyliperda* Fabr., which makes round holes, sometimes reaching the seed, which makes dates unmarketable. The injured fruits usually fall prematurely. Its incidence is variable in Spain and no particular management is reported to be applied. The nitidulid beetles *Carpophilus hemipterus* L. and *C. dimidiatus*

Fabr. can attack ripe fruits, preferably the decayed or damaged ones. Protection of fruit bunches with dense plastic nets and early harvesting is suggested as efficient prevention methods. Larvae of *Arenipses sabella* Hamp., the greater date moth, feed upon young leaves, inflorescences, and immature dates. Although biological control with *Braconidae* and *Bacillus thuringiensis* Berl. is used in other countries, no particular management is reported to be used in Spain.

As far as diseases are concerned, new data on incidences have been reported, but they do not cause significant damages in Spain. Inflorescence rot disease has been recently detected in Elche (Abdullah et al. 2005; Gómez 1999). Brownish or rusty-colored lesions appear on unopened spathes, producing destruction of the flowers and strands. The pathogen was identified as *Mauginiella scaettae* (Cav.) Maire but *Fusarium* spp. or *Botrytis* sp. were also present in some samples. Cutting and burning affected bunches and applications of cupric products are control strategies. A high incidence of *Thielaviopsis paradoxa* (de Seynes) Hohn. and *T. punctulata* (Hennebert) Paulin, Hamington, and McNew was found in soil from date plantations at Elche. These fungi are involved, among others, in root rot and heart bud rot diseases which suggest the possibility of infection of newly transplanted offshoots (Abdullah et al. 2009). *Graphiola phoenicis* (Moug.) Poit. can be found forming subepidermal spots on young leaves in nurseries (pers obs and Laboratory of Plant Protection, Balearic Islands). Propagation of these fungi is favored by humid conditions; however, damage to date palms is relatively limited.

15.2.5 Agroforestry Utilization and Potential

Rivera et al. (1997) described *Phoenix iberica* as a type of date palm growing spontaneously in ravines and salt marshes of Murcia and Alicante provinces. It has a large trunk with numerous basal offshoots (Fig. 15.5), short glaucous leaves with strong and

Fig. 15.5 Group of palms, *Phoenix iberica*, Rambla de la Parra, Abanilla, Murcia (Photo: C. Obón)

relatively short acanthophylls, rigid leaflets, and small fruits with very thin flesh. In addition, date seeds of local *P. dactylifera* types spontaneously germinate in natural and seminatural habitats of the littoral and in the valleys of Almeria, Murcia, Alicante, and Valencia, where there is sufficient groundwater. Therefore, the date palm is a relevant tree species of areas in the Chicamo, Segura, Vinalopó, and Almanzora river valleys. The date palms suffer no great damage, overcoming frequent recurrent fires which occur in those areas. But these spontaneous date palms are threatened by red palm weevil since the palms are not subject to any preventive or curative treatment.

Phoenix dactylifera (perhaps with some proportion of hybrids with *P. iberica*) was used for forest restoration and stabilization of the sand dunes of Guardamar del Segura, Alicante, in the early years of the twentieth century. In afforestation, it was the main species for planting in saline depressions between dunes where, besides helping to stop dune migration, helped drain temporary standing water, at that time important to help prevent the spread of malaria.

In the early 1950s, the issue of dealing with spontaneous *Phoenix* palms fell to the forest service because the palms were growing in no man's land (agricultural and forestry) for which a specific forest category, palm forests (*montes palmerales*), was created. This category included a few examples such as the palm groves of Elche, Orihuela, and Alicante. However, *Phoenix* palms were difficult to manage in administrative terms as crops, as they did not function according to the standards of orange groves or orchards. Furthermore, when date palm production was focused on white palm leaves for religious ritual use, the palms were even more difficult to relate to fruit or vegetable production.

15.2.6 Date Palm Mycorrhization

Mycorrhization has successfully been tested in *Phoenix* to improve hardiness and yields. *Phoenix dactylifera* has been shown to be mycorrhizal (Dreyer 2004; Khaliel and Abou-Heilah 1985; Oihabi et al. 1993; St. John 1988). The mycorrhizal potential to enhance plant growth, shown for palms under low-nutrient conditions (Janos 1977), is of great interest for the management of cultivated date palms.

Arbuscular mycorrhizal (AM) colonization in date palm is restricted to the inner cortex of the third-order roots, except for the pneumatorhizas and the short thick roots (Dreyer et al. 2010). The first- and almost all the second-order roots of this palm species were not colonized by AM fungi. In terms of mycorrhizal colonization, the root system presents a division of two different specialized types of third-order roots: (1) mycorrhizal thickened roots where the arbuscules form and (2) mycorrhiza fine roots with only intraradical hyphae and spores, but without arbuscules, and pseudomantles of spores anchored in the pneumatorings of the second-order roots, which were described for the first time by Dreyer et al. (2010). The mycorrhiza formed by *Phoenix dactylifera* is intermediate between the Arum and the Paris types and is characterized by intercalary arbusculate coils and not only by intracellular but also by intercellular fungal growth (Dreyer et al. 2010).

The date palm is native to arid and semiarid regions and is characterized by a very low number of AM fungal spores in the soil (Dreyer 2004). A low spore number is also typical in the rhizosphere of other plants of arid or semiarid Mediterranean ecosystems (Azcón-Aguilar et al. 2003). Indeed, in these soils, the main source of inocula is extraradical mycelium (Requena et al. 1996). The strategy developed by *Phoenix* spp. to increase propagule numbers by developing different root types and structures, such as the pseudomantle, is a notable example of adaptation between host and fungal partner in response to such conditions (Dreyer et al. 2010).

The introduction of AM fungi into date palm nursery production systems guarantees a higher growth than non-mycorrhizal date palms, decreasing the time spent in nursery and therefore reducing costs. The AM fungi *Glomus mosseae* (Nicol. and Gerd.) Gerd. and Trap. (≡ *Funneliformis mosseae*) and *G. intraradices* Schen. and Sm. (≡ *Rhizophagus irregularis*) have been very effective in promoting date palm growth at low fertilization levels (Dreyer 2004). Date palm responds to the inoculation of many AM fungal species and is less selective in fungal terms than other palm species like *Phoenix canariensis,* the Mexican blue palm (*Brahea armata*), or the European fan palm (*Chamaerops humilis*).

15.3 Genetic Resources and Conservation

15.3.1 Chronological Account of Research in Genetics, Breeding, and Conservation

Although accounts of historical cultivation of date palm can be traced back to the fifteenth century, the earliest systematic research into date palm cultivation is found in Muñoz-Palao (1929), where a few types grown in the southeastern provinces of Spain are mentioned, such as Candits, Verdales, Tenaos, and Negros. Further research late in the twentieth century revealed local names for date palm ethnovarieties, such as Moscatel, Rojos, and Largo (Rivera et al. 1997). The status of these ethnovarieties is questionable and research under way suggests that the names may allude only to the fruit characteristics in a particular year, rather than fixed characteristics and genetic identity (unpublished results).

In Spain, date palm improvement has relied on conventional methods, particularly obtaining seedlings from selected female palm individuals. From the 1940s to the 1980s, conventional breeding programs were developed in Elche at the Estación Experimental Agraria; however, no definite result in terms of relevant seedlings or cultivars was reported. In the early 2000s, a pilot study was conducted to select the seedling female date palms bearing the best quality edible fresh fruit. One hundred local ethnovarieties were selected for further study, based on fruit flavor, aroma, and texture, as a fresh perishable fruit at the khalal or rutab stages. Once superior ethnovarieties were ascertained they should be propagated by offshoots to create new cultivar palm groves to be established outside the boundary of the historic *Palmeral de Elche*. The overall objective was to produce attractive fresh dates for the European gourmet food market (Johnson et al. 2013; Orts and Johnson 2007). Since 2000, the Phoenix Research Station in Elche has produced about 30,000 tissue-cultured plants for commercial fruit cultivation, for use on surrounding farms not subject to regulations governing conservation of the historic *Palmeral*. Date palm cultivation beyond the protected *Palmeral* was expected to form a beneficial buffer zone and to extend the palm landscape of Elche. Plantlets produced are of the introduced cv. Medjool and cvs. Confitera and Léon both allegedly derived from local ethnovarieties (Gómez and Ferry 2010). All three produce soft dates. Fruits are harvested by cutting the entire bunches at the rutab stage for artificial ripening before being sold (Johnson et al. 2013).

15.3.2 Current Status and Prospect of Genetic Resources

To analyze the threats to genetic resources of the date palm in Spain, a distinction must be made between wild and cultivated populations. Up to three species have been recognized (*Phoenix canariensis, P. iberica,* and *P. dactylifera*), each with specific threats and conservation solutions. Although the specific rank of *P. iberica* is not accepted by Govaerts et al. (2011), independent of the taxonomic level, the conservation status of its populations is pitiful. In addition, for the feral/wild species (*P. canariensis* and *P. iberica*), the distribution data and conservation risks are very different. Both wild and cultivated specimens for agricultural purposes (excluding clear ornamental use) have had a modest amount of protection given since 1956, by the former Law on Forests, which declared them to be forest trees. In the case of date palms as crops—for fruits, leaves, and other products—the larger plantations have been considered as palm tree forests, even though of artificial origin.

The Canary Islands hold the entire native population of the endemic *Phoenix canariensis*, mainly concentrated on La Gomera and Gran Canaria (Fig. 15.6), forming natural palm groves in ravines and deep valleys and on rocky slopes. A census revealed more than 140,000 individuals, and some sites hold exceptionally rich populations of old palms (Rigueiro 2005). The natural groves of this species declined markedly after the Spanish occupation in the fifteenth century, and most of their ancient sites have been considerably transformed or destroyed because of land reclamation for agricultural and expanding urban uses, overgrazing, fuel extraction, and recurrent fires (Naranjo et al. 2009; Sosa et al. 2013). Locally some palm tree groves have suffered in the past decades from overexploitation, caused by the intensification of the traditional extraction of sap, obtained by making cuts in the leaf bases near the apical meristem. The sap is used to produce *guarapo* or Canarian palm tree honey. A major threat to the species in its natural areas is the historical hybridization with the introduced *P. dactylifera* (González-Pérez and Sosa 2002; Gonzalez-Perez et al. 2004a, b). To regulate and protect the natural populations, in 2006 the Canarian regional government issued a Decree which provides for strict protection of the wild palms and also forbids the transfer of native palm trees among the different islands to preserve the genetic identity of

Fig. 15.6 Palms in a *barranco* (ravine), *Phoenix canariensis*, Sorrueda, Gran Canaria, Canary Islands (Photo: E. Carreño)

the local races. *Phoenix canariensis* is the main indicator species of the habitats protected under the European Union Directive of Habitats.

In Spain, wild or feral specimens of *Phoenix* aggr. *dactylifera*—including isolated individuals of *P. iberica*—were considered as forest trees beginning in the late nineteenth century. The Spanish Forest Law (Ley de Montes) in force from 1956 to 2003 maintained this designation which forbade the cutting and extraction of wild trees, as well as those in large historic plantings. During the first half of the twentieth century, the most representative artificial palm tree crops in Spain, the so-called *palmerales*, were declared artistic gardens or picturesque sites, loose designations for seminatural protected areas. For these sites, specific rules to maintain the traditional uses (dates for food and leaves for handcrafts) were passed, forbidding tree removal and requiring the replacement of dead trees. The two most extensive palm groves in Alicante province, the *Palmeral de Elche* and the *Palmeral de Orihuela*, were protected. This protection was maintained only until 1975, replaced by a new Spanish law on protected areas. The Spanish Constitution empowered the Autonomous Communities (ancient cultural regions) to protect and manage a variety of protected sites for natural or cultural reasons. In 1986, the Valencian Parliament passed a regional law to protect the large *Palmeral de Elche*; in 2000, this site was officially inscribed as a World Heritage Site (Cremades 2009).

Some *Phoenix dactylifera* palms studied at Elche and Orihuela were found to be highly valued for their dates and receive individual names other than those already mentioned. These plants, both male and female, are the basis for further breeding and selection. The field research associated with building a National Germplasm Collection of *Phoenix* revealed the enormous seedling diversity within the *palmerales* of Elche, Orihuela, Valle de Ricote, Murcia, Alicante, and others. This rich diversity is not only restricted to the Spanish mainland. Formations of *P. canariensis* (wild, feral, and cultivated) on Fuerteventura (Canary Islands) are extraordinarily diverse both from the morphological and genetic viewpoints.

15.3.3 Threats and Degradation

Outside its natural habitat, *Phoenix canariensis* is a major species for ornamental purposes, represented by hundreds of remarkable trees in Spanish gardens. The main present threat is the red palm weevil. At this time, most of the cultivated *P. canariensis* at low elevations in Spain are severely affected, and many majestic Canarian date palms survive thanks to the regular application of specific biocides.

Phoenix iberica, a taxon virtually extinct in wild, is mostly maintained through isolated individuals scattered through *ramblas* (dry ravines), gardens, or cultivated palms and is suspected to house introgressions with *P. dactylifera* (according to the variability of offspring). Several private orchards managed by horticulturists linked to nongovernmental conservation organizations began some years ago to produce new specimens in Murcia and Albacete, with seeds harvested from the last wild individuals found in the Chícamo river valley near Murcia. However, recovery of pure forms of this species, after selected crosses among the newly obtained individuals, could take 20–30 years. Good remaining sites with stands of historic date palms such as small coastal bays, saline rivers, and wadis must be selectively monitored in the future to find new representative specimens.

For cultivated specimens of *Phoenix*, the protection of older trees depends on legislation of the Autonomous Communities, which has established, in some cases, lists of protected specimens or minimal sizes, in order to declare them as strictly protected trees. For instance, the Law on Monumental Trees in the Valencian Community, passed in 2006, protects all the date palm trees more than 12 m in height. The main problems facing Spanish *Phoenix* palm tree populations are as follows:

(a) Introduction of any foreign palm trees (not only *Phoenix* species) for ornamental purposes which are vectors of pests and diseases such as *Rhynchophorus* and *Paysandisia*
(b) The control of already-introduced existing pest populations of *Rhynchophorus* and *Paysandisia*
(c) Clonal propagation for large date palm plantations using only a few cultivars, which can lead to a significant reduction of the rich diversity of local ethnovarieties or seedlings
(d) Loss of traditional economic uses of palms and palm products (Picó 1997)
(e) Urban developments on traditional *Palmeral* properties (Gracia 2006)

15.3.4 Conservation Efforts

For 80 years, governmental actions have been taken at the national and provincial levels to protect Spain's historic palm groves and the *Phoenix* palms they contain. A Decree of March 8, 1933, declared the Elche palm grove to be of social interest for conservation and entrusted implementation of the provisions to the

Ministry of Agriculture, Industry, and Commerce. A Board was established by an order dated March 28, 1942, and subsequently restructured by another order dated October 18, 1967. The latter gave the Ministry of Agriculture authority to include the historic palms on a list of species in Article 228 of the Rules of Forestry. The rules regulated use of the groves and the issuance of felling licenses and made them subject to inspection and supervision by the Forest District Headquarters.

As far as the cultural viewpoint is concerned, Decrees of July 31, 1941, and July 27, 1943, declared the palm grove of Elche to be an official artistic garden. Under the auspices of the Ministry of National Education and the Board for the Protection of Artistic Gardens, state supervision was to be exercised by the Ministry of Education under the Artistic Treasure Act and the initial Decree. Broader urban aspects were considered which led to enactment of municipal ordinances in 1951 which were included in the General Urban Plan of Elche in 1962 and thereafter pursuant to the Act Regulating Land Use and Urban Planning. As a result, a Special Development Plan was drafted for the palm tree gardens and an order of October 11, 1972 included in the 1973 revision of the Plan.

Phoenix dactylifera became a protected species within the municipality of Elche under Generalitat Valenciana Law 1/1986, of May 9, which regulates and protects the *Palmeral de Elche*. The objective of this Act was the protection and promotion of Elche's date palms and their areas of growth, by regulating their use, purpose, and consumption, in order to ensure the historical continuity of the natural and cultural values they represent and to promote their cultivation. The Board of the Elche Palm Grove was created as the implementing body for the provisions of the Act. The Board depends, structurally and functionally, on the Valencia Council of Culture, Education, and Science and is based in Elche. Composition of the Board is as follows: Chairman, the Counselor of Culture, Education, and Science; Vice Chairman, the Mayor of Elche; four additional members from public agencies; and one date-grower representative. The Board meets at least once per year.

Date palm groves are considered an important historical legacy and deeply ingrained in the landscape and culture of Alicante and Murcia. In Alicante, the governmental actions taken in the twentieth century set the stage for the *Palmeral de Elche*'s inscription by UNESCO as a World Heritage Site in 2000. This designation recognized the unique historical landscape created in Europe by the exotic date palms and the associated irrigated agricultural system introduced from North Africa by the Moors during the period of Arab domination of Iberia. It also strengthened the legal framework for sustainable management and protection of the palm grove which typifies the city of Elche. In the Autonomous Community of Murcia, *Phoenix dactylifera* has the status of a protected species and is subject to obtaining authorization for certain uses. These policies have led to the transformation of palm groves into gardens. Little effort is focused on protecting traditional management practices, which carries with it the risk of losing the essence of the complete agricultural system with its associated crops and the subsequent degradation of the landscape (Gracia 2006; Larrosa 2003).

15.3.5 Germplasm Banks of Genetic Resources

At the beginning of the current century, the lack of a germplasm bank of date palm genetic resources in Spain became evident. Authorities of the Valencia Regional Government promoted field activities by Miguel Hernandez and Murcia universities to focus on the study and conservation of date palm populations related to *Phoenix iberica*. This work provided core genetic material for the beginning of the National Germplasm Phoenix bank. For 6 years, INIA (National Institute for Agrarian Research) and FEDER (European Fund for Regional Development) of the European Commission sponsored activities to collect *Phoenix* palm accessions (fruits and seeds) within the territory of Spain and to produce seedlings. At present this bank includes over 700 living accessions of *Phoenix* (almost all seedlings) which are maintained at the Escuela Politécnica Superior, Campus de Desamparados, Universidad Miguel Hernández, Orihuela, and in 25 acres of the park *Soto I6* on a meander of Segura river, with the cooperation of Confederación Hidrográfica del Segura and Ayuntamiento de Orihuela.

The *Phoenix* living germplasm bank has accessions represented by plants derived from selected seeds collected in traditional *palmerales* of Almería, Murcia, Alicante, and Valencia, as well as in the Balearic and Canary Islands; included are the 25 different seedling types known from southeastern Spain (Rivera et al. 1997). Accessions also include specimens germinated from seeds collected from important *Phoenix* individuals (selected for their age or singularity) cultivated in gardens, streets, and parks of Spain, France, and Italy. Attention was paid to the diversity of *Phoenix canariensis* under cultivation. A protocol for desiccation of seeds and conservation at low temperatures (4 °C) was developed; the germinative capacity of seeds stored under these conditions was tested with positive results.

15.3.6 Quarantine Regulations

Spain has taken several legal measures relative to the phytosanitary status of *Phoenix* palms. At the national level, Decree 131/2003, of July 11, established comprehensive phytosanitary protection for palm groves of historic, economic, social, and cultural relevance. A subsequent Decree 58/2005 of January 21 (BOE 19, of January 22, 2005, pp. 2583–2665) adopted protective measures against the introduction and spread in the country of organisms harmful to plants or plant products, as well as for export and transit to third countries. Under this measure, all imports of *Phoenix* species are prohibited, although seed and fruit imports are permitted. Decree B.O.E. 247 (of 13/10/2004) prohibited the import of all kinds of *Phoenix* to the Canary Islands (APA/3281/2004).

In Valencia, an order dated December 22, 2009, empowered the Department of Agriculture, Fisheries, and Food of the Valencian Council to establish mandatory phytosanitary measures for the control and eradication of the pest *Rhynchophorus ferrugineus*. Article 11 provides specific rules on palm movements in historic *palmerales*. Article 11.1, in accordance with the provisions of Decree 131/2003, of July 11 of the Council, established a comprehensive security plan for *palmerales* of historic, economic, social, and cultural relevance in the Valencian area. For the purposes of all the provisions in the Decree, it recognized the palm groves of Monforte del Cid, Aspe, Crevillente, Albatera, Catral, Dolores, San Fulgencio, Santa Pola, and Guardamar del Segura, as well as those of Elche, Orihuela, and Alicante.

15.4 Plant Tissue Culture

15.4.1 Role and Importance

Date palm has conventionally been propagated either through seeds or offshoots. In Spain, however, propagation has been based seeds. Due to the lack of sufficiently high temperatures in the fall, dates do not generally ripen fully or well, and therefore, date palms have been cultivated for several uses other than just for fruit. This explains why there has been no need to select and propagate date palms by offshoots. Nevertheless, due to the date palm's long life cycle and its strong heterozygous nature, seed propagation might be undesirable. This is the case of the variable field performance or the poorer quality of fruits. Offshoot propagation is also limited due to the small number of offshoots produced during the tree's lifetime, resulting in low propagation potential. Another characteristic of the palm groves in Spain, particularly in Elche, is the age of trees, since the majority of them is more than 50 years old and includes rare genotypes (Ferry et al. 2002; Johnson et al. 2013).

Since 1970, extensive efforts have been made worldwide to improve mass propagation of date palm through tissue culture, in order to develop efficient and adequate propagation methods (Al-Khayri 2007; Jain et al. 2011). Because of the presence of rare and interesting genotypes for commercial date production and the scarcity of offshoots due to pruning at the palm base, the research and development of tissue-culture technology makes sense in Spain (Ferry et al. 2002).

15.4.2 Chronological Account of Research and Development

Research on palm tissue culture in Spain has been developed mainly at the Phoenix Station in Elche. Initial work was to study the structural biology of adult date palms, particularly the production of axillary buds from the shoot tip (Ruiperez and Ferry 1996). Then, in vitro propagation by organogenesis was carried out, and high percentages of explants were obtained by culturing them in liquid media (Ruiperez et al. 1995) although the majority of them led to the development of floral organs. Propagation through other initial explants such as young spikelets or small leaves was also attempted. In vitro plants were obtained either by indirect or direct somatic embryogenesis or by adventitious organogenesis with the proliferation stage maintained without callus (Navarro et al. 1999). A mixed embryogenesis/organogenesis method (Ferry et al. 2000) was also suggested to be a very effective procedure. Indeed, research on in vitro propagation by organogenesis was also carried out with 11 date palm cultivars (Medjool, Zahidi, Thoory, Bouffegous, etc.) provided by the GRFP (French Date Palm Research Group).

15.4.3 Scale-Up Production and Other Tissue-Culture Applications

In vitro–cultured palms have been effectively produced in recent years in response to local demand. Several cultivars with high economic potential (Medjool, Confitera, Lucerga, and Léon) have been propagated, the latter three from quality adult date palms grown in the *Palmeral de Elche* (VitroPalm Technology 2013).

The first plantation of about 1,500 palms was established in 2001 and planting increased steadily in the following years. In 2004, around 2,200 plants were distributed among local farmers, and in 2006, the first flowerings occurred and the initial harvest of tissue-culture-derived palms achieved. In 2008, the COOPELCHE and SAT ELX cooperatives participated in the marketing and sale of dates from in vitro-produced palms.

Since 2000, the Phoenix Station has produced and distributed about 30,000 tissue-cultured plants for use on surrounding farms outside the limits of the World Heritage Site and for export to other countries such as Morocco, Mali, Niger, Mauritania, and Djibouti. However, the Elche City Council prohibited to sale these plants in Spain to individuals who were not native to or living in Elche.

In vitro palms have also been used for research. For instance, Phoenix Station provided in vitro plants to Alicante University for research on the identification of palm cultivars based on amplified fragment length polymorphism (AFLP) markers (Diaz et al. 2003) and on date palm responses to colonization by entomopathogenic fungi (Gómez et al. 2009).

15.4.4 Survey of Research and Commercial Labs

Research on date palm micropropagation in Spain was initiated and developed by the Phoenix Station at Elche. The Station was established in 1991 with the support of the Elche City Council and in collaboration with Alicante University, Miguel Hernández University, the Generalitat Valenciana, INRA-France, and CIRAD-France (Centre for International Cooperation in Agronomic Research for Development). After functioning for two decades, in March 2012, the Phoenix Station was closed by the Elche City Council. Researchers moved to the Phoenix Station Association in nearby Aspe, where they continue some of the former activities.

VitroPalm Technology is a company in Elche devoted to date palm tissue culture. The technical team, field experience, and tissue-culture protocols were originated from the Phoenix Station. The company offers consulting in the selection of cultivars and genotypes based on commercial interest, design of laboratory facilities to produce in vitro culture plants, staffing requirements, laboratory establishment, training services for technicians, and so on. They offer the application of the acquired technology to the multiplication of any cultivar of date palm derived from adult or juvenile palms (VitroPalm Technology 2013). However, the 2010 economic crisis led to a discontinuity in its activities.

15.4.5 Recommended Protocols

Several studies have reported protocols for date palm propagation through somatic embryogenesis or organogenesis (Al-Khayri 2007; Jain et al. 2011; Zaid and Arias 2002). The protocols used in Spain are mainly based on organogenesis (VitroPalm Technology 2013). The technique has been improved over several years and adjusted to ensure true-to-type plants and good productivity. Propagation is made from offshoots, using portions of the terminal apex as explants. A few months later, the first shoots appear, from which the proliferation explants are obtained. Procedures have also been optimized to achieve the multiplication of adult date palms using their terminal apex as explants. In this sense, new local cultivars (e.g., Confitera, Lucerga, and Léon) have been created from high-quality selected samples.

A new propagation protocol has been developed to be applied when there are insufficient numbers of offshoots for regeneration. In this case, the terminal apex of the date palm with between four and six leaves is used to renew the proliferation series. Inflorescences are also used as explants both for the organogenesis and embryogenesis techniques. The elongation, development, and rooting phases do not present major problems and high parameters are usually achieved. The acclimatization phase is achieved in 8 weeks with about 90 % survival rate.

In vitro multiplication of date palm has both advantages and disadvantages. High-quality plants, a large quantity of plants in a short time, early date production, and production of healthy plants that avoid spreading damaging organisms are some of the advantages. In the particular case of the seedling date palm groves of Elche, tissue-culture techniques allow propagation of elite ethnovarieties, rare quality genotypes, or genotypes without offshoots (Ferry 2011).

However, there is also a risk of accelerating agrodiversity erosion by cloning a small number of cultivars selected only for their interest in date-fruit production and to replace autochthonous or local cultivars with exotic materials. This could increase susceptibility to exotic pests and diseases or to adverse climatic conditions.

Another factor limiting tissue culture is the production of abnormal plants obtained with somatic embryogenesis; therefore, this technique must be used very carefully. For that reason, some authors recommend avoiding use of the embryogenesis approach, whereas organogenesis assures genetic stability (Ferry 2011).

15.5 Cultivar Identification

15.5.1 Role and Importance

It is noteworthy that the traditional selection of date palm trees in Spain has been through the organoleptic and preservation features of fruits that were considered of good quality. Their seeds were sown, but *palmereros* agree that seeds from a good date palm may not yield seedlings of the same type and quality as the mother palm. What they usually did was find a use for each type of date fruit. If these were small or had very large seeds and thin flesh, they were used as animal feed; if strongly astringent (high polyphenol contents and others) they were artificially ripened with vinegar (Obón et al. 2009). Dates that matured on the palm were directly consumed fresh; these were the most sought after and propagated. Vegetative characteristics were less influential, although in the most important *palmerales* at Elche and Orihuela, offshoot production was a problem and they were systematically destroyed; palms producing few or no offshoots were preferred for sowing.

15.5.2 Research in Morphological Descriptors

Hernández-Bermejo et al. (2012) recorded from medieval writers of Al-Andalus (Moorish Iberia) the names of different date ethnovarieties such as Aywaa, Barni, Sahriz, Wadiyya, Yabbara, etc. Cavanilles (1797) distinguished in Elche two ethnovarieties: Candits, sweet, becoming wrinkled when they ripen on the palm, and eaten without artificial ripening techniques (*adobo*), and Ásperos, harsh, which need to be sprinkled with vinegar and keep covered for 2 days to be artificially ripened; at 6 days they become fermented and sour. The first modern descriptions of date palms from southeastern Spain were provided by Escribano (1884) who distinguished six types. Later, Muñoz-Palao (1929) and more recently Orts (2004) comprehensively described the known date types of Elche. Complementing those accounts, Rivera et al. (1997) described date palm seedlings grown in the Segura and Chicamo river basins in Murcia and Alicante.

A total of 112 descriptors/characteristics (qualitative, quantitative, and allometric) totalizing 517 states are used in the National Phoenix germplasm collection for describing Spanish date palms. These characteristics are organized into physiological (2); phenological (2); vegetative, stem (9); vegetative, leaves (32); reproductive, male inflorescences (12); reproductive, female inflorescences (16); and fruits (39). These are applied to the description of the mother plants and the offspring because the collection consists almost exclusively of seedlings.

15.5.3 Research on Molecular Descriptors

The earliest works focus on varietal identification, and testing of the integrity of tissue-culture-developed palms through AFLP markers was based on material from the palm grove of Elche (Diaz et al. 2003) at the Phoenix Station. The development of simple sequence repeat (SSR) markers (Akkak et al. 2009; Billotte et al. 2004) stimulated interest in the study of diversity in date palm groves throughout southeastern Spain (Carreño 2012; García-Martínez et al. 2010, 2012) as a basis for selection. High diversity within the sampled palm groves of Alicante, Murcia, and Almería provinces is the common denominator, due to seed propagation rather than by offshoots, as well as the evidence of imported material within the different areas surveyed, be it adult plants, dates, or pollen (Rivera et al. 2008). Recent works attest to a growing interest in the study of date palm diversity with molecular markers in southeastern Spain (Agulló 2009; Carreño 2012; Sánchez 2009) (Fig. 15.7). Molecular characterization of *Phoenix canariensis* in the Canary Islands and the search for specific markers for the species was developed by Gonzalez-Perez et al. (2004a, b).

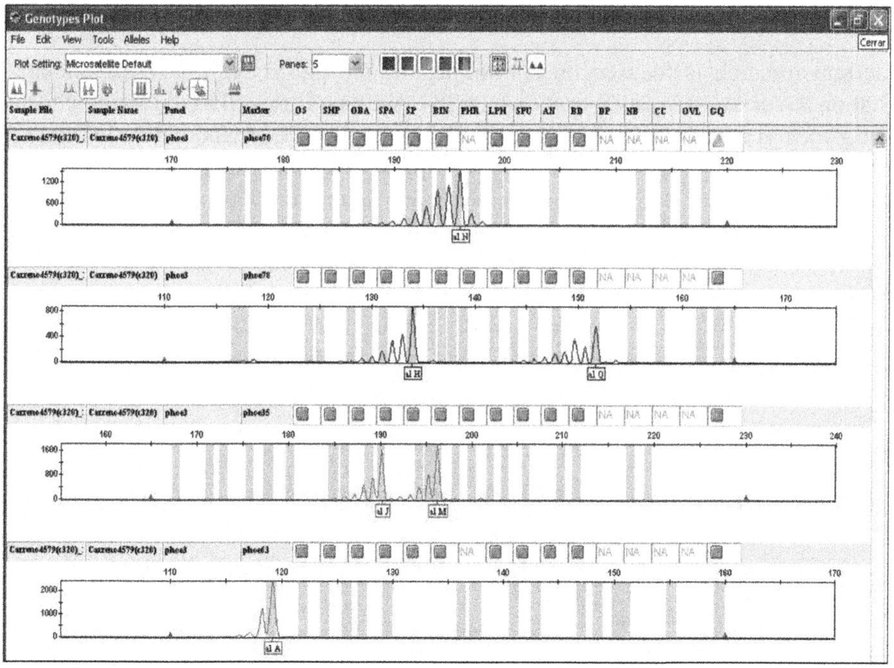

Fig. 15.7 Electropherogram of *Phoenix* samples. An electropherogram is a plot of results from an analysis done by electrophoresis automatic sequencing (Photo: E. Carreño)

Research into genetic diversity and identification of genetic resources of interest for agriculture has barely scratched the surface in Spain. More extensive fieldwork together with identification with molecular markers is needed. Although some genotypes of interest for breeding have already been identified, their characteristics are not widely known by the public, making commercialization difficult. Protection measures can be a double-edged sword, since they do not usually allow for substitution of old palms with selected ethnovarieties from local genotypes, which could be an incentive for farmers to maintain orchards. Joining forces with local administrations and producers and pooling resources would be of great interest for the development of future breeding programs as well as for conservation.

To maintain the traditional diversity of palms, it is recommended to continue the planting of seeds and rejuvenate traditional and historic *palmerales*, which are aging and are suffering from infestations of the red palm weevil and the palm borer, already described. The future of those date palm groves within urban environments such as Elche is linked to revitalization of agriculture for fruit production in harmony with the ornamental and landscape value (Picó 1997).

15.5.4 Survey of Research and Commercial Labs

Current research is focused on controlling the red palm weevil, by the Phoenix Station Association in collaboration with INRA-France, as well as the study of local genotypes as a basis for selection, carried out at Miguel Hernández University and Murcia University. Local names for date palm ethnovarieties, such as Candíos, Moscatel, Rojos, or Largo, have been previously identified (Rivera et al. 1997). The status of these ethnovarieties is questionable and research under way with SSR markers (Akkak et al. 2009; Billotte et al. 2004) points out that some among these names may allude only to the fruit characteristic in a particular year, rather than fixed characteristics and genetic identity (unpublished results). VitroPalm is a new business venture aimed at commercial production of plantlets of cv. Medjool as well as local genotypes, such as Confitera, Lucerga, and Léon.

15.6 Cultivar Description

15.6.1 Growth Requirements

Traditional local date palm seedlings were planted on the margins of agricultural fields along the canals. Seedlings only need care in their early years such as weeding and digging the soil in autumn and spring to add some fertilizer, along with annual pruning at the waning January moon. When the palms reach a height of 60–80 cm, they only require watering in summer and annual pruning of three to four whorls of their fronds for its development (Escribano 1884).

15.6.2 Cultivar Distribution

Modern cultivars as previously described are restricted to areas surrounding the historic *Palmeral de Elche*. Traditional seedlings/cultivar distribution is strongly overlapping with the maximum diversity in the *Palmeral de Elche* as a World Heritage Site (Fig. 15.8). The most relevant seedlings are Candits (candíos, maduros de la palmera, maduros), Tenats (tenaicos), and Tendres (tiernos); these are frequent in the orchards of Elche and Orihuela in Alicante and Abanilla in Murcia. Traditional seedling dates can also be found in Alicante, in Valle de Ricote, and even in Almería and Valencia in the form of isolated individuals.

15.6.3 Nutritional Aspects

Date fruits of the Negros type from the *Palmeral de Orihuela* have maturity index values of 45.33, due to an increase in solid soluble contents and a decrease in acidity (Serrano et al. 2001). Date fruits from the *Palmeral de Elche* in the rutab stage have a sucrose concentration of 1–5/100 g, but these dates have a similar amount of fructose and glucose of 17–28/100 g and therefore are a good energy source (Amorós et al. 2009). These were compared with samples of other *Phoenix* species by Amorós et al. (2014).

Besides the fruits, terminal buds (hearts) are sometimes consumed fresh in salads and called *palmito*. Commercial exploitation of *palmito* is limited to local festivals (San Antón, c. 17 January) in Elche and Orihuela and has a very low impact on date palm populations. Local informants from Abanilla and Murcia report that *palmito* is eaten when harvested from palms felled by strong winds.

15.6.4 Description of Spanish Seedling Dates

Candíos is a group of seedling dates sharing fruit characteristics, especially the ability to naturally ripen fruit on the tree without artificial processes, under the average climatic conditions of the Spanish *palmerales*.

Phenology Flowering (March) April–May (June). Ripening (September) November–December (February).

Vegetative Morphology Stem 5–20 m tall; 22–40 cm in diameter. Offshoots below ground level 1–4. Offshoots on the trunk above the ground level rarely present. Crown with 50–90 leaves in adults; leaf 310–500 cm long; leaf rachis 20–30 cm wide at base. Leaf base green (bright or dull), or somewhat glaucous,

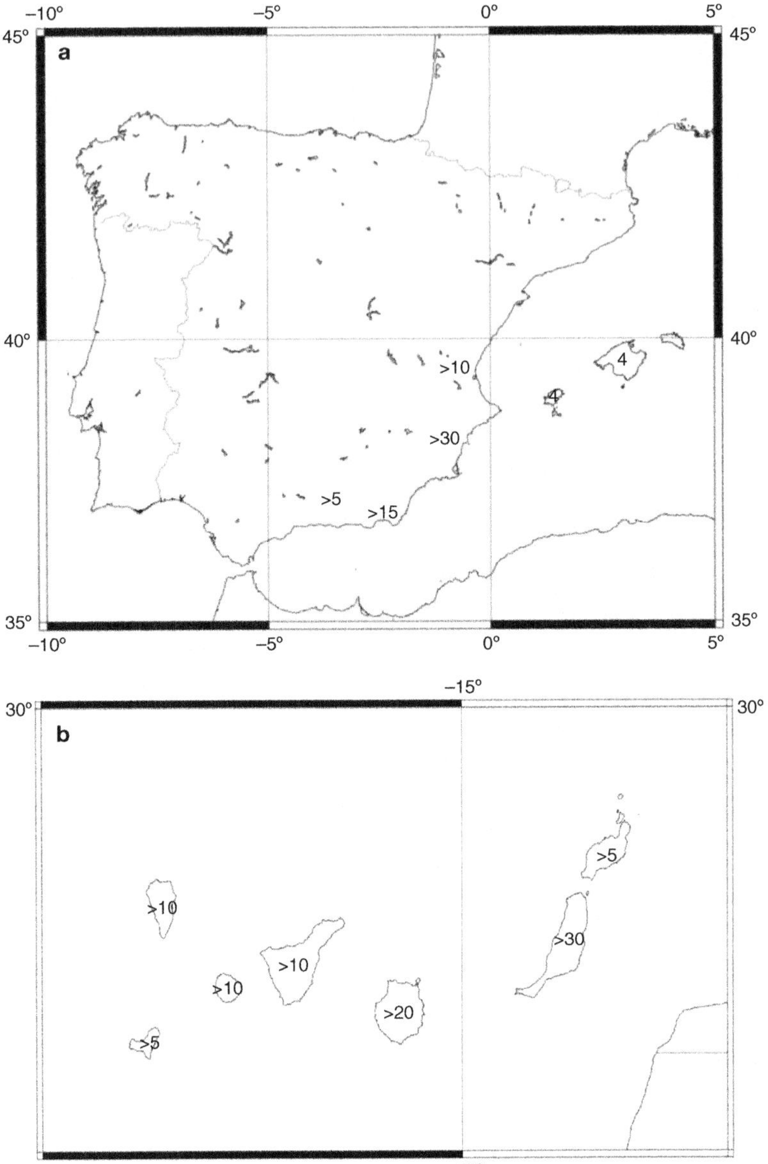

Fig. 15.8 Assessment of palm diversity in terms of number of different cultivars or seedlings in palm groves and forms in wild populations including *Phoenix canariensis*. (**a**) Iberian Peninsula. (**b**) Canary Islands. Maps (Maps were drawn by F. Alcaraz using Quantum GIS (2014) for digitizing polygons and GMT (2014) to generate the final maps)

with a glaucous cast (Fig. 15.9), or with faint yellowish-red tinge, old leaves with reddish-maroon discoloration regularly on edges, or with slight mottling of maroon appearing vertically in the center. Leaf sub-glaucous; haut brown; pseudopetiole 50–150 cm long. P/L (pseudopetiole vs. leaf length): 0.2–0.4. The *haut* is defined as a structure unique to the genus *Phoenix*, thus to the Palm family, formed by the coalescence in early development of the adaxial folds in the leaf (Fig. 15.10). Just before sword leaf emergence, the haut atrophies and then disintegrates on emergence, detaching itself from the V-shaped folds that form the leaflets (Dransfield et al. 2008). Basal pulvinula of spines not noticeably swollen. Divergence between basal pulvinula of spines 30–90°; basal neck in the upper spines of adult leaves 0–25 mm long; spines solitary or paired; 2–15 spines on each side of the rachis; spines 4–20 cm long and 0.4–0.8 cm wide; spines green; distal part of spines concolor. Basal pulvinula of leaflets not noticeably swollen; divergence between basal pulvinula of leaflets 60–90°; leaflets solitary, or in pairs, or in groups of (3) or (4); 40–90 leaflets on each side of

Fig. 15.9 Candíos palm tree grown by Francisco Serrano at Elche (Photo: D. Rivera)

Fig. 15.10 The remnants of the brown haut on a Candit seedling exhibiting a brown coloration. The haut color varies in other *Phoenix* spp. and is an aid to species identification (Photo: D. Rivera) (Color figure online)

the rachis; leaflets at the midpoint of fully developed leaves 40–66 cm long; leaflets 1.7–4 cm wide; leaflets arranged along the rachis regularly and spreading in the same plane or irregularly in clusters and spreading in different planes (quadrifarious); leaflet consistence feathery or stiff. Leaflet apex very sharp needlelike; apical leaflet similar in shape and dimensions to the subapical leaflets. Leaf blade anatomy. Prominent marginal veins in leaflets lacking; tannin-filled sclerotic cells in the margin of leaflets lacking. Abaxial ramenta lacking in young and adult leaves.

Fruit Morphology Perianth in ripe fruits 2.5–4.5 mm high × 7.5–9 mm wide. Remains of perianth (in fresh fruits) yellow or orange. At khalal stage, fruit shape in side view oblong-elliptical and fruit shape in section circular. Fruit apex obtuse. Apical stigmatic remains inconspicuous or very short. Fruit base rounded or oblique. Fruit 25–45 × 15–30 mm. Ratio fruit breadth to fruit length 0.5–0.6. Fresh crunchy fruit yellow or orange; soft ripe fruit (rutab) black or amber brown; fruit flesh 3–5 mm thick. Fruit consistency soft (moisture over 50 %) or semidry (moisture 20–50 %). Epicarp (peel) not adherent to the flesh. Fruit quality as food good. Seed 15.5–25 × 7–9.5 × 6–8.5 mm. Ratio seed breadth to seed length 0.35–0.56. Ratio seed thickness to seed breadth 0.75–1. Seed weight 0.5–1.22 g. Seed shape elliptical-oblong or cylindrical-narrow, apex obtuse, base obtuse, mucronate, or oblique, surface irregular.

Tenats mostly differ from the above for their late ripening and lower water content, which permits longer conservation. These also differ by the epicarp adherent (Fig. 15.11). Vegetatively, the date palm tree is similar. Tendres are characterized by their lower astringency and larger and more globose dimensions, which permits their consumption at an earlier ripening stage. Furthermore, these are late ripening and can be found fresh in the markets from late December to mid-January (Fig. 15.12).

Fig. 15.11 Tenats dates, from Tabalón Alto, Elche (Photo: D. Rivera)

Fig. 15.12 Tendres dates, commercialized at Elche (Photo: D. Rivera)

15.7 Date Production and Marketing

15.7.1 Practical Approaches

The date production and marketing approach taken in Elche was to create cooperatives for the production and marketing of dates. These were created as a result of introducing new cultivars selected at the Phoenix Station. A facility was created to carry out post-harvest artificial aging and to establish distribution channels through gourmet product lines such as El Corte Inglés, the large Spanish department store group.

In parallel the local company Huerto de Elche (formerly Huerto del Cura) is focused on commercialization of local Candit dates produced in Elche and ripened naturally on the tree and served fresh, initially at the restaurant of the Millennium Garden Hotel in Elche and later in various renowned Spanish restaurants such as Martin Berasategui's (Michelin 2 stars) and Rodrigo de la Calle's, inspired by an initiative termed *gastrobotany* and led by Santiago Orts (González 2011). In 2012, fresh Elche dates were presented at the most important gastronomic fair in the Middle East, Gourmet Abu Dhabi, by Santiago Orts (Romero 2012).

15.7.2 Optimization of Yield

Performance is optimized by each individual date palm due to the great diversity of palms in the Spanish *palmerales*. Planting of selected individuals, cloned, began in Spain with the distribution of the material obtained by the Phoenix Station after the year 2000.

Fruit size is very important in the wholesale price received by farmers; therefore, the *palmereros* reduce the number of infructescences on the tree and the number of dates per infructescence to produce larger fruits, a fairly common practice in date growing. Usually, superior male date palms are selected to artificially pollinate the female palms to achieve large, good-quality dates.

15.7.3 Harvest Mechanization

In Spain, mechanization is rare or lacking in the harvest of dates. Instead, harvesting and other manual cultivation tasks are carried out by skilled workers, the *palmereros*. They climb palm trees and manage leaves and infructescences and cut bunches of ripe dates. The *palmereros* climb the palms using traditional ropes looped around the trunk, modern climbing techniques, or ladders.

15.7.4 Postharvest Operations

The processing which follows the date harvest depends greatly on the characteristics of the fruits, whether these are soft, semidry (Candíos), or dry (Tenaicos). At present, the usual form of conservation is freezing at −1 to −5 °C, removing dates from the freezers according to market demand. Once thawed, the dates become stale in 2–3 days. Candits dates left on the fruit branch may be stored at 20 °C for 2 weeks and Tenats for 2 months or even more, without losing their organoleptic characteristics, but they do gradually lose moisture.

A joint venture between INRA (France), IVIA (Valencian Institute for Agricultural Research), and the Elche City Council led to the development of a treatment procedure for dates, patented in 2005 (Vilella et al. 2005). This process comprises one or more of the following steps: (a) cold storage of immature dates, (b) ripening of dates, and (c) cold storage of wet ripe dates. The procedure allows, on the one hand, improvement of current methods of harvesting and processing of dates in the factory and, on the other hand, offers to the market a fresh fruit, i.e., moist rather than the date dry, or even preserved as a confit, which is the essence of the current date fruit supply. This method was associated with the development of cultivation of fruit bearing clones such as cvs. Medjool and Confitera.

Researchers of the Polytechnic University of Cartagena have developed techniques with NaClO, UV-C, ozonated water, and alkaline and neutral electrolyzed water to decrease carob moth (*Ectomyelois ceratoniae*) infestation and maintain date-fruit quality and give a longer shelf life (Jemni et al. 2014). The treatment of dates with NaClO, UV-C, O_3, and electrolyzed water showed a positive effect for lowering their natural infestation by *E. ceratoniae* as well as the microbial growth after 30 days of storage at 20 ° C. In particular, UV-C and neutral electrolyzed water were the most effective against moth proliferation without an adverse effect to objective and subjective quality attributes. Independent of sanitizing treatments, the phenolics concentration significantly increased throughout storage, while the antioxidant activity determined by the DPPH method remained quite constant. The quality attributes of dates were maintained after 30 days at 20 °C.

15.7.5 Marketing Status and Research

In southeastern Spain, especially in Elche, traditional exploitation has been and still is conducted by *palmerero* families, among them the largest producer is Francisco Serrano (*Sopascures*), who manages about 2,000 date palms. Numerous *palmereros* are not owners of the palm groves they care for (Pomata 1984), but are instead hired for their services.

Since at least the Middle Ages, Spanish dates have been a good commodity of maritime trade in both the Mediterranean Sea and across the Atlantic Ocean (Bover and Roselló 2008; Rivera et al. 2013). However, presently Spanish date marketing is very local and restricted to nearby markets and fairs. Recently the advertising and promotion of dates reached prestigious restaurants and gourmet establishments. Naturally ripened Elche dates were presented at well-known gastronomic fairs such as Madrid Fusion.

Date-fruit marketing is being associated with symbolic images: the historic artistic garden, Huerto del Cura; the *Palmeral de Elche* as a World Heritage Site; and the Iberian figure of the Lady of Elche all provide images which identify with the dates of Elche.

15.7.6 Current Import and Export

Spanish dates have long been a traditional food resource to combat hunger in the region of the *palmerales*. Currently the Spanish date is becoming a food consumed by people with high purchasing power, when, traditionally, it has been a resource for poor people.

Due to the minimal economic importance of date consumption in Spain, statistics are not reported individually for dates; they are included under *Other Goods* within the category of dry fruits in the available import/export statistical information. Market surveys show that dates in Spain are mainly imported from Tunisia, Deglet Noor cv. dates, and from Israel as fresh large Salomon or Jumbo Medjool cv. dates as well as yellow (khalal) Barhi cv. dates. Other frequent sources are California (USA), Algeria, and, to a lesser extent, Iraq and Namibia (cf. Mili 1993 for older data).

15.8 Processing and Novel Products

15.8.1 Industrial Processing Activities

Researchers of Miguel Hernández University of Elche are developing technologies to use date fiber as an ingredient in food and nutraceuticals. Of particular importance was presentation in 2012 of a date paté that helps lower cholesterol and combat anemia. The addition of 10 % Confitera date paste in the formulation of champagne-type pork liver pâté leads to the enhancement of moisture, fiber, and phenolic compounds, and it was enough to avoid lipid oxidation during 4 days after the elaboration process. Color was the most affected parameter; however, in terms of overall acceptability, panelists preferred samples with added date fruit (Martín-Sánchez et al. 2013).

Another novel product under investigation by researchers at Miguel Hernández University is a date paste made with by-products from fresh dates. The addition of up to 15 % date paste in the formulation of Bologna-type meat products leads to the enhancement of the nutritional (lower fat content and higher fiber content than control) and technological quality (redder-colored and less hard, chewy and cohesive product than the control) together with a satisfactory sensory quality (Sánchez-Zapata et al. 2011).

15.8.2 Survey of Commercial Date Processers

The earliest and latest maturing ethnovariety fruits fetch a higher price in the market, especially those sold in local festivals and fairs, for these are preserved dates of better quality, both in size, color, and sweetness. Festivals where the dates are sold in southeastern Spain are St. Anthony Day (January 17 at Orihuela, Albacete, and Elche), San Sebastian Day (January 30 at Orihuela), Candlemas Day (February 2, at Murcia), St. Blaise Day (February 3 at Albal), St. Agatha Day (February 5 at Catral), All Saints' Day (November 1), and Holy Face Day (second Thursday after Easter, at Alicante) (Jaén 1994), this being the last day in which fresh dates are widely sold. In all cases, processers are small enterprises and family businesses. However, Tunisian dates, e.g., Deglet Noor cv., have been commercialized fresh, dried, or candied, in Spain in large quantities since at least 1919, by medium-sized enterprises such as El Monaguillo, located in El Campello, Alicante.

15.8.3 Secondary Metabolites and Health Benefits

The medicinal uses of date palm and its products were relatively common in Spain; it reached top of diversity and importance during the Middle Ages, both in Al-Andalus and the Christian territories of the northern border, but have been lost progressively. Rivera et al. (2014) reviewed the historical evolution of medicinal uses of date palm and the Canary Island date palm in both the Iberian Peninsula and the Canary and Balearic Islands, from first century AD until present times. Main categories of uses are summarized in Fig. 15.13 for *P. dactylifera* and Fig. 15.14 for *P. canariensis* in terms of percentage of records in the data base.

At present there are very few persisting uses. Dates were the palm product most commonly used, but the sap, pollen, and tender buds are also used. The dates of

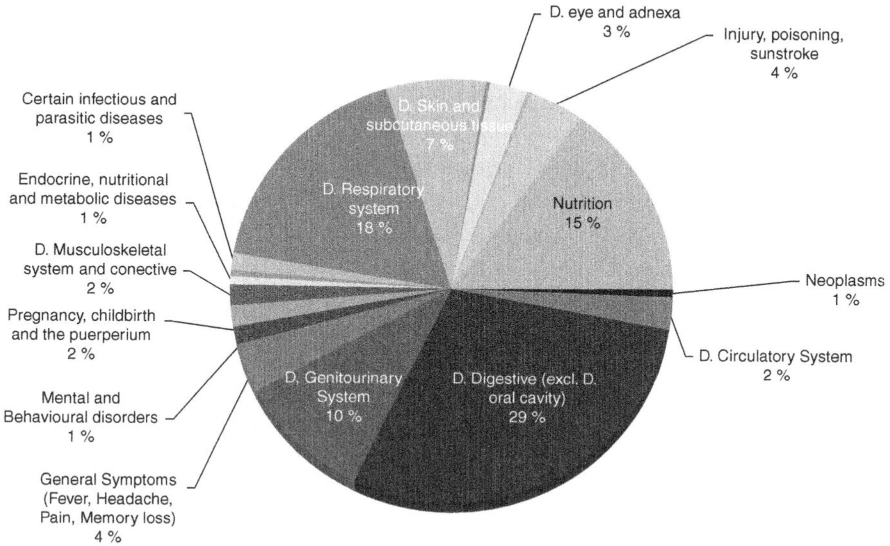

Fig. 15.13 Medicinal uses traditionally adopted in Spain for *Phoenix dactylifera* including fruits, seeds, inflorescences, and spathe. Classes of diseases and related health problems according to Version 2010 (ICD-10 2010). *D* diseases (*Source*: Rivera et al. (2014))

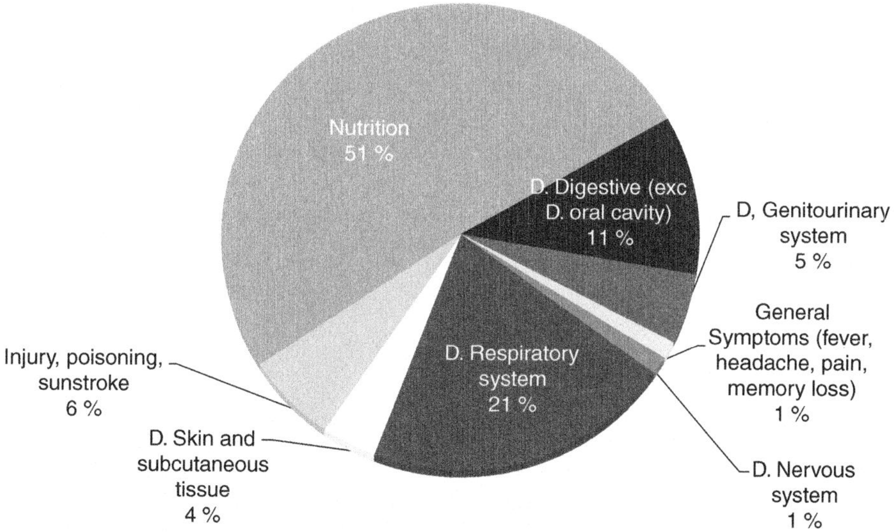

Fig. 15.14 Medicinal uses traditionally adopted in Spain for *Phoenix canariensis* including fruits, sap, seeds, inflorescences, and spathe. Classes of diseases and related health problems according to Version 2010 (ICD-10 2010). *D* diseases (*Source*: Rivera et al. (2014))

Phoenix dactylifera were used as an analgesic, to treat anemia and digestive disorders or to strengthen the gums, in the treatment of erectile dysfunction and as an aphrodisiac, to facilitate childbirth, and to soothe postpartum pains and treat matrix prolapse or excessive menstrual flow. Dates were also used as diuretics and in dysuria and bladder disorders and are still used for the treatment of various respiratory problems. Externally they were used to treat skin problems, wounds, bleeding, and hemorrhoids. Canary palm (*Phoenix canariensis*) raw sap, or concentrated sap, or fermented sap (palm wine) are consumed, especially in the island of La Gomera, as food and are also used as diuretic; remedy for genitourinary, digestive, and oral infections; expectorant; antitussive; and treatment of oral cavity and throat irritation disorders.

In Toledo of Al-Andalus, spathes of *P. dactylifera* were used, almost a thousand years ago, in the treatment of weakness, pain, nephritis, bladder diseases, liver disorders, diarrhea, digestive disorders, pain in the abdomen and stomach, excessive menstrual bleeding, skin ulcers and scabies, joint pain, and heart disorders. Rational phytotherapy should pay attention to this resource, consider the available scientific evidence (pharmacological and even clinical), and incorporate it into our modern therapeutic repertoire.

Date fruits from the *Palmeral de Elche* could provide a good source of natural antioxidants, since they have high hydrophilic total antioxidant activity (HTAA) as compared to other fruits, with a maximum value in khalal state of 200–1,000 mg Trolox eq./100 g depending on the cultivar. These dates also have a high phenolic content, with values of 125–400 gallic acid eq./100 g, which is correlated to HTAA content (Amorós et al. 2009).

The processing of dates yields high volumes of blanching water. Blanching water from Confitera date processing has an important content of phenols and flavonoids, which confer interesting antioxidant properties, as well as organic acids and sugars extracted during blanching. The use of blanching water for reconstituting skim milk powder showed that it is suitable for direct use in the manufacture of yogurt by increasing the content of natural antioxidants and organic acids and has a promising future as a functional food ingredient (Trigueros et al. 2012).

15.8.4 Bioenergy and Other Uses

Several studies are ongoing at Miguel Hernandez University to develop composite building materials from *Phoenix* palm leaves. For example, the Canary Islands palm is pruned up to twice a year in Spain, producing huge amounts of biomass now disposed of in landfills. García-Ortuño et al. (2012) have investigated the performance of particleboards made from pruning residues and a commercial potato starch as a natural binder. The average values of modulus of rupture, modulus of elasticity, internal bond strength, and thermal conductivity suggest that it is completely feasible to manufacture acceptable eco-friendly particleboards for general purposes using the leaf bases of Canary Islands palm (*P. canariensis*) as an alternative lignocellulosic raw material. Very likely, the same applications are possible with leaves of the date palm.

Increasing use of the date palm as an ornamental species in the gardens of Spanish cities, like Elche and Orihuela, generates a large amount of waste. The date palm leaves are a recalcitrant material which is difficult to compost. Martinez et al. (2011) have developed a method of composting this waste along with grass trimmings, vegetable crop residues, and sheep manure as a source of inoculum to facilitate the composting process. The compost obtained has macronutrients and micronutrients in amounts similar to that found in other materials frequently used as organic amendments, such as urban organic waste and manure, and compost quality was verified as adequate for plant development (Moral et al. 2012).

15.9 Conclusions and Recommendations

Efforts underway to improve upon the traditional seedling date area in Elche represent a significant step forward for the utilization of seedling date palms and may well serve as a model in other locations such as Marrakech, Morocco. Unfortunately, excessive focus on clonal germplasm, irruption of introduced pests, funding cuts, and subsequent closure in 2011 of the Phoenix Station at Elche seriously disrupted the commercial development of novel cultivars.

It is also important to assess environmental, economic, and social issues before any massive tissue-culture production program is to be initiated. In many countries, the low productivity of palms is due to agronomic causes such as deficient irrigation or inadequate fruit storage techniques, more than genetic limitations. On the other hand, an effort on mass propagation should be accompanied by efficient trade and marketing that ensures date sale and consumption.

Research into genetic diversity and identification of *Phoenix* genetic resources of interest for agriculture in Spain is at its starting point. Although some genotypes of interest for breeding have already been identified, their characteristics are not widely known and limited to a locality or a particular farmer, making commercialization difficult.

Phoenix dactylifera and *P. canariensis* palms are socioeconomically important for local populations in southeastern Spain and in the Canary Islands, not only because of the nutritive value of dates but also for the ecological services of palm shelter for other crops and the multiple uses of palm leaves and trunks.

Innovative approaches combining agroecological perspectives, rural tourism, and gastronomy may focus on stratified mixed crops and recovery of local and heirloom vegetable and fruit cultivars traditionally associated with date palm cultivation to offer future opportunities to farmers to increase their incomes and preserve the sustainable use of the Spanish *palmeral*.

References

Abdullah SK, Asensio L, Monfort E et al (2005) Occurrence in Elx, SE Spain of inflorescence rot disease of date palm caused by *Mauginiella scaettae*. J Phytopathol 153:1–6

Abdullah SK, Asensio L, Monfort E et al (2009) Incidence of the two date palm pathogens, *Thielaviopsis paradoxa* and *T. punctulata* in soil from date palm plantations in Elx, South-East Spain. J Plant Prot Res 49(3):276–279

Abdullah SK, Lopez Lorca LV, Jansson HB (2010) Diseases of date palms (*Phoenix dactylifera* L.). Basrah J Date Palm Res 9(2):1–44

Agulló R (2009) Estudio de la variabilidad genética del Palmeral de Elche mediante marcadores moleculares. Trabajo Fin de Carrera EPSO-Universidad Miguel Hernández, Orihuela

Akkak A, Scariot V, Torello Marinoni D et al (2009) Development and evaluation of microsatellite markers in *Phoenix dactylifera* L. and their transferability to other *Phoenix* species. Biol Plant 53:164–166

Al-Khayri JM (2007) Date palm *Phoenix dactylifera* L. micropropagation. In: Jain SM, Higgmann H (eds) Protocols for micropropagation of woody trees and fruits. Springer, Dordrecht, pp 509–526

Amorós A, Pretel MT, Almansa MS et al (2009) Antioxidant and nutritional properties of date fruit from Elche grove as affected by maturation and phenotypic variability of date palm. Food Sci Technol Int 15:65–72

Amorós A, Rivera D, Larrosa E, Obón C (2014) Physico-chemical and functional characteristics of date fruits from different *Phoenix* species (*Arecaceae*). Fruits 69:315–323

Azcón-Aguilar C, Palenzuela J, Roldán A et al (2003) Analysis of the mycorrhizal potential in the rhizosphere of representative plant species from desertification threatened Mediterranean shrublands. Appl Soil Ecol 22:29–37

Billotte N, Marseillac N, Brottier P et al (2004) Nuclear microsatellite markers for the date palm (*Phoenix dactylifera* L.): characterization, utility across the genus *Phoenix* and in other palm genera. Mol Ecol Notes 4:256–258

Blumberg D (2008) Date palm arthropod pests and their management in Israel. Phytoparas 36(5):411–448

Bover J, Roselló R (2008) Palmes i datils a Mallorca. Segles XIII-XV: Phoenix dactylifera i Phoenix canariensis. BSAL 64:311–320

Brotons B (1989) Los palmerales de Elche desde sus orígenes. Ed. B. Brotons, Crevillente

Brotons B (2001) El cultiu de la palmera datilera a Elx. Instituto Municipal de Cultura, Ajuntament d'Elx, Elche

Carreño E (2012) La diversidad genética en los palmerales del sureste ibérico como recurso agroambiental. Master thesis, Universidad Miguel Hernández

Cavanilles AJ (1793) Icones et descriptiones plantarum quae aut sponte in Hispnia crescunt aut in hortis hospitantur, vol 2. Imprenta Real, Madrid

Cavanilles AJ (1797) Observaciones sobre la historia natural, geografia, agricultura, población y frutos del reyno de Valencia. Imprenta Real, Madrid

Cremades VJ (2009) Protección y tutela normativa de "El Palmeral de Elche". Rev Fac Cien Soc Juríd Elche 1(4):82–109

Dembilio O, Quesada-Moraga E, Santiago-Álvarez C, Jacas JA (2010) Potential of an indigenous strain of the entomopathogenic fungus *Beauveria bassiana* (Ascomycota; Hypocreales) against the red palm weevil, *Rhynchophorus ferrugineus* (Coleoptera: Curculionidae). J Invertebr Pathol 104:214–221

Diaz S, Pire C, Ferrer J, Bonete MJ (2003) Identification of *Phoenix dactylifera* L. varieties based on amplified fragment length polymorphism (AFLP) markers. Cell Mol Biol Lett 8(4):891–899

Dransfield J, Uhl NW, Asmussen CB et al (2008) Genera palmarum: the evolution and classification of palms. Kew Publishing, Kew

Dreyer B (2004) Estudios de caracterización y eficiencia de las micorrizas arbusculares de las palmeras *Brahea armata* S. Watson, *Chamaerops humilis* L., *Phoenix canariensis* Chabaud y *P. dactylifera* L. PhD thesis, Universidad de Murcia, Murcia

Dreyer B, Morte A, Lopez JA, Honrubia M (2010) Comparative study of mycorrhizal susceptibility and anatomy of four palm species. Mycorrhiza 20:103–115

El-Shafie HAF (2012) List of arthropod pests and their natural enemies identified worldwide on date palm, *Phoenix dactylifera* L. Agric Biol J N Am 3(12):516–524

Escribano JM (1884) Pomona de la Provincia de Murcia ó sed descripcion cientifica y cultivo de los árboles frutales conocidos en esta localidad. Mem Real Acad Cienc Exact Físic Nat Madrid 10:1–224

Esteban-Duran J, Yela JL, Beitia CF, Jimenez AA (1998) Biology of red palm weevil, *Rhynchophorus ferrugineus* (Olivier) (Coleoptera: Curculionidae: Rhynchophorinae), in the laboratory and field, life cycle, biological characteristics in its zone of introduction in Spain, biological method of detection and possible control. Bol San Veg Plagas 24:737–748

FAO (2013) Production Dates Spain. www.Faostat3.fao.org/faostat-gateway/go/to/download/Q/QC/E. Last accessed 12 Mar 2014

Ferry M (2011) Potential of date palm micropropagation for improving small farming systems. In: Jain SM, Al-Khayri JM, Johnson DV (eds) Date palm biotechnology. Springer, Dordrecht, pp 15–28

Ferry M, Gómez S (2002) The red palm weevil in the Mediterranean area. Palms 46(4):172–178

Ferry M, Navarro J, Ruiperez E (2000) An embryogenesis-organogenesis mixed system for in vitro date palm durable mass propagation. In: Proceedings of date palm international symposium, Windhoek. p 27

Ferry M, Gómez S, Jimenez E et al (2002) The date palm grove of Elche, Spain: research for the sustainable preservation of a World heritage site. Palms 46(3):139–148

García-Martínez S, Agulló R, Sánchez JM et al (2010) Estudio de la variabilidad genética en el palmeral de Elche utilizando marcadores SSR. Actas Hort 55:267–268

García-Martínez S, Sánchez JM, Alonso A et al (2012) Estudio de la variabilidad genética en el palmeral de Orihuela utilizando marcadores SSR. Actas Hort 60:54–57

García-Ortuño T, Ferrández-García MT, Andreu-Rodríguez J. et al (2012) Valorization of pruning residues: the use of *Phoenix canariensis* to elaborate eco-friendly particleboards. Structures and environmental technologies. International Conference of Agricultural Engineering – CIGR-AgEng 2012, Valencia

GMT (2014) The generic mapping tools. Release 5.1.0. http://gmt.soest.hawaii.edu. Last accessed 24 Jan 2014

Gómez S (1999) Plagas y enfermedades de la palmera datilera (*Phoenix dactylifera* L.) en España. Phytoma 114:188–191

Gómez S (2002) Cría masiva de *Rhyzobius lophanthae* Blaisdell (Coleoptera: Coccinellidae) depredador de la cochinilla roja de las palmeras (*Phoenicococcus marlatti* Cockerell). Bol San Veg Plagas 28:167–176

Gómez S, Ferry M (2010) La palmeraie historique d'Elche. In: Aberlenc-Bertossi F (ed) Biotechnologies du palmier dattier. IRD Ėditions, Paris, pp 95–103

Gómez S, Capilla MA, Ferry M (1996) Una nueva plaga en España: la cochinilla roja de la palmera datilera, *Phoenicococcus marlatti* Ckll. (Cocc: Phoenicococcidae). Phytoma 82:28–36

Gómez S, Muñoz C, Ferry M, Martínez MM (2008) Primeros resultados sobre el uso de *Steinernema carpocapsae* (Rhabditida: Steinernematidae) asociado a quitosano para el control de *Rhynchophorus ferrugineus*, Olivier en palmeras datileras. Bol San Veg Plagas 34:147–149

Gómez S, Salinas J, Tena M, Lopez-Llorca LV (2009) Proteomic analysis of date palm (*Phoenix dactylifera* L.) responses to endophytic colonization by entomopathogenic fungi. Electrophoresis 30:2996–3005

González E (2011) Gastrobotánica: todo al verde. http://jamasvolvereapasarhambre.wordpress.com/2011/12/05/gastrobotanica-todo-al-verde/. Last accessed 10 Apr 2013

González-Pérez MA, Sosa P (2002) La palmera canaria (*Phoenix canariensis*). Diversidad genética e hibridación. Primera evidencia molecular de la existencia de híbridos entre *Phoenix canariensis* y *P. dactylifera*. Rev Medio Amb (Gobierno de Canarias) 23. http://www.grancanaria.com/eventos/medioambiente/docs/palmre.pdf. Last accessed 10 Oct 2013

Gonzalez-Perez MA, Caujapé-Castells J, Sosa PA (2004a) Allozyme variation and structure of the Canarian endemic palm tree *Phoenix canariensis* (Arecaceae): implications for conservation. Heredity (Edinb) 93:307–315

Gonzalez-Perez MA, Caujapé-Castells J, Sosa PA (2004b) Molecular evidence of hybridisation between the endemic *Phoenix canariensis* and the widespread *P. dactylifera* with Random Amplified Polymorphic DNA (RAPD) markers. Plant Syst Evol 247:165–175

Govaerts R, Dransfield J, Zona SF et al (2011) World checklist of Arecaceae. Royal Botanic Gardens, Kew. http://apps.kew.org/wcsp/. Last accessed 6 June 2013

Gracia L (2006) Indicadores Ambientales y Paisajísticos del Palmeral de Elche. PhD thesis, Universidad Miguel Hernández, Elche

Güerri-Agulló B, Gomez-Vidal S, Asensio L et al (2010) Infection of the red palm weevil (*Rhynchophorus ferrugineus*) by the entomopathogenic fungus *Beauveria bassiana*: A SEM Study. Microsc Res Tech 73:714–725

Güerri-Agulló B, López-Follana R, Asensio L et al (2011) Use of a solid formulation of *Beauveria bassiana* for biocontrol of the red palm weevil (*Rhynchophorus ferrugineus*) (Coleoptera: Dryophthoridae) under field conditions in SE Spain. Fla Entom 94(4):737–747

Gutiérrez A, Ruiz V, Moltó E, Tapia G, Téllez MM (2010) Development of a bioacoustic sensor for the early detection of red palm weevil (*Rhynchophorus ferrugineus* Olivier). Crop Prot 29(7):617–676

Hernández-Bermejo E, García E, Carabaza J (2012) Flora agrícola y forestal de Al-Andalus, vol 1. Ministerio de Agricultura, Alimentación y Medio Ambiente, Madrid

Howard FW, Moore D, Giblin-Davis RM, Abad RG (2001) Insects on palms. CABI Publishing, Wallingford

ICD-10 (2010) International statistical classification of diseases and related health problems 10th revision. http://apps.who.int/classifications/icd10/browse/2010/en. Last accessed 11 Mar 2014

INE (2013) Censo agrario 2009. www.ine.es/CA/informe.do. Last accessed 9 June 2013

Jaén G (1994) Les palmeres del migjorn Valencia. Generalitat Valenciana, Valencia

Jain SM, Al-Khayri JM, Johnson DV (eds) (2011) Date palm biotechnology. Springer, Dordrecht

Janos D (1977) Vesicular-arbuscular mycorrhizae affect the growth of *Bactris gasipaes*. Principes 21:12–18

Jemni M, Otón M, Ramirez JG et al (2014) Conventional and emergent sanitizers decreased *Ectomyelois ceratoniae* infestation and maintained quality of date palm after shelf-life. Postharv Biol Tech 87:33–41

Johnson DV, Al-Khayri JM, Jain SM (2013) Seedling date palms (*Phoenix dactylifera* L.) as genetic resources. Emir J Food Agric 25(11):809–830

Khaliel AS, Abou-Heilah AN (1985) Formation of vesicular-arbuscular mycorrhizae in *Phoenix dactylifera* L. cultivated in Qassim region, Saudi Arabia. Pak J Bot 17:267–270

Larrosa JA (2003) El palmeral de Elche: patrimonio, gestión y turismo. Invest Geogr 30:77–96

Llácer E, Martínez De Altube MM, Jacas JA (2009) Evaluation of the efficacy of *Steinernema carpocapsae* against the red palm weevil, *Rhynchophorus ferrugineus* in *Phoenix canariensis*. BioControl 54:559–565

Martinez FJ, Quiles A, Galvez L et al (2011) Co-compostaje de residuos de poda de palmera junto con residuos verdes y estiércol de oveja. Actas del I Congreso Estatal de Agricultura Ecológica Urbana y Periurbana sobre Huertos Urbanos y Desarrollo sostenible. Sociedad Española de Agricultura Ecológica, Elche, pp 1–11

Martínez-Tenedor J, Gómez Vives S, Ferry M, Díaz Espejo G (2008) Ensayos en túnel de viento para la mejora de la eficacia de las trampas de feromona de *Rhynchophorus ferrugineus* (Colepotera: Dryophthoridae), picudo rojo de la palmera. Bol San Veg Plagas 34:151–161

Martín-Sánchez AM, Ciro-Gómez G, Sayas E et al (2013) Date palm by-products as a new ingredient for the meat industry: application to pork liver pâté. Meat Sci 93:880–887

Mili S (1993) El mercado Español de dátiles: situación actual y perspectivas. Agricultura 727:155–161

Montagud S (2004) *Paysandisia archon* (Burmeister 1880) (Lepidoptera, Castniidaes), nuevas localizaciones en la Península Ibérica y su gestión. Bol SEA 34:237–246

Moral R, Martínez-Teruel F, Quiles A et al (2012) Desarrollo de compost a partir de residuos de jardinería urbana y huertos ecológicos urbanos de tipo mediterráneo. Agr Ganad Ecol 8:28–29

Muñoz-Irles C, Gómez-Vives S, Ferry M (2008a) Estudio de campo de la eficacia de *Rhyzobius lophanthae*, Blaisdell en el control biológico de *Phoenicococcus marlatti,* Cockerell, cochinilla roja, de la palmera datilera, mediante suelta inoculativa. Bol San Veg Plagas 34(1): 117–127

Muñoz-Irles C, Gómez-Vives S, López Fernández JA, Cantús Talens JM (2008b) Ensayo de campo sobre la capacidad de migración del insecticida thiamethoxam 25% inyectado en el tronco de la palmera datilera y sobre su eficacia en el control de la cochinilla roja (*Phoenicococcus marlatti* Cockerell). Bol San Veg Plagas 34(1):129–134

Muñoz-Palao JM (1929) La palmera datilera. Confederación Sindical Hidrográfica del Segura, Murcia

Naranjo A, Sosa P, Márquez M (2009) Habitat 9370: Palmerales de Phoenix canariensis endémicos canarios. In: VV.AA. Bases ecológicas preliminares para la conservación de los tipos de hábitats de interés comunitario en España. Ministerio de Medio Ambiente, y Medio Rural y Marino, Madrid

Navarro J, Ruiperez E, Ferry M (1999) Regeneración de plantas por embriogénesis somática a partir de inflorescencieas de palmera datilera (*Phoenix dactylifera*) de Elche (España). VIII Congreso Nacional de Ciencias Hortícolas, Murcia, pp 152–157

Obón, C, Rivera D, Alonso A et al (2009) Etnobotánica de la palmera datilera y especies próximas (*Phoenix*, Arecaceae) en la Comunidad Valenciana. In: Guillem-Llobat X, Garcia Frasquet G (eds) VI Trobades, Seminari d'Estudis sobre la Ciència. Ed. CEIC Alfons el Vell, Gandia

Oihabi A, Perrin R, Marty F (1993) Effet des mycorrhizes V.A. sur la croissance et la nutrition minerale du palmier dattier. Rev Rés Amél Prod Agr Milieu Aride 5:1–9

Orts F (2004) Antología de palabras, dichos y refranes de la comarca de Elche. Francisco Orts Serrano, Elche

Orts S, Johnson DV (2007) Commercial date fruit production from local genotypes in Elche, Spain. Paper presented at 4th symposium on date palm in King Faisal University, Saudi Arabia, Al Hassa, 5–8 May 2007

Picó F (1997) El palmeral histórico de Elche. Ayuntamiento de Elche, Elche

Pomata AG (1984) Partidas, personajes y cosas del Elche rural. Sdad. Cooperativa del Campo y Caja Rural de Elche, Elche

Quantum GIS (2014) Release 2.0. http://www.qgis.org/es/site. Last accessed 24 Jan 2014

Requena N, Jeffries P, Barea J (1996) Assessment of natural mycorrhizal potential in a desertified semiarid ecosystem. Appl Environ Microbiol 62:842–847

Rigueiro A (2005) Bosques monumentales de España. Mundi-Prensa, Madrid

Rivera D, Obon C, Asencio A (1988) Arqueobotánica y Paleoetnobotánica en el Sureste de España, datos preliminares. Trab Prehist 45:317–334

Rivera D, Obón C, Ríos S et al (1997) Frutos secos, oleaginosos, frutales de hueso, almendros y frutales de pepita. Universidad de Murcia, Murcia

Rivera D, Obón de Castro C, Carreño E et al (2008) Morphological systematics of date-palm diversity (*Phoenix*, Arecaceae) in western Europe and some preliminary molecular results. Acta Hort 799:97–104

Rivera D, Johnson D, Delgadillo J et al (2013) Historical evidence of the Spanish introduction of date palm (*Phoenix dactylifera* L., Arecaceae) into the Americas. Genet Res Crop Evol 60:1433–1452

Rivera D, Obón C, Verde A et al (2014). La palmera datilera y la palmera canaria en la fitoterapia tradicional de España. Revista de Fitoterapia 14:67–81

Romero JC (2012) El dátil ilicitano se promociona en Abu Dhabi con la gastrobotánica. http://www.laverdad.es/alicante/v/20120216/elche/datil-ilicitano-promociona-dhabi-20120216.html. Last accessed 10 May 2013

Ruiperez E, Ferry M (1996) Morphological study of the axilar productions of the adult date palm tree potentiality for in vitro culture. Options Méd A/28:176

Ruiperez E, Ferry M, Casas JL (1995) Comportamiento de yemas axilares de palmera adulta en medio líquido. IV congreso luso-espanhol de Fisiología Vegetal

Sánchez JM (2009) Estudio de la variabilidad genética en el Palmeral de Orihuela mediante marcadores microsatélites. Trabajo Fin de Carrera EPSO-Universidad Miguel Hernández, Orihuela

Sánchez-Zapata E, Fernández-López J, Peñaranda M et al (2011) Technological properties of date paste obtained from date by-products and its effect on the quality of a cooked meat product. Food Res Int 44:2401–2407

Sansano-Javaloyes MP, Gómez-Vives S, Ferry M, Díaz-Espejo G (2008) Ensayos de campo para la mejora de la eficacia de las tramas de captura de *Rhynchophorus ferrugineus* Olivier (Coleóptera: Dryophthoridae), picudo rojo de la palmera. Bol San Veg Plagas 34:135–145

Serrano M, Pretel MT, Botella MA, Amorós A (2001) Physicochemical changes during date ripening related to ethylene production. Food Sci Tech Int 7:31–36

Sosa P, Saro I, González-Pérez MA et al (2013) Patrón de colonización y conquista de la palmera canaria (*Phoenix canariensis*) en el archipiélago canario. Evidencias genéticas e históricas. In: Sánchez-Gómez P, Torrente P (eds) Libro de Resúmenes, 6° Congreso de Biología de Conservación de Plantas. Universidad de Murcia, Murcia, p 25

Soto A, Duart M (2013) Incidencia de *Paysandisia archon* (Burmeister 1880) en la Comunidad Valenciana, medidas de control biológico. www.ivia.es/nuevaweb/jornadas/ornamental/DAAOANTONIA_SOTO-UPV.pdf. Last accessed 17 Dec 2013

St. John TV (1988) Prospects for application of vesicular–arbuscular mycorrhizae in the culture of tropical palms. Adv Econ Bot 6:50–55

Trigueros L, Sayas-Barberá E, Pérez-Alvarez JA, Sendra E (2012) Use of date (*Phoenix dactylifera* L.) blanching water for reconstituting milk powder: yogurt manufacture. Food Bioprod Proc 90:506–514

Vilella J, Río M, Ferry M (2005) Procedimiento de tratamiento de dátiles. http://www.oepm.es/pdf/ES/0000/000/02/24/25/ES-2242545_A1.pdf. Last accessed 30 Sep 2013

VitroPalm Technology (2013) Information dossier. www.vitropalm.com. Last accessed 11 Oct 2013

Zaid A, Arias EJ (eds) (2002) Date palm cultivation. Rev.1. FAO plant production and protection paper. 156. FAO, Rome

Zaid A, de Wet PF, Djerbi M, Oihabi A (2002) Diseases and pests of date palm. In: Zaid A, Arias EJ (eds) Date palm cultivation. Rev.1. FAO Plant Production and Protection Paper. 156. FAO, Rome, pp 223–287

Appendixes

© Springer Science+Business Media Dordrecht 2015

J.M. Al-Khayri et al. (eds.), *Date Palm Genetic Resources and Utilization: Volume 2: Asia and Europe*, DOI 10.1007/978-94-017-9707-8

Appendix A

Fruit characteristics of major date palm cultivars grown in various locations along with estimated tree yield and the number of trees per cultivar of some date-producing countries in Asia and Europe

Country	Cultivar	Yield per tree (kg)	Number of planted trees	Fruit characteristics	Location
Bahrain	Amari	60	535	Poor	Capital, Muharraq, central, northern, southern
Bahrain	Ashhal	60	535	Very good	Capital, Muharraq, central, northern, southern
Bahrain	Banat-Alabade	45	535	Poor	Capital, Muharraq, central, northern, southern
Bahrain	Banat-Alssyid	70	1,069	Medium	Capital, Muharraq, central, northern, southern
Bahrain	Barhee	70	5,881	Excellent	Capital, Muharraq, central, northern, southern
Bahrain	Brismi	55	1,069	Medium	Capital, Muharraq, central, northern, southern
Bahrain	Buchairah	70	6,950	Medium	Capital, Muharraq, central, northern, southern
Bahrain	Fardh	65	1,069	Good	Capital, Muharraq, central, northern, southern
Bahrain	Gharrah	55	11,227	Excellent	Capital, Muharraq, central, northern, southern
Bahrain	Hallaw	75	6,950	Good	Capital, Muharraq, central, northern, southern
Bahrain	Hallaw/Taroot	65	2,138	Medium	Capital, Muharraq, central, northern, southern
Bahrain	Hatimi	65	1,069	Good	Capital, Muharraq, central, northern, southern

Bahrain	Hilali	70	6,415	Excellent	Capital, Muharraq, central, northern, southern
Bahrain	Humri	65	535	Medium	Capital, Muharraq, central, northern, southern
Bahrain	Jabiri	70	1,604	Very good	Capital, Muharraq, central, northern, southern
Bahrain	Khasbat-Asfoor	70	15,503	Very good	Capital, Muharraq, central, northern, southern
Bahrain	Khawaja	65	5,346	Excellent	Capital, Muharraq, central, northern, southern
Bahrain	Khalas	60	18,711	Excellent	Capital, Muharraq, central, northern, southern
Bahrain	Khunaizi	80	58,806	Very good	Capital, Muharraq, central, northern, southern
Bahrain	Merziban	75	70,567	Good	Capital, Muharraq, central, northern, southern
Bahrain	Mubashir	40	1,604	Medium	Capital, Muharraq, central, northern, southern
Bahrain	Mudallal	45	535	Excellent	Capital, Muharraq, central, northern, southern
Bahrain	Muwaji	60	32,076	Very good	Capital, Muharraq, central, northern, southern
Bahrain	Nabtat Saif	60	2,673	Excellent	Capital, Muharraq, central, northern, southern
Bahrain	Rzaiz	65	1,069	Good	Capital, Muharraq, central, northern, southern
Bahrain	Sabo	65	2,138	Good	Capital, Muharraq, central, northern, southern
Bahrain	Selmi	65	23,522	Poor	Capital, Muharraq, central, northern, southern

(continued)

Country	Cultivar	Yield per tree (kg)	Number of planted trees	Fruit characteristics	Location
Bahrain	Setrawi	60	535	Poor	Capital, Muharraq, central, northern, southern
Bahrain	Shabibi	65	1,069	Good	Capital, Muharraq, central, northern, southern
Bahrain	Shambari	65	3,208	Good	Capital, Muharraq, central, northern, southern
Bahrain	Shishi	65	3,742	Good	Capital, Muharraq, central, northern, southern
Bahrain	Sils	50	1,069	Poor	Capital, Muharraq, central, northern, southern
Bahrain	Tanjoob	60	535	Medium	Capital, Muharraq, central, northern, southern
Bahrain	Tayer	50	535	Poor	Capital, Muharraq, central, northern, southern
Bahrain	Um Rahim	70	2,673	Good	Capital, Muharraq, central, northern, southern
India	Barhee	120	na	Nonastringent, suitable for raw dates, late maturity, golden yellow fruit, oval, tree is medium, semivigorous	Kutch, Bhuj, western India (leading cultivar in India)
India	Halawy	100	na	Suitable for raw dates, early maturity, light orange with yellow shade fruit, oblong elliptical with obtuse apex, tree is tall, vigorous	Jodhpur
India	Khadrawy	60	na	Astringent, suitable for soft and dry dates, medium maturity, greenish-yellow fruit, oblong with broad apex, tree is dwarf	Bikaner
India	Khalas	75	na	Suitable both for raw and soft dates, early maturity, yellow at *doka*, golden fruit brown at *pind*, oblong, tree is semivigorous	Small scattered areas
India	Khuneizi	50	na	Suitable for raw dates, early maturity, dark red at *doka*, dark brown fruit at *pind*, oval with pointed apex, tree is semivigorous	Abohar, Ganganagar

India	Medjool	90	na	Astringent, suitable only for dry dates, late maturity, golden yellow fruit, irregular ridges on seed, tree is tall, vigorous	Bikaner, Jodhpur
India	Sewi	50	na	Suitable for raw and soft dates, late maturity, yellowish green fruit, oblong, tree is tall, vigorous	Small scattered areas
India	Shamran	100	na	Astringent, medium maturity, yellow, pink-shaded fruit, base oblong with obtuse apex, tree is semivigorous	Bikaner, Jodhpur
India	Zaghloul	120	na	Astringent, suitable for soft dates, medium maturity, red fruit, oblong oblique and asymmetric, tree is vigorous, erect	Kutch, Jaisalmer
India	Zahidi	125	na	Astringent, suitable for soft dates, late maturity, yellow fruit, ovate with broader end toward apex, tree is semivigorous	Jaisalmer
Iran	Almehtari	75	150,000	Soft, the earliest in ripening, recommended for early-season markets	Hormozgan
Iran	Barhee	350	300,000	Soft, edible in khalal, recommended for early-season markets	Khuzestan
Iran	Breim	95	50,000	Semidry, edible in khalal, recommended for early-season markets	Khuzestan
Iran	Dayri	110	110,000	Dry, long shelf life in storage rooms	Khuzestan
Iran	Dehdar Moradi	110	50,000	Semidry, candidate for export	Khuzestan
Iran	Haji Mohammadi	85	50,000	Soft, candidate for export	Khuzestan
Iran	Halawy	135	100,000	Semidry, recommended for early-season markets	Khuzestan
Iran	Kabkab	95	2,000,000	Soft, recommended for export	Khuzestan, Bushehr
Iran	Khasouee	90	350,000	Soft, recommended for by-product industries (liquid sugar)	Hormozgan
Iran	Mordaseng	130	400,000	Soft, recommended for by-product industries	Hormozgan
Iran	Mozafati	195	1,800,000	Soft, recommended for export	Kerman
Iran	Piarom	95	95,000	Semidry, the best for export	Hormozgan
Iran	Rabbi	60	1,600,000	Semidry	Baluchistan, Kerman
Iran	Sayer	100	1,900,000	Semidry, recommended for export for by-product industries	Khuzestan
Iran	Shahani	175	600,000	Soft, recommended for export	Fars

(continued)

Country	Cultivar	Yield per tree (kg)	Number of planted trees	Fruit characteristics	Location
Iran	Zahidi	130	550,000	Semidry to dry, recommended for export, long shelf life in storage	Mostly Khuzestan
Iraq	Barhee	92	147,463	Soft, with thick flesh and rich flavor; of superb quality. Nearly cylindrical, light amber to dark brown when ripe, excellent in quality	Southern areas
Iraq	Dayri	58	976,535	Semidry, medium in size, sweet with few fibers of flesh, oval shape, good quality	Southern areas
Iraq	Halawy	57	569,874	Semidry, extremely sweet, small to medium in size. Thick flesh, caramel taste, and sweet, somewhat wrinkled in appearance, with a yellow color ripening to a light amber and then to a golden brown, good in quality	Southern areas
Iraq	Khadrawy	64	683,877	Soft, very dark date. It has many desirable qualities. It cures well; it ripens to amber, cured to a reddish brown, with a caramel-like texture and a sweet flavor, good quality	Middle, east, southern areas
Iraq	Khastawi	68	1,361,020	Soft, small to medium in size, cylindrical shape, sweet with few fibers of flesh, good in quality	Middle, western areas
Iraq	Sayer	55	903,735	Semidry, dark orange-brown, medium size, soft and syrupy, oval curved to one side, medium in quality	Southern areas
Iraq	Zahidi	73	5,966,423	Semidry, medium size, cylindrical, light golden brown, very sugary, and sold as soft, medium-hard, and hard, medium in quality	All date palm-growing areas
Israel	Ameri		10,000	Large fruit, with a hard texture	Jordan Valley
Israel	Barhee	200–300	27,000	Round-shaped. Harvested and consumed at yellow khalal	Jordan Valley, Valley of Springs
Israel	Dayri	100	36,000	Small, elongated, and dark at tamar. Planted mainly as a source for Lulav (spear leaf for Jewish religious ceremony)	Jordan Valley, Valley of Springs
Israel	Deglet Noor	100–150	28,000	Harvested either as soft, almost rutab natural fruit on its spikelets or as a dry fruit. Has a special taste and quality	Arava, Dead Sea regions
Israel	Halawy	140	13,000	7–10 g. Yellow at khalal and brown at tamar stage	Jordan Valley, Valley of Springs
Israel	Hayany	150–200	36,000	Red at khalal. Harvested at khalal but consumed at rutab	Jordan Valley, Valley of Springs

Israel	Khadrawy	100	6,500	Extremely slow growth habit. Fruit averaging 9 g	Jordan Valley, Valley of Springs
Israel	Medjool	90–130	425,000	Very large. Harvested selectively in succulent, semidry condition	All regions. Higher qualities in southern regions
Israel	Zahidi	150–170	12,000	Fruit has a light amber honey-like color. Fruit size is small to moderate with a unique taste due to high sucrose fraction. Long storability of dried fruit	Jordan Valley, Valley of Springs
Kuwait	Anbarah	75	500	Soft and excellent	Abdally, Wafra, home gardens
Kuwait	Awaidi	100	1,000	Soft and very good	Abdally, Wafra, home gardens
Kuwait	Barhee	250	318,785	Excellent, eaten at khalal, rutab, and tamar stages	Abdally, Wafra, Sulaibiah, Jahra, home gardens
Kuwait	Braim	100	1,000	Soft, excellent, and eaten at all stages	Abdally, Wafra, home gardens
Kuwait	Hilali	120	1,000	Late, soft, and good	Abdally, Wafra, home gardens
Kuwait	Jouzi	100	1,000	Acceptable and eaten at all stages	Abdally, Wafra, home gardens
Kuwait	Khalas	100	48,906	Excellent, eaten at rutab and tamar	Abdally, Wafra, Sulaibiah, home gardens
Kuwait	Khasab	150	1,000	Late and good	Abdally, Wafra, home gardens
Kuwait	Khyarah	75	1,000	Excellent	Abdally, Wafra, home gardens
Kuwait	Lolwi	200	1,000	Acceptable	Abdally, Wafra, home gardens
Kuwait	Maktoomi	100	1,000	Late and excellent	Abdally, Wafra, home gardens
Kuwait	Medjool	150	4,598	Excellent, eaten at tamar stage	Abdally, Wafra, Sulaibiah, home gardens
Kuwait	Nebut Seif	100	1,841	Excellent, eaten at tamar stage	Abdally, Wafra, Sulaibiah, home gardens
Kuwait	Qantar	100	500	Soft and good	Abdally, Wafra, home gardens
Kuwait	Samaran	150	1,000	Good, early ripening cultivar	Home gardens
Kuwait	Shishi	100	1,000	Soft and good	Abdally, Wafra, home gardens
Kuwait	Suckari	100	10,929	Excellent, eaten at tamar stage	Abdally, Wafra, Sulaibiah
Kuwait	Um Al-Dehn	150	1,000	Soft and excellent	Abdally, Wafra, home gardens

(continued)

Country	Cultivar	Yield per tree (kg)	Number of planted trees	Fruit characteristics	Location
Oman	Barni	33.17	152,415	Yellowish green, elongated at khalal stage, light brown at tamar stage, mild sweetness, mid-season cultivar	Sharqiyah, Dakhlia
Oman	Bu Narenjah	20.6	88,288	Golden at khalal stage, brown at tamar stage, sweet, mid-season processing cultivar	Sharqiyah, Dakhlia, Muscat
Oman	Fard	47.07	435,120	Orange at khalal stage, dark brown/black at tamar stage, mild sweetness, mid-season commercial processing cultivar	Dhahira, Dakhlia, Sharqiyah, Buraimi
Oman	Handhal	45.36	83,165	Golden at khalal stage, brown at tamar stage, mild sweetness, mid-season cultivar consumed mainly as tamar	Dakhlia, Dakhlia, Batinah
Oman	Jabri	22.84	92,328	Golden at khalal stage, light brown at tamar stage, sweet, late-season premium cultivar	Batinah, Sharqiyah
Oman	Khalas	50.34	251,422	Yellow at khalal stage, reddish yellow/golden at tamar stage, mild sweetness, mid-season premium cultivar	Muscat, Dhahira, Dakhlia
Oman	Khasab	49.3	566,797	Red at khalal stage, black at tamar stage, mild sweetness, late-season cultivar	Batinah, Dhahira, Dakhlia, Buraimi
Oman	Khunaizi	36.74	306,591	Scarlet red at khalal stage, black at tamar stage, very sweet, mid-season cultivar, commonly used for date syrup	Muscat, Musandam, Dhahira, Buraimi
Oman	Mabsli	46.18	675,027	Yellow at khalal stage, sweet, early-season cultivar, boiled at khalal stage for export	Batinah, Dakhlia, Sharqiyah
Oman	Madloki	37.58	137,079	Bright golden at khalal stage, brown at tamar stage, sweet, mid-season premium cultivar	Sharqiyah, Dhofar
Oman	Masli	49.7	83,791	Dark red at khalal stage, dark at tamar stage, extra sweet, mid-season cultivar	Dakhlia, Batinah
Oman	Naghal	36.28	679,211	Yellow at khalal stage, light brown at tamar stage, mild sweetness, early-season cultivar	Muscat, Batinah, Dhahira, Dakhlia, Buraimi
Oman	Qash	20.89	134,316	Reddish yellow at khalal stage, dark brown at tamar stage, sweet, mid-season cultivar (Qash Abu Saif)	Musandam, Dhofar, Muscat
Oman	Salani	36.24	63,122	Golden at khalal stage, brown at tamar stage, sweet, early-season cultivar	Batinah, Muscat

Oman	Shahl	38.74	325,299	Bright red at khalal stage, black at tamar stage, mild sweetness, mid-season cultivar	Batinah, Musandam
Oman	Um Sella	35.99	978,542	Dark red at khalal stage, mild sweetness, mid-season cultivar consumed mostly at rutab stage	Batinah, Muscat
Oman	Zabad	25.75	89,854	Yellow-pink round at khalal stage, light brown at tamar stage, sweet, mid- to late-season cultivar	Dakhlia, Batinah
Pakistan	Aseel	80–90	na	Medium size, yellow, mid-season	Khairpur, Sindh
Pakistan	Begum Jangi	80–100	na	Small size, yellow, mid-season	Kech, Baluchistan
Pakistan	Dedhi	90–100	na	Medium size, yellow, mid-season	Khairpur, Sindh
Pakistan	Dhakki	90–100	na	Large size, yellow, late	D. I. Khan, Dhakki, Khyber Pakhtunkhwa
Pakistan	Fasli	70–80	na	Small size, yellow, early	Khairpur, Sindh
Pakistan	Gajjar	90–100	na	Large size, yellow, early	Sukkur, Sindh
Pakistan	Goknah	80–120	na	Large size, yellow, mid-season	Kech, Baluchistan
Pakistan	Gulistan	70–100	na	Large size, yellow, mid-season	D. I. Khan, Khyber Pakhtunkhwa
Pakistan	Hallini	60–70	na	Medium size, yellow, late	Kech, Baluchistan
Pakistan	Hillawi	70–100	na	Medium size, yellow, early	Jhang, Multan, Bahawalpur, D.G. Khan, Muzafer, Garh, Punjab
Pakistan	Hussaini	70–80	na	Medium size, yellow, early	Kech, Baluchistan
Pakistan	Jawan Sour	70–100	na	Medium size, yellow	Panjgur, Baluchistan
Pakistan	Karbaline	80–90	na	Medium size, yellow, mid-season	Khairpur, Sindh
Pakistan	Kasho Wari	90–100	na	Large size, yellow, early	Sukkur, Sindh
Pakistan	Kehraba	80–90	na	Small size, yellow, late	Panjgur, Baluchistan
Pakistan	Kupro	70–80	na	Small size, red, early	Khairpur, Sindh
Pakistan	Muzawati	80–100	na	Medium size, red, mid-season	Panjgur, Mashkel, Baluchistan
Pakistan	Otakin	90–100	na	Large size, orange, mid-season	Khairpur, Sindh
Pakistan	Pashpag	50–60	na	Medium size, yellow, early	Kech, Baluchistan

(continued)

Country	Cultivar	Yield per tree (kg)	Number of planted trees	Fruit characteristics	Location
Pakistan	Rabai	50–60	na	Medium size, red, mid-season	Panjgur, Mashkel, Baluchistan
Pakistan	Roghni	60–70	na	Small size, yellow, early	Kech, Baluchistan
Palestine	Barhee	80–120	2,000	High quality consumed mainly as *balah*, very sweet	Jordan Valley, Jericho
Palestine	Deglet Noor	60–80	1,500	Medium	Jericho, Gaza
Palestine	Khadrawy	40–60	500	Soft round date with dark mahogany-toned skin	Jericho, Gaza
Palestine	Halawy	40–60	500	Very sweet, soft, light brown skin-colored date	Jericho, Gaza
Palestine	Hayany	75–150	4,200	High quality consumed as rutab, medium size, dark red	Gaza
Palestine	Khidri	40–60	300	Large maroon-red date with a firm chewy texture	Jericho
Palestine	Kustawy	150–160	1,300	One of the premier dates with a rich sweet flavor	Jericho, Gaza
Palestine	Medjool	70–80	80,000	High	Jericho, Jordan Valley
Palestine	Zahidi	40–60	2,800	Soft golden brown date, also known as the *golden date*	Jericho, Gaza
Qatar	Aswad	na	na	na	na
Qatar	Azat	na	na	na	na
Qatar	Barhee	na	na	na	na
Qatar	Bin Saif	na	na	na	na
Qatar	Berz	na	na	na	na
Qatar	Bashbak	na	na	na	na
Qatar	Hallini	na	na	na	na
Qatar	Hilali	na	na	na	na
Qatar	Hitmi	na	na	na	na
Qatar	Jabri	na	na	na	na
Qatar	Khalas	na	na	na	na
Qatar	Nebut Seif	na	na	na	na
Qatar	Niqal	na	na	na	na
Qatar	Qashmak	na	na	na	na
Qatar	Rotana	na	na	na	na

Qatar	Shishi	na	na	na	na
Qatar	Tarahim	na	na	na	na
Qatar	Ward	na	na	na	na
Saudi Arabia	Ajwa	na	na	Ovoid, medium, dark red at rutab stage, dark at tamar stage, mid-season cv. consumed at tamar stage	Madinah
Saudi Arabia	Anbara	na	na	Falcoid elongated, large, maroon red at rutab stage, maroon red at tamar stage, late-season cv. consumed at tamar stage	Madinah
Saudi Arabia	Barhee	145	na	Ovoid, medium to small, apricot yellow at rutab stage, light brown at tamar stage, mid-season cv. consumed at rutab and tamar stages	Al-Hasa, Riyadh, Qassim
Saudi Arabia	Barni Madinah	na	na	Ovoid elongated, medium, yellowish brown at rutab stage, brown at tamar stage, late-season cv. consumed at rutab and tamar stages	Madinah
Saudi Arabia	Bayad	na	na	Ovoid, medium, golden yellow at rutab stage, mid-season cv. consumed at rutab stage	Najran
Saudi Arabia	Beid	na	na	Ovoid, medium to small, apricot yellow at rutab stage, light brown at tamar stage, mid-season cultivar consumed at rutab and tamar stages	Madinah
Saudi Arabia	Buraimi	na	na	Ovoid elongated, medium to small, apricot at rutab stage, brown at tamar stage, early-season cv. consumed at rutab and tamar stages	Qatif
Saudi Arabia	Deglet Noor	na	na	Ovoid elongated, medium to large, light brown at rutab stage, light brown at tamar stage, late-season cv. consumed at rutab and tamar stages	Madinah, Riyadh
Saudi Arabia	Ghur	90	na	Falcoid elongated, medium, yellow at bser stage, yellowish brown at rutab stage, early-season cv. consumed at rutab stage	Al-Hasa, Qatif
Saudi Arabia	Hilali	83	na	Ovoid, large, yellowish brown at rutab stage, brown at tamar stage, late-season cv. consumed at rutab stage	Al-Hasa
Saudi Arabia	Hulwa	na	na	Ovoid elongated, medium, dark red at rutab stage, dark at tamar stage, mid-season cv. consumed at rutab and tamar stages	Madinah, Jouf, Hail

(continued)

Country	Cultivar	Yield per tree (kg)	Number of planted trees	Fruit characteristics	Location
Saudi Arabia	Khalas	94	na	Ovoid elongated, medium, yellowish brown at rutab stage, brown at tamar stage, mid-season cv. consumed at rutab and tamar stages	Al-Hasa, Riyadh, Qassim
Saudi Arabia	Khenaizi	92	na	Ovoid elongated, medium, dark red at rutab stage, dark brown at tamar stage, mid-season cv. consumed at rutab and tamar stages	Qatif, Al-Hasa
Saudi Arabia	Khodry	na	na	Cylindrical elongated, large, dark red at rutab stage, dark brown at tamar stage, late-season cv. consumed at tamar stage	Riyadh, Qassim, Jazan
Saudi Arabia	Mabroom (Barni Al Ola)	na	na	Elongated, medium to large, yellowish brown at rutab stage, brown at tamar stage, early-season cv. consumed at rutab and tamar stages	Madinah
Saudi Arabia	Maktoumi	na	na	Cylindrical, medium, apricot yellow at rutab stage, brown at tamar stage, mid-season cv. consumed at bser, rutab, and tamar stages	Qassim, Al-Hasa, Madinah, Northern Borders
Saudi Arabia	Medjool	na	na	Ovoid elongated; small, medium, and large; yellowish brown at rutab stage; reddish brown at tamar stage; early-season cv. consumed at tamar stage	Madinah
Saudi Arabia	Meneifi	na	na	Ovoid elongated, medium, apricot brown at rutab stage, light brown at tamar stage, late-season cv. consumed at rutab and tamar stages	Riyadh
Saudi Arabia	Miskani	na	na	Ovoid, medium, yellow red at bser stage, brown at rutab stage, mid-season cv. consumed at tamar stage	Riyadh
Saudi Arabia	Nabtat Ali	na	na	Ovoid elongated, medium to large, maroon at rutab stage, maroon at tamar stage, mid-season cv. consumed at rutab and tamar stages	Qassim
Saudi Arabia	Nabtat Saif	na	na	Ovoid to global, medium, golden at rutab stage, golden brown at tamar stage, mid-season cv. consumed at rutab and tamar stages	Riyadh, Qassim
Saudi Arabia	Nabtat Sultan	na	na	Ovoid to global, medium to large, golden at rutab stage, golden brown at tamar stage, late-season cv. consumed at rutab and tamar stages	Riyadh

Saudi Arabia	Qatarah	na	na	Ovoid elongated, medium, dark red at rutab stage, dark at tamar stage, mid-season cv. consumed at rutab stage	Riyadh, Qassim
Saudi Arabia	Rabia	na	na	Ovoid elongated, large, yellowish brown at rutab stage, dark brown at tamar stage, early-season cv. consumed at rutab and tamar stages	Madinah
Saudi Arabia	Ruthana	na	na	Ovoid, medium, yellowish brown at rutab stage, yellowish brown at tamar stage, early-season cv. consumed at rutab stage	Madinah, Riyadh
Saudi Arabia	Ruzeiz	112	na	Ovoid, small, yellowish brown at rutab stage, dark brown at tamar stage, mid-season cv.	Al-Hasa
Saudi Arabia	Sabaka	na	na	Ovoid elongated, medium, golden brown at rutab stage, light brown at tamar stage, mid-season cv. consumed at rutab and tamar stages	Qassim
Saudi Arabia	Safawi	na	na	Ovoid elongated, medium, red at rutab stage, dark red at tamar stage, late-season cv. consumed at tamar stage	Madinah
Saudi Arabia	Sari	na	na	Ovoid elongated, medium to large, yellowish red at bser stage, reddish brown at rutab stage, early-season cv. consumed at rutab stage	Al-Aflaj, Wadi Al-Dawasir
Saudi Arabia	Sefri	na	na	Cylindrical, medium, golden yellow at rutab stage, brown at tamar stage, early-season cv. consumed at tamar stage	Al-Baha, Jazan, Jouf, Tabouk, Qassim
Saudi Arabia	Segae	na	na	Cylindrical elongated, medium to large, yellowish brown at rutab stage, reddish brown at tamar stage, mid-season cv. consumed at tamar stage	Riyadh, Qassim, Northern Boarders, Asir
Saudi Arabia	Shahal	118	na	Ovoid, medium, reddish yellow at rutab stage, reddish brown at tamar stage, late-season cv. consumed at rutab stage	Al-Hasa
Saudi Arabia	Shaishi	75	na	Long cylindrical, medium to large, yellowish brown at rutab stage, reddish brown at tamar stage, mid-season cv. consumed at rutab and tamar stages	Al-Hasa, Riyadh
Saudi Arabia	Shebebi	85	na	Ovoid, medium to large, amber at rutab stage, dark amber at tamar stage, mid-season cv. consumed at rutab and tamar stages	Al-Hasa, Qatif

(continued)

Country	Cultivar	Yield per tree (kg)	Number of planted trees	Fruit characteristics	Location
Saudi Arabia	Sukkari	na	na	Ovoid elongated, medium, yellowish brown at rutab stage, brown at tamar stage, mid-season cv. consumed at rutab and tamar stages	Qassim, Riyadh
Saudi Arabia	Sullaj	na	na	Cylindrical, medium, golden yellow at rutab stage, maroon at tamar stage, mid-season cv. consumed at rutab stage	Riyadh, Asir
Saudi Arabia	Tayyar	71	na	Ovoid elongated, small to medium, yellowish brown at rutab stage, dark brown at tamar stage, early-season cv. consumed at rutab stage	Al-Hasa
Saudi Arabia	Um Raheem	116	na	Ovoid, small to medium, yellowish brown at rutab stage, light brown at tamar stage, late-season cv. consumed as bser and rutab	Al-Hasa
Saudi Arabia	Wannana	na	na	Ovoid elongated, large, brown at rutab stage, dark brown at tamar stage, mid-season cv. consumed at rutab and tamar stages	Qassim, Madinah
Spain	Bollior	60–120	5,000	Semidry. Quality is improved by heat, using boiling water or steam. Fresh dates are usually yellow and harsh	Elche
Spain	Candits, Cándidos, Maduros	60–120	30,000	Semidry. Fruits are of average size, 4–4.5 × 2 cm, cylindrical. Skin is smooth yellowish, at full maturity dark reddish brown. Flesh is firm, smooth, granulose, creamy, and farinaceous and tastes sweet and pleasant. Used in desserts when fully mature but spoils easily. Ripe from early November	Elche, Segura, Chicamo valleys
Spain	Candíos Puntiagudos	60–120	5,000	Fruits are of average size, 3.5–4.5 × 2 cm, cylindrical but narrowing toward the apex. Skin is smooth yellowish turning reddish brown at maturity. Flesh is firm, smooth, granulose, and farinaceous and tastes sweet and pleasant. Harvested from mid-November but if left until tamar stage becomes brownish	Segura, Chicamo valleys
Spain	Confitera	100–150	5,000	Soft. Thick, fleshy, elongated, with good size, seed markedly fusiform. Artificially ripened, but also ripens naturally and gradually depending on the year	Exclusively in Elche

Spain	Daurat, Dorado	60–120	10,000	Soft. Yellow, sweet, and savory, consumed fresh but also ripe, good but quickly ferments	Elche, Segura, Chicamo valleys, Murcia, Almería
Spain	De Adobo	60–120	30,000	Soft. Fruits ripen slowly, extending availability for table consumption. Edible only after being marinated in vinegar. Trees being replaced by more desirable cvs. Becoming scarce	Segura, Chicamo valleys, Elche
Spain	De Berberia	60–120	1,000	Fruits are small, ovoid, 2 cm in length. Skin is smooth to rugose, very dark reddish brown. Flesh is dark, farinaceous, sweet, and aromatic. Ripe from early November	Segura, Chicamo valleys
Spain	De Espiga	60–100	5,000	Fruits' average size is 4×2 cm. Skin is yellow, straw colored to shiny reddish, smooth shiny and consistent, and dense and strongly adhering to the flesh, which is firm, fibrous, somewhat dry, and sweet-sour. Ripe from early December. Resembles cv. Fard of Oman	Segura, Chicamo valleys
Spain	De Rambla	30–60	500	Fruits are small, 2.4–2.6×1.5–1.8 cm. Skin is yellow, straw colored shiny reddish, smooth, consistent, firm, and strongly adhering to the flesh, which is firm, fibrous, somewhat dry, and sweet-sour	Segura, Chicamo valleys
Spain	De Sol	60–100	1,000	Fruits are only suitable for livestock feed, although after freezing they are fit for human consumption	Segura, Chicamo valleys
Spain	Gros, Cavirots, Redondos	60–120	10,000	Semidry. Grown in Elche since the late sixteenth century. A variety of large, thick-fleshed dates. In Ojós (Murcia) it becomes very large when thinned	Elche, Segura, Chicamo valleys
Spain	Largos	80–120	1,000	Fruits are large, 5×2 cm, cylindrical, smooth brownish yellow. Skin is smooth, shiny, strongly adhering to the mesocarp. Flesh is white, fibrous, sweet, and slightly sour. Consumed fresh and in dry sweet confectionery. Ripe from mid-December. Similar to cv. Ghars (Rhars) of North Africa (Algeria)	Segura, Chicamo valleys
Spain	Léon	100–120	500	Soft	Elche
Spain	Marraner	60–120	5,000	Semidry	Elche

(continued)

Country	Cultivar	Yield per tree (kg)	Number of planted trees	Fruit characteristics	Location
Spain	Moscatel	60–120	1,000	Fruit are of average size, cylindrical to 4 cm. Skin is yellow, shiny, turning reddish brown when mature, rugose. Flesh is farinaceous and consistent, yellowish green, very sweet when mature. Ripe from mid-November. Dubious resemblance to cv. Deglet Noor of Tunisia and cv. Halawy of Iraq. Also, could be related to cv. Khalasa from the Arabian Gulf because the flesh is amber and without fibers	Segura, Chicamo valleys
Spain	Rojo	60–120	500	Fruits are large 4.5–5 × 1.8–2.5 cm, elongated oval. Skin is smooth and shiny red, turning brownish at maturity. Flesh is white, fibrous fleshy, juicy, and sweet and tastes a bit sour. Eaten fresh. Ripe from early November. Similar to cv. Aruri of Egypt, known for its large fruit size	Segura, Chicamo valleys
Spain	Tenat	80–150	5,000	Dry. Medium, 3.5–4, 5 × 2 cm, cylindrical but tapered toward the apex. Yellow. Flesh is thin, soft, grainy and mealy, sweet and pleasant. Collected in December but then left to mature on the bunch, can remain on the bunch for 2 months without deterioration	Elche, Ricote Valley
Spain	Tendre Dolz	60–100	1,000	Semidry	Elche
Spain	Tiernos, Tendre	60–100	5,000	Soft. Fruits spoil easily starting from the end; hence, they are eaten before fully mature. At the time consumed color can be green, yellow, or red and accordingly their name changes: Tendres Verts, Tiernos Rojos	Segura, Chicamo valleys, Elche
Spain	Verdal	60–120	1,000	Fruit is small oval, elongated, 2.8–3.5 × 2 cm or more, apiculate. Skin is smooth and fine, green, turning brownish green when mature. Ripe from mid-November	Segura, Chicamo valleys
Syria	Ashrasi	60–90	12,000	Excellent	Deir Al-Zor, Homs (Palmyra), Hasaka, Raqqa, Damascus countryside
Syria	Barhee	150–300	21,000	Excellent	Deir Al-Zor, Homs (Palmyra), Hasaka, Raqqa, Damascus countryside

Syria	Birbin	300–400	28,000	Good	Deir Al-Zor, Homs (Palmyra), Hasaka, Raqqa, Damascus countryside
Syria	Deglet Noor	75–100	15, 000	Excellent	Deir Al-Zor, Homs (Palmyra), Hasaka, Raqqa, Damascus countryside
Syria	Gish Rabi	150–250	10,000	Good	Deir Al-Zor, Homs (Palmyra), Hasaka, Raqqa, Damascus countryside
Syria	Kabkab	100–150	7,300	Very good, yellow fruit	Deir Al-Zor, Homs (Palmyra), Hasaka, Raqqa, Damascus countryside
Syria	Kabkab	100–150	7,000	Very good, red fruit	Deir Al-Zor, Homs (Palmyra), Hasaka, Raqqa, Damascus countryside
Syria	Khadrawy	150	2,500	Very good	Deir Al-Zor, Homs (Palmyra), Hasaka, Raqqa, Damascus countryside
Syria	Khalas	150–250	22,000	Excellent	Deir Al-Zor, Homs (Palmyra), Hasaka, Raqqa, Damascus countryside
Syria	Khastawi	250–300	40,000	Good	Deir Al-Zor, Homs (Palmyra), Hasaka, Raqqa, Damascus countryside
Syria	Khineze	200–350	4,300	Very good	Deir Al-Zor, Homs (Palmyra), Hasaka, Raqqa, Damascus countryside
Syria	Lolo	150–200	9,500	Very good	Deir Al-Zor, Homs (Palmyra), Hasaka, Raqqa, Damascus countryside

(continued)

Country	Cultivar	Yield per tree (kg)	Number of planted trees	Fruit characteristics	Location
Syria	Maktoom	80–120	8,500	Very good	Deir Al-Zor, Homs (Palmyra), Hasaka, Raqqa, Damascus countryside
Syria	Medjool (Mujhoolah)	70–90	20,000	Excellent	Deir Al-Zor, Homs (Palmyra), Hasaka, Raqqa, Damascus countryside
Syria	Nabtat Saif	70–130	20,700	Excellent	Deir Al-Zor, Homs (Palmyra), Hasaka, Raqqa, Damascus countryside
Syria	Shahabi	300	4,000	Excellent	Deir Al-Zor, Homs (Palmyra), Hasaka, Raqqa, Damascus countryside
Syria	Zaghloul	150–250	8,000	Very good	Deir Al-Zor, Homs (Palmyra), Hasaka, Raqqa, Damascus countryside
Syria	Zahidi	250–350	46,000	Good	Deir Al-Zor, Homs (Palmyra), Hasaka, Raqqa, Damascus countryside
Yemen	Abdel Rahman	34	na	Very good quality. Consumed at rutab and tamar; khalal (bser) yellow, rutab brown, fruit oval	Abdel Rahman
Yemen	Abyadh (Bathri, seeded)	120	na	Fruits good–excellent, harvest time August	Marib
Yemen	Al Sahagi	42	na	Excellent quality if harvested July	Shabwah
Yemen	*Asabia el Aroos*	38	na	Very good quality. Consumed at rutab and tamar; khalal (bser) red, rutab black, fruit elongated oval	Hadhramout, Shabwah
Yemen	Azzani	45	na	Excellent quality if harvested in July–August	Shabwah
Yemen	Baqal	49	na	Very good quality, if harvested in August–September	Wadi Hadhramout, Al Mahrah
Yemen	Barbosa	30	na	Excellent quality if harvested in July–August	Sokotra

Yemen	Gabiley	48	na	Excellent quality if harvested in July–August	Shabwah
Yemen	Ghudairey	200	na	Excellent quality if harvested in August	Al Jawf
Yemen	Gzaz	52	na	Excellent quality if harvested in July–September	Hadhramout
Yemen	Hajri	37	na	Very good quality. Used at rutab and tamar; khalal (bser) yellow, rutab black, fruit elongated oval	Hadhramout, Shabwah
Yemen	Hamrah	47	na	Good quality, if harvested in July–September	Hadhramout, Al Mahrah
Yemen	Madini	41	na	Excellent quality if harvested in August–September	Hadhramout
Yemen	Mijraf	47	na	Excellent quality if harvested in August–September	Hadhramout
Yemen	Momeg	40	na	Very good quality, harvested in August	Al Mahrah
Yemen	Serfateh	16	na	Excellent quality if harvested in July–August	Sokotra
Yemen	Sokotri	49	na	Good quality, matures early in August–September	Hadhramout, Shabwah
Yemen	Tha'al (Manasif)	35	na	Excellent quality if harvested in June–July	Tihama, Shabwah
Yemen	Tubaig	32	na	Excellent quality if harvested in May–June	Tihama
Yemen	Zurghi (Makkawy)	100	na	Inferior quality if harvested in June	Tihama

This list is not inclusive of all cultivars grown. All entries were provided by the authors of the corresponding country chapter *na* not available

Appendix B

Commercial sources of dates in some date-producing countries in Asia and Europe

Country	Name	Contact information	Website link
India	Gorasiya Farm	Kutch, Gujarat, India	na
India	Ministry of Commerce (controls all date imports)	Government of India, New Delhi, India	na
Iran	AHT	2 Ramin St., North Valiasr Ave., Tehran 1966843341, Iraninfo@aht.ir	www.aht.ir
Iran	ARAT CO.	61, Mir Emad Ave. Tehran 15878, Iraninfo@aratco.com	www.aratco.com
Iran	Dombaz	No. 1 Industrial Town, Bandar Abbas, Iraninfo@dombaz.com	www.dombaz.com
Iran	Iran Dates Exporting Co. (IDEC)	Suite#21, 1729 Sina St. Dr. Shariati Ave. Tehran 19338–13841, Iraninfo@irandatesco.com	www.irandatesco.com
Iran	KDA Group (Khezri Group)	Sadi Ave, 5th Area Industrial, Shadegan, Iraninfo@khezrigroup.com	www.khezrigroup.com
Iran	Middle East Products Export Co. (MEPE CO.)	45 West Nahid Blvd, Africa Ave, Tehran 19669, Iranmepe@mepeco.com, ipc@mepeco.com, doc.mepeco.com	www.mepeco.com
Iraq	Iraqi Date Processing and Marketing	Shalchia, Baghdad, Iraq.iraqiandate1@yahoo.com	
Iraq	Al Moosawi Dates	Abulkaseeb, Basrah, Iraq.Almoosawi_dates@hotmail.com	www.almoosawigroup.com
Iraq	Al-Basra Dates	Abulkasseb,Basrah,Iraq.basra_dates@yahoo.com	www.basradatesjeeran.com
Israel	Hadiklaim	hadiklaim@hadiklaim.co.il	www.Hadiklaim.com
Israel	Mehadrin	export@mtex.co.il	www.mtex.co.il
Kuwait	Gulf Palm Company	P.O. Box 27971, Safat 13136, Kuwait, palms@gulfpalms.co	www.gulfpalms.com
Kuwait	House of Development Company	P.O. Box 28870, Safat 13150, Kuwait. hodag@qualitynet.net	www.hodagri.com
Oman	Samail Dates Factory	PO Box 34 Lizgh, Postal Code 615	
Oman	Agricultural Association for Al-batinah Region Farmers	PO Box 553 Suwaiq, Postal Code 315, aabrfsuq@omantel.om	na
Oman	United Date Processing Co.	PO Box 1161, Azaibah, Oman, udfll@omantel.net.om	www.goldendatesoman.com
Oman	Bright Sun Dates	PO Box 3208, Oman, Hzk1951@hotmail.com	www.brightsundates.com

Pakistan	Al-Ghazi Fruit Farm	Salahuddin Phulpoto, Khairpur, Sindh, Pakistan. +92-300-8313732	na
Pakistan	Amanullah Khan	Gramkan, Panjgur, Baluchistan, Pakistan. Ph.: +92-8293-631093/631367	na
Pakistan	Ispeeza Enterprises	Jama Masjid Rood, Turbat, Baluchistan, Pakistan. Ph.: +92-861-413308/411154, Fax: +92-861-9409	na
Pakistan	Khairpur Foods	Therhi, Khairpur, Sindh, Pakistan. Ph.: +92-300-3290789, khairpurfoods@gmail.com	na
Pakistan	Meno Date Farm	1-College Road, 149-G/3, Block-5, Satellite Town, Quetta, Baluchistan, Pakistan. Ph.: +92-81-412795/442780	na
Pakistan	Sardar Foods	Mr. Mohammad Anis, Karamabad, Near Khairpur, Sindh, Pakistan. Ph.: +92-300-2007767, sardarfoods@yahoo.com	na
Pakistan	Shahzad & Co.	Dyial Road, D.I. Khan, KPK, Pakistan. Ph.: +92-961-740138/740095	na
Pakistan	Sindh Foods International	Mr. Zameer Ahmed Memon, NeaMadarsa Dar-ul-Huda, Therhi, Khairpur Mirs, Sindh, Pakistan, Ph.: +92-24-3514766, Mob.:+92-0305-3125410	na
Pakistan	Zamindar Date Farm Products	Abdul Haleem Memon, Therhi Near Rohrri Canal, Khairpur, Sindh, Pakistan. Ph.: +92-300-3143622, zamindar_dates@yahoo.com	na
Palestine	Manasrah Development and Investment Company (MADICO)	Jericho, info@madico.ps	http://www.madico.ps/products.html
Palestine	Nakheel Palestine for Agricultural Investment (Nakheel)	Jericho, info@Nakheelpal.ps	http://www.nakheelpal.com/
Saudi Arabia	Stock Market: Riyadh Dates Market	Riyadh Municipality, Riyadh 11146, Saudi Arabia	http://www.alriyadh.gov.sa
Saudi Arabia	Stock Market: Buraidah Dates City	PO Box 1119, Qassim Municipality, Buraidah, Qassim, Saudi Arabia	http://www.qassim.gov.sa/AR/MunicipalityDepts/eh/Pages/da_dept.aspx
Saudi Arabia	Stock Market: King Abdullah City for Dates	Al-Hasa Municipality, PO Box 1790, Al-Hasa 31982, Saudi Arabia	http://www.alhasa.gov.sa/SitePages/Home.aspx

(continued)

Country	Name	Contact information	Website link
Saudi Arabia	Stock Market: Madinah Dates Market	Madinah Regional Municipality, PO Box 4952, Madinah 41412, Saudi Arabia	http://www.amana-md.gov.sa/
Saudi Arabia	Al-Ahli Ideal Dates Packing Factory	P.O. Box 188 Madinah 41455, Saudi Arabia, Phone +966-1846-1900	na
Saudi Arabia	Al-Ansar Factory for Dates & Sweets, Packing and Wrapping	P.O. Box 20847, Madinah 41455, Saudi Arabia, Phone +966-14-827-4767	www.alansardates.com
Saudi Arabia	Al-Babtain Factory for Dates Filling & Packing	P.O. Box 60802, Riyadh 11555, Saudi Arabia, Phone +966-11-551-2424	na
Saudi Arabia	Al-Faisaliah Dates Factory	P.O. Box 231, Al-Kharj 11942, Saudi Arabia Phone +966-11-545-8899	na
Saudi Arabia	Al-Sharq Dates Food Manufacturing Co.	P.O. Box 1055 Al-Hasa, 31982, Saudi Arabia Phone +966-13-531-4258: ali@tomoors.com	www.tomoors.com
Saudi Arabia	Al-Hasa Company for Food Industry Date Processing Plant	P.O. Box 3531 Al-Hasa, 31982, Saudi Arabia Phone +966-13-586-9456: food@yahoo.com	http://www.ahsa-foodind.com/index.htm
Saudi Arabia	Al Jawhara Dates Factory	P.O. Box 15533 Al-Hasa 31982, Saudi Arabia Phone +966-13-595-0216: a.alhabdan@hotmail.com	
Saudi Arabia	Al-Jazirah Dates and Food Factory	P.O. Box 5998 Al-Hasa, 31982, Saudi Arabia Phone 0+966-13-532-2121: analajaji@aljazirahdates.com.sa	www.aljazirahdates.com.sa
Saudi Arabia	Al-Kharj Dates Factory	P.O. Box 1518, Al-Kharj 11942, Saudi Arabia Phone +966-11-550-0224	na
Saudi Arabia	Al-Mohamadia Dates Company	P.O. Box 7045, Al-Kharj 11942, Saudi Arabia Phone +966-11-545-8970: dates@foma.com.sa	www.mohamadia.com.sa
Saudi Arabia	Al-Rajhia Dates Factory	P.O. Box 17793 Riyadh 11494, Saudi Arabia Phone +966-11-410-6753	na
Saudi Arabia	Al-Salehaiah Dates Factory	P.O. Box 63254 Riyadh 11516, Saudi Arabia Phone +966-11-424-2043	www.salehia.com.sa
Saudi Arabia	Al-Yamamah Dates Packing Factory	P.O. Box 60573 Riyadh 11555, Saudi Arabia Phone +966-11-544-7584	na
Saudi Arabia	Amal Al-Khair Dates Packing & Packaging Factory	P.O. Box 124264 Riyadh 11761, Saudi Arabia Phone +966-11-415-9426: dates@amalalkhair.com.sa	http://www.amalalkhair.com.sa

Saudi Arabia	Barakat Al-Madinah Dates and Sweets Factory	P.O. Box 18241 Jeddah 21415, Saudi Arabia Phone +966-1840-9232	http://www.madinahdatefactory.com
Saudi Arabia	Bateel Sweets and Chocolates Factory	P.O. Box 10071 Riyadh 11433, Saudi Arabia Phone +966-11-265–2477: rashid@badrahn.net	http://bateel.com
Saudi Arabia	Kingdom Dates Factory	P.O. Box 222, Al-badaya, 51951, Qassim, Saudi Arabia Phone +966-16-331-1111, info@kingdomdates.com.sa	http://www.kingdomdates.com.sa/
Saudi Arabia	Madaen Star Dates Factory	P.O. Box 224, Al-Kharj 11942, Saudi Arabia Phone +966-11-550-1111	na
Saudi Arabia	Al-Madina Munawarah Dates Company	P+E102.O. Box 5090 Madinah 41455 Saudi Arabia Phone +966-14-845-7777: info@tomoor.com	http://www.tomoor.com/
Saudi Arabia	Nadheed Dates Factory	P.O. Box 5022 Buraidah 5142, Saudi Arabia, Phone +966-16-322-2222	http://www.nadheed.com/jnad/
Saudi Arabia	Nakhail Al-Watan Dates Factory	P.O. Box 371999 Riyadh 11325, Saudi Arabia Phone +966-11-439–9999,: nakhailalwatan@hotmail.com	http://nakhealalwatan.com/en/
Saudi Arabia	Qassim Agricultural Company Dates Packaging Factory	P.O. Box 4000 Buraidah 51461, Saudi Arabia Phone +966-16-380-0777	www.gaco-agri.com
Saudi Arabia	Quba Dates Factory	P.O. Box 539 Madinah 41455, Saudi Arabia Phone +966-14-840-3153	na
Saudi Arabia	Taiba Dates Packing Factory	P.O. Box 2779 Madinah 41455, Saudi Arabia Phone +966-14-840-2027	na
Spain	Associació de Palmerers d'Elx (APELX)	na	na
Spain	Associaçió per al Desenvolupament Rural del Camp d'Elx (ADRCE)	na	na
Spain	Cooperativa del Campo de Elche (Coopelche)	Partida La Hoya, 03294 Elche, Alicante, Spain, coopelche@yahoo.es	na
Spain	Sociedad Agraria de Transformación (DATELX)	na	na
Yemen	Seiyun Packing Factory	Seiyun, Hadhramout, Yemen	na
Yemen	Ettahaitah Packing Factory	Al Hudaidah, Yemen	na

na not available

Appendix C

Commercial sources of offshoots in some date-producing countries in Asia and Europe

Country	Name	Contact information	Website link
Cameroon	Agricultural Research Institute for Development	IRAD, Garoua P.O. Box: 415 Garoua, Cameroon	www.iradcameroon.org
India	Swami Keshwanand Rajasthan Agricultural University	Beechwal, Bikaner, Rajasthan, India	http://www.raubikaner.org
India	Fruit Research Station	Mundra, Bhuj, Gujarat, India	na
Iran	Agricultural and Natural Resources Research Centers	Kerman, Bushehr, Fars, Hormozgan, Baluchistan	na
Iran	Date Palm and Tropical Fruits Research Institute	Km 10 Coastal Ahvaz-Khorramshahr Road, Ahvaz, Irandptfrii@yahoo.com	http://khorma.areo.ir
Iraq	Horticulture Directorate, Ministry of Agriculture	Abu-Graib, Baghdad, Iraq.hort_and_forests@moagr.org	na
Iraq	Private sector	Most date palm governorates	na
Israel	Produced by many local growers	na	na
Kuwait	Gulf Palm Company	P.O. Box 27971, Safat 13136, Kuwait, palms@gulfpalms.co	www.gulfpalms.com
Kuwait	Palm Agro Production Company	P.O. Box 1976, Safat 13020, Kuwait. palms@qualitynet.net	na
Pakistan	Al-Badar Fruit Farm	Sadruddin Phulpoto, Khairpur, Sindh, Pakistan. Ph.: +92-300-3115211	na
Pakistan	Date Palm Research Institute (DPRI)	Shah Abdul Latif Univ., Khairpur, Sindh, Pakistan. Ph.:+92-243-9280344, egypak06@gmail.com	http://www.salu.edu.pk/
Pakistan	Horticulture Research Station	Bahawalpur, Punjab, Pakistan. Ph.: +92-301-17771602	http://bzu.academia.edu/Departments/Horticultural_Research_Station_Bahawalpur
Pakistan	Jiskani Fruit Farm	Post Office Jiskani, Kot Deeji, Khairpur, Sindh, Pakistan. Ph.: +92-300-9318343	na

Palestine	Alahlieh for date palm development	Gaza	http://www.gazaark.org
Palestine	Khan Yunis Date Nursery	Khan Yunis, Gaza	http://paltoday.ps/ar/post/68954
Palestine	Palestinian palm association for development	Khan Yunis, Gaza, palmtrees1945@yahoo.com	na
Palestine	Palestinian Agricultural Relief Committee (PARC)	West Bank and Gaza, info@parc.ps	http://www.pal-arc.org/
Palestine	National association for date palm development	Gaza	na
Spain	Asociación de Cultivadores de Palmeras Datileras	C/Luis Gonzaga 24-6 s, 03202 Elche, Alicante, Spain	na
Spain	Asociación de Empresas Productoras de Plantas de Viveros de la Provincia de Alicante (VAME)	C/Pueste Ortices 13 Bajo, 03201 Elche, Alicante, Spain info@asoc-vame.es	www.asoc-vame.es/vame/vame
Spain	Asociación Profesional de Flores, Plantas y Tecnología Hortícola de la Comunidad Valenciana (ASFPLANT)	C/Guillem de Castro 79 4°, 46008 Valencia, Spain asfplant@asfplant.com	www.asfplant.com/index.php
Syria	Al-Jalaa Date Palm Propagation Center in Al-Bukamal	Al-Jalaa town, Al-Bukamal, Deir Al-Zor Province	na
Syria	Date Palm Propagation Center in Palmyra	Palmyra, Homs Province	na
Syria	Date Palm Propagation Center in Sabkhat Al Moh	Palmyra, Homs Province	na
Yemen	Public Corporation for Agricultural Services	Ministry of Agriculture and Irrigation, Sana'a, Yemen.	www.yemen.gov.YE

na not available

Appendix D

Commercial source of in vitro plants in some date-producing countries in Asia and Europe

Country	Name	Contact information	Website link
India	Atul Rajasthan Date Palms Ltd	Chaupasani, Jodhpur, Rajasthan, India	http://www.ardp.co.in
India	Kutch Crop Ltd. c/o Excel Crop Care Ltd	Bhuj, Gujarat, India	na
Iran	Agricultural Biotechnology Research Institute of Iran	Seed and Plant Improvement Institutes Campus, Mahdasht Road, P. O. Box, 31535–1897, Karaj, Iranmail@abrii.ac.ir	http://abrii.areo.ir
Iran	RANA Agro-Industry Corporation	22, Atlasee St, Mirdamad Blvd, Tehran 19116, Iran, P.O. Box: 15875/3997info@ranagro.com	www.ranagro.com, www.rana.ir
Iraq	Iraqi Center for Plant Tissue Culture Ltd.	District, 28 Al-Tajiat, Baghdad-Iraq. iraqicptc@gmail.com	na
Iraq	Uruk Centre for Date Palm Tissue Culture	North gate, Nasiriyah City, Iraq.urukpalm@yahoo.com	na
Israel	Zemach Tissue Culture Laboratory, Zvieli Nursery	zvieli@zvieli.co.il	www.zvieli.co.il
Kuwait	Kuwait Institute for Scientific Research (KISR)	P.O. Box 24885, Safat 13109, Kuwait. kisrdg@kisr.edu.kw	www.kisr.edu.kw
Oman	Tissue Culture Laboratory of the Ministry of Agriculture and Fisheries	P.O Box 467 P.C. 100, Tel: + 96824696300, Fax: + 96824696271, infonet@maf.gov.om	http://maf.gov.om
Pakistan	Date Palm Research Institute (DPRI)	Shah Abdul Latif Univ., Khairpur, Sindh, Pakistan. Ph.: +92-243-9280344, egypak06@gmail.com	http://www.salu.edu.pk/
Saudi Arabia	Al-Rajhi Tissue Culture Laboratory (Clone Biotech)	P.O. Box 55155, Riyadh 11534, Saudi Arabia, info@clonebiotech.com	http://www.clonebiotech.com/
Saudi Arabia	SAPAD Tissue Culture Date Palm Co.	PO Box 1806, Dammam 31441, Saudi Arabia, Tel: +966 13 822 3850, Fax: +966 13 821 0385	http://sapad.com.sa/
Spain	Vitropalm Technology	Partida Peña de las Aguilas 1, 217. Elche (Alicante) Spain, vitropalm@vitropalm.com	www.vitropalm.com

United Arab Emirates	Al Wathba Marionnet LLC	P.O. Box: 41522 Abu Dhabi, UAE, Tel: +971 2 6661522, Fax: +971 2 6663850, Mobile: +97150 6428951, datepalm@emirates.net.ae; info@awm-datepalm.com	www.awm-datepalm.com
United Arab Emirates	Date Palm Tissue Culture Laboratory, United Arab Emirates University	P.O. Box 81908, Al Ain, UAE, Tel: +971 3 7832334, Fax: +971 3 7832472	http://datepalm.uaeu.ac.ae/subpages/Laboratory.html
Yemen	Date Palm Developments (D. P. D. Ltd) in UK	Date Palm Developments (D. P. D. Ltd) in UK	www.datepalm.co.uk

na not available

Appendix E

Research institutes concerned with date palm research in some date-producing countries in Asia and Europe

Country	Name	Contact information	Website link
Bahrain	Arabian Gulf University	Desert and Arid Land Sciences Program, College of Graduate Studies Arabian Gulf University, Kingdom of Bahrain	www.agu.edu.bh
Bahrain	Ministry of Municipalities Affairs and Urban Planning, Directorate of Plant Wealth	Agriculture Affairs, Ministry of Municipalities Affairs and Urban Planning, Budaiya, Kingdom of Bahrain	http://websrv.municipality.gov.bh/agri/
Bahrain	University of Bahrain	College of Science, University of Bahrain, P. O. Box 32038, Sakhir, Kingdom of Bahrain	www.uob.edu.bh/
India	Fruit Research Station	Mundra (Sardar Dantiwada Agriculture University Campus), Bhuj, Gujarat, India	http://www.sdau.edu.in
India	Central Institute for Arid Horticulture	Ganganagar Highway, Beechwal, Bikaner, Rajasthan, India	http://www.ciah.ernet.in
India	Anand Agriculture University	Department of Horticulture and Biotechnology, College of Agriculture, Anand, Gujarat, India	http://www.aau.in
India	Chaudhary Charan Singh Haryana Agricultural University	Department of Biotechnology, Chaudhary Charan Singh Haryana Agricultural University, Hisar, Haryana, India	http://www.hau.ernet.in
India	Date Palm Research Station, Swami Keshwanand Rajasthan Agricultural University	Beechwal, Bikaner, Rajasthan, India	http://www.raubikaner.org
Iran	Agricultural Biotechnology Research Institute of Iran	Seed and Plant Improvement Institutes Campus, Mahdasht Road, P. O. Box, 1535–1897, Karaj, Iranmail@abrii.ac.ir	http://abrii.areo.ir
Iran	Date Palm and Tropical Fruits Research Institute	Km 10 Coastal Ahvaz-Khorramshahr Road, Ahvaz, Irandptfrii@yahoo.com	http://khorma.areo.ir
Iran	Seed and Plant Certification and Registration Institute	Corner of Koleksion St, Nabovvat Blvd, P.O. Box 31535–1516 Karaj, Iraninfo@spcir.ir	http://spcri.areo.ir
Iraq	Date Palm Research Center, University of Basrah	Basrah, Iraq.Datepalm-center@yahoo.com	www.uobasrah.edu.iq

Iraq	Date Palm Research Unit, College of Agriculture, University of Baghdad	Al-Jadriyah, Baghdad, Iraq, dpru@coagri.uobaghdad.edu.iq	www.coagri.uobaghdad.edu.iq
Iraq	Horticulture Directorate, Ministry of Agriculture	Abu-Graib, Baghdad, Iraq.hort_and_forests@moagr.org	www.zeraa.gov.iq
Iraq	Plant Protection Directorate, Ministry of Agriculture	Abu-Graib, Baghdad, Iraq. crop_prot@moagr.org	www.zeraa.gov.iq
Iraq	Agriculture Research Directorate, Ministry of Science and Technology	Zaafrania, Baghdad, Iraq	www.most.gov.iq
Israel	Faculty of Agriculture, Food and Environment	The Hebrew University of Jerusalem	http://www.agri.huji.ac.il/english/
Israel	Southern Arava R&D	na	http://www.aravard.org.il/
Israel	Volcani Research Center	na	www.agri.gov.il
Kuwait	Kuwait Institute for Scientific Research (KISR)	P.O. Box 24885, Safat 13109, Kuwait. kisrdg@kisr.edu.kw	www.kisr.edu.kw
Kuwait	Public Authority for Agriculture and Fish Resources (PAAFR)	PO Box 21422 Safat, 13075 Kuwait City – Kuwait	www.paaf.gov.kw
Oman	Ministry of Agriculture and Fisheries Wealth	P.O Box 467 P.C. 100, tel: + 96824696300, Fax: + 96824696271, infonet@maf.gov.om	http://maf.gov.om
Oman	Sultan Qaboos University	PO Box 34, Al-Khod 123, Oman, cams@squ.edu.om	http://www.squ.edu.om
Pakistan	Agriculture, Research Institute	Dera Ismail Khan, KPK Pakistan	
Pakistan	Date Palm Research Institute	Shah Abdul Latif Univ., Khairpur, Sindh, Pakistan. Ph.: +92-243-9280344, egypak06@gmail.com	http://www.salu.edu.pk/
Pakistan	HEJ Research Institute of Chemistry	Karachi University, Karachi, Pakistan	http://www.iccs.edu/
Pakistan	National Institute of Genomics and Advanced Biotechnology	NARC, Park Road, Islamabad, 44000 Pakistan, Ph.: +92-51 9255061, drgmali@yahoo.ca	http://www.parc.gov.pk/narc/Nigab/Genom.html
Pakistan	University College of Agriculture and Environmental Sciences	Islamia University of Bahawalpur, Bahawalpur, Punjab, Pakistan, vc@iub.edu.pk	http://www.iub.edu.pk/department.php?id=MzY=

(continued)

Country	Name	Contact information	Website link
Pakistan	University of Agriculture	Faisalabad, Punjab, Pakistan. Ph, 92–41 9200161	http://uaf.edu.pk/new/
Palestine	Al-Aqsa University	Gaza	http://www.alaqsa.edu.ps
Palestine	An-Najah National University	West Bank	http://www.najah.edu
Palestine	Omega Cooperation for Consultancy and Development	Gaza, info@omegapal.com	http://www.omegapal.com
Qatar	Ministry of Environment	Government of Qatar, Doha Qatar	www.moe.gov.qa/ENGLISH
Qatar	Qatar University	Doha, Qatar	www.qu.edu.qa/
Saudi Arabia	FAO Project, Al-Hasa Date Palm Research Center	Ministry of Agriculture, P O Box 43, Al-Hasa 31982, Saudi Arabia	
Saudi Arabia	Date Palm Research Center for Excellence, King Faisal University	P. O. Box 400, Al-Hasa 31982, Saudi Arabia	http://www.dprckfu.org
Saudi Arabia	Chair of Date Palm Research	College of Food and Agricultural Sciences, King Saud University, Riyadh, KSA	http://www.rpwrc-ksu.org
Saudi Arabia	Saleh Kamel Chair for Palm Research	Qassim University, Almulyda, Qassim, Saudi Arabia	http://www.skcdr.rc.qu.edu.sa
Saudi Arabia	National Center for Palm and Dates	Kingdom of Saudi Arabia, Al-Masharq Tower, King Fahad Road, P.O. Box 16166, Riyadh 11464, Saudi Arabia	www.ncpd.org.sa
Saudi Arabia	Joint Center for Genomics Research (JCGR)	King Abdulaziz City for Science and Technology (KACST), Riyadh 11442, Saudi Arabia	http://www.kacst.edu.sa
Spain	Cátedra Palmeral d'Elx, Universidad Miguel Hernández	Universidad Miguel Hernández de Elche, Avda de la Universidad s/n. Elche, info@umh.es	palmeralelx.umh.es
Spain	Colección Nacional de Phoenix	Formulary at, www.phoenix-spain.org	www.phoenix-spain.org, www.inia.es/webcrf/CRFesp/Paginaprincipal.asp
Syria	Date Palm Department	Ministry of Agriculture	na
Yemen	Seiyun Agricultural Research Centre	Seiyun, Hadhramout, Yemen, agr.res.seiyun@y.net.ye	www.area.gov.net.ye
Yemen	Nasir's College of Agriculture, University of Aden	Al Hawtah, Lahej Governorate	www.aden-univ.net

Yemen	Date Palm Center, University of Hadhramout for Science and Technology	Seiyun, Hadhramout, Yemen	na
Yemen	Science and Technology Centre; University of Aden	University of Aden, Khormaksar, Aden, Yemen, sctec@yahoo.com	www.aden-univ.net

na not available

Appendix F

Scientific societies concerned with date palm research in some date-producing countries in Asia and Europe

Country	Name	Contact information	Website link
Iran	Agricultural Biotechnology Research Institute of Iran	Seed and Plant Improvement Institutes Campus, Mahdasht Road, P. O. Box, 31535–1897, Karaj, Iranmail@abrii.ac.ir	http://abrii.areo.ir
Iran	Date Palm and Tropical Fruits Research Institute	Km 10 Coastal Ahvaz-Khorramshahr Road, Ahvaz, Irandptfrii@yahoo.com	http://khorma.areo.ir
Iran	Seed and Plant Certification and Registration Institute	Corner of Koleksion St, Nabovvat Blvd, P.O. Box 31535–1516 Karaj, Iraninfo@spcir.ir	http://spcri.areo.ir
Iraq	Iraqi Date Palm Network	ijboory@iraqi-datepalms.net	www.iraqi-datepalms.net
Kuwait	Kuwait Science Club	Kuwait	www.ksclub.org/site/
Pakistan	Pakistan Botanical Society	Department of Botany, University of Karachi, Karachi-75270, Pakistan. Ph.:+92-21-36073119, pakjbot@yahoo.com	http://www.pakbs.org/
Pakistan	The Chemical Society of Pakistan	The Chemical Society of Pakistan, Country Coordination Centre in Chemical Sciences (C4S), Department of Chemistry, Quaid-i-Azam University, Islamabad. Ph.: +92-51-2601055, +92-051-9064-2007, ccccsqau@gmail.com, c4scsp@gmail.com	http://www.chemicalsocietyofpakistan.com
Saudi Arabia	The Arab League for Palm And Date Research	Hosted by KACST, P.O. Box 6086 Riyadh 11442 Saudi Arabia, dp-rabitah@kacst.edu.sa	http://www.kacst.edu.sa/ar/depts/palm/Pages/Home.aspx
Saudi Arabia	Saudi Society of Agricultural Sciences	College of Food and Agricultural Sciences, King Saud University, Riyadh, Saudi Arabia	http://www.ssas.org.sa
Spain	Asociación Estación Phoenix	Av. Padre Ismael, 58, 03680 Aspe, Alicante, Spain	na
Yemen	Friends of Date Palm	Seiyun, Hadhramout, Yemen	na
Yemen	Yemeni Biological Society	Sana'a, Yemen, ybsoc@yahoo.com	www.yemenbios.org

na not available

Index

© Springer Science+Business Media Dordrecht 2015
J.M. Al-Khayri et al. (eds.), *Date Palm Genetic Resources and Utilization: Volume 2: Asia and Europe*, DOI 10.1007/978-94-017-9707-8

E

F

G

R

S

T

CPSIA information can be obtained
at www.ICGtesting.com
Printed in the USA
LVOW04*1624030516
486513LV00008B/93/P

9 789401 797061